Principles of
Molecular Mechanics

Principles of Molecular Mechanics

by

Katsunosuke Machida

A KODANSHA LTD. and JOHN WILEY & SONS, INC. co-publication

KODANSHA Tokyo

WILEY New York·Chichester·Weinheim·Brisbane·Singapore·Toronto

Katsunosuke Machida
Professor Emeritus
Kyoto University
Kyoto 606-8501, Japan

Published jointly by Kodansha Ltd., Tokyo (Japan) and John Wiley & Sons, Inc. (USA).

Kodansha Ltd.
12-21, Otowa 2-chome
Bunkyo-ku, Tokyo 112-8001
Japan

Telephone: 81 3 3946-6201

ISBN Kodansha: 4-06-208637-9

John Wiley & Sons, Inc.
605 Third Avenue
New York, N. Y. 10158-0012
USA

Telephone: (212) 850-6000

ISBN Wiley: 0-471-35727-8

Library of Congress Cataloging-in-Publication Data is available upon request from the Library of Congress

Printed in Japan

Contents

Preface

Nowadays molecular mechanics is understood to be a branch of computational chemistry that aims to predict various molecular properties as deduced in analytical expressions from empirically designed molecular force fields. Here a molecular force field traditionally implies a force field for atomic nuclei and not for electrons. The force field for electrons is purely Coulombic and axiomatic. No arbitrariness enters in writing down the force field for electrons. The force field for atomic nuclei must also be axiomatic in principle, but it is too complicated and has too many unknown parts to be deduced from a set prescription given the present status of physics and chemistry. Hence, in order to find a faithful model of the true force field for atomic nuclei, we must refer to as many types of experimental data as possible whenever the data contain any information on the force field.

Spectroscopy and crystallography offer many data related to molecular force fields, but the relationships are not straightforward in general. The analytical expressions used in calculating molecular properties from the force field are derived from a wide variety of branches of mathematics and physics. They include, for example, linear algebra, vector analysis, Fourier transforms, complex functions, group theory, mechanics of rigid bodies, statistical mechanics, quantum mechanics and perturbation theory. Fundamental knowledge of these subjects is indispensable to anyone who is engaged in the design or the improvement of computer programs for molecular mechanics.

Even a routine user of a molecular mechanics program cannot be entirely ignorant of the theoretical aspects of what he or she is going to do. Due to the constant addition of new functions and new objects of calculation, recent molecular mechanics programs are becoming larger and larger. Although the contents of a large program cannot be interpreted easily, it is misleading to use the program as a black box and to trust the output uncritically. The algorithms used in molecular mechanics programs involve more or less approximations which may be accurate enough for one purpose but too rough for another. No user of molecular mechanics programs can become aware of such pitfalls without understanding the mutual relationship between the physical quantities to be calculated. For chemists interested in the possible outcome of a molecular force field, it is helpful to have an introductory guide covering the theoretical procedures of deducing as much from the force field as possible.

This volume is a comprehensive theoretical treatise on the mathematical and physical background of molecular mechanics calculation. A few formulae expressing the transition probabilities are given without proof in Chapters 8 and 9, because the derivation of these formulae requires too lengthy discussion based on the theory of interaction between electromagnetic waves and matter to be contained in this monograph. The other important formulae to be used in the calculation are derived from fundamental principles in primitive, though not very elegant, ways. Practical algorithms are outlined with emphasis on speeding up the calculation, saving computer memory and keeping significant digits in numerical analyses. Examples of calculated data have been taken from rather simple molecules which can be treated with the highest level of approximation presently attainable in each problem. This work is not intended to be an exhaustive survey of the literature on the application of molecular mechanics to various chemical species. The reader interested in the application to complex molecules is referred to the well-organized monographs and reviews cited at the end of Chapters 1 and 2. Selected references to the recent development in this field are listed in Addendum.

The author wishes to express his sincere thanks to Professor Eiji Osawa whose suggestion was the direct motivation for writing this book. Thanks are also due to Dr. Yoshihisa Miwa for his valuable comments on thermodynamics, to Professor Tooru Taga for information on crystallography and to Professor Teizo Kitagawa for his expertise in lattice dynamics. The author hopes that the publication of this book will encourage communication between different fields of chemistry and physics through a common interest in molecular mechanics.

January 1999

Katsunosuke Machida

Chapter 1

Introduction

From the viewpoint of chemists, a molecule is an assembly of two kinds of charged particles, electrons and atomic nuclei. If molecular mechanics is literally interpreted as the mechanics of molecules, both electrons and atomic nuclei may be taken as the object of molecular mechanics, but this is not the case. According to the widely accepted definition in the literature,[1–7] molecular mechanics is the mechanics of atomic nuclei moving around in a molecule or in an assembly of molecules. No attention is paid explicitly to the movements of electrons in molecular mechanics. Atomic nuclei are often denoted by an abbreviated expression, "nuclei". This convention will also be used in this book, as long as no ambiguity arises.

The standpoint of molecular mechanics in which the electrons are not taken as the subject of motion is based on the Born–Oppenheimer approximation. Born and Oppenheimer have shown that the motions of electrons and nuclei are "separable" from each other to the fourth order of approximation with respect to the quantity $(m_e / \overline{m})^{1/4}$, where m_e is the mass of electron and $\overline{m}$ is the average of the masses of nuclei in the molecule.[8] Electrons are statistically distributed around nuclei according to the principle of quantum mechanics, and offer a part of the force field which regulates the motions of nuclei. The total force on each nucleus is the sum of the force generated by the electrons and the electrostatic force between the positively charged nuclei themselves. A nucleus is regarded to be a material point. The inner structure of atomic nuclei is not a subject of molecular mechanics. In this respect, we must pay attention to the distinction between the two concepts, an atomic nucleus and an atom. Here, an atomic nucleus is a concept to which no extension is attributed, while an atom means a composite particle consisting of an atomic nucleus and a certain number of electrons around it. Thus, the term "atom" is not clearly defined in a molecule, although such terms as "atomic coordinates" and "interatomic distances" are often used. In principle, the terms "atomic nuclei" (or "nuclei") and "atoms" should be distinguished rigorously from each other.

In this volume, the term "nuclei" stands for the material points which are the subjects of motion in molecular mechanics, while the term "atom" is used as a somewhat fuzzy concept which represents an atomic nucleus and its surrounding electrons. Examples of the former include expressions such as "nuclear coordinates" and "internuclear distances", while examples of the latter include "nonbonded atom–atom interaction" and "effective atomic charges".

The energy eigenvalue of a stationary electronic state of an isolated N-atomic molecule is a continuous function of the $3N$ nuclear coordinates. The $3N$-dimensional space spanned by $3N$ Cartesian coordinates of an N-particle system is called the configuration space of the system. A point in a configuration space is called a configuration point. The force acting on a nucleus is clearly defined at every configuration point of the molecule as minus the gradient of the energy eigenvalue evaluated at the point. This force is a conservative force which depends neither on the history of motion nor on the velocity of that nucleus, but is uniquely determined by the relative positions of the nuclei constituting the system. Thus the energy eigenvalue of a stationary electronic state including the nuclear repulsion terms plays the role of the potential energy by which the entire motion of the nuclei in the system is controlled.

Since the potential energy depends solely on the relative positions of nuclei, it can be described as a function of appropriate variables which uniquely represent relative nuclear positions at any given time. This function is called the potential energy function, or simply, the "potential function" of the system. The force field being generated from this potential function and acting on the moving nuclei is called the molecular force field. The relation between the molecular force field and the potential function is the same as the relation between an electrostatic field and the electrostatic potential, the gradient of which is a vector quantity representing the electrostatic force.

Once the potential function of a molecule is given, many physical quantities characteristic of the molecule can be derived from it by simply expanding the function as a power series of appropriate variables or by solving the quantum mechanical or the classical equations of motion which incorporate the power series as the potential energy. In the case where the system is an isolated molecule, the physical quantities to be derived include the thermodynamic functions (enthalpies, entropies, free energies and heat capacities), the equilibrium and the thermally averaged structures and various spectroscopic constants. The last items include the normal frequencies, the centrifugal distortion constants, the Coriolis coupling constants, the anharmonicity constants and the vibration–rotation interaction constants. For assemblies of a few molecules taken as molecular compounds, the energies of association, the dissociation constants, the structures and the spectroscopic properties are deducible from the potential function. Finally, the potential function of a molecular crystal gives the equilibrium structure (cell constants and atomic coordinates in the asymmetric unit), the elastic constants, the lattice energies, the surface energies, the normal frequencies, the anisotropic temperature factors in X-ray and neutron diffractions, the temperature diffuse scattering patterns and the inelastic neutron scattering spectra.

By using any reasonable model of the potential function, all these physical quantities are given certain analytical expressions which can be evaluated with the help of appropriate computer programs. Any physical quantity deducible from the potential function in this way can be a subject of molecular mechanics. For a random assembly of molecules, many physical quantities cannot be expressed analytically. To evaluate such a physical quantity, the motion of the whole system under an assumed potential function is simulated over a time course according to the classical or the quantum mechanical equation of motion. The ensemble average of the physical quantity is recorded as a time sequence, and the time average of this sequence is taken as the observable physical quantity of the system under the given condition. This approach has been taken to form a large branch of computational molecular science, *i.e.*, molecular dynamics,[9–12] which is traditionally distinguished from molecular mechanics. Molecular dynamics covers the physical properties of liquids and transcendental phenomena which are not tractable by molecular mechanics, but the time required for this sort of computation is about 10^4–10^5 times greater than those required for molecular mechanics. Obviously, molecular mechanics and molecular dynamics are closely related to each other. From the standpoint of molecular dynamics, molecular mechanics plays the role of setting up and calibrating the potential function.

In earlier works, the potential function to be used in molecular mechanics was constructed exclusively by referring to experimental data.[13–16] With the progress of molecular orbital methods, useful information on the potential function has become available from sophisticated quantum mechanical calculations.[17–20] The theoretical computation has the advantage that any quantities obtained in the course of calculation are available as the output at any stage of calculation. Thus we can partition the energy into several parts according to the physical origin, or follow details of the molecular distortion along a reaction path. However, the theoretical calculation at the present stage cannot be free from certain approximations which are inherently associated with systematic error. On the other hand, any observed data collected through a properly designed experimental work should be equivalent to the results of an exact quantum mechanical calculation. While experiments are superior to theoretical calculations in not being biased by any approximations, it is an essential limitation that experimental data are available only for observable quantities. Thus, the roles of the experimental and the theoretical methods are complementary as the source of information on the potential function. In setting up the potential function, we must understand well the advantages and the shortcomings of both methods, and utilize them properly. In general, we must rely much on theory when the form of the potential model is to be set up, while experimental data must be used when determining individual potential parameters.

A good potential function cannot be set up without understanding the physical meaning of what is to be calculated. The physical meaning of a molecular property expressed by a mathematical formula is closely related to the logical process of deriving the formula from the fundamental concepts underlying the principles of molecular mechanics. This is the reason why mathematical procedures of deriving various properties of molecules and crystals related to the nuclear motion have been collected in this book. Topics not yet fully explored in current molecular mechanics programs are included if any information on the potential function is expected to be available therefrom in the future. The contents are arranged in the following order.

The subjects related to the coefficients of Taylor series expansion of the potential function are taken up in Chapters 2 through 5 according to the order of power in the series. Chapter 2 deals with the potential function itself, that is, the zeroth-order term of the power series expanded at a configuration point of nuclei. Chapter 3 is concerned with the search for the potential minima through the calculation of first-order terms followed by their elimination. In Chapter 4, normal coordinate analyses based on second-order terms calculated at a potential minimum are taken up as an essential source of information on molecular vibrations. Chapter 5 deals with the role of third- and fourth-order terms in the second-order perturbation theory of vibrational anharmonicity and the vibration–rotation interaction. The molecular structures and constants treated in Chapters 2 through 5 are used in the statistical mechanical and the empirical methods of calculation of thermodynamic functions described in Chapter 6.

In Chapter 7, empirical methods of describing the electric properties of molecules in terms of the nuclear coordinates are discussed. The charge distribution in a molecule is related not only to the electrostatic part of the potential function but also to such observable physical properties as dipole moments, polarizabilities and vibrational spectra. Construction of any reliable models of the intramolecular charge distribution is an important problem to be solved by molecular mechanics. Chapter 8 describes how to simulate infrared absorption and Raman spectra by using the potential function, the spectral intensity parameters and the bandwidth parameters. Generally, vibrational spectra contain much information on the potential function and the electric properties of molecules in comparison with the labor and cost required for measurement. Progress in this field is thus expected to improve the reliablility and applicability of molecular mechanics calculations.

Finally, molecular crystals are taken up in Chapter 9. Although a crystal is an assembly of an uncountable number of molecules, the periodicity enables to give analytical expression to some of its physical properties. Since the potential function regulating the nuclear motion arises from a common principle irrespective of whether the system is an isolated molecule or an assembly of molecules, a potential function which can predict the properties of isolated molecules should predict crystal properties equally well. As the molecular size increases, the experimental data available in the crystal phase become much more abundant than those available in the gas phase. For polar molecules, gas phase data are difficult to measure in general. In this respect, extension of molecular mechanics to crystals is particularly important in the application to molecules showing biological activities.

References

1) L.S. Bartell, *J. Chem. Phys.*, **32**, 827 (1960).
2) K.B. Wiberg, *J. Am. Chem. Soc.*, **87**, 1070 (1965).
3) D.N.J. White, *Struct. Diffr. Methods*, **6**, 38 (1978).
4) U. Burkert and N.L. Allinger, *Molecular Mechanics*, The American Chemical Society, Washington D.C. (1982).
5) Kj. Rasmussen, *Potential Energy Functions in Conformational Analysis*, Springer, Berlin (1985).
6) J.P. Bowen and N.L. Allinger, *Molecular Mechanics : The Art and Science of Parametrization, Reviews in Computaitonal Chemistry II*, (ed. K.B. Lipkowitz and D. B. Boyd) VCH Publishers, New York (1991), p.81.
7) P.M. Ivanov, *JCPE Newsletter*, **3** (4), 16 (1992); *ibid.*, **4** (1), 9 (1992); *ibid.*, **7** (3), 3 (1995).
8) M. Born and R. Oppenheimer, *Ann. Phys.* **84**, 457 (1927); M. Born and K. Huang, *Dynamical Theory of Crystal Lattices*, App. VII, VIII, Oxford University Press, London (1954).
9) A. Rahman and F.H. Stillinger, *J. Chem. Phys.*, **55**, 3336 (1971).
10) M.P. Allen and D. J. Tildesley, *Computer Simulation of Liquids*, Oxford University Press, London (1989).
11) H.C. Andersen, *J. Chem. Phys.*, **72**, 2384 (1980).
12) S. Nose, *Mol. Phys.*, **52**, 255 (1984); *idem.*, *J. Chem. Phys.*, **81**, 511 (1984).
13) F.H. Westheimer and J.E. Meyer, *J. Chem. Phys.*, **14**, 733 (1946).
14) N.L. Allinger, *Adv. Phys. Org. Chem.*, **13**, 1 (1976).
15) O. Ermer, *Struct. Bonding (Berlin)*, **27**, 161 (1976).
16) E. M. Engler, J.D. Andose and P.v.R. Schleyer, *J. Am. Chem. Soc.*, **95** 8005 (1973).
17) P. Pulay and W. Meyer, *J. Mol. Spectrosc.*, **40**, 59 (1971).
18) J.A. Pople, R. Krishnan, H.B. Schlegel and J.S. Binkley, *Int. J. Quantum Chem.*, **S13**, 225 (1979).
19) Y. Osamura, Y. Yamaguchi and H.F. Schaefer, III, *J. Chem. Phys.*, **75**, 2919 (1981); *idem., ibid.*, **77**, 383 (1982).
20) P. Saxe, Y. Yamaguchi and H.F. Schaefer, III, *J. Chem. Phys.*, **77**, 5647 (1982).

Chapter 2

Molecular Force Field

2.1 Outline of the Method of Calculation

The potential function of an isolated molecule in the electronic ground state is uniquely determined if the coordinates of all the constituent nuclei are given. Let X_i, Y_i and Z_i be the coordinates of the ith nucleus of an N-atomic molecule in a space-fixed Cartesian coordinate system, and let the potential function of the molecule in this system be denoted by

$$V \equiv V(X_1, Y_1, Z_1, X_2, Y_2, Z_2, \cdots, X_N, Y_N, Z_N) \tag{2-1}$$

Then the components of the force on the ith nucleus are given by

$$(f_i)_X = -\partial V/\partial X_i,\ (f_i)_Y = -\partial V/\partial Y_i,\ (f_i)_Z = -\partial V/\partial Z_i \tag{2-2}$$

The use of different symbols X, Y and Z for the Cartesian coordinates along the three axes is required for clarifying the discussion in some cases, but leads to a tedious description in those cases where axes X, Y and Z need not be distinguished from one another. In the latter cases we shall use a single symbol x with the consecutive numbering from 1 to $3N$ for the Cartesian coordinates with the convention that

$$X_i \equiv x_{3i-2},\quad Y_i \equiv x_{3i-1},\quad Z_i \equiv x_{3i},\qquad (i = 1\text{–}N). \tag{2-3}$$

The potential functions expressed in terms of Cartesian coordinates of the nuclei are useful for mathematical treatment in general theory, but are not so convenient from the chemical point of view. The relationship between the internal energy and the shape of a molecule is more intuitively represented in terms of a potential function with arguments not affected by the position and the orientation of the whole molecule.[1)] The bond lengths, bond angles and other structure parameters, which are ordinarily used for describing the molecular structure, are suitable examples of such arguments. They are not only independent of the molecular position and orientation, but appear so commonly in a wide variety of molecules that detailed comparison of potential functions between different molecules is greatly facilitated. These parameters will be called "the internal variables" and distinguished from "the external variables", which will be introduced in Chapter 3 for describing the amounts of translational and rotational motions of a molecule as a whole. By denoting the ith internal variable by R_i, the potential function is written as

$$V = V(R_1, R_2, \ldots). \tag{2-4}$$

In this case, the force exerted on individual atoms can be calculated by expressing each internal variable, say R_i, as a differentiable function of the Cartesian coordinates of nuclei,

$$R_i = R_i(x_1, x_2, \ldots, x_{3N}), \tag{2-5}$$

and using the chain formula for the partial derivatives of multivariable functions in the form

$$\frac{\partial V}{\partial x_k} = \sum_i \frac{\partial V}{\partial R_i} \frac{\partial R_i}{\partial x_k} . \tag{2-6}$$

Evaluation of the potential function for a given nuclear configuration is the most fundamental process of all molecular mechanics calculations. To understand the mathematical background of the calculation, it is convenient to divide the process into the following three steps. The first step is the calculation of the Cartesian coordinates of the nuclei, the second is the calculation of the structure parameters to be used as the arguments of the potential function, and the third is the calculation of the potential function itself. In most computer programs of molecular mechanics, numerical calculation of the potential function is actually carried out in these three steps, although the first step is omitted if the nuclear coordinates are already known. Details of the first and the second steps will be given in sections 2.2 and 2.3, respectively. The third step is straightforward if the potential function has been defined as a sum of simple functions of the structure parameters to be determined in section 2.2. General properties of a potential function given in the form of such a sum will be discussed in section 2.4, and section 2.5 will deal with parametric methods of describing the potential function in this way.

2.2 Generation of Initial Nuclear Coordinates

There are a great number of molecules for which the Cartesian coordinates of the constituent nuclei are available from various experiments in spectroscopy and crystallography. One must calculate the nuclear coordinates, however, when reliable experimental data are lacking or when a non-existing molecular form is to be dealt with. The orientation of a more-than-three atomic molecule with a nonlinear bond angle θ_{123} in space is determined if the positions of nuclei 1, 2 and 3 are fixed. The Cartesian coordinates of the other nuclei of this molecule can be calculated from the data of appropriate internal variables by carrying out a series of translational and rotational transformations of the coordinates axes.[2] The following is a variation of this method in which a transformation of the coordinate axes is replaced by a certain expression of the vector along a newly formed bond in terms of the unit vectors along the preceding bonds.[3] The use of a fixed coordinate system throughout is convenient when a local group of a large molecule is to be replaced by another group for which only the internal variables are known.

Suppose that the Cartesian coordinates of three atomic nuclei i, j and k bonded by two chemical bonds i–j and j–k are known. The bond angle θ_{ijk} is assumed to be smaller than 180°, so that the plane through i, j and k is uniquely determined. Let us consider how to add another nucleus l to this system and how to specify the three internal variables necessary for determining its coordinates uniquely. There are two cases, serial and branched, as to the choice of the three internal variables, depending on which of the nuclei j and k is bonded to l. The serial case is shown in Fig. 2.1(a), where l is bonded to k, and the bond length r_{kl}, the bond angle θ_{jkl} and the dihedral angle τ_{ijkl} are given. In the branched case, shown in Fig. 2.1(b), l is bonded to j, and the bond length r_{jl} are given together with two bond angles θ_{ijl} and θ_{kjl}. In the serial case, which is far more frequently used than the branched case, the

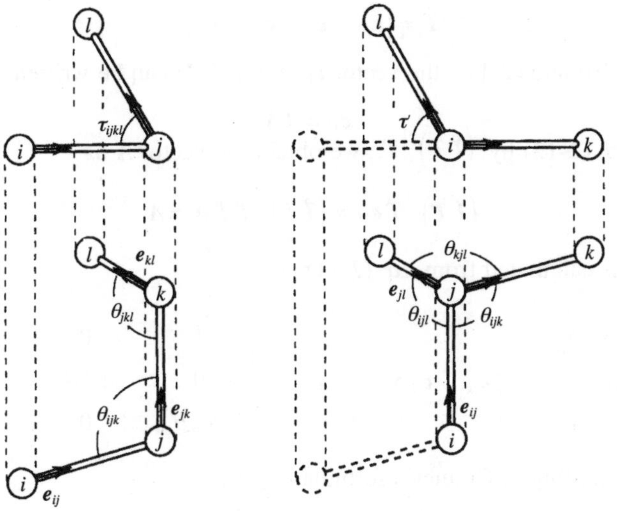

(a) (b)

Fig. 2.1 Generation of nuclear Cartesian coordinates.

Cartesian coordinates of l are calculated as follows. Let the unit vectors along the bonds i–j, j–k and k–l be denoted by $\boldsymbol{e}_{ij}$, $\boldsymbol{e}_{jk}$ and $\boldsymbol{e}_{kl}$, respectively. Then the vector $\boldsymbol{e}_{kl}$ is expressed as a linear combination of three vectors $\boldsymbol{e}_{ij}$, $\boldsymbol{e}_{ij} \times \boldsymbol{e}_{jk}$ and $\boldsymbol{e}_{jk}$ in the form

$$\boldsymbol{e}_{kl} = a\boldsymbol{e}_{jk} + b\,(\boldsymbol{e}_{ij} \times \boldsymbol{e}_{jk}) + c\boldsymbol{e}_{ij} \,. \tag{2-7}$$

Once the coefficients a, b and c are given, the components of the vector $\boldsymbol{e}_{kl}$ can be calculated from Eq. (2-7) in which the vectors $\boldsymbol{e}_{ij}$ and $\boldsymbol{e}_{jk}$ are determined from the coordinates of the nuclei i, j and k. On defining the bond vector $\boldsymbol{r}_{kl}$ as a vector directed from k to l in the form

$$\boldsymbol{r}_{kl} = [X_{kl} \quad Y_{kl} \quad Z_{kl}]^{\mathrm{T}} = r_{kl}\boldsymbol{e}_{kl} = [X_l - X_k \quad Y_l - Y_k \quad Z_l - Z_k]^{\mathrm{T}} \,, \tag{2-8}$$

the coordinates of the nucleus l are calculated by

$$X_l = X_k + X_{kl}, \qquad Y_l = Y_k + Y_{kl}, \qquad Z_l = Z_k + Z_{kl} \,. \tag{2-9}$$

The symbol $[\ \ ]^{\mathrm{T}}$ will represent hereafter the transpose of a vector or a matrix whenever the elements are explicitly given or the upper wave ˜ cannot be used as a standard symbol for the transpose. If a vector or a matrix is given as a single letter, e.g., $\boldsymbol{a}$, its transpose will be denoted by $\tilde{\boldsymbol{a}}$. A bold type symbol for a vector will implicitly mean a column vector unless stated otherwise, but the definition of a vector will be given using $[\ \ ]^{\mathrm{T}}$ for a row vector whenever the space can be saved by the symbol.

To evaluate the coefficients a, b and c in Eq. (2-7), we may introduce a vector $\boldsymbol{A}$ consisting of the components a, b and c in the form

$$\boldsymbol{A} = [a \quad b \quad c]^{\mathrm{T}} \tag{2-10}$$

and a 3×3 matrix $\boldsymbol{T}$ consisting of the three column vectors [(3×1) matrices] in Eq. (2-7) arranged rowwise, *i.e.*,

$$\boldsymbol{T} = \left[\boldsymbol{e}_{jk} \quad \boldsymbol{e}_{ij} \times \boldsymbol{e}_{jk} \quad \boldsymbol{e}_{ij}\right] . \tag{2-11}$$

By using Eqs. (2-10) and (2-11), the vector $\boldsymbol{e}_{kl}$ in Eq.(2-7) can be written as

$$\boldsymbol{e}_{kl} = \boldsymbol{T}\boldsymbol{A} \tag{2-12}$$

Premultiplying Eq. (2-12) by $\left(\tilde{\boldsymbol{T}}\boldsymbol{T}\right)^{-1}\tilde{\boldsymbol{T}}$, we obtain the vector $\boldsymbol{A}$ as

$$\left(\tilde{\boldsymbol{T}}\boldsymbol{T}\right)^{-1}\tilde{\boldsymbol{T}}\boldsymbol{e}_{kl} = \left(\tilde{\boldsymbol{T}}\boldsymbol{T}\right)^{-1}\tilde{\boldsymbol{T}}\boldsymbol{T}\boldsymbol{A} = \boldsymbol{A} . \tag{2-13}$$

The matrix $\tilde{\boldsymbol{T}}\boldsymbol{T}$ is calculated from Eq. (2-11) as

$$\tilde{\boldsymbol{T}}\boldsymbol{T} = \begin{bmatrix} \tilde{\boldsymbol{e}}_{jk} \\ \left(\boldsymbol{e}_{ij} \times \boldsymbol{e}_{jk}\right)^{\mathrm{T}} \\ \tilde{\boldsymbol{e}}_{ij} \end{bmatrix} \left[\boldsymbol{e}_{jk} \quad \boldsymbol{e}_{ij} \times \boldsymbol{e}_{jk} \quad \boldsymbol{e}_{ij}\right] = \begin{bmatrix} 1 & 0 & -\cos\theta_{ijk} \\ 0 & \sin^2\theta_{ijk} & 0 \\ -\cos\theta_{ijk} & 0 & 1 \end{bmatrix} \tag{2-14}$$

and is inverted according to Cramer's formula to give

$$\left(\tilde{\boldsymbol{T}}\boldsymbol{T}\right)^{-1} = \frac{1}{\sin^2\theta_{ijk}} \begin{bmatrix} 1 & 0 & \cos\theta_{ijk} \\ 0 & 1 & 0 \\ \cos\theta_{ijk} & 0 & 1 \end{bmatrix} . \tag{2-15}$$

Premultiplying $\boldsymbol{e}_{kl}$ by $\tilde{\boldsymbol{T}}$ leads to

$$\tilde{\boldsymbol{T}}\boldsymbol{e}_{kl} = \begin{bmatrix} \tilde{\boldsymbol{e}}_{jk}\boldsymbol{e}_{kl} \\ \left(\boldsymbol{e}_{ij} \times \boldsymbol{e}_{jk}\right)^{\mathrm{T}}\boldsymbol{e}_{kl} \\ \tilde{\boldsymbol{e}}_{ij}\boldsymbol{e}_{kl} \end{bmatrix} = \begin{bmatrix} -\cos\theta_{jkl} \\ \sin\theta_{ijk}\sin\theta_{jkl}\sin\tau_{ijkl} \\ \cos\theta'_{ij,kl} \end{bmatrix} , \tag{2-16}$$

where $\theta'_{ij,kl}$ is the angle formed by the unit vectors $\boldsymbol{e}_{ij}$ and $\boldsymbol{e}_{kl}$.

According to the cosine theorem in spherical trigonometry (Appendix 1), cos $\theta'_{ij,kl}$ can be written in terms of θ_{ijk}, θ_{jkl} and τ_{ijkl}, as

$$\cos\theta'_{ij,kl} = \cos\theta_{ijk}\cos\theta_{jkl} + \sin\theta_{ijk}\sin\theta_{jkl}\cos\left(\pi - \tau_{ijkl}\right). \tag{2-17}$$

From Eqs. (2-13) and (2-15–17), the vector $\boldsymbol{A}$ is calculated to be

$$\boldsymbol{A} = \left(\tilde{\boldsymbol{T}}\boldsymbol{T}\right)^{-1}\tilde{\boldsymbol{T}}\boldsymbol{e}_{kl} = \begin{bmatrix} -\cos\theta_{jkl} - s_1\cos\theta_{ijk}\cos\tau_{ijkl} \\ s_1\sin\tau_{ijkl} \\ -s_1\cos\tau_{ijkl} \end{bmatrix} , \tag{2-18}$$

where $s_1 = \sin\theta_{jkl}/\sin\theta_{ijk}$.

The branched case may be reduced to the serial case by transferring the bond vector $\boldsymbol{r}_{jk}$ into the position indicated by the dotted line in Fig. 2.1(b). The angle τ' in Fig. 2.1(b) corresponds to the torsion angle τ_{ijkl} in Fig. 2.1(a), so that the relation between angles θ_{kjl} and τ' in Fig. 2.1(b) is the same as the relation between angles $\theta'_{ij,kl}$ and τ_{ijkl} in Fig. 2.1(a). In the branched case, angles θ_{ijl} and θ_{kjl} are usually known but τ' is not. In that case, cos τ' can be calculated from θ_{kjl} by

$$\cos\tau' = \frac{-\cos\theta_{kjl} + \cos\theta_{ijk}\cos\theta_{ijl}}{\sin\theta_{ijk}\sin\theta_{ijl}}, \tag{2-19}$$

opposite to the serial case in which Eq. (2-17) is used. In analogy with Eq. (2-7), the unknown unit vector $\boldsymbol{e}_{jl}$ can be written in the form

$$\boldsymbol{e}_{jl} = a\,\boldsymbol{e}_{jk} + b(\boldsymbol{e}_{ij} \times \boldsymbol{e}_{jk}) + c\,\boldsymbol{e}_{ij} = \boldsymbol{TA}', \tag{2-20}$$

and the vector $\boldsymbol{A}'$ is obtained as

$$\boldsymbol{A}' \equiv \begin{bmatrix} a \\ b \\ c \end{bmatrix} = \begin{bmatrix} -s_2\cos\tau' \\ -s_2\sin\tau' \\ -\cos\theta_{ijl} - s_2\cos\theta_{ijk}\cos\tau' \end{bmatrix}, \tag{2-21}$$

where

$$s_2 = \sin\theta_{ijl}/\sin\theta_{ijk}. \tag{2-22}$$

In Eq. (2-21), $\sin\tau'$ is calculated by

$$\sin\tau' = \pm(1-\cos^2\tau')^{1/2}, \tag{2-23}$$

where the double sign is taken to be positive if $0 < \tau' < \pi$, that is, if the plane i–j–l viewed from the side of i is brought to the position of plane i–j–k by a clockwise rotation. The negative sign is taken otherwise.

In the standard input data for computer programs calculating the molecular structure, the information necessary for generating the nuclear coordinates is usually written in a line per nucleus with a properly designed format. The collection of such lines for all the nuclei in a molecule is often called the Z matrix, and is used in many *ab initio* MO programs.[4)] Once the Z matrix is given, one can construct the vector $\boldsymbol{A}$ from two successive bond vectors as given by Eq. (2-18) or (2-21), and thereby calculate all the nuclear coordinates successively. There is an arbitrariness in determining the positions of the first three nuclei 1, 2 and 3. A simple choice for the case where $\theta_{123} < \pi$ is to place the nucleus 1 at the origin, 2 on the x-axis, and 3 on the XY-plane. Then the coordinates of 1, 2 and 3 are given by

$$X_1 = 0, \qquad Y_1 = 0, \qquad Z_1 = 0, \tag{2-24a}$$

$$X_2 = r_{12}, \qquad Y_2 = 0, \qquad Z_2 = 0 \tag{2-24b}$$

and

$$X_3 = r_{12} + r_{23}\cos\theta_{123}, \quad Y_3 = r_{23}\sin\theta_{123}, \quad Z_3 = 0. \tag{2-24c}$$

In Eqs. (2-24a–c), there are six conditions, each of which fixes a coordinate to zero, and three structure parameters r_{12}, r_{23} and θ_{123}. The position of any other nucleus l ($l \geq 4$) is determined by giving three structure parameters r_{kl}, θ_{jkl} and τ_{ijkl} for the serial case, and r_{jl}, θ_{ijl} and θ_{kjl} for the branched case. If the equilibrium value of the valence angle θ_{jkl} is 180°, we can define neither the plane j–k–l nor the torsion angle τ_{ijkl}. In this case, the constraint of the nucleus l on the straight line passing j and k is attained by giving a single condition, $\theta_{jkl} = \pi$. The position of l on this line is determined by the bond length r_{kl}. Thus, the coordinates of the nucleus l seem to be determined by two structure parameters θ_{jkl} and r_{kl}.

However, since there are two degrees of freedom of motion on a plane perpendicular to the bond j–k, the constraint to keep the linearity of the sequence j–k–l implicitly requires two conditions. In fact, if the nucleus l deviates from the straight line connecting j and k by an infinitesimal amount, its position cannot be described uniquely in terms of a single angle θ_{jkl}. The total number of conditions and parameters is always three per nucleus, giving the number of degrees of freedom of an N-particle system, $3N$.

We have specified the orientation of a nonlinear molecule in a Cartesian frame by introducing three conditions in Eq. (2-24a), two in Eq. (2-24b) and one in Eq. (2-24c). Thus, the number of internal variables necessary for determining the nuclear configuration of an N-atomic nonlinear molecule is $3N - 6$. In the case of a linear molecule, the orientation in a Cartesian frame is already determined when we introduce five conditions in Eqs. (2-24a) and (2-24b). Specification of $3N - 5$ internal variables is thus necessary and sufficient for describing the nuclear configuration of an N-atomic linear molecule.

2.3 Calculation of Internal Variables

2.3.1 Bond Lengths, Bond Angles and Torsion Angles

The potential function of a polyatomic molecule cannot be determined unless the nuclear configuration is uniquely given. Thus, three types of internal variables, bond lengths, bond angles and torsion angles, which are necessary parameters for generating the nuclear coordinates, are also necessary as arguments of the potential function. Besides these three, the potential function used in molecular mechanics includes several types of internal variables, nonbonded internuclear distances, out-of-plane angles and linear bond angles. The nonbonded internuclear distances are used for describing the force acting between any pairs of nonbonded nuclei. The necessity of the other two types of variables will be discussed later in connection with the transformation of the potential function into the Cartesian coordinate system.

The internal variables commonly used in describing the potential function are illustrated in Fig. 2.2. These internal variables are expressed in terms of the bond vectors defined by Eq. (2-8), and can be calculated from the Cartesian coordinates of relevant nuclei as follows.[5)]

The bond length r_{ij} is the absolute value of the bond vector $\boldsymbol{r}_{ij}$, and is given by

$$r_{ij} = \left(\boldsymbol{r}_{ij} \cdot \boldsymbol{r}_{ij}\right)^{1/2} = \left(X_{ij}^{\ 2} + Y_{ij}^{\ 2} + Z_{ij}^{\ 2}\right)^{1/2} \ . \tag{2-25}$$

The complete Z matrix of an acyclic compound should contain all the bond lengths, while some bond lengths do not necessarily appear in the Z matrix of a ring compound. From the mathematical viewpoint, the nonbonded internuclear distances are not distinguished from the bond lengths, since they are expressed in the same form as Eq. (2-25) in terms of the distance vector between the two nonbonded nuclei i and j.

The bond angle θ_{ijk} spanning the two bonds j–i and j–k can be calculated by using the well-known formula for the scalar product of two bond vectors,

$$r_{ij}\, r_{jk} \cos\theta_{ijk} = \boldsymbol{r}_{ji} \cdot \boldsymbol{r}_{jk} = X_{ji}X_{jk} + Y_{ji}Y_{jk} + Z_{ji}Z_{jk} \ . \tag{2-26}$$

From Eq. (2-26), we have

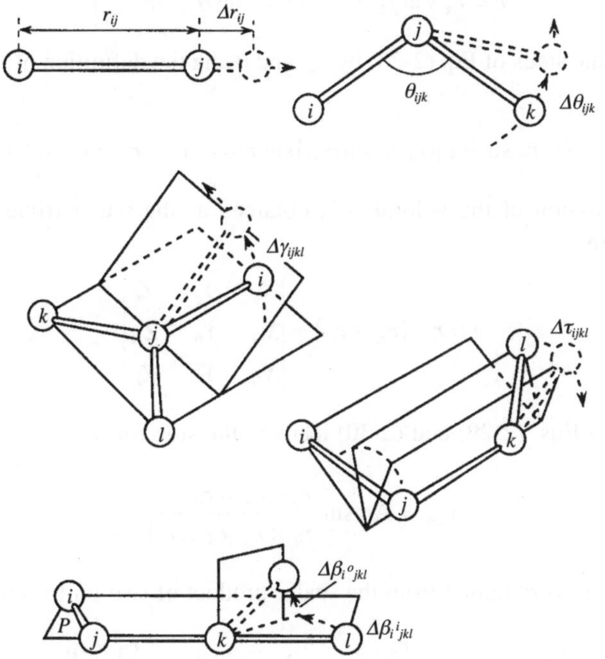

Fig. 2.2 Types of internal variables and internal coordinates.

$$\theta_{ijk} = \text{Arccos}\left(\boldsymbol{r}_{ji} \cdot \boldsymbol{r}_{jk} / r_{ij}\, r_{jk}\right) \tag{2-27}$$

Except for the cases where θ_{ijk} is near π, Eq. (2-27) can be used unconditionally in a computer program, since the bond angles are always defined within the variable range of the principal values of arccosines.

It is recommended that the torsion angle τ_{ijkl} be calculated in two ways through formulae for sines and cosines.[6] Let v be the volume of the parallelepiped formed by three bond vectors $\boldsymbol{r}_{ij}$, $\boldsymbol{r}_{jk}$ and $\boldsymbol{r}_{kl}$, taken as positive when the vectors $\boldsymbol{r}_{ij}$, $\boldsymbol{r}_{jk}$ and $\boldsymbol{r}_{kl}$ form a right-handed system in this order. Projecting this parallelepiped on a plane perpendicular to the bond j–k as shown in Fig. 2.3, we obtain parallelogram ABCD, the area s of which is related to v by

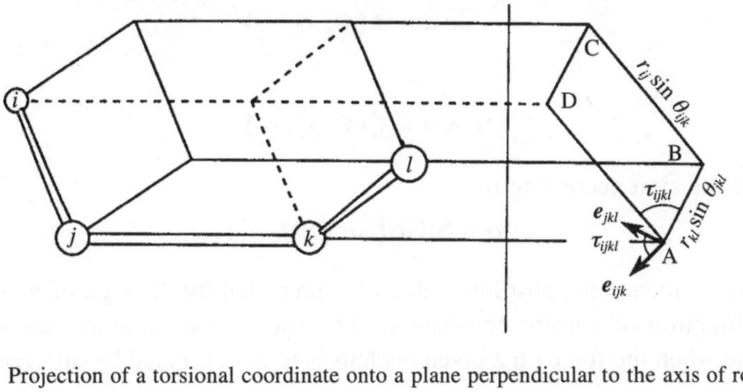

Fig. 2.3 Projection of a torsional coordinate onto a plane perpendicular to the axis of rotation.

$$v = r_{jk}\, s = r_{jk}\left(r_{ij}\, r_{kl} \sin\theta_{ijk} \sin\theta_{jkl} \sin\tau_{ijkl}\right) . \tag{2-28}$$

Multiplying both the sides of Eq. (2-28) by r_{jk}, and using the definition of the vector product, we have

$$r_{jk}\, v = \left(r_{ij}\, r_{jk} \sin\theta_{ijk}\right)\left(r_{jk}\, r_{kl} \sin\theta_{jkl}\right)\sin\tau_{ijkl} = \left|\boldsymbol{r}_{ij} \times \boldsymbol{r}_{jk}\right|\left|\boldsymbol{r}_{jk} \times \boldsymbol{r}_{kl}\right| \sin\tau_{ijkl} \tag{2-29}$$

Another expression of the volume v is obtained as the scalar triple products of bond vectors in the form

$$v = \boldsymbol{r}_{ij} \cdot \left(\boldsymbol{r}_{jk} \times \boldsymbol{r}_{kl}\right) = \begin{vmatrix} X_{ij} & Y_{ij} & Z_{ij} \\ X_{jk} & Y_{jk} & Z_{jk} \\ X_{kl} & Y_{kl} & Z_{kl} \end{vmatrix} . \tag{2-30}$$

Eliminating v from Eqs. (2-29) and (2-30) leads to the sine formula[7]

$$\tau_{ijkl} = \operatorname{Arcsin} \frac{\boldsymbol{r}_{ij} \cdot \left(\boldsymbol{r}_{jk} \times \boldsymbol{r}_{kl}\right) r_{jk}}{\left|\boldsymbol{r}_{ij} \times \boldsymbol{r}_{jk}\right|\left|\boldsymbol{r}_{jk} \times \boldsymbol{r}_{kl}\right|} . \tag{2-31}$$

The cosine formula[1] is obtained from the scalar product of two unit vectors defined by

$$\boldsymbol{e}_{ijk} = -\frac{\boldsymbol{r}_{ij} \times \boldsymbol{r}_{jk}}{\left|\boldsymbol{r}_{ij} \times \boldsymbol{r}_{jk}\right|} \quad \text{and} \quad \boldsymbol{e}_{jkl} = -\frac{\boldsymbol{r}_{jk} \times \boldsymbol{r}_{kl}}{\left|\boldsymbol{r}_{jk} \times \boldsymbol{r}_{kl}\right|} .$$

Since the vectors $\boldsymbol{e}_{ijk}$ and $\boldsymbol{e}_{jkl}$ are perpendicular to the planes formed by the atom sequences i–j–k and j–k–l, respectively, we have $\cos\tau_{ijkl} = \boldsymbol{e}_{ijk} \cdot \boldsymbol{e}_{jkl}$, or

$$\tau_{ijkl} = \operatorname{Arccos} \frac{\left(\boldsymbol{r}_{ij} \times \boldsymbol{r}_{jk}\right) \cdot \left(\boldsymbol{r}_{jk} \times \boldsymbol{r}_{kl}\right)}{\left|\boldsymbol{r}_{ij} \times \boldsymbol{r}_{jk}\right|\left|\boldsymbol{r}_{jk} \times \boldsymbol{r}_{kl}\right|} . \tag{2-32}$$

On calculating an angle variable from a given value of its trigonometric functions, the principal value is usually obtained by using an inverse trigonometric function installed as a standard built-in function in current computers. Accordingly, to determine the value of an internal angle variable α within the range $-\pi \le \alpha \le \pi$, one should calculate both the sines and cosines as the functions of the nuclear coordinates in the forms

$$\sin\alpha = f_s(x_1, x_2, \cdots) \tag{2-33a}$$

and

$$\cos\alpha = f_c(x_1, x_2, \cdots), \tag{2-33b}$$

and determine the sign according to

$$\alpha = \mathrm{SIGN}(\operatorname{Arccos} f_c, f_s) ,$$

where SIGN(a, b) means the absolute value of a preceded by the sign of b, and is available as a built-in function of current computers. The use of both sine and cosine formulae is convenient also when the force on a given nucleus is to be estimated by differentiating internal variables with respect to the Cartesian coordinates according to Eq. (2-6).

Differentiation of Eqs. (2-33a) and (2-33b) gives

$$\partial\alpha/\partial x_i = (1/\cos\alpha)(\partial f_s/\partial x_i) \tag{2-34}$$

and

$$\partial\alpha/\partial x_i = -(1/\sin\alpha)(\partial f_c/\partial x_i), \tag{2-35}$$

respectively. The denominator of Eq. (2-34) vanishes when α approaches either 0° or 180°, while that of Eq. (2-35) vanishes when α approaches ±90°. In both these cases, the numerical calculation becomes unstable.

The first derivatives of a torsion angle τ_{ijkl} with respect to the Cartesian coordinates of the nuclei i, j, k and l can be expressed in a non-divergent form by eliminating the vanishing factors from both the denominator and the numerator in the right side of either Eq. (2-34) or (2-35). These are known as the components of the ***s*** vectors widely used in the normal coordinate analysis.[1] The use of successive differentiations of Eqs. (2-34) and (2-35) is more compactly programed, however, for estimation of the higher derivatives as shown in Chapter 3.

In dealing with changes of internal variables due to any small changes of molecular structure, the change of a structure parameter from a certain standard value is more suitable as the argument of the potential function than the structure parameter itself. Similarly, the changes of the Cartesian coordinates of nuclei from their initial positions are more convenient than the Cartesian coordinates themselves as the arguments for discussing the displacements of nuclei due to any small changes of the molecular structure. The merit of using small arguments is that the potential function can be expanded in powers of these arguments. We can then terminate the power series at any order according to the required accuracy. Usually, only the terms of the lowest power are retained in the first-order approximation.

Now we define the Cartesian displacement coordinates, or simply the Cartesian displacements, of the nucleus i of an N-atomic molecule as

$$\left.\begin{aligned} \Delta x_{3i-2} &\equiv \Delta X_i = X_i - X_i^0 \\ \Delta x_{3i-1} &\equiv \Delta Y_i = Y_i - Y_i^0 \\ \Delta x_{3i} &\equiv \Delta Z_i = Z_i - Z_i^0 \end{aligned}\right\} \quad (i = 1 - N), \tag{2-36}$$

where X_i^0, Y_i^0 and Z_i^0 are the initial values of the Cartesian coordinates of the nucleus i. Any infinitesimal motion of atomic nuclei around their equilibrium positions in a molecule can be described by the well-established methods of normal coordinate analyses,[1] the details of which are given in Chapter 4. A useful type of variable for the normal coordinate analysis is the internal coordinate.[1] The kth internal coordinate, ΔR_k, is defined as the difference between the instantaneous value of the kth internal variable R_k and its equilibrium value R_k^0,

$$\Delta R_k = R_k - R_k^0. \tag{2-37}$$

The internal coordinates defined for the three basic types of internal variables, the bond length, the bond angles and the dihedral angles, are called the (bond) stretching, the (bond angle) bending and the torsional coordinates, respectively. In these expressions, the terms in parentheses are omitted unless there is risk of confusion. In calculations for molecular mechanics, we do not restrict R_k^0 to be the equilibrium value of R_k for individual molecules, but regard it as the initial value of R_k calculated from the starting set of the nuclear Cartesian coordinates. They are often taken commonly throughout a given type of local structure of molecules. In this extension of the definition, the use of internal coordinates can be extended to any molecules for which no equilibrium structures are known. We must note in this case, however, that linear terms may appear on expanding the potential function in powers of internal coordinates.

Generally, the internal coordinate ΔR_k is a function of the Cartesian displacement coordinates, Δx_1, Δx_2, ...[1] When both ΔR_j and Δx_p are small quantities, the former can be expanded as a Taylor series of the latters in the form

$$\Delta R_j = \sum_p B_{jp}\,\Delta x_p + \tfrac{1}{2}\sum_p \sum_q B_{j,pq}\,\Delta x_p \Delta x_q + \cdots, \tag{2-38}$$

where

$$B_{jp} = \left(\partial R_j / \partial x_p\right)_0 \tag{2-39a}$$

and

$$B_{j,pq} = \left(\partial^2 R_j / \partial x_p \partial x_q\right)_0. \tag{2-39b}$$

The subscript 0 in Eqs. (2-39a,b) indicates that the derivatives are to be evaluated at the nuclear positions specified by X_i^0, Y_i^0 and Z_i^0. The expansion in the form of Eq. (2-38) plays the key role in the transformation of the potential function from the internal coordinate space to the Cartesian coordinate space.

2.3.2 Out-of-plane Angles

Some types of internal variables other than the basic three (bond length, bond angles and torsion angles) are necessary for particular molecules. Imagine a molecule in which the nucleus j is surrounded by three co-planar bonds j–i, j–k, and j–l formed, for example, by the sp^2-hybridization. If this plane is taken to be the XY plane, the Z-components of the bond vectors $\boldsymbol{r}_{ji}$, $\boldsymbol{r}_{jk}$ and $\boldsymbol{r}_{jl}$ all vanish at the equilibrium position. In this case, no first order changes of the internal variables, r_{ji}, r_{jk}, r_{jl}, θ_{ijk}, θ_{ijl} and θ_{kjl} can be brought about by infinitesimally displacing the nuclei i, j, k and l along the Z-axis. Then the Cartesian displacement coordinates ΔZ_a with $a = i, j, k$ and l disappear from the linear part of the expansion in the right side of Eq. (2-38), and the transformation from the Cartesian displacements into the internal coordinates becomes incomplete within the framework of the first order approximation.

To avoid this difficulty, the out-of-plane angle γ_{ijkl} is introduced as an internal variable linearly related to the Cartesian displacements, ΔZ_a ($a = i, j, k$ and l). The variable γ_{ijkl} is defined as the angle between the bond j–i and the plane formed by the nuclei j, k and l, taken to be positive when the change of the bond vector $\boldsymbol{r}_{ji}$ satisfies the condition $\Delta\boldsymbol{r}_{ji} \cdot (\boldsymbol{e}_{jk} \times \boldsymbol{e}_{jl}) > 0$ as shown in Fig. 2.2. The change of the out-of-plane angle, $\Delta\gamma_{ijkl}$, is then used as an internal coordinate called the out-of-plane bending coordinate. When the equilibrium value of the out-of-plane angle is zero, the out-of-plane angle γ_{ijkl} and the out-of-plane bending coordinate $\Delta\gamma_{ijkl}$ mean the same variable as each other. The out-of-plane angle γ_{ijkl} can be expressed as a function of the Cartesian coordinates of the relevant nuclei *i, j, k* and *l* by writing the volume of the parallelepiped formed by the three bond vectors $\boldsymbol{r}_{ji}$, $\boldsymbol{r}_{jk}$ and $\boldsymbol{r}_{jl}$ in two ways as

$$v = \boldsymbol{r}_{ji} \cdot (\boldsymbol{r}_{jk} \times \boldsymbol{r}_{jl}) \qquad (2\text{-}40a)$$

and

$$v = |\boldsymbol{r}_{jk} \times \boldsymbol{r}_{jl}|(r_{ji} \sin\gamma_{ijkl}) \qquad (2\text{-}40b)$$

Eliminating v from Eqs. (2-40a,b) and solving the resulting equation for γ_{ijkl}, we obtain

$$\gamma_{ijkl} = \text{Arcsin}\frac{\boldsymbol{r}_{ji} \cdot (\boldsymbol{r}_{jk} \times \boldsymbol{r}_{jl})}{r_{ji}|\boldsymbol{r}_{jk} \times \boldsymbol{r}_{jl}|}. \qquad (2\text{-}41)$$

The out-of-plane angle formed by three nuclei *i*, *k* and *l* bonded to a planar trivalent nucleus *j* may be described in three ways according to which of the three anchor nuclei is regarded to be displaced from the plane defined by the other two anchor nuclei and the apex nucleus *j*. As seen from Eq. (2-40b), the three out-of-plane angles are related to one another by

$$\sin\theta_{kjl}\sin\gamma_{ijkl} = \sin\theta_{lji}\sin\gamma_{kjli} = \sin\theta_{ijk}\sin\gamma_{ljik} \qquad (2\text{-}42a)$$

Taking the differential of Eq. (2-42a) for the planar configuration, we have

$$\sin\theta_{kjl}\,\Delta\gamma_{ijkl} = \sin\theta_{lji}\,\Delta\gamma_{kjli} = \sin\theta_{ijk}\,\Delta\gamma_{ljik}\,. \qquad (2\text{-}42b)$$

Then, if we choose $\sin\theta_{kjl}\,\Delta\gamma_{ijkl}$ instead of $\Delta\gamma_{ijkl}$ as the internal variable representing the out-of-plane deformation of the four-atomic system *ijkl*, we need not explicitly specify the anchor nucleus deviating from the plane of the other three, as long as the linear approximation is employed. The scaled variable $\sin\theta_{kjl}\,\Delta\gamma_{ijkl}$ is recommended as the standard internal coordinate for describing the out-of-plane deformation of an sp^2 system by IUPAC.[8)]

Another internal variable used for describing any deviation of the three bonds formed by an sp^2 hybridization from the planar structure is the dihedral angle between the plane formed by the bonds *j–i* and *j–k* and that formed by the bonds *j–i* and *j–l*, $\tau'_{ij(kl)}$. The dihedral angle $\tau'_{ij(kl)}$ can be calculated from the Cartesian coordinates of the nuclei *i, j, k* and *l* by using a formula analogous to that used for the torsion angle τ_{ijkl}, Eq. (2-31), that is,

$$\tau'_{ij(kl)} = \text{Arcsin}\frac{\boldsymbol{r}_{ji} \cdot (\boldsymbol{r}_{jk} \times \boldsymbol{r}_{jl})\, r_{ji}}{|\boldsymbol{r}_{ji} \times \boldsymbol{r}_{jk}||\boldsymbol{r}_{ji} \times \boldsymbol{r}_{jl}|}, \qquad (2\text{-}43)$$

where the vectors $\boldsymbol{r}_{ji}$, $\boldsymbol{r}_{jk}$ and $\boldsymbol{r}_{jl}$ are used instead of $\boldsymbol{r}_{ij}$, $\boldsymbol{r}_{jk}$ and $\boldsymbol{r}_{kl}$ in Eq. (2-31), respectively. In Eq. (2-43), a positive $\tau'_{ij(kl)}$ is taken to be anticlockwise from the plane *i–j–k* to the plane *i–j–l* when observed from the side of *i*. Angle deformations of this type were used as internal coordinates in the calculation of anharmonic potential of ethylene.[7)] The four torsional coordinates of an ethylene molecule, which lies on the *XY*-plane as shown in Fig. 2. 4, lead to four linear combinations in the forms

$$\tau = (\tau_{3125} + \tau_{3124} + \tau_{6124} + \tau_{6125})/2 \qquad (2\text{-}44a)$$

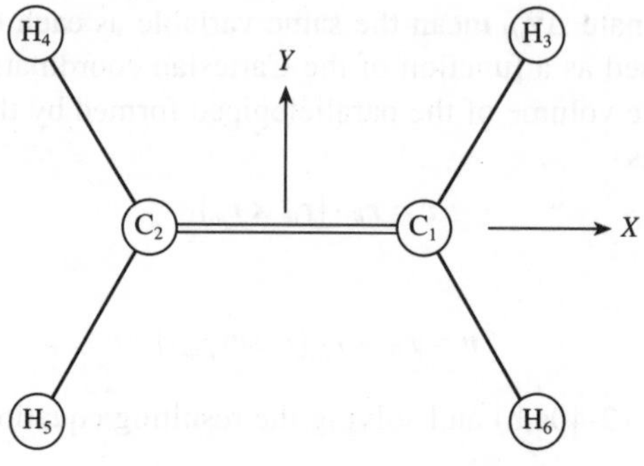

Fig. 2.4 Ethylene molecule on the *XY*-plane.

$$\tau_1' = (\tau_{3125} + \tau_{3124} - \tau_{6124} - \tau_{6125})/2 \qquad (2\text{-}44b)$$

$$\tau_2' = (\tau_{3125} - \tau_{3124} - \tau_{6124} + \tau_{6125})/2 \qquad (2\text{-}44c)$$

$$S_{red} = (\tau_{3125} - \tau_{3124} + \tau_{6124} - \tau_{6125})/2 \qquad (2\text{-}44d)$$

On the increase of these four coordinates, the four hydrogen atoms move along the *Z*-axis with the signs listed in Table 2.1. The net movement of the hydrogen atoms indicates that τ is twice the angle of mutual torsion of the two methylene groups, while τ'_1 and τ'_2 are minus twice the dihedral angles formed by the H–C=C planes around the nuclei C_1 and C_2, respectively. The quantity S_{red}, which does not change on any nuclear displacement, is an example of redundant coordinates which are explained in Chapter 4.

Distortion of the planar trivalent structure in Fig. 2.2 can also be described by the dihedral angle between the planes of the nuclei *i–k–l* and *j–k–l*, $\tau''_{kl(ij)}$. This angle is regarded as the torsion angle of the sequence *j–k–l–i* in which the real bond is only *j–k*. Analogously, the angle $\tau'_{ij(kl)}$ in Eq. (2-43) may be defined as the torsion angle of the sequence *k–i–j–l*. These angles are called improper torsions. The angles $\tau''_{kl(ij)}$ and $\tau'_{ij(kl)}$ are employed in the widely used program systems CHARMm[9a–c] and AMBER[10], respectively, both particularly adapted for large molecules of biological interest. The merit of using the improper torsion is that the torsion angles and the out-of-plane deformation can be expressed by a common

Table 2.1 Displacements of hydrogen nuclei of ethylene on the increase of torsion angles.[a]

No.	H_3	H_4	H_5	H_6
τ_{3124}[b]	+	–	0	0
τ_{3125}[b]	+	0	+	0
τ_{6124}[b]	0	–	0	–
τ_{6125}[b]	0	0	+	–
τ	2+	2–	2+	2–
τ_1'	2+	0	0	2+
τ_2'	0	2+	2+	0
S_{red}	0	0	0	0

[a] Movement along the *Z*-axis (+ for the positive direction).
[b] Sign convention taken from Ref. 8.

formula and the relevant computer program can be simplified accordingly.

A relation between the angles γ_{ijkl} and $\tau'_{ij(kl)}$ is obtained by combining Eqs. (2-40b) and (2-43), rewritten respectively as

$$\sin\gamma_{ijkl} = v/\left(r_{ji}\left|\boldsymbol{r}_{jk}\times\boldsymbol{r}_{jl}\right|\right) = v/\left(r_{ji}\,r_{jk}\,r_{jl}\sin\theta_{kjl}\right) \tag{2-45a}$$

and

$$\sin\tau'_{ij(kl)} = v\,r_{ji}/\left(\left|\boldsymbol{r}_{ji}\times\boldsymbol{r}_{jk}\right|\left|\boldsymbol{r}_{ji}\times\boldsymbol{r}_{jl}\right|\right) = v/\left(r_{ji}\,r_{jk}\,r_{jl}\sin\theta_{ijk}\sin\theta_{ijl}\right)\,, \tag{2-45b}$$

and eliminating v from them, as shown in the form

$$\sin\gamma_{ijkl} = \sin\theta_{ijk}\sin\theta_{ijl}\sin\tau'_{ij(kl)}/\sin\theta_{kjl}\,. \tag{2-46}$$

Within the framework of the first-order approximation in which the relation between the internal coordinates and the Cartesian displacements is taken to be linear, it follows from Eq. (2-46) that the internal coordinates $\Delta\gamma_{ijkl}$ and $\Delta\tau'_{ij(kl)}$ are related to each other by

$$\Delta\gamma_{ijkl} = f\Delta\tau'_{ij(kl)}$$

where

$$f = \left(\sin\theta_{ijk}\sin\theta_{ijl}\cos\tau'_{ij(kl)}/\sin\theta_{kjl}\cos\gamma_{ijkl}\right)_0\,.$$

For the planar structure at the equilibrium, the factor f is simplified as

$$f = \left(-\sin\theta_{ijk}\sin\theta_{ijl}/\sin\theta_{kjl}\right)_0\,.$$

The potential function written in terms of $\Delta\gamma_{ijkl}$ can be converted to that in terms of $\Delta\tau'_{ij(kl)}$ by using this conversion factor. In the approximation with a higher order than the first, however, we must note that the relation between $\Delta\gamma_{ijkl}$ and $\Delta\tau'_{ij(kl)}$ as well as that among $\Delta\gamma_{ijkl}$, $\Delta\gamma_{kjli}$ and $\Delta\gamma_{ljik}$ in Eq. (2-42b) are no longer linear, and the nuclei are displaced along different paths on the increase of these variables. As seen from Eqs. (2-45a,b), the out-of-plane angle γ_{ijkl} becomes indefinite if θ_{kjl} approaches π, while the dihedral angle $\tau'_{ij(kl)}$ becomes indefinite if either θ_{ijk}or θ_{ijl} approaches π.

2.3.3 Linear Angle Bending

For a sequence of three nuclei, j–k–l, which lie exactly or nearly on a straight line at the equilibrium, neither the bond angle θ_{jkl} in a usual sense nor its change $\Delta\theta_{jkl}$ from the equilibrium value is appropriate as an explicit variable for describing the nuclear motion. There are two reasons for this difficulty. First, since θ_{jkl} takes the maximum value π radian at the equilibrium, the variable range of $\Delta\theta_{jkl}$ is limited by an upper bound 0. Second, since θ_{jkl} is related to the Cartesian displacements of the nuclei by an Arccosine formula, Eq. (2-27), the derivatives given by Eq. (2-34) cannot be evaluated near the limit where θ_{jkl} approaches π. Furthermore, the torsion angle τ_{ijkl} cannot be defined when θ_{jkl} is π. Thus, neither of the two internal variables of the already known types, τ_{ijkl} and θ_{jkl}, can be used for describing any motion of the nuclei j, k and l, which are linearly arranged at the equilibrium, in the direction perpendicular to the line joining j and l.

There are many ways of choosing two internal variables to be assigned to the angular deformations of the linear bond sequence j–k–l. If the sequence j–k–l can be chosen to lie on a Cartesian coordinate axis, e.g., the Z-axis by virtue of the molecular symmetry, the two linear bending variables will be best defined in the forms recommended by Hoy, Mills and Strey,[6)] *i.e.*,

$$R_{jkl(x)} = \frac{\boldsymbol{e}_y \cdot (\boldsymbol{r}_{kj} \times \boldsymbol{r}_{kl})}{r_{kj}\, r_{kl}} \quad \text{and} \quad R_{jkl(y)} = -\frac{\boldsymbol{e}_x \cdot (\boldsymbol{r}_{kj} \times \boldsymbol{r}_{kl})}{r_{kj}\, r_{kl}} , \tag{2-47}$$

where $\boldsymbol{e}_y$ and $\boldsymbol{e}_x$ are the unit vectors along the Y- and the X-axes, respectively. The advantage of Eq. (2-47) is that the linear bending potential can be expressed in terms of the quantity $R^2_{jkl(x)} + R^2_{jkl(y)}$ in such a way as to be independent of the orientation of the X- and the Y-axes.[5)]

On the other hand, some device is necessary for a linear sequence j–k–l in a general position where the unit vectors corresponding to $\boldsymbol{e}_y$ and $\boldsymbol{e}_x$ in Eq. (2-47) cannot be chosen *a priori*. If another nucleus i is bonded to j and the bond angle θ_{ijk} is not 180°, it is convenient to choose two angles $\beta^o_{i\,jkl}$ and $\beta^i_{i\,jkl}$ shown in Fig. 2.2. The former is the angle between the bond vector $\boldsymbol{r}_{kl}$ and the plane formed by the bonds i–j and j–k, while the latter is the angle between the projection of $\boldsymbol{r}_{kl}$ onto the i–j–k plane and the extension of the bond j–k. The angles β^o_{ijkl} and β^i_{ijkl} may be called the out-of-plane and the in-plane linear bending coordinates, respectively. Both angles $\beta^i_{i\,jkl}$ and $\beta^o_{i\,jkl}$ can be calculated from the relevant bond vectors by

$$\beta^i_{i\,jkl} = \text{Arcsin}\frac{(\boldsymbol{r}^0_{ij} \times \boldsymbol{r}_{jk}) \cdot (\boldsymbol{r}_{jk} \times \boldsymbol{r}_{kl})}{r_{jk}\left|(\boldsymbol{r}_{ij} \times \boldsymbol{r}_{jk}) \times \boldsymbol{r}_{kl}\right|} , \tag{2-48}$$

and

$$\beta^o_{i\,jkl} = \text{Arcsin}\frac{(\boldsymbol{r}^0_{ij} \times \boldsymbol{r}_{jk}) \cdot \boldsymbol{r}_{kl}}{\left|\boldsymbol{r}^0_{ij} \times \boldsymbol{r}_{jk}\right| r_{kl}} \tag{2-49}$$

The formula for the in-plane linear angle bending, Eq. (2-48), is derived by writing the distance between the points A and B in Fig. 2.2 in two ways as

$$\overline{\text{AB}} = r_{kl} \sin\theta_{jkl} \cos\tau_{ijkl} = r_{kl} \cos\beta^o_{i\,jkl} \sin\beta^i_{i\,jk.}$$

and multiplying both sides by $r_{ij}\, r_{jk}{}^2 \sin\theta_{ijk}$. Rearrangement of the result after substituting the relations

$$(\boldsymbol{r}_{ij} \times \boldsymbol{r}_{jk}) \cdot (\boldsymbol{r}_{jk} \times \boldsymbol{r}_{kl}) = r_{ij}\, r_{jk}{}^2\, r_{kl} \sin\theta_{ijk} \sin\theta_{jkl} \cos\tau_{ijkl}$$

and

$$\left|(\boldsymbol{r}_{ij} \times \boldsymbol{r}_{jk}) \times \boldsymbol{r}_{kl}\right| = r_{ij}\, r_{jk}\, r_{kl} \sin\theta_{ijk} \cos\beta^o_{i\,jkl}$$

leads to Eq. (2-48). To derive the formula for the out-of-plane linear angle bending, the area of the parallelogram formed by $\boldsymbol{r}_{ij}$ and $\boldsymbol{r}_{jk}$ is multiplied by the length of $\overline{\text{AC}}$ in Fig. 2.5, giving the volume of the parallelpiped formed by the bond vectors $\boldsymbol{r}_{ij}$, $\boldsymbol{r}_{jk}$ and $\boldsymbol{r}_{kl}$ in the form

$$v = r_{ij}\, r_{jk}\, r_{kl} \sin\theta_{ijk} \sin\beta^o_{i\,jkl} = \left|\boldsymbol{r}^0_{ij} \times \boldsymbol{r}_{jk}\right| r_{kl} \sin\beta^o_{i\,jkl} \tag{2-50}$$

Eliminating v from Eqs. (2-30) and (2-50) leads to Eq. (2-49).

By using Eqs. (2-25, 27, 31, 32, 41, 43, 47– 49), the numerical values of any internal

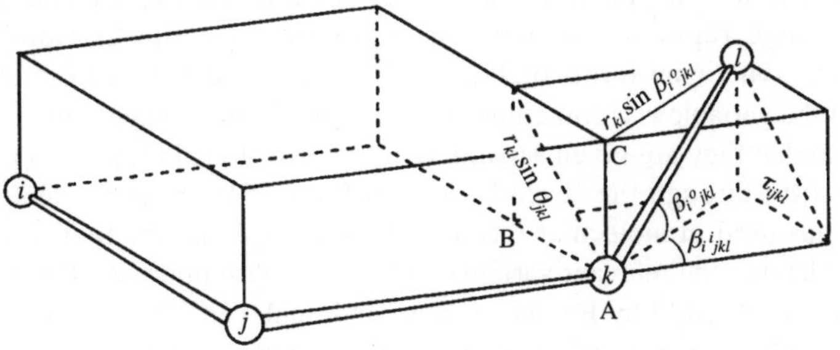

Fig. 2.5 In-plane and out-of-plane linear angle bending coordinates.

variables necessary for evaluating the potential function can be calculated from the Cartesian coordinates of the atomic nuclei. The derivatives of these internal variables with respect to the Cartesian coordinates of nuclei are necessary for expanding the potential function in powers of the nuclear Cartesian displacements. An efficient method of calculating these derivatives is described in Chapter 3.

2.4 General Properties of Potential Energy Functions

2.4.1 Partition of the Potential Energy

In order for the energy of a molecule to be computable at any point in the nuclear configuration space, the potential function should be given explicitly as a single-valued function of a set of properly chosen internal variables. To make the correct choice, the set should include at least the same number of mutually independent internal variables as the degrees of freedom of the internal motion of the nuclei in the molecule. The use of an excess number of internal variables over the degrees of freedom is not forbidden, but one should note that in such a case certain variables are not independent of one another.

The most commonly used types of internal variables in molecular mechanics are the bond length, r_{ij}, the bond angle, θ_{ijk}, the linear bond angle, β_{jkl}, the out-of-plane angle, γ_{ijkl}, the torsion angle, τ_{ijkl}, and the nonbonded internuclear distance, r'_{ij}. Sometimes, the improper torsion angle, τ'_{ijkl}, or other types of variables defined in different ways are used instead of the out-of-plane angle. Usually, the potential function is written as the sum of various types of simple functions each of which contains only a small number (mostly one and rarely exceeding three) of internal variables in the form

$$V = \sum_r V_r(r) + \sum_\theta V_\theta(\theta) + \sum_\beta V_\beta(\beta) + \sum_\gamma V_\gamma(\gamma) + \sum_\tau V_\tau(\tau) + \sum_{r'} V_{\mathrm{ED}}(r')$$

$$+ \sum_{r'} V_{\mathrm{C}}(r') + \sum_{i,j,\cdots} V_{ij\cdots}(R_i, R_j, \cdots), \qquad (2\text{-}51)$$

where R_i, R_j, $\cdots$ stand for any variables of the types, r, θ, β, γ, and τ.

Each of the first through the seventh sums in the right side of Eq. (2-51) is taken over the single variable functions of the variable type shown in parentheses. The sixth and the seventh sums arise from two types of through-space interaction between nonbonded atoms,[12,13] called the exchange repulsion–dispersion interaction and the Coulomb interaction, respectively. The functional forms of V_{ED} and V_C are quite different from each other, and the definition of the variable r' is sometimes modified in different ways between the two types of interaction, so that they are given separately in the two sums in Eq. (2-51). The first, the second and the fifth through the seventh sums in Eq. (2-51) are contained in most of the potential functions used in molecular mechanics. The third and the fourth sums are added if the molecules having the relevant variable types are to be supported. The earlier potential functions written by Boyd,[14] by Engler, Andose and Schleyer,[15] by Andose and Mislow,[16] by White and Bovill[17] and by Jacob, Thompson and Bartell (MUB1)[18] consist of such single variable terms. The generic potential functions, SYBYL[19], Chem-X[20], DREIDING[21], UFF[22], GROMOS[23] and COSMIC[24a–c] belong to this category. Here a generic potential means a potential in which the parameters for each variable are generated by applying a certain general rule to the atom types of those nuclei by which the variable is defined. The potential functions in the first versions of CFF,[25,26] which aim to predict the structures, thermodynamic quantities and vibrational spectra from a single potential consistently and AMBER[10,11] are constructed also from single variable terms only. In the program ECEPP[27,28](revised later as ECEPP/2[29,30] and ECEPP/3[31,32]), which has been widely used in searching for stable conformers of peptides and proteins, bond lengths, bond angles and out-of-plane angles are all fixed at certain standard values, so that the potential function consists of the fifth through the seventh sums in Eq. (2-51).

The eighth (and last) sum consists of multivariable functions representing the through-bond interactions among any internal variables not far apart. The quadratic two-variable terms of a few limited types appear in the potential functions MUB2[33], FLEX[34], MM1[35,36], MM2[37–39] and its extensions and variations, MM2*[40], MM2′,[41] and MMP2.[42] The MM2 potential has been used extensively for predicting the structures and the heats of formation of stable conformers of organic compounds. The program CHARMm[9a] and the later versions of CFF (CVFF[43–47] and CFF93[48,49]) and AMBER[50] include various types of interaction terms. Many more types of quadratic interaction terms are introduced in the programs MM3[51a–l], MM4[52a–e], UTAH5[53], RISE[54a–d], EFF[55], SPASIBA[56a–d], SDFF[57a–c] and MMFF94[58a,b], which aim to closer fitting of normal frequencies.

The essential strategy of molecular mechanics is to regard the single variable terms as the main part of the potential function, retaining as a small number of multivariable terms as possible according to the purpose of the calculation. On the basis of this partitioning of the potential function in molecular mechanics, we can deal with the energies, structures and many other properties of an infinite number of compounds by using a finite number of parameters which reflect the local environment of individual internal variables. Each term in Eq. (2-51) may thus be called a local potential function.

Any internal variable appearing in the single variable local potential functions involved in Eq. (2-51) has its own optimum values specific to the variable type, the participating atomic species and the scheme of hybridization. The number of the optimum values may be up to a few for a variable of the type τ, but is always one for each of the other types. Under the condition that the molecule remains stable, each variable of the types r, θ, β and γ rarely takes a value outside a small range around the optimum. Then a single variable function in the first through the fourth sums in Eq. (2-51) may be expanded as a power series of the deviation of an instantaneous value of the variable, R_i, from its optimum value, R_i^0, *i.e.*,

$$V_i(R_i) = \sum_{n=2}^{N} (1/n!) k_i^{(n)} (R_i - R_i^0)^n \,, \tag{2-52}$$

Note that the expansion in Eq. (2-52) should begin with the quadratic term in order for the energy gradient to vanish when $R_i = R_i^0$. Similarly, a multivariable function may be expanded as

$$V_{ij\cdots}(R_i, R_j, \cdots) = \sum_{n\geq 1, m\geq 1, \cdots} [1/(n+m+\cdots)!] k_{ij\cdots}^{(n,m,\cdots)} (R_i - R_i^0)^n (R_j - R_j^0)^m \cdots \tag{2-53}$$

In practical calculations, the expansion is terminated at a finite power according to the required accuracy.

The minimum number of internal variables necessary for describing the potential function of a stable N-atomic molecule is the number of degrees of freedom of the internal motion, *i.e.*, $3N-5$ for a linear molecule and $3N-6$ for a nonlinear molecule. If the potential function is expressed by the minimum number of internal variables, the energy of the molecule reaches the minimum value 0 when each variable R_i takes its optimum value R_i^0. In this case, the quantity $R_i - R_i^0$ coincides with the internal coordinate ΔR_i defined by Eq. (2-37), and the potential energy in terms of the latter is written as

$$V(\Delta R_1, \Delta R_2, \cdots) = \tfrac{1}{2} \sum_{i,j} F_{ij}\, \Delta R_i\, \Delta R_j + \tfrac{1}{3!} \sum_{i,j,k} F_{ijk}\, \Delta R_i\, \Delta R_j\, \Delta R_k + \cdots \tag{2-54}$$

where

$$F_{ij} = \left(\partial^2 V / \partial R_i\, \partial R_j\right)_0 \tag{2-55a}$$

and

$$F_{ijk} = \left(\partial^3 V / \partial R_i\, \partial R_j\, \partial R_k\right)_0 \tag{2-55b}$$

The coefficients of expansion defined in Eqs. (2-55a,b) are called the force constants. The suffix 0 indicates that the derivatives are evaluated at the potential minimum. If the displacements of the nuclei from their initial positions are small, it may be allowed as the first-order approximation to retain only the quadratic terms in the expansion in Eq. (2-54). In this case, by introducing the vector of the internal coordinates,

$$\boldsymbol{\Delta R} = [\Delta R_1 \quad \Delta R_2 \quad \cdots \quad \Delta R_n]^{\mathrm{T}} \tag{2-56}$$

and a real symmetric matrix defined by

$$\boldsymbol{F} = \begin{bmatrix} F_{11} & F_{12} & \cdots & F_{1n} \\ F_{21} & F_{22} & \cdots & F_{2n} \\ \vdots & \vdots & \ddots & \vdots \\ F_{n1} & F_{n2} & \cdots & F_{nn} \end{bmatrix}, \tag{2-57}$$

the potential energy is given a simple expression in matrix algebra,

$$V(\Delta R_1, \Delta R_2, \cdots) = \tfrac{1}{2} \sum_{i,j} F_{ij}\, \Delta R_i\, \Delta R_j = \tfrac{1}{2} \boldsymbol{\Delta \tilde{R}}\, \boldsymbol{F}\, \boldsymbol{\Delta R} \,. \tag{2-58}$$

The matrix $\boldsymbol{F}$ is well known as the force constant matrix, or simply as the $\boldsymbol{F}$ matrix, in the theory of molecular vibrations.[1)]

The simplest possible potential function for a stable molecular structure consists only of the quadratic diagonal terms, $\frac{1}{2} F_{ii} \Delta R_i^2$ in Eq. (2-58). The variable R_i may be any of those involved in the first five sums in the right side of Eq. (2-51). A force field generated from a potential function of this type is called a simple valence force field (SVFF). The SVFF was used in earlier normal coordinate analyses, but failed in most cases to give satisfactory results because of its too simple form.[59)]

If the potential function for an SVFF is supplemented with all possible interaction terms of the form $F_{ij}\, \Delta R_i\, \Delta R_j$, the resulting force field is called a general valence force field (GVFF).[60)] A GVFF corresponds to the most general representation of the quadratic force field in terms of internal coordinates used for the normal coordinate analysis of a molecule. Since the number of parameters in a GVFF increases roughly in proportion to the square of the number of degrees of freedom of the motion of the system, it is not straightforward even for very simple molecules to determine all the required parameters uniquely from spectroscopic data. In general, the search for a reasonable compromise between SVFF and GVFF is an important but controversial problem in molecular mechanics. It is essential in this process to estimate properly the significance of interaction force constants. It should be kept in mind that the interaction force constants affect not only the vibrational levels but also the transfer of local strain through the bonds in a molecule.

2.4.2 Interaction Force Constants and Intramolecular Strains

Any local strain in a part of a molecule gives rise to more or less structural change in the other parts of the molecule. The extent of transfer of a strain from one part to another depends roughly on the sign and magnitude of the interaction force constants connecting the two parts of the molecule.[61–63)] Eventually, there must be some relationship between the interaction force constants and the changes in the geometry of molecules on dissociation or isomerization reactions.[64,65)] Let the *p*th internal coordinate be kept at a given value, ΔR_p, and all the other coordinate be relaxed to take their own optimum values corresponding to the structure which minimizes the energy under the given constraint on ΔR_p. Then the instantaneous equilibrium value of an internal coordinate ΔR_i $(i \neq p)$, $(\Delta R_i)_p$, may be regarded as a continuous function of the specified value of ΔR_p, and can be expanded in powers of ΔR_p as long as the latter is within a small range, *i.e.*,

$$(\Delta R_i)_p = A_{ip}\, \Delta R_p + A_{ipp}\, \Delta R_p^{\,2} + A_{ippp}\, \Delta R_p^{\,3} + \cdots . \qquad (2\text{-}59)$$

The coefficients of this expansion are determined as follows.[64)] Since the potential energy is stationary with respect to all coordinates except ΔR_p, the derivative of Eq. (2-54) with respect to an internal variable R_q $(q \neq p)$ should vanish if $\Delta R_i = (\Delta R_i)_p$ $(i \neq p)$, *i.e.*,

$$\partial V / \partial R_q = \sum_i F_{qi}\, \Delta R_i + \tfrac{1}{2} \sum_{i,j} F_{qij}\, \Delta R_i\, \Delta R_j + \cdots = 0 \, . \qquad (2\text{-}60)$$

Substituting Eq. (2-59) into Eq. (2-60), we obtain

$$\begin{aligned}
&F_{pq}\,\Delta R_p + \sum_{i\neq p} F_{iq}\left(A_{ip}\,\Delta R_p + A_{ipp}\,\Delta R_p^{\,2} + A_{ippp}\,\Delta R_p^{\,3} + \cdots\right) \\
&+ \tfrac{1}{2} F_{ppq}\,\Delta R_p^{\,2} + \sum_{i\neq p} F_{piq}\,\Delta R_p\left(A_{ip}\,\Delta R_p + A_{ipp}\,\Delta R_p^{\,2} + \cdots\right) \\
&+ \tfrac{1}{2} \sum_{i,j\neq p} F_{ijq}\left(A_{ip}\,\Delta R_p + A_{ipp}\,\Delta R_p^{\,2} + \cdots\right)\left(A_{jp}\,\Delta R_p + A_{jpp}\,\Delta R_p^{\,2} + \cdots\right) \\
&+ \tfrac{1}{6} F_{pppq}\Delta R_p^{\,3} + \tfrac{1}{2}\sum_{i\neq p} F_{ppiq}\,A_{ip}\,\Delta R_p^{\,3} + \tfrac{1}{2}\sum_{i,j\neq p} F_{pijq}\,A_{ip}\,A_{jp}\,\Delta R_p^{\,3} \\
&+ \tfrac{1}{6} \sum_{i,j,k\neq p} F_{ijkq}\,A_{ip}\,A_{jp}\,A_{kp}\,\Delta R_p^{\,3} + \cdots = 0
\end{aligned} \tag{2-61}$$

where the terms with higher powers in ΔR_p than the third are omitted. In order for Eq. (2-61) to hold identically for any values of ΔR_p, the coefficients of ΔR_p, ΔR_p^2, ΔR_p^3, ... must vanish separately. From the linear through the cubic terms, we have

$$F_{pq} + \sum_{i\neq p} F_{iq}\,A_{ip} = 0\,, \tag{2-62a}$$

$$\tfrac{1}{2} F_{ppq} + \sum_{i\neq p}\left(F_{iq}\,A_{ipp} + F_{piq}\,A_{ip}\right) + \tfrac{1}{2}\sum_{i,j\neq p} F_{ijq}\,A_{ip}\,A_{jp} = 0 \tag{2-62b}$$

and

$$\begin{aligned}
&\tfrac{1}{6} F_{pppq} + \sum_{i\neq p}\left(F_{iq}\,A_{ippp} + F_{piq}\,A_{ipp} + \tfrac{1}{2} F_{ppiq}\,A_{ip}\right) \\
&+ \tfrac{1}{2} \sum_{i,j\neq p}\left\{F_{ijq}\left(A_{ip}\,A_{jpp} + A_{ipp}\,A_{jp}\right) + F_{pijq}\,A_{ip}\,A_{jp}\right\} \\
&+ \tfrac{1}{6} \sum_{i,j,k\neq p} F_{ijkq}\,A_{ip}\,A_{jp}\,A_{kp} = 0.
\end{aligned} \tag{2-62c}$$

The restriction on the running indices in the sums in Eqs. (2-62a–c) may be removed if we adopt the convention that $A_{pp} = 1$ and $A_{ppp} = A_{pppp} = 0$, which is obtained by comparing Eq. (2-59) with the nominal relation, $(\Delta R_p)_p = \Delta R$. The terms in Eqs. (2-62a–c) without the running indices are then brought into the sums as given by

$$\sum_i F_{iq}\,A_{ip} = 0 \tag{2-63a}$$

$$\sum_i F_{iq}\,A_{ipp} + \tfrac{1}{2}\sum_{i,j} F_{ijq}\,A_{ip}\,A_{jp} = 0 \tag{2-63b}$$

and

$$\sum_i F_{iq}\,A_{ippp} + \tfrac{1}{2}\sum_{i,j} F_{ijq}\left(A_{ip}\,A_{jpp} + A_{ipp}\,A_{jp}\right) + \tfrac{1}{6}\sum_{i,j,k} F_{ijkq}\,A_{ip}\,A_{jp}\,A_{kp} = 0\,. \tag{2-63c}$$

In order to solve each of Eqs. (2-63a–c) for the corresponding unknowns A_{ip}, A_{ipp} or A_{ippp} simultaneously, we introduce the asymmetric matrices $\boldsymbol{A}^{(i)}$ and $\boldsymbol{U}^{(i)}$ the elements of which are defined as given in the first and the second columns of Table 2.2, and a diagonal matrix $\boldsymbol{T}^{(i)}$ with the pth element

Table 2.2 The elements of the matrices used for calculating the coefficients of internal coordinates at instantaneous equilibria.

i	$A_{ip}^{(i)}$	$U_{ip}^{(i)}$	$T_{pp}^{(i)}$
1	A_{ip}	0	$\sum_{j} F_{jp} A_{jp}$
2	A_{ipp}	$\frac{1}{2}\sum_{j,k} F_{jkl} A_{jp} A_{kp}$	$\frac{1}{2}\sum_{j,k,l} F_{jkl} A_{jp} A_{kp} A_{lp}$
3	A_{ippp}	$\sum_{j,k} F_{jkl} A_{jpp} A_{kp} + \frac{1}{6}\sum_{j,k,l} F_{jkl} A_{jp} A_{kp} A_{lp}$	$\sum_{j,k,l} F_{jkl} A_{jpp} A_{kp} A_{lp} + \frac{1}{6}\sum_{j,k,l,m} F_{jklm} A_{jp} A_{kp} A_{lp} A_{mp}$

$$T_{pp}^{(i)} = \sum_{i} F_{ip} A_{ip}^{(i)} + U_{pp}^{(i)} \, , \quad (i = 1, 2 \text{ and } 3). \tag{2-64}$$

Equations (2-63a–c) can then be rewritten in a common matrix form,

$$\boldsymbol{F}\boldsymbol{A}^{(i)} + \boldsymbol{U}^{(i)} = \boldsymbol{T}^{(i)} . \tag{2-65}$$

The inverse of an $\boldsymbol{F}$ matrix is called a compliance matrix, and is denoted by $\boldsymbol{C}$.[66] In terms of the compliance matrix, Eq. (2-65) is written as

$$\boldsymbol{A}^{(i)} = \boldsymbol{C}(\boldsymbol{T}^{(i)} - \boldsymbol{U}^{(i)}) . \tag{2-66}$$

By comparing the pth diagonal elements of the matrices in the right and the left sides of Eq. (2-66), the elements of $\boldsymbol{T}^{(i)}$ are given by

$$T_{pp}^{(i)} = \left(A_{pp}^{(i)} + \sum_{l} C_{pl}\, U_{lp}^{(i)} \right) \Big/ C_{pp} \, ,$$

which, after appropriate substitutions, gives the expressions in the third column of Table 2.2. Then the off-diagonal elements of Eq. (2-66) yield the desired coefficients of the expansion in Eq. (2-59) as

$$A_{ip} = C_{ip} / C_{pp} \tag{2-67a}$$

$$A_{ipp} = \frac{1}{2}\left(C_{ip} \sum_{j,k,l} F_{jkl} A_{jp} A_{kp} A_{lp} - C_{ii} \sum_{j,k,l} F_{jkl} A_{jp} A_{kp} A_{li} \right) \tag{2-67b}$$

and

$$\begin{aligned} A_{ippp} &= C_{ip} \sum_{j,k,l} F_{jkl} A_{jpp} A_{kp} A_{lp} - C_{ii} \sum_{j,k,l} F_{jkl} A_{jpp} A_{kp} A_{li} \\ &+ \frac{1}{6}\left(C_{ip} \sum_{j,k,l,m} F_{jklm} A_{jp} A_{kp} A_{lp} A_{mp} - C_{ii} \sum_{j,k,l,m} F_{jklm} A_{jp} A_{kp} A_{lp} A_{mi} \right) . \end{aligned} \tag{2-67c}$$

The linear coefficient A_{ip} has also been obtained by Jones and Ryan by solving Eq. (2-63a) directly from Cramer's rule.[67] According to Eqs. (2-67b,c), the higher-order coefficients in the expansion of $(\Delta R_i)_p$ in Eq. (2-60) can be calculated from the higher-order force constants if the compliance constants are available. Unfortunately, the higher-order force constants necessary for calculating the coefficients A_{ipp} and A_{ippp} have been obtained for only a small number of simple molecules. General procedures of estimating the cubic and quartic force constants are described in detail in Chapter 5.

The coefficient A_{ip} is the rate of change of R_i at the beginning of the increase of R_p. Consider the case where R_p is a bond length. If the initial direction of the change of R_i on the increase of R_p is kept throughout the change, *i.e.*, if R_i changes monotonically on the dissociation of the bond R_p, the quantity $\Delta R_i = R_i(\text{product}) - R_i(\text{reactant})$ must have the same sign as A_{ip}. In Table 2.3, the changes of R_i on a variety of dissociation reactions of simple molecules are compared with the relevant coefficient A_{ip}. According to Eq. (2-67a), the signs of A_{ip} and A_{pi} must be the same as each other since the compliance matrix is symmetric and positive definite.

In those cases where both R_i and R_p are bond lengths, the change of R_i on the dissociation

Table 2.3 Changes in bond lengths on dissociation reaction.

		Reactant		Product			
R_i	R_p	Formula	R_i^e [a)]	Formula (State)	R_i^e [a)]	ΔR_i^e [a)]	A_{ip}
r(O–H)	O–H	H_2O	95.6[b)]	OH($^2\Pi$)	97.06[c)]	1.46	0.029[d)]
r(N–H)	N–H	NH_3	101.24[b)]	NH_2($^2B_1$)	102.4[b)]	1.16	0.047[e)]
θ(HNH)	N–H	NH	107.8[b)]	NH_2($^2B_1$)	103.4[b)]	–4.3	–0.339[e)]
r(C–H)	C≡N	HCN	106.57[b)]	CH($^2\Pi$)	111.98[c)]	5.41	0.032[f)]
r(C≡N)	C–H	HCN	115.30[b)]	CN($^2\Sigma^+$)	117.18[c)]	1.88	0.011[f)]
r(C=O)	C=O	CO_2	116.21[b)]	CO($^1\Sigma^+$)	112.82[c)]	–3.39	–0.157[g)]
r(C=S)	C=S	CS_2	155.45[b)]	CS($^1\Sigma^+$)	153.44[c)]	–2.01	–0.164[g)]
r(S=O)	S=O	SO_2	143.08[b)]	SO($^3\Sigma^-$)	148.11[h)]	5.03	–0.002[i)]
r(N≡N)	N=O	N_2O	112.82[b)]	N_2($^1\Sigma_g^+$)	109.4[c)]	–3.42	–0.056[j)]
r(N=O)	N≡N	N_2O	118.42[b)]	NO($^2\pi$)	115.08[c)]	–3.34	–0.085[g)]
r(C–H)	C=O	HCHO	110.2[b)]	CH_2($^3\Sigma^-$)	102.9[b)]	–7.3	–0.107[k)]
θ(HCH)	C=O	HCHO	121.1[b)]	CH_2($^3\Sigma^-$)	180.0[b)]	58.9	0.552[k)]
r(C–H)	C–H	HCHO	110.2[b)]	HCO($^2A'$)	109.0[b)]	–1.2	–0.007[k)]
r(C=O)	C–H	HCHO	121.1[b)]	HCO($^2A'$)	119.8[b)]	–1.3	–0.043[k)]
r(C–H)	C≡C	C_2H_2	106.0[b)]	CH($^2\Sigma^+$)	111.98[c)]	5.98	0.015[l)]
r(C–H)	C=C	C_2H_4	108.69[m)]	CH_2($^3\Sigma^-$)	102.9[b)]	–5.79	–0.026[n)]
θ(HCH)	C=C	C_2H_4	117.44[m)]	CH_2($^3\Sigma^-$)	180.0[b)]	62.56	0.353[n)]
r(C–H)	C–C	C_2H_6	108.8[o)]	CH_3($^2A_2''$)	107.9[b)]	–0.9	–0.022[o)]
θ(HCH)	C–C	C_2H_6	107.4[o)]	CH_3($^2A_2''$)	120.0[b)]	12.6	0.233[o)]

[a)] bond length in pm, bond angles in degrees.
[b)] G. Herzberg, *Molecular Spectra and Molecular Structure* I., van Nostrand, Princeton (1950)
[c)] G. Herzberg, *Molecular Spectra and Molecular Structure* III., van Nostrand, Princeton (1966)
[d)] P. Jensen, *J. Mol. Spectrosc.*, **133**, 438 (1989)
[e)] Y. Morino, K. Kuchitsu and S. Yamamoto, *Spectrochim. Acta*, **24A**, 335 (1968)
[f)] I. Suzuki, M. A. Pariseau and J. Overend, *J. Chem. Phys.*, **44**, 3561 (1966)
[g)] I. Suzuki, *Bull. Chem. Soc. Jpn.*, **48**, 3565 (1975)
[h)] K. Kuchitsu and Y. Morino, *Bull. Chem. Soc. Jpn.*, **38**, 814 (1965)
[i)] T. Amano, E. Hirota and Y. Morino, *J. Phys. Soc. Jpn.*, **22**, 399 (1967)
[j)] I. Suzuki, *J. Mol. Spectrosc.*, **32**, 54 (1969)
[k)] L.B. Harding and W.C. Ermler, *J. Comput. Chem.*, **6**, 13 (1985)
[l)] G. Strey and I.M. Mills, *J. Mol Spectrosc.*, **59**, 103 (1976)
[m)] E. Hirota, Y. Endo, S. Saito, K. Yoshida, I. Yamaguchi, and K. Machida, *J. Mol. Spectrosc.*, **89**, 223 (1981)
[n)] K. Machida and Y. Tanaka, *J. Chem. Phys.*, **61**, 5040 (1974)
[o)] L.S. Bartell, S. Fitzwater and W.J. Hehre, *J. Chem. Phys.*, **63**, 4750 (1975)

of R_p must have the same sign as the change of R_p on the dissociation of R_i, and the sign of the changes of both R_i and R_p is given as the sign of the compliance constant C_{ip}. This relation holds for most reactions listed in Table 2.3, but there are a few exceptions for which the monotonicity of the geometry change breaks down somewhere along the dissociation path. Such an anomaly may occur if the potential surface involves an avoided crossing or a conical intersection, and may thus include certain information on the potentail surfaces of low-lying electronic excited states. Note that the sign inconsistency between C_{ij} and ΔR indicates necessarily a breakdown of the monotonicity in the geometry change, but the latter does not always give rise to the former. Actually, a contraction following the elongation of a bond does not necessarily bring the bond length back to its original value.

2.4.3 Invariance of Compliance Constants

The compliance constants have an advantage over the force constants in their independence of choice of internal coordinates as the arguments of the potential function.[68] Suppose that the potential function of a molecule is written in terms of two sets of internal coordinates, $\boldsymbol{\Delta R}$ and $\boldsymbol{\Delta R}'$, which are partially different from each other. Let us write the common parts of the two sets as $\boldsymbol{\Delta R}_c$ and $\boldsymbol{\Delta R}'_c$, and the different parts as $\boldsymbol{\Delta R}_d$ and $\boldsymbol{\Delta R}'_d$. If each internal coordinate in $\boldsymbol{\Delta R}_d$ can be written as a linear combination of those in $\boldsymbol{\Delta R}'_c$ and $\boldsymbol{\Delta R}'_d$, the interconversion relations between the two sets may be written in the matrix form as

$$\begin{bmatrix} \boldsymbol{\Delta R}_c \\ \boldsymbol{\Delta R}_d \end{bmatrix} = \begin{bmatrix} \boldsymbol{E} & \boldsymbol{0} \\ \boldsymbol{a}_c & \boldsymbol{a}_d \end{bmatrix} \begin{bmatrix} \boldsymbol{\Delta R}'_c \\ \boldsymbol{\Delta R}'_d \end{bmatrix} \tag{2-68a}$$

and

$$\begin{bmatrix} \boldsymbol{\Delta R}'_c \\ \boldsymbol{\Delta R}'_d \end{bmatrix} = \begin{bmatrix} \boldsymbol{E} & \boldsymbol{0} \\ \boldsymbol{b}_c & \boldsymbol{b}_d \end{bmatrix} \begin{bmatrix} \boldsymbol{\Delta R}_c \\ \boldsymbol{\Delta R}_d \end{bmatrix} \tag{2-68b}$$

where $\boldsymbol{b}_c = \boldsymbol{a}_d^{-1}\boldsymbol{a}_c$ and $\boldsymbol{b}_d = \boldsymbol{a}_d^{-1}$. The potential functions expanded in powers of $\boldsymbol{\Delta R}$ and $\boldsymbol{\Delta R}'$ are then given by

$$V = \begin{bmatrix} \tilde{\boldsymbol{F}}_c & \tilde{\boldsymbol{F}}_d \end{bmatrix} \begin{bmatrix} \boldsymbol{\Delta R}_c \\ \boldsymbol{\Delta R}_d \end{bmatrix} + \tfrac{1}{2} \begin{bmatrix} \boldsymbol{\Delta \tilde{R}}_c & \boldsymbol{\Delta \tilde{R}}_d \end{bmatrix} \begin{bmatrix} \boldsymbol{F}_{cc} & \boldsymbol{F}_{cd} \\ \boldsymbol{F}_{dc} & \boldsymbol{F}_{dd} \end{bmatrix} \begin{bmatrix} \boldsymbol{\Delta R}_c \\ \boldsymbol{\Delta R}_d \end{bmatrix}$$

$$= \begin{bmatrix} \tilde{\boldsymbol{F}}'_c & \tilde{\boldsymbol{F}}'_d \end{bmatrix} \begin{bmatrix} \boldsymbol{\Delta R}'_c \\ \boldsymbol{\Delta R}'_d \end{bmatrix} + \tfrac{1}{2} \begin{bmatrix} \boldsymbol{\Delta \tilde{R}}'_c & \boldsymbol{\Delta \tilde{R}}'_d \end{bmatrix} \begin{bmatrix} \boldsymbol{F}'_{cc} & \boldsymbol{F}'_{cd} \\ \boldsymbol{F}'_{dc} & \boldsymbol{F}'_{dd} \end{bmatrix} \begin{bmatrix} \boldsymbol{\Delta R}'_c \\ \boldsymbol{\Delta R}'_d \end{bmatrix}, \tag{2-69}$$

where the vector of the linear coefficients and the $\boldsymbol{F}$ matrix in the $\boldsymbol{\Delta R}'$ system are represented in terms of those in the $\boldsymbol{\Delta R}$ system as

$$\begin{bmatrix} \tilde{\boldsymbol{F}}'_c & \tilde{\boldsymbol{F}}'_d \end{bmatrix} = \begin{bmatrix} \tilde{\boldsymbol{F}}_c & \tilde{\boldsymbol{F}}_d \end{bmatrix} \begin{bmatrix} \boldsymbol{E} & \boldsymbol{0} \\ \boldsymbol{a}_c & \boldsymbol{a}_d \end{bmatrix} = \begin{bmatrix} \tilde{\boldsymbol{F}}_c + \tilde{\boldsymbol{F}}_d \boldsymbol{a}_c & \tilde{\boldsymbol{F}}_d \boldsymbol{a}_d \end{bmatrix} \tag{2-70}$$

and

$$\begin{bmatrix} \boldsymbol{F}'_{cc} & \boldsymbol{F}'_{cd} \\ \boldsymbol{F}'_{dc} & \boldsymbol{F}'_{dd} \end{bmatrix} = \begin{bmatrix} \boldsymbol{E} & \tilde{\boldsymbol{a}}_c \\ \boldsymbol{0} & \tilde{\boldsymbol{a}}_d \end{bmatrix} \begin{bmatrix} \boldsymbol{F}_{cc} & \boldsymbol{F}_{cd} \\ \boldsymbol{F}_{dc} & \boldsymbol{F}_{dd} \end{bmatrix} \begin{bmatrix} \boldsymbol{E} & \boldsymbol{0} \\ \boldsymbol{a}_c & \boldsymbol{a}_d \end{bmatrix}$$
$$= \begin{bmatrix} \boldsymbol{F}_{cc} + \tilde{\boldsymbol{a}}_c \boldsymbol{F}_{dc} + \boldsymbol{F}_{cd}\boldsymbol{a}_c + \tilde{\boldsymbol{a}}_c \boldsymbol{F}_{dd}\boldsymbol{a}_c & \boldsymbol{F}_{cd}\boldsymbol{a}_d + \tilde{\boldsymbol{a}}_c \boldsymbol{F}_{dd}\boldsymbol{a}_d \\ \tilde{\boldsymbol{a}}_d \boldsymbol{F}_{dc} + \tilde{\boldsymbol{a}}_d \boldsymbol{F}_{dd}\boldsymbol{a}_c & \tilde{\boldsymbol{a}}_d \boldsymbol{F}_{dd}\boldsymbol{a}_d \end{bmatrix}, \quad (2\text{-}71)$$

respectively. On the transformation from the set $\Delta\boldsymbol{R}$ to the set $\Delta\boldsymbol{R}'$, the submatrix of $\boldsymbol{F}$ corresponding to the unchanged internal coordinates, $\boldsymbol{F}_{cc}$, changes if $\boldsymbol{a}_c \neq \boldsymbol{0}$, that is, if the changed internal coordinates in the old system $\Delta\boldsymbol{R}_d$ are related linearly to the unchanged coordinates in the new system $\Delta\boldsymbol{R}'_c$. On the other hand, the compliance matrix in the new system is calculated by inverting Eq. (2-71) as

$$\begin{bmatrix} \boldsymbol{C}'_{cc} & \boldsymbol{C}'_{cd} \\ \boldsymbol{C}'_{dc} & \boldsymbol{C}'_{dd} \end{bmatrix} = \begin{bmatrix} \boldsymbol{E} & \boldsymbol{0} \\ \boldsymbol{b}_c & \boldsymbol{b}_d \end{bmatrix} \begin{bmatrix} \boldsymbol{C}_{cc} & \boldsymbol{C}_{cd} \\ \boldsymbol{C}_{dc} & \boldsymbol{C}_{dd} \end{bmatrix} \begin{bmatrix} \boldsymbol{E} & \tilde{\boldsymbol{b}}_c \\ \boldsymbol{0} & \tilde{\boldsymbol{b}}_d \end{bmatrix}$$
$$= \begin{bmatrix} \boldsymbol{C}_{cc} & \boldsymbol{C}_{cc}\boldsymbol{b}_c + \boldsymbol{C}_{cd}\boldsymbol{b}_d \\ \boldsymbol{b}_c\boldsymbol{C}_{cc} + \boldsymbol{b}_d\boldsymbol{C}_{dc} & \boldsymbol{b}_c\boldsymbol{C}_{cc}\tilde{\boldsymbol{b}}_c + \boldsymbol{b}_d\boldsymbol{C}_{dc}\tilde{\boldsymbol{b}}_c + \boldsymbol{b}_c\boldsymbol{C}_{cd}\tilde{\boldsymbol{b}}_d + \boldsymbol{b}_d\boldsymbol{C}_{dd}\tilde{\boldsymbol{b}}_d \end{bmatrix}. \quad (2\text{-}72)$$

According to Eq. (2-72), the submatrix of $\boldsymbol{C}$ corresponding to the unchanged part of the internal coordinate vector is invariant on a partial transformation of the coordinates.

As seen from the definition, the reciprocal of a diagonal compliance constant, $1/C_{ii}$, has the same dimension as the diagonal force constant F_{ii}. An important difference in the physical meaning between $1/C_{ii}$ and F_{ii} has been pointed out by Swanson.[65] On truncating the expansion of the potential function in Eq. (2-54) at the quadratic terms and setting the values of internal coordinates as $\Delta R_q = 1$ and ΔR_i $(i \neq q) = 0$, Eq. (2-61) turns out to be

$$(\partial V/\partial R_q) = F_{qq}\,. \quad (2\text{-}73)$$

On the other hand, on substituting Eq. (2-60) with $p = q$ into Eq. (2-61), retaining only the linear terms and using Eq. (2-67), we have

$$(\partial V/\partial R_q) = \left(\sum_i F_{qi}C_{iq}\right)\Delta R_q \Big/ C_{qq} = \Delta R_q / C_{qq}\,. \quad (2\text{-}74)$$

Thus the reciprocal of a diagonal compliance constant represents the force required to keep the internal coordinate ΔR_q at unity under the condition that $(\partial V/\partial R_p) = 0$ for all the other internal coordinates ΔR_p. Accordingly, F_{qq} must be greater than $1/C_{qq}$ by the amount of energy required to constrain ΔR_i $(i \neq q)$ at zero. In fact, as proved below, the quantity $F_{qq} - 1/C_{qq}$ cannot be negative if $\boldsymbol{F}$ is symmetric and positive definite. Since the order in the array of the internal coordinates is arbitrary, we may choose ΔR_q to be ΔR_n in an n-dimensional problem without loss of generality. From the definition of the compliance matrix, $\boldsymbol{C} = \boldsymbol{F}^{-1}$, the (n, n) element of the matrix $\boldsymbol{CFC} = \boldsymbol{C}$ is written as

$$C_{nn} = \sum_{i=1}^{n}\sum_{j=1}^{n} C_{ni}F_{ij}C_{jn} = \sum_{i=1}^{n-1} C_{ni}\sum_{j=1}^{n} F_{ij}C_{jn} + C_{nn}\sum_{j=1}^{n} F_{nj}C_{jn}$$

$$= \sum_{i=1}^{n-1} C_{ni}\left(\sum_{j=1}^{n-1} F_{ij}C_{jn} + F_{in}C_{nn}\right) + C_{nn} = \sum_{i=1}^{n-1} C_{ni}\sum_{j=1}^{n-1} F_{ij}C_{jn} + C_{nn}\sum_{i=1}^{n-1} C_{ni}F_{in} + C_{nn}$$

$$= \sum_{i=1}^{n-1} C_{ni}\sum_{j=1}^{n-1} F_{ij}C_{jn} + C_{nn}(1 - C_{nn}F_{nn}) + C_{nn}. \tag{2-75}$$

The fifth (and last) equality of Eq. (2-75) results from the relation

$$\sum_{i=1}^{n} C_{ni}F_{in} = \sum_{i=1}^{n-1} C_{ni}F_{in} + C_{nn}F_{nn} = 1.$$

Rearranging Eq. (2-75) and dividing by C_{nn}^2, we have

$$F_{nn} - 1/C_{nn} = \sum_{i=1}^{n-1}\sum_{j=1}^{n-1} C_{ni}F_{ij}C_{jn} \Big/ C_{nn}^2. \tag{2-76}$$

Since the compliance matrix $\boldsymbol{C}$ is symmetric, *i.e.*, $C_{jn} = C_{nj}$, the numerator of the right side of Eq. (2-76) is a quadratic form with respect to $n - 1$ variables, C_{n1}, C_{n2}, ..., $C_{n,n-1}$. The coefficients F_{11} through $F_{n-1,n-1}$ are the elements of the $(n - 1)$th principal submatirx of $\boldsymbol{F}$. Since $\boldsymbol{F}$ is positive defnite, all of its principal submatrices should be positive definite. Accordingly, the right side of Eq. (2-76) is positive or zero, and it has been proved that

$$F_{nn} \geq 1/C_{nn}. \tag{2-77}$$

As obvious from the definition of the compliance matrix, the equality in Eq. (2-77) holds only when $\boldsymbol{F}$ is diagonal, *i.e.*, when all the C_{ni}'s vanish. The force required to distort an internal coordinate is thus independent of the number and types of the other internal coordinates of the molecule if the $\boldsymbol{F}$ matrix is diagonal. A potential function consisting of single variable potential terms corresponds to a diagonal $\boldsymbol{F}$ matrix in the approximation in which the terms higher than second-order are neglected. Accordingly, the parameters in the partitioned potential function in the form of Eq. (2-51) are expected to show better transferability among different molecules than the GVFF force constants in the internal coordinate space without redundancy.

2.5 Modeling of Local Potential Functions

2.5.1 Potential Models for Non-infinitesimal Variables

The power series expansion of the potential energy of a molecule works well as long as the change in each internal variable is restricted within a small range around its equilibrium value. The assumption of an infinitesimal change around a particular value is usually valid for such internal variables as the bond lengths, r, the bond angles, θ and β and the out-of-plane angles, γ. In order to follow the changes in the energy and the structure of a molecule up to a certain dissociation limit, however, no upper bound can be assumed for the relevant bond length. Even if no dissociation reactions take place, the torsion angles around a single bond may change over a wide range between 0 and 2π in the course of a conformational isomerization. Furthermore, the potential functions used in molecular mechanics always involve a number of single-variable functions of the internuclear distances between two atoms not directly bonded to each other, $V_{ER}(r'_{ij})$ and $V_C(r'_{ij})$ (see Eq. (2-51)). The argument r'_{ij} of such a function may take a widely varying value from the lower bound roughly given by the sum of the van der Waals radii of the atoms i and j to the upper bound comparable with the molecular size. To overcome the failure of the power series expansion in these cases, we must assume some apopropriate model potential functions in closed forms which can be used over the whole variable range of particular types of internal variables.

2.5.2 Bond Stretching Potential

Since the development of quantum mechanics, it has become possible to carry out theoretical calculations of the bond energies of diatomic molecules as functions of the bond distances quantitatively. On the other hand, the accurate shapes of the experimental bond energy curves of diatomic molecules have become available from the spectroscopically determined energy levels assigned on the basis of quantum mechanical treatment of the nuclear motion. In an attempt to represent the experimentally determined bond energy curves of diatomic molecules by simple analytical functions, the Morse function

$$V_b = D_{ij}^e\left[\exp\left\{-2a_{ij}\left(r_{ij}-r_{ij}^0\right)\right\}-2\exp\left\{-a_{ij}\left(r_{ij}-r_{ij}^0\right)\right\}\right] \tag{2-78}$$

was proposed.[69)]

Since it follows from Eq. (2-78) that $V_b\left(r_{ij}^0\right)=-D_{ij}^e$ and $V_b(\infty)=0$, D_{ij}^e represents the energy required to stretch the bond r_{ij} from its equilibrium distance to infinity, corresponding to the dissociation energy from the potential minimum. Differentiating Eq. (2-78) n times with respect to r_{ij}, and putting $r_{ij}=r_{ij}^0$, one has

$$\left(\partial^n V_b/\partial r_{ij}{}^n\right)_0=(-1)^n(2^n-2)a_{ij}{}^n D_{ij}^e\ . \tag{2-79}$$

Equation (2-78) is then expanded in powers of $\left(r_{ij}-r_{ij}^0\right)$ in the form

$$V_b\left(r_{ij}\right)=\tfrac{1}{2}k_{ij}\left(r_{ij}-r_{ij}^0\right)^2\left\{1-a_{ij}\left(r_{ij}-r_{ij}^0\right)+\frac{7a_{ij}{}^2}{12}\left(r_{ij}-r_{ij}^0\right)^2-\cdots\right\}, \tag{2-80}$$

where $k_{ij}=2a_{ij}{}^2D_{ij}^e$ according to Eq. (2-79). The expanded form in Eq. (2-80) truncated at the quartic term is used in MM3[51)], MM4[52)] and MMFF94[58)] potential functions.

Historically, the Morse function was derived as the analytical function which gave the vibrational energy levels in the form of a polynomial of the vibrational quantum number terminated at the quadratic term.[69)] According to precise spectroscopic measurements, the fitting of the experimental energy levels to the power series of the vibrational quantum number requires cubic or higher-order terms although they are usually very small, as shown in Table 2.4. A number of empirical functions have been proposed to elucidate these higher-order terms. One of these empirical functions, the Lippincott function,[70)]

$$V_b=D_{ij}^e\left[1-\exp\left\{-k_{ij}r_{ij}^0\left(r_{ij}-r_{ij}^0\right)^2/2D_{ij}^e r_{ij}\right\}\right], \tag{2-81}$$

Table 2.4 Coefficients of expansion of vibrational energy in powers of vibrational quantum number[a] of diatomic molecules.[b]

Molecule	State	ω(cm^{-1})	x(cm^{-1})	$100x/\omega$	y(cm^{-1})	$100y/x$
H_2	$^1\Sigma_g^+$	4395.24	117.91	2.68	0.293	0.249
D_2	$^1\Sigma_g^+$	3118.46	94.96	3.045	1.457	1.534
CO	$^1\Sigma^+$	2170.21	13.46	0.620	0.031	0.230
$H^{35}Cl$	$^1\Sigma^+$	2989.74	52.05	1.741	0.056	0.108
HF	$^1\Sigma^+$	4138.52	90.07	2.176	0.980	1.008
I_2	$^1\Sigma_g^+$	214.57	0.6127	0.286	−0.0009	−0.146
KBr	$^1\Sigma^+$	231	0.7	0.303	0.0011	0.157
KCl	$^1\Sigma^+$	280	0.9	0.321	0.0011	0.122
Li_2	$^1\Sigma_g^+$	351.44	2.592	0.738	−0.0058	0.224
LiH	$^1\Sigma^+$	1405.65	23.200	1.650	0.1633	0.704
N_2	$^1\Sigma_g^+$	2359.61	14.46	0.613	0.0075	0.052
NO	$^2\Pi_{1/2}$	1904.03	13.97	0.734	−0.0012	−0.009
Na_2	$^1\Sigma_g^+$	159.23	0.726	0.456	−0.0027	−0.372
NaBr	$^1\Sigma^+$	315	1.15	0.365	0.0008	0.070
O_2	$^3\Sigma_g^-$	1580.36	12.07	0.764	0.0546	0.452
P_2	$^1\Sigma_g^+$	780.43	2.804	0.359	−0.0053	−0.189
Se_2	$^1\Sigma_g^+$	391.77	1.06	0.271	0.002	0.189
SiO	$1\Sigma^+$	1242.03	6.047	0.487	0.0033	0.054

[a] The energy E_v/hc (cm^{-1}) is expanded in powers of (v + 1/2) as

$$E_v/hc = \omega(v+\tfrac{1}{2}) + x(v+\tfrac{1}{2})^2 + y(v+\tfrac{1}{2})^3 + \cdots \;;$$

see Chapter 5.

[b] G. Herzberg, *Spectra of Diatomic Molecules*, van Nostrand, Toronto (1950)

has been used mainly for describing the stretching potential of hydrogen bonds. The parameters k_{ij}, D_{ij}^e and r_{ij}^0 in Eqs. and (2-81) have the same meaning as the corresponding parameters in Eqs. (2-78) and (2-80). It should be noted that this function has the value $V_b(0) = 0$ at $r_{ij} = 0$, and should always be used with a function for which $V(0) = \infty$.

In most of the potential functions used in molecular mechanics, the energies of interaction between the hydrogen and the acceptor atoms in hydrogen bonds are calculated from certain analytical functions in closed forms. In some programs, the coordinate bonds between metals and ligands in coordination compounds can be treated in a similar way. The use of analytical functions is essential for these bonds if the search for equilibrium structures is to be carried out for a system in which the formation or the dissociation of such a bond may take place automatically in the course of energy minimization. An inverse power-type potential function in the form

$$V_{HB}(r_{H\cdots A}) = \frac{A'}{r_{H\cdots A}^{\;m}} - \frac{B'}{r_{H\cdots A}^{\;n}}, \tag{2-82}$$

where $m = 12$ and $n = 10$, was proposed by McGuire *et al.* for use in the conformational study of peptides and proteins in an earlier stage of the development of molecular mechanics.[71] Since then this 12–10 potential has been employed in many empirical potential functions aiming at the calculation of energies and structures, and has proved to be fairly successful in spite of its rather simple structure.

It is well known that a stable hydrogen bond is formed in various systems where the donor, the hydrogen and the acceptor atoms lie on a straight line. Typical examples of such a system are found in carboxylic acid dimers. In the cases of α-helix and β-sheet structures of proteins, the bond angle subtended at the acceptor oxygen atom is linear, too. The selectivity of particular directions in a hydrogen bond can be taken into account in the potential function by introducing an angle-dependent factor in the hydrogen bond stretching potential[34,51h]. The multivariable function of the form

$$V'_{\mathrm{HB}}(r_{\mathrm{H\cdots A}},\theta_{\mathrm{D\text{-}H\cdots A}},\theta'_{\mathrm{H\cdots A\text{-}X}}) = V_{\mathrm{HB}}(r_{\mathrm{H\cdots A}})\cos^m(\theta_{\mathrm{D\text{-}H\cdots A}})\cos^n(\theta'_{\mathrm{H\cdots A\text{-}X}}) \qquad (2\text{-}83)$$

is adopted in the program for biological macromolecules, CHARMm[9a], with m = 0, 2 or 4 and n = 0 or 2, as well as in the program for complexes of proteins with small molecules, YETI[72] which has been derived from AMBER,[10,11] with m = 2 and n = 4. The program YETI deals also with the coordinate bonds using the function

$$V'_{\mathrm{MC}}(r_{\mathrm{M\cdots L}},\theta_{\mathrm{L\cdots M\cdots L'}}) = V_{\mathrm{ML}}(r_{\mathrm{M\cdots L}})\prod\cos^2(\theta_{\mathrm{L\cdots M\cdots L'}} - \theta^0_{\mathrm{L\cdots M\cdots L'}}) \qquad (2\text{-}84)$$

where L and L′ are the ligand atoms, θ^0 is the optimum bond angle and the product is taken over the independent angles around the central metal atom, M. On the other hand, Eq. (2-82) with m = 12 and n = 6 is used without the angle-dependent factor for the hydrogen bonds in the program OPLS[73], which has also been derived from AMBER. This potential has the same form as the 12 – 6 potential for the exchange repulsion–dispersion interaction described in the next section, so that the programing can be made much simpler than the case of Eq. (2-83). Both methods have been reported to reproduce the experimental protein structures quite successfully[72,73]. An elaborate model may represent the true feature of the potential function more faithfully than a simple model, but the latter is more advantageous in reducing the load on the computer, so that it can deal with larger systems than the former. The choice between elaborateness and simplicity depends on what aspect of the individual problem is considered most important.

2.5.3 Exchange Repulsion and Dispersion Potential

There are two types of force acting in opposite directions between a pair of neutral atoms which are not bonded to each other. One is repulsive and is called the exchange repulsion force, while the other is attractive and is called the dispersion force.[74] These types of force are also called the short range force, since they fall off at much shorter distances than the Coulomb force between opposite charges, *i.e.*, the long range force. Irrespective of the relevant atomic species, the exchange repulsion force is far more dominant at a very short internuclear separation. Both types of force diminish on the increase of the internuclear distance, whereby the rate of diminishing of the dispersion force is always lower than that of the exchange repulsion force. Thus the net force acting between a pair of nonbonded neutral atoms is weakly attractive at a large internuclear distance, and turns out to be strongly repulsive on the approach of the two atoms. This effect prevents abnormal shortening of the distance between any two nuclei i and j not directly bonded to each other, r'_{ij}, and plays an important role in determining the molecular conformation. Hence the empirical potential functions used in molecular mechanics are always designed to include a certain function of

r'_{ij} representing the energy from which the exchange repulsion force and the dispersion force are deduced as minus the gradients. Because of the large variable range of r'_{ij}, this function should be given in a closed analytical form. Two types of functions have been employed by many authors, the exp – 6 (Buckingham) type[75)]

$$V_{\exp-6}(r'_{ij}) = B_{ij} \exp(-C_{ij}\, r'_{ij}) - A_{ij}\, r'^{-6}_{ij} \tag{2-85a}$$

and the n – 6 (Lennard–Jones) type[76)]

$$V_{n-6}(r'_{ij}) = \frac{B_{ij}}{r'^{n}_{ij}} - \frac{A_{ij}}{r'^{6}_{ij}}\,, \tag{2-85b}$$

both of which converge to zero at the limit $r'_{ij} \to \infty$. The origin of the attractive and the repulsive short range force acting between two neutral atoms is elucidated as follows.

A neutral atom is composed of a nucleus and a number of electrons moving around the nucleus. On taking the time average of trajectory, the center of the mass of electrons may coincide with the position of the nucleus, but the nucleus and the electrons never occupy the same position as each other at a given moment. Hence the center of the mass of electrons at a moment does not necessarily coincide with the nuclear position, and various patterns of momentary charge distributions may be formed within two nonbonded atoms in contact. In the absence of any restriction on the movement of electrons, the center of the mass of electrons of one atom is liable to occupy a position as close to the nucleus of the other atom as possible. In Fig. 2.6, this situation of oriented charge distribution is schematically shown in (a) and (b). The oppositely oriented charge distribution shown in (c) and (d) is energetically less favorable. The attractive distribution ((a), (b)) occurs with a higher probability than the repulsive distribution ((c), (d)). Eventually, an attractive force acts between two nonbonded atoms. This attractive force is called the dispersion force. The quantum mechanical treatment based on the second-order perturbation theory shows that the energy gain by this force is inversely proportional to the sixth power of the distance between the two atoms.[77)]

On the other hand, if no bond formation takes place between two atoms approaching each other, the electrons which enter the newly formed antibonding orbitals give rise to a very strong repulsive force between the two atoms due to the exchange repulsion. This repulsive energy is given by the exchange integral, which generally includes a factor exhibiting an

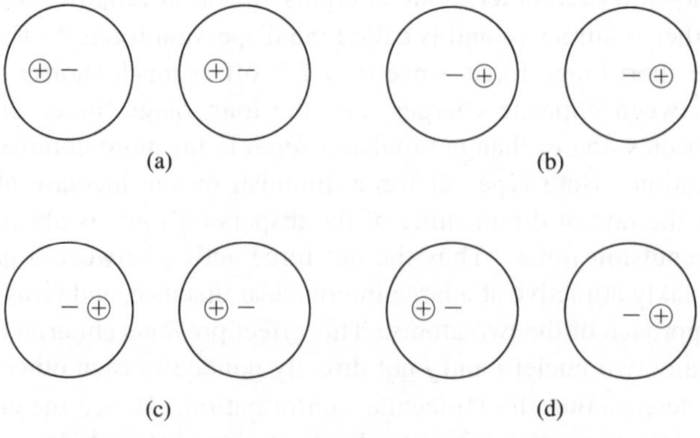

Fig. 2.6 Orientation of electron clouds of two contacting atoms.

exponential dependence on the internuclear distance.[78] The net energy of interaction between a nonbonded atom pair is the sum of the two types of energies leading to the attractive and the repulsive force, and is expected to depend on the internuclear distance with the closed form of the exp – 6 type given by Eq. (2-85a).

The numerical calculation of the exponential function involved in the repulsive term of Eq. (2-85a) is much more complicated and time consuming than the inverse power function involved in the attractive term. Furthermore, since exp(0) = 1, the exp – 6 type potential falls into the negative infinity when the two nuclei approach each other abnormally. Once an extreme shortening of r'_{ij} takes place during the search for energy minima, a normal internuclear distance cannot be recovered at all. Such a pitfall is avoided by assuming the inverse power type dependence of the repulsive potential on the distance, r^{-n} ($n > 6$) as given by Eq. (2-85b). The distance dependent factor of the repulsive term in Eq. (2-85b) with $n = 12$ can be obtained by a single multiplication from that of the attractive term, so that this potential is quite efficient for the speed-up of computation.

It is not easy to imagine the shape of a potential function directly from the potential parameters in Eqs. (2-85a,b). In order for the functional form to be more quickly recognizable, it is recommended that Eqs. (2-85a,b) be rewritten in terms of the equilibrium internuclear distance, r^e_{ij}, and the potential depth at the equilibrium distance, D^e_{ij}, in the forms

$$V_{\exp-6}(r'_{ij}) = D^e_{ij}\left\{\frac{6}{c_{ij}-6}\exp c_{ij}\left(1-\frac{r'_{ij}}{r'^e_{ij}}\right) - \frac{c_{ij}}{c_{ij}-6}\left(\frac{r'^e_{ij}}{r'_{ij}}\right)^6\right\}, \tag{2-86a}$$

and

$$V_{n-6}(r'_{ij}) = D^e_{ij}\left\{\frac{6}{n-6}\left(\frac{r'^e_{ij}}{r'_{ij}}\right)^n - \frac{n}{n-6}\left(\frac{r'^e_{ij}}{r'_{ij}}\right)^6\right\}, \tag{2-86b}$$

respectively. The parameter c_{ij} in Eq. (2-86a) is related to C_{ij} in Eq. (2-85a) by $c_{ij} = C_{ij}\, r'^e_{ij}$. It follows immediately from Eq. (2-86a) that

$$V_{\exp-6}(r'^e_{ij}) = -D^e_{ij} \quad \text{and} \quad \left(\partial V_{\exp-6}/\partial r'_{ij}\right)_{r'^e_{ij}} = 0\ .$$

The same relations for V_{n-6} are obtained from Eq. (2-86b). It is seen from Eqs. (2-86a,b) that the dispersion terms of $V_{\exp-6}$ and V_{n-6} having the same r'^e_{ij} and D^e_{ij} differ from each other by the constant factor

$$A_{\exp-6}/A_{n-6} = c_{ij}(n-6)/n(c_{ij}-6)\ . \tag{2-87}$$

If the parameter c_{ij} in Eq. (2-86a) is near 12, the behaviors of $V_{\exp-6}$ and V_{12-6} are quite similar except for the strongly repulsive region at short distances, where V_{12-6} increases more rapidly than $V_{\exp-6}$ on the shortening of r'_{ij}. The hardness of a potential function may be estimated roughly from the second derivative at the equilibrium. Differentiating twice Eqs. (2-86a,b) with respect to r', we have

$$\left(\partial^2 V_{\exp-6}/\partial r'^2\right)_{r'^e} = \left(6c\,D^e/r'^{e2}\right)\{(c-7)/(c-6)\} \tag{2-88a}$$

and

$$(\partial^2 V_{n-6}/\partial r'^2)_{r'^e} = 6n D^e / r'^{e2} \tag{2-88b}$$

respectively. From Eqs. (2-88a,b), V_{12-6} is estimated to be about 20 % harder than $V_{\mathrm{exp}-6}$ if $c \sim 12$. The failure of $V_{\mathrm{exp}-6}$ falling into negative infinity at $r' = 0$ can be avoided, without affecting the values of the potential itself and its second derivative at the equilibrium, by a modification in which the exponential repulsive term in Eq. (2-85a) is replaced by a product of a Lorentzian function and an inverse power of r_{ij} in the form[54a)]

$$V_N(r_{ij}) = \frac{b^2 B_{ij}}{(r_{ij} - r_{ij}^e)^2 + b^2} \left(\frac{r_{ij}^e}{e\, r_{ij}} \right)^n - \frac{A_{ij}}{r_{ij}^{\,6}} . \tag{2-89}$$

This expression is derived from the identity

$$(r_m / r_{ij})^n \equiv \exp\{-n \ln(r_{ij}/r_m)\} \tag{2-90}$$

in which the logarithmic part is expanded around r_m as

$$\ln\left(\frac{r_{ij}}{r_m} \right) = \frac{1}{r_m}(r_{ij} - r_m) - \frac{1}{2\, r_m^{\,2}}(r_{ij} - r_m)^2 + \cdots . \tag{2-91}$$

Note that V_{n-6} in Eq. (2-86b) is derived from the first-order approximation of Eq. (2-91) in which only the linear term in the right side is retained. In the second-order approximation, we retain the linear and the quadratic terms in Eq. (2-91), and substitute it into Eq. (2-90). The result is written, after some rearrangements, in the form

$$\exp(-C_{ij}\, r_{ij}) = \left(\frac{r_m}{e\, r_{ij}} \right)^n \exp\left\{ -\frac{n}{2\, r_m^{\,2}} (r_{ij} - r_m)^2 \right\} , \tag{2-92}$$

where

$$C_{ij} = n / r_m .$$

To save computing time, the Gaussian distribution function in Eq. (2-92) is replaced by the Lorentzian function with the same curvature at the equilibrium

$$b = (2 \ln 2 / n)^{1/2}\, r_m , \tag{2-93}$$

leading to the repulsive potential in Eq. (2-89). The modified Buckingham type potential is used in RISE, in which the power n is taken to be 8 if the atoms i and j form a geminal (1–3) pair, and 12 otherwise[54a)]. The form of Eq. (2-89) is somewhat complicated, but computation using it can be done faster than if the original exp–6 type potential is used, since it does not involve evaluation of the exponential function.

In the MM2 potential[37)], the exchange repulsion–dispersion potential with the minimum at $r_{ij}'^e$ is expressed as the exp–6 type potential in the form

$$V_{\mathrm{exp}-6}(r_{ij}') = \varepsilon_{ij} \left\{ b_{ij} \exp\left(-c_{ij} \frac{r_{ij}'}{r_{ij}'^e} \right) - a_{ij} \left(\frac{r_{ij}'^e}{r_{ij}'} \right)^6 \right\} , \tag{2-94}$$

where the dimensionless parameters a_{ij} and c_{ij} are fixed at 2.25 and 12.5, respectively, irrespective of the atomic types of the nuclei i and j. The corresponding parameters in MM3 are $a_{ij} = 1.86$ and $c_{ij} = 12$. In Eq. (2-94), the parameter b_{ij} is fixed to be so related to a_{ij} and c_{ij} by

$$b_{ij} = 6a_{ij}\exp(c_{ij})/c_{ij}$$

as to satisfy the equilibrium condition

$$\frac{dV}{dr_{ij}} = \frac{\varepsilon_{ij}}{r_{ij}'^{e}}\left\{-b_{ij}c_{ij}\exp\left(-c_{ij}\frac{r_{ij}'}{r_{ij}'^{e}}\right) + 6a_{ij}\left(\frac{r_{ij}'^{e}}{r_{ij}'}\right)^{7}\right\} = 0 \qquad (2\text{-}95)$$

at $r_{ij}' = r_{ij}'^{e}$.

The potential involves therefore two adjustable parameters instead of three in the usual exp-6 potential. The negative divergence of the exchange repulsion–dispersion potential at $r_{ij}' = 0$ is avoided by replacing the exp–6 function with an inverse quadratic function of the form

$$V_{-2}(r_{ij}') = \varepsilon_{ij}\, d_{ij}\left(r_{ij}'^{e}/r_{ij}'\right)^{2} \qquad (2\text{-}96)$$

whenever r_{ij}' is within a certain threshold $r_{ij}'^{e}/P_{ij}$.

The constants d_{ij} and P_{ij} are determined from the relations $V_{\exp-6} = V_{-2}$ and $\left(dV_{\exp-6}/dr_{ij}'\right) = \left(dV_{-2}/dr_{ij}'\right)$ to be satisfied simultaneously at $r_{ij}' = r_{ij}'^{e}/P_{ij}$. Differentiating Eq. (2-96) gives

$$\frac{dV_{-2}}{dr_{ij}'} = -\frac{2\varepsilon_{ij}\, d_{ij}}{r_{ij}'^{e}}\left(\frac{r_{ij}'^{e}}{r_{ij}'}\right)^{3}. \qquad (2\text{-}97)$$

Equating Eq. (2-94) with Eq. (2-96) and Eq. (2-95) with Eq. (2-97), and eliminating ε_{ik}, r^{0}_{ik} and b_{ik} from the resulting relations, we obtain

$$\left(\frac{2P_{ik}}{c_{ik}} - 1\right)\exp c_{ik}\left(1 - \frac{1}{P_{ik}}\right) = -\frac{2P_{ik}^{\;7}}{3} \qquad (2\text{-}98)$$

and

$$d_{ik} = a_{ik}\, P_{ik}^{\;4}\left(\frac{6P_{ik}}{c_{ik}} - 1\right)\Big/\left(1 - \frac{2P_{ik}}{c_{ik}}\right). \qquad (2\text{-}99)$$

The constant P_{ik} is obtained by solving Eq. (2-98) numerically, and is substituted into Eq. (2-99) to give d_{ik}. The results for $c_{ik} = 12.5$ and $a_{ik} = 2.25$ are : $P_{ik} = 3.311$ and $d_{ik} = 336.176$. The parameters for the heterogeneous atom pair A···B are evaluated from those for the homogeneous pairs A···A and B···B by the relations $r_{AB}'^{e} = (r_{AA}'^{e} + r_{BB}'^{e})/2$ and $\varepsilon_{AB} = (\varepsilon_{AA}\varepsilon_{BB})^{1/2}$.

The nonbonded atom–atom interaction energy differs from the bond energy not only in the magnitude but also in the dependence on the direction. If the valence of an atom A forming a bond with B is not saturated, another atom C approaching A receives an attractive force toward the direction favorable for the formation of a stable bond angle C–A–B. Thus, the driving force for the formation of chemical bonds is strongly anisotropic. On the contrary, the force due to the nonbonded atom–atom interaction is isotropic in principle as long as the nucleus is at the center of mass of the surrounding electrons. There acts only an attractive or a repulsive force between two nonbonded nuclei, and no force arises in the direction perpendicular to the line connecting the two nuclei.

The electron cloud of a hydrogen atom bonded to another atom is liable to be so attracted from the bond that its center of mass is shifted more or less from the nuclear position towards the bonded atom. Accordingly, any atoms approaching a bonded hydrogen atom from the direction opposite to the bond can contact the hydrogen nucleus more closely than those approaching from the direction perpendicular to the bond. A method of taking account of this situation is to replace the internuclear distance $|\boldsymbol{r}_{\mathrm{Y\cdots H}}|$ in the calculation of the nonbonded atom-atom potential by $|\boldsymbol{r}_{\mathrm{Y\cdots H}} + f_r\,\boldsymbol{r}_{\mathrm{H\text{–}X}}|$. The effective center of interaction is then shifted by the factor f_r from the nuclear position of the hydrogen atom along the bond as illustrated in Fig. 2.7. Such an apparent foreshortening of the C–H bond was proposed by Williams in the calculation of the structures and the lattice energies of molecular crystals.[79)] Allinger *et al.* applied this method to the intramolecular nonbonded interaction involving hydrogen atoms. The factor f_r is taken to be 0.085 and 0.077 for MM2[37)] and MM3[51a)], respectively.

In most of the model potential functions, the nonbonded atom pair bonded to a common atom, *i.e.*, the atoms at the geminal position of each other, are treated somewhat differently from the other nonbonded atom pairs. Three schemes of the modeling have been proposed in the literature. In the first scheme, the same form of the potential function as Eqs. (2-85a) is used with different values of the parameters, *A*, *B* and *C*. The second scheme of the modeling is to omit the potential term $V_{\mathrm{ED}}(r'_{ij})$ entirely for the geminal atom pairs. In this case, the geminal nonbonded interaction energy is considered to be implicitly incorporated in the potential terms involving the relevant bond lengths and bond angles. In the third scheme, the quadratic repulsive term of the form

$$V_N(r'_{ij}) = \tfrac{1}{2} F_{ij}\left(r'_{ij} - r'^{0}_{ij}\right)^2 \qquad \text{(2-100a)}$$

is used for each geminal nonbonded internuclear distance, r'_{ij}. The potential functions of this type were employed in the first version of CFF.[25,26)] The program SPASIBA,[56)] in which

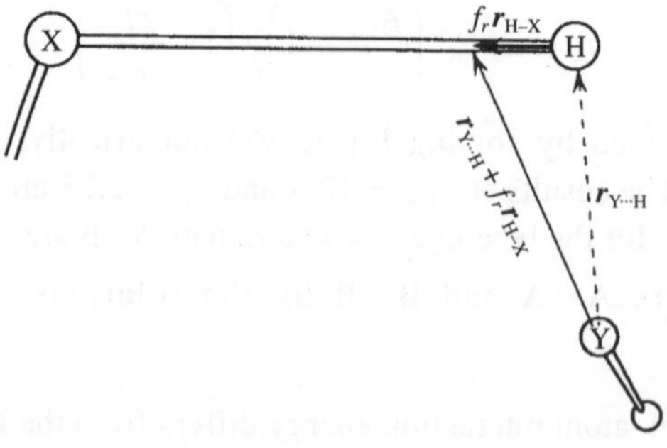

Fig. 2.7 Foreshortening of the X–H bond on calculating nonbonded atom–atom interaction energy involving a hydrogen atom.

the frequency fitting is taken as the main subject of calculation, takes account of the 1-3 interaction of this type.

Among the quadratic force fields used in the normal coordinate analysis, the SVFF supplemented with the geminal nonbonded interaction terms in the form of Eq. (2-100a) is called the Urey–Bradley force field (UBFF).[80] The UBFF and its various modifications have been successful in reproducing the normal frequencies of a wide variety of compounds with reasonable accuracy.[81a,b] In the UBFF defined for an equilibrium structure to be used in the vibrational spectroscopy, the 1–3 repulsive terms are written as

$$V_N\left(r'_{ij}\right) = \tfrac{1}{2} F_{ij}\left(r'_{ij} - r'^{0}_{ij}\right)^2 + F'_{ij}\left(r'_{ij} - r'^{0}_{ij}\right) \tag{2-100b}$$

and the relevant bond stretching and the angle bending terms as

$$V_r\left(r_{pk}\right) = \tfrac{1}{2} K_{pk}\left(r_{pk} - r^0_{pk}\right)^2 + K'_{pk}\left(r_{pk} - r^0_{pk}\right), \qquad (p = i \text{ or } j)$$

and

$$V_\theta\left(\theta_{ikj}\right) = \tfrac{1}{2} H_{ikj}\left(\theta_{ikj} - \theta^0_{ikj}\right)^2 + H'_{ikj}\left(\theta_{ikj} - \theta^0_{ikj}\right),$$

and the coefficients K'_{pk} and H'_{ikj} are so written in terms of F'_{ij} as to make the first derivatives of the total potential vanishing. In the molecular mechanics potential defined for general structure, addition of linear terms is not necessary because the gradients vanish automatically after the potential minimization.

A new type of exchange–repulsion–dispersion potential named the buffered-*n*–*m* potential

$$V_{\text{buf}-n-m}\left(r'_{ij}\right) = D^e_{ij}\left\{\frac{1+\delta}{\left(r'_{ij}/r'^e_{ij}\right)+\delta}\right\}^{n-m}\left\{\frac{1+\gamma}{\left(r'_{ij}/r'^e_{ij}\right)^m+\gamma}\right\}, \tag{2-101}$$

was proposed by Halgren[82] by fitting the functional form to the vibrational levels of rare gas dimers, virial coefficient and viscosity, and was implemented in MMFF94 force field.[58a,b] The parameters δ and γ were fixed to be 0.07 and 0.12, respectively, and the exponents n and m to be 14 and 7, leaving the function as a two-parameter potential.

2.5.4 Coulomb Interaction Potential

The seventh sum in Eq. (2-51) represents the Coulomb interaction energy between the charges distributed within a molecule. Since a molecule is an assembly of positively charged nuclei and negatively charged electrons, this type of interaction must exist between localized charges. Since the coordinates of electrons are not used in molecular mechanics, the effect of electronic charges must be implicitly incorporated in certain functions of nuclear coordinates. The simplest way of describing the Coulomb interaction energy within a molecule is to use a certain effective charge placed at each nuclear position. This "effective nuclear charge," which will be denoted by q_i hereafter, is the algebraic sum of the positive charge of the nucleus i and the negative charges of electrons distributed in the neighborhood of i. Because of the ambiguity in defining "the neighborhood" of a nucleus, the values of

q_i in the literature are quite divergent. Various methods of empirically estimating the effective atomic charges are described in Chapter 7.

Once the effective atomic charges are estimated according to a certain convention, the Coulomb interaction energy is given by

$$V_C(r'_{ij}) = q_i\, q_j / \varepsilon_{ij}\, r'_{ij} \,, \tag{2-102}$$

where ε_{ij} is the effective dielectric constant of the medium surrounding the nuclei i and j. The Coulomb interaction within a molecule may also be described in terms of the effective dipole moment $\boldsymbol{\mu}_a$ of each bond a instead of the effective nuclear charge q_i. In this case, the energy of the Coulomb interaction between the bond dipoles $\boldsymbol{\mu}_a$ and $\boldsymbol{\mu}_b$ is given by

$$V_C(R_{ab}) = \frac{1}{\varepsilon_{ab}} \left\{ \frac{\boldsymbol{\mu}_a \cdot \boldsymbol{\mu}_b}{|\boldsymbol{R}_{ab}|^3} - \frac{3(\boldsymbol{\mu}_a \cdot \boldsymbol{R}_{ab})(\boldsymbol{\mu}_b \cdot \boldsymbol{R}_{ab})}{|\boldsymbol{R}_{ab}|^5} \right\}, \tag{2-103}$$

where ε_{ab} is the effective dielectric constant of the medium and $\boldsymbol{R}_{ab}$ is the vector connecting the midpoints of the bonds a and b with the direction from a to b. The Coulomb potential in the form of Eq. (2-103) is used in the MM series potential functions developed by Allinger and his coworkers.[34–42,51,52] The effective dielectric cosntants in Eqs. (2-102) and (2-103) are taken to be a constant in some models and to be a function of the internuclear or the interbond distance in others. If an odd power dependence on the distance is assumed in the latter case, no square roots need to be computed in calculating the denominators in Eqs. (2-102) and (2-103), and appreciable computing time is saved. In MMFF94, the Coulomb potential is expressed in the buffered form

$$V_{\text{buf-C}}(r'_{ij}) = q_i q_j \big/ D(r'_{ij} + \delta)^n \,. \tag{2-104}$$

The exponent n is normally 1, but the use of a distance-dependent dielectric constant ($n = 2$) is also supported.[58a,b)]

In some potential functions, negative point charges or point dipoles are introduced as representatives of electrons localized around special regions in a molecule, and the electrostatic potential arising therefrom is added to $V_C(r'_{ij})$. The effect of lone-pair electrons is taken into account in MM2 by using point dipoles. In the EPEN force field, electrons in bonding orbitals, π-electrons and lone-pair electrons are expressed as negative point charges.[83a,b)] In these cases, the coordinates of the point dipoles or the point charges are determined from the coordinates of the nuclei of the donor atoms of the electrons. Occasionally, the coordinates of neighboring nuclei are also referred to. Thus the independent variables in these potential functions are exclusively the nuclear coordinate as defined for the molecular mechanics potential.

2.5.5 Angle Bending Potential

A bond angle does not change much from its equilibrium value except for the case of small ring formation or extreme congestion of bulky groups, so that the angle bending potential is usually written as a single quadratic term or a power series in the form of Eq. (2-52). The use of powers higher than 2 is effective for preventing too much deviation of the angle from its equilibrium value. A single additional term with $n = 6$ is used in MM2.[37)]

Various analytical forms of the angular potential function, which may cover a wider range of the variable than the power series, have also been tested. In most of these cases, trigonometric functions of an angular variable is used instead of the change in the angle itself. In his pioneering work on the anharmonic potential of polyatomic molecules, Pliva used a three-parameter function for the valence angle θ_m in the form

$$V_\theta(\theta_m) = D_m \frac{\sin^2(\theta_m - \theta_m^0)}{\sin(\theta_m/2)} \left\{ 1 + b_m \frac{\sin^2(\theta_m - \theta_m^0)}{\sin^2(\theta_m/2)} \right\}, \tag{2-105}$$

which includes only two adjustable parameters except for the equilibrium angle θ_m^0, and can be used over the whole variable range $0 \le \theta_m \le 2\pi$.[84] For a nonlinear bond angle, this potential gives a maximum at $\theta_m = \pi$ with the barrier height

$$V_\theta(\pi) = D_m \sin^2 \theta_m^0 (1 + b_m \sin^2 \theta_m^0) . \tag{2-106}$$

Differentiating Eq. (2-105) twice and thrice with respect to θ_m and putting $\theta_m = \theta_m^0$, the corresponding quadratic and cubic force constants are obtained as

$$F_{mm} = \left(\partial^2 V_m / \partial \theta_m{}^2\right)_0 = 2D_m / \sin(\theta_m^0/2) \tag{2-107a}$$

and

$$F_{mmm} = \left(\partial^3 V_m / \partial \theta_m{}^3\right)_0 = -D_m \cos(\theta_m^0/2)/\sin^2(\theta_m^0/2) \tag{2-107b}$$

respectively. Although no adjustable parameter enters the relation between F_{mm} and F_{mmm}, the anharmonicity of water molecule calculated from Eqs. (2-107a,b) agrees well with the observed data. The expressions for the second and the third derivatives of $V_m(\theta_m)$ at a non-equilibrium position are very complicated.

To amend the failure of the harmonic potential in reflecting the periodicity inherent to any angular variables, the molecular mechanics program DREIDING[21] supports the functional form

$$V_\theta = \tfrac{1}{2} C_{ijk} \left(\cos\theta_{ijk} - \cos\theta_{ijk}^0\right)^2 . \tag{2-108}$$

In the program UFF[22,23], the angular potential functions are written in the form

$$V(\theta_j) = K_{ijk} \sum_{n=0}^{m} C_n \cos n\theta_j . \tag{2-109}$$

The coefficients C_n are so determined as to give the minimum of the function at $\theta_m = \theta_m^0$.

Allured *et al.* proposed a new type of angular potential in a closed form suitable to transition-metal complexes of the ML_n type which often exhibits largely distorted L(ligand)–M(metal)–L angles.[85] The functional form is

$$V_\phi(\phi) = k_\phi \left\{1 + \cos(n\phi + \psi_\phi)\right\} \tag{2-110}$$

where ψ_ϕ is the phase shift determining the positions of potential minima.

The angular variable ϕ is defined in terms of the bond vectors $\boldsymbol{r}_{ML}$ and their projections onto a plane or an axis which is uniquely determined from the positions of the nuclei M, L_1, L_2, $\cdots$. The potential of this type was reported to fit more closely to results of the *ab initio* calculation over a wide variable range than the usual harmonic potential. Note further that the derivatives of the sine or cosine of an angular variable with respect to the Cartesian displacements can be obtained more easily than the derivatives of the angle itself, because the inverse trigonometric functions in the expressions of these variables (Eqs. (2-27, 31, 32, 41, 43, 47–49)) need not be calculated.

An analytical angle bending potential based on the hybrid orbital theory has recently been proposed by Root *et al.*[86] These authors calculated the overlap integral between two sp^{m}d^n-type hybrid orbitals, one lying along the Z-axis and the other on the ZX-plane forming an angle θ with the former, in the form

$$\Delta = \frac{1 + m\cos\theta + n(3\cos^2\theta - 1)/2}{1 + m + n}. \tag{2-111}$$

The angle bending potential is expressed in the form

$$V_\theta(\theta) = \sum_i^{\text{all ligand}} \sum_{j \neq i} k_i \{S_i^{\max} - S_i(\theta_{ij})\}, \tag{2-112}$$

where $S_i(\theta_{ij})$ is a function of θ_{ij} called the orbital strength. The forms of $S_i(\theta_{ij})$ for various hybrid orbitals were given in terms of Δ by Pauling.[87] The number of parameters in the right side of Eq. (2-112) can be reduced considerably since the parameter k_i depends only on the atom type of the apex nuclei i. In this sense, the potential $V_\theta(\theta)$ is generic. Nevertheless, this potential was reported to reproduce the bond angles and the angle bending frequencies of a wide variety of compounds. The potential functions SHAPES[85,88] and VALBOND,[86] which include Eq. (2-112) as the angle-bending potential, have been implemented in CHARMm.

2.5.6 Torsional Potential

The torsion angles defined by Eq. (2-31) or (2-32) are indispensable arguments of the potential functions of molecules containing at least one sequence of four successive atoms, e.g., i–j–k–l. Any change in the torsion angle τ_{ijkl} causes a change in the nonbonded internuclear distance r'_{il} according to the relation

$$\begin{aligned} r'_{il} &= \{(\boldsymbol{r}_{ij} + \boldsymbol{r}_{jk} + \boldsymbol{r}_{kl}) \cdot (\boldsymbol{r}_{ij} + \boldsymbol{r}_{jk} + \boldsymbol{r}_{kl})\}^{1/2} \\ &= \{r_{ij}{}^2 + r_{jk}{}^2 + r_{kl}{}^2 - 2r_{ij}r_{jk}\cos\theta_{ijk} - 2r_{jk}r_{kl}\cos\theta_{jkl} \\ &\quad + 2r_{ij}r_{kl}(\cos\theta_{ijk}\cos\theta_{jkl} - \sin\theta_{ijk}\sin\theta_{jkl}\cos\tau_{ijkl})\}^{1/2} \end{aligned} \tag{2-113}$$

where the quantity in the parentheses in the last formula arises from Eq. (2-17). The bond vectors $\boldsymbol{r}_{ij}$ and $\boldsymbol{r}_{kl}$ in Eq. (2-113) may be replaced by general distance vectors between nonbonded nuclei. In this case, an internuclear distance between any two groups separated by the bond j–k is given as a function of the rotational angle of one group against the other around the bond. Then the dependence of the total nonbonded interaction energy within a molecule on the angle of internal rotation around a skeletal bond can be quickly estimated by combining Eq. (2-113) with Eqs. (2-85a,b), (2-101) or their modifications. Generally, the nonbonded interaction energies calculated in this way are not sufficient for elucidating experimental data for the height of barriers to the internal rotation or the energy differences between rotational isomers. In molecular mechanics, such a difficulty is avoided by introducing a torsional potential term $V_\tau(\tau_{ijkl})$ given in the fifth sum in Eq. (2-51). Generally, the force required for distorting a torsion angle around a single bond is much weaker than the force required for distorting either bond lengths or bond angles. Then the torsion angle τ may deviate largely from the initially assumed value during the process of structure optimization. In fact, the equilibrium torsion angles of a given type are known to vary from one molecule to another according to the environment of the single bond. Such a wide variable range cannot be covered by a power series in $\Delta\tau$, the increment of a certain torsion angle from its standard value. Accordingly, the torsional potential is given as the sum of trigonometric functions in the closed form

$$V_\tau(\tau_{ijkl}) = \tfrac{1}{2}\sum_{n=1}^{L} V_n^0(1 + P_n \cos n\tau_{ijkl}) \,, \qquad (2\text{-}114)$$

where $V_n^0 \geq 0$ and P_n is +1 or − 1 according to whether the term containing V_n^0 takes a maximum or a minimum value at $\tau_{ijkl} = 0$. The upper limit of the sum, L, is mostly less than 7. The introduction of the torsional potential in the form of Eq. (2-114) is not only requested empirically but justified theoretically. Quantum mechanical calculations show that the energy of stabilization due to the partial overlap between the bond orbitals of the vicinal bonds i–j and k–l is actually affected by the torsion angle τ_{ijkl}.[88] The magnitude of such an overlap may be affected by the changes in the bond length r_{jk} and the bond angles θ_{ijk} and θ_{jkl}, and the resulting energy change may be manifested in any structure-dependent barrier heights. This effect can be taken into account in the versatile molecular mechanics program UTAH5 in which torsional potentials in the forms

$$V_\tau(\tau_{ijkl}, r_{jk}) = \tfrac{1}{2} V_3^0(r_{jk} - r_{jk}^0)(1 + \cos 3\tau_{ijkl}) \qquad (2\text{-}115)$$

and

$$V_\tau(\tau_{ijkl}, r_{jk}) = \tfrac{1}{2} V_3^0(\theta_{ijk} - \theta_{ijk}^0)(\theta_{jkl} - \theta_{jkl}^0)(1 + \cos 3\tau_{ijkl}) \qquad (2\text{-}116)$$

are supported.[53] The r-dependent potential in Eq. (2-115) is also employed in MM3,[51a] MM4[52a] and MUB2.[33] The total energy constituting the barrier to the internal rotation around the C–C bond of ethane can be partitioned into three components arising from V_τ, V_{ED} and V_C. The partitioned energies calculated for the model potential functions in MM2, MM3 and RISE are listed in Table 2.5. In all cases, the change in the nonbonded interaction energy can elucidate at most 20% of the observed barrier height. For the internal rotation

Table 2. 5 Torsional barrier of ethane.

Potential	MM2	MM3	RISE
r_{C-H}(pm)[c]	102.02[a]	102.73[b]	110.28
r_{C-C}(pm)[c]	153.2	153.1	153.0
θ_{CCH}(°)[c]	111.0	111.4	109.72
V_{ED}	2.634	1.543	2.229
V_C	0.0	0.0	0.017
V_τ	8.924	8.962	9.913
Barrier height	11.56	10.51	12.16

[a] 1.115 pm × 0.915 (Ref. 37).
[b] 1.113 pm × 0.923 (Ref. 51a).
[c] Fixed at the optimized values for the staggered conformation.

of a methyl group, each term in Eq. (2-114) may be omitted unless n is an integer multiple of 3, since the contributions from the three equivalent hydrogen atoms cancel one another out according to the relation

$$\cos n\tau + \cos n(\tau + 2\pi/3) + \cos n(\tau + 4\pi/3) = 0 \,, \qquad (2\text{-}117)$$

in the case where $n \neq 0$ and mod $(n, 3) \neq 0$. On the other hand, the difference in the free energy between the *trans* and the *gauche* conformers of *n*-butane is reproduced well by adding the terms with $n = 1$ and 2 in Eq. (2-114) to the torsional potential for the sequence C–C–C–C.[90,91] Since molecular mechanics is the study of how to describe and how to use the force field for nuclear motion, the search for efficient functions which represent the torsional potential in closed forms is an important objective. Extensive collection of spectroscopic data on torsional levels and thermodynamic data on equilibria among rotational isomers as well as detailed quantum mechanical calculations is necessary for further progress in this direction.

References

1) E. B. Wilson, Jr., J. C. Decius and P. C. Cross, *Molecular Vibrations*, McGrow-Hill, New York (1955).
2) H. B. Thompson, *J. Chem. Phys.*, **47**, 3407 (1967).
3) T. Miyazawa, *Kagaku to Denshikeisanki*, (in Japanese) p. 48, ed. by T. Yonezawa and K. Osaki, Nankodo, Kyoto (1968).
4) T. Clark, *A Handbook of Computational Chemistry*, p. 99, John Wiley and Sons, New York (1985).
5) M. Pariseau, I. Suzuki and J. Overend, *J. Chem. Phys.*, **42**, 2335 (1965).
6) A. R. Hoy, I. M. Mills and G. Strey, *Mol. Phys.*, **24**, 1265 (1972).
7) K. Machida, *J. Chem. Phys.*, **44**, 4186 (1966).
8) Y. Morino and T. Shimanouchi, *Pure Appl. Chem.*, **50**, 1707 (1978).
9) a) B. R. Brooks, R. E. Bruccoleri, B. D. Olafson, D. J. States, S. Swaminathan and M. Karplus, *J. Comput. Chem.*, **4**, 187 (1983); b) A. D. MacKerell, Jr., J. Wiorkiewicz and M. Karplus, *J. Am. Chem. Soc.*, **117**, 11946 (1995); c) J. J. Pavelites, J. Gao, P. A. Bash and A. D. Mackerell, Jr., *J. Comput. Chem.*, **18**, 221 (1997); M. C. Nicklaus, *ibid.*, **18**, 1056 (1997); D. Yin and A. D. Mackerell, Jr., *ibid.*, **19**, 334 (1998); O. Donini and D. F. Weaver, *ibid.*, **19**, 1515 (1998)..
10) S. J. Weiner, P. A. Kollman, D. A. Case, U. C. Singh, C. Ghio, G. Alagona, S. Profeta and P. Weiner, *J. Am. Chem. Soc.*, **106**, 765 (1984) (United atom version).
11) S. J. Weiner, P. A. Kollman, D. T. Nguyen and D. A. Case, *J. Comput. Chem.*, **7**, 230 (1986) (All atom version).
12) T. L. Hill, *J. Chem. Phys.*, **14**, 465 (1946).
13) J. O. Hirschfelder, C. F. Curtiss and R. B. Bird, *The Molecular Theory of Gases and Liquids*, John Wiley & Sons, New York (1954).
14) R. H. Boyd, *J. Chem. Phys.*, **49**, 2574 (1968).
15) E. M. Engler, J. D. Andose and P. R. Schleyer, *J. Am. Chem. Soc.*, **95**, 8005 (1973).
16) J. D. Andose and K. Mislow, *J. Am. Chem. Soc.*, **96**, 216 (1974).
17) D. N. J. White and M. J. Bovill, *J. Mol. Struct.*, **33**, 273 (1976); *idem., J. Chem. Soc.* (*Perkin II*), **1977**, 1610.
18) J. Jacob, H. B. Thompson and L. S. Bartell, *J. Chem. Phys.*, **47**, 3736 (1967).
19) M. Clark, R. D. Cramer III and N. Van Opdenbosch, *J. Comput. Chem.*, **10**, 982 (1989).
20) E. K. Davies and N. W. Murrel, Comput. Chem., 13, 149 (1989); D. Galisteo, J. A. Lopez-Sastre and H. Marginez-Garcia, *J. Mol. Struct.*, **384**, 25 (1996).
21) S. L. Mayo, B. D. Olafson and W. A. Goddard III, *J. Phys. Chem.*, **94**, 8897 (1990).
22) A. K. Rappe, C. J. Casewit, K. S. Colwell, W. A. Goddard III, and W. M. Skiff, *J. Am. Chem. Soc.*, **114**, 10024 (1992); C. J. Casewit, K. S. Colwell, and A. K. Rappe, *ibid.*, **114**, 10035, 10046 (1992).
23) J. Hermans, H. J. C. Berendsen, W. F. van Gunsteren and J. P. M. Postma, *Biopolymers*, **23**, 1513 (1984); K. H. Ott and B. Meyer, *J. Comput. Chem.*, **17**, 1068 (1996).
24) a) J. C. Vinter, A. Davis and M. R. Saunders, *J. Comput.-Aided Mol. Design*, **1**, 31 (1987); b) R. J. Abraham and I. S. Howorth, *ibid.*, **2**, 125 (1988); c) S. D. Morley, R. J. Abraham, I. S. Howorth, D. E. Jackson and M. R. Saunders, *ibid.*, **5**, 475 (1991); d) R. Grifith, J. B. Brenner and S. J. Titmuss, *J. Comput. Chem.*, **18**, 1211 (1997); e) M. J. Szabo, R. K. Szilagyi and L. Bencz, *J. Mol. Struct.*, **427**, 55 (1998).
25) S. Lifson and A. Warshel, *J. Chem. Phys.*, **49**, 5116 (1968).
26) A. Warshel and S. Lifson, *J. Chem. Phys.*, **53**, 582 (1970).

27) F. A. Momany, L. M. Caruthers, R. F. McGuire and H. A. Scheraga, *J. Phys. Chem.*, **78**, 1595, 1621 (1974).
28) F. A. Momany, R. R. McGuire, A. W. Burgess and H. A. Scheraga, *J. Phys. Chem.*, **79**, 2361 (1975).
29) S. S. Zimmerman, M. S. Pottle, G. Nemethy and H. A. Scheraga, *Macromolecules*, **10**, 1 (1977).
30) H. Chuman, F. A. Momany and L. Schafer, *Int. J. Pept. Protein Res.*, **24**, 233 (1984); Z. Wang and R. Pachtor, *J. Comput. Chem.*, **18**, 323 (1997).
31) G. Nemethy, K. D. Gibson, K. A. Palmer, C. N. Yoon, G. Parterlini, A. Zagari, S. Rumsey and H. A. Scheraga, *J. Phys. Chem.*, **96**, 6472 (1992).
32) M. Vasquez, G. Nemethy and H. A. Scheraga, *Chem. Rev.*, **94**, 2183 (1994).
33) S. Fitzwater and L. S. Bartell, *J. Am. Chem. Soc.*, **98**, 5107 (1976).
34) R. Lavery, H. Sklenar, K. Zakrzewska, and B. Pullman, *J. Biol. Struct. Dynam.*, **3**, 989 (1986); R. Lavery, K. Zakrzewska and H. Sklenar, *Comput. Phys. Commun.*, **91**, 135 (1995); T. H. Duong and K. Zakrzewska, *J. Comput. Chem.*, **18**, 796 (1997).
35) N. L. Allinger, M. T. Tribble, M. A. Miller and D. H. Wertz, *J. Am. Chem. Soc.*, **93**, 1637 (1971); N. L. Allinger and J. T. Sprague, *ibid.*, **95**, 3893 (1973); *idem.*, *Tetrahedron*, **31**, 21 (1975).
36) D. H. Wertz and N. L. Allinger, *Tetrahedron*, **30**, 1579 (1994).
37) N. L. Allinger, *J. Am. Chem. Soc.*, **99**, 3279, 8127 (1977).
38) D. R. Ferro, P. Pumilia and M. Raggazi, *J. Comput. Chem.*, **18**, 351 (1997); M. C. A. Costa and Y. Takahata, *ibid.*, **18**, 712 (1997); I. Komaroni and J. M. J. Trouchet, *J. Mol. Struct.*, **395/396**, 15 (1997); H. M. Marques, C. Warden, M. Monye, M. S. Shongwe and K. L. Brown, *Inorg. Chem.*, **37**, 2578 (1998).
39) P. M. Ivanov, *JCPE Newsletter*, **4** (1), 9; *idem.*, *ibid.*, **4** (2), 9(1992).
40) F. Mohamadi, N. G. J. Richards, W. C. Guida, R. Liskamp, M. Lipton, C. Caufield, G. Chang, T. Hendrickson and W. C. Still, *J. Comput. Chem.*, **11**, 440 (1990).
41) C. Jaime and E. Osawa, *Tetrahedron*, **39**, 2769 (1983); L. P. Burke, A. D. Debellis, H. Fuhrer, H. Meier, S. D. Pastor, G. Rihs, G. Rist, R. K. Rodebaugh and S. P. Shum, *J. Am. Chem. Soc.*, **119**, 8313 (1997).
42) J. T. Sprague, J. C. Tai, Y. Yuh and N. L. Allinger, *J. Comput. Chem.*, **8**, 581 (1987); B. Reindl, T. Clark and P. von R. Schleyer, *J. Comput. Chem.*, **17**, 1406 (1996); *idem.*, *ibid.*, **18**, 28, 533 (1997), E. L. Eliel, *J. Org. Chem.*, **62**, 9154 (1998).
43) O. Ermer and S. Lifson, *J. Am. Chem. Soc.*, **95**, 4121 (1973).
44) O. Ermer, *Struct. Bonding* (*Berlin*), **27**, 161 (1976).
45) A. T. Hagler, E. Huler and S. Lifson, *J. Am. Chem. Soc.*, **96**, 5319 (1974).
46) A. T. Hagler, P. S. Stern, S. Lifson and S. Ariel, *J. Am. Chem. Soc.*, **101**, 813 (1979); A. T. Hagler, S. Lifson and P. Dauben, *ibid.*, **101**, 5122, 5131 (1979).
47) D. R. Black, C. G. Parker, S. S. Zimmerman and M. L. Lee, *J. Comput. Chem.*, **17**, 531 (1996); U. Eichler, C. M. Kolmel and J. Sauer, *ibid.*, **18**, 463 (1997); K. Gaedt and H.-D. Holtje, *ibid.*, **19**, 935 (1998)
48) S. Lifson and P. S. Stern. *J. Chem. Phys.*, **77**, 4542 (1982).
49) P. M. Ivanov, *JCPE Newsletter*, 7, No. 4, 13 (1996).
50) Y. Sun, J. W. Caldwell and P. A. Kollman, *J. Phys. Chem.*, **99**, 10081 (1995) ; P. M. Ivanov and C. Jaime, *J. Mol. Struct.*, **377**, 157 (1997).
51) a) N. L. Allinger, Y. H. Yuh and J. H. Lii, *J. Am. Chem. Soc.*, **111**, 8551 (1989); J. H. Lii and N. L. Allinger, *ibid.*, **111**, 8566, 8576 (1989); b) J. H. Lii and N. L. Allinger, *J. Comput. Chem.*, **12**, 186 (1991); c) G. Liang, P. C. Fox and J. P. Bowen, *ibid.*, **17**, 940 (1996); S. Li and C.-H. Huang, *ibid.*, **17**, 1013 (1996); d) S. W. Carrigan, J.-H. Lii and J. P. Bowen, *J. Comput.-Aided Mol, Des.*, **11**, 61 (1997); G. Liang, X. Chen, J. A. Dustman, A. H. Lewin and J. P. Bowen, *J. Comput. Chem.*, **18**, 1371 (1997); J.-Y. Shim and J. P. Bowen, *ibid.*, **19**, 1387 (1998); e) G. Liang, J. P. Bays and J. P. Bowen, *J. Mol. Struct.*, **403**, 165 (1997); B. P. Hay, L. Yang, D. Zhang, J. R. Rustad and E. Wasserman, *ibid.*, **417**, 19 (1997); B. P. Hay, L. Yang, J.-H. Lii and N. L. Allinger, *ibid.*, **428**, 203 (1998); X. Sanchez-Rives, M. Ramos and C. Jaime, *ibid.*, **442**, 93 (1998); f) B. P. Hay, O. Clement, G. Sandrone and D. A. Dixon, *Inorg. Chem.*, **37**, 6408 (1998); g) J. C. Tai and N. L. Allinger, *J. Comput. Chem.*, **19**, 475 (1998); h) J.-H. Lii and N. L. Allinger, *J. Phys. Org. Chem.*, **7**, 591 (1994), *idem.*, *J. Comput. Chem.*, **19**, 1001 (1998); i) P.-O. Norrby and T. Liljefors, *ibid.*, **19**, 1146 (1998); j) J.-Y. Shim and J. P. Bowen, *ibid.*, **19**, 1370, 1387 (1998); k) C. R. Landis, T. Cleveland and T. K. Firman, *J. Am. Chem. Soc.*, **120**, 2641 (1998); I Columbus and S. E. Biali, *ibid.*, **120**, 3060 (1998); D. M. Power and E. A. Noe, *ibid.*, **120**, 5312 (1998); l) M. H. Langoor, L. M. J. Kroon-Batenburg and J. H. Van der Maas, *J. Chem. Soc.*, *Faraday Trans.*, **93**, 4107 (1997).
52) a) N. L. Allinger, K. Chen and J.-H. Lii, *J. Comput. Chem.*, **17**, 642 (1996); b) N. Nevins, K. Chen, and N. L. Allinger, *J. Comput. Chem.*, **17**, 669 (1996); c) N. Nevins, J.-H. Lii and N. L. Allinger, *J. Comput. Chem.*, **17**, 695 (1996); d) N. Nevins and N. L. Allinger, *J. Comput. Chem.*, **17**, 730 (1996); e) N. L. Allinger, K. Chen, J. A. Katzenellenbogen, S. R. Wilson and G. M. Anstead, *J. Comput. Chem.*, **17**, 747 (1996); N. L. Allinger and Y. Fan, *ibid.*, **18**, 1829 (1997).
53) D. H. Faber and C. Altona, *Comput. Chem.*, **1**, 203 (1977).
54) a) Y. Miwa and K. Machida, *J. Am. Chem. Soc.*, **110**, 5183 (1988); *ibid.*, **111**, 7733 (1989); b) I. Yokoyama, Y. Miwa and K. Machida, *ibid.*, **113**, 6459 (1991), *J. Phys. Chem.*, **95**, 9740 (1991); *Bull. Chem. Soc. Jpn.*, **65**,

746 (1992); I. Yokoyama, Y. Miwa, K. Machida, J. Umemura and S. Hayashi, *ibid.*, **66**, 400 (1993); c) Y. Miwa, N. Mimura, K. Machida, T. Nakagawa, J. Umemura and S. Hayashi, *Spectrochim. Acta*, **50A**, 1629 (1994); d) E. Ganeshsrinavas, D. N. Sathyanarayana, K. Machida and Y. Miwa, *J. Mol. Struct.*, **361**, 217 (1996), *ibid.*, **403**, 153 (1997).

55) J. L. M. Dillen, *J. Comput. Chem.* **11**, 1125 (1990), *ibid.*, **13**, 257 (1992), *ibid.*, **16**, 595, 610 (1995).
56) a) P. Derreumaux, P. Lagant and G. Vergoten, *J. Mol. Struct.*, **295**, 203, 223 (1993); b) F. Tristram, V. Durier and G. Vergoten, *ibid.*, **377**, 47 (1996), *idem., ibid.*, **378**, 249 (1997); c) M. Chhiba and G. Vergoten, *ibid.*, **384**, 55 (1997); d) V. Durier, F. Tristram and G. Vergoten, *ibid.*, **395/396**, 81 (1997); P. Lagant and G. Vergoten, *ibid.*, **412**, 59 (1997).
57) a) K. Palmo, L.-O. Pietila and S. Krimm, *J. Comput. Chem.*, **12**, 385 (1991); b) K. Palmo, N. G. Mirkin. L.-O. Pietila and S. Krimm, *Macromolecules*, **26**, 6831 (1993); c) K. Palmo, N. G. Mirkin and S. Krimm, *J. Phys. Chem.*, **A102**, 6448 (1998).
58) a) T. A. Halgren, *J. Comput. Chem.* **17**, 490, 520, 553, 616 (1996); b) T. A. Halgren and R. B. Nachbar, *J. Comput. Chem.*, **17**, 587 (1996).
59) N. Bierrum, V*erh. Dtch. Phys. Ges.*, **16**, 737 (1914), Ref. 1, p. 174.
60) G. Herzberg, *Moleculara Spectra and Molecular Structure, II, Infrared and Raman Spectra of Polyatomic Molecules*, p. 186, van Nostrand, Princeton; Tronto (1945).
61) D. F. Heath and J. W. Linnet, *Trans. Faraday Soc.*, **44**, 556, 873 (1948).
62) J. W. Linnet and M. F. Hoare, *Trans. Faraday Soc.*, **45**, 844 (1949).
63) I. M. Mills, *Spectrochim. Acta*, **19**, 1585 (1963).
64) K. Machida and J. Overend, *J. Chem. Phys.*, **50**, 4437 (1969).
65) B. I. Swanson, *J. Am. Chem. Soc.*, **98**, 3067 (1976).
66) J. C. Decius, *J. Chem. Phys.*, **38**, 241 (1963).
67) L. H. Jones and R. R. Ryan, *J. Chem. Phys.*, **52**, 2003 (1970).
68) S. J. Cyvin, *Molecular Vibrations and Mean Square Amplitudes*, Elsevier, Amsterdam (1968).
69) P. M. Morse, *Phys. Rev.*, **34**, 57 (1929).
70) E. R. Lippincott and Schroeder, *J. Chem. Phys.*, **23**, 1099 (1955).
71) R. F. McGuire, F. A. Momany, and H. A. Scheraga, *J. Phys. Chem.*, **76**, 375 (1972).
72) A.Vedani, *J. Comput. Chem.*, **9**, 269 (1988).
73) W. L. Jorgensen and J. Tirado-Rives, *J. Am. Chem. Soc.*, **110**, 1657 (1988); W. Damm, A. Frontera, J. Tirado-Rives and W. L. Jorgensen, *J. Comput. Chem.*, **18**, 1955 (1997).
74) A. I. Kitaigorodsky, *Molecular Crystals and Molecules*, Academic Press, New York (1973).
75) R. A. Buckingham, *Faraday Soc. Discuss.*, **22**, 75 (1956).
76) J. E. Lennard-Jones, *Proc. Roy. Soc. Ser. A*, **106**, 463 (1924); *ibid.*, **112**, 214 (1926).
77) F. London, *Z. Phys.*, **63**, 245 (1930).
78) T. L. Hill, *J. Chem. Phys.*, **16**, 399 (1948).
79) D. E. Williams, *J. Chem. Phys.*, **43**, 4424 (1965); *ibid.*, **45**, 3770 (1966); *ibid.*, **47**, 4680 (1967).
80) H. C. Urey and C. A. Bradley, *Phys. Rev.*, **38**, 1969 (1931).
81) a) T. Shimanouchi, *The Molecular Force Field, Physical Chemistry*, Vol.4, ed. by H. Eyring, D. Henderson and W. Jost, Academic Press (1970); pp.233~306; b) T. Shimanouchi, *J. Phys. Chem. Ref. Data*, **1**, 189 (1972); *ibid.*, **2**, 121, 225 (1973); *ibid.*, **3**, 269 (1974).
82) T. A. Halgren, *J. Am. Chem. Soc.*, **114**, 7827 (1992).
83) a) L. L. Shipman, A. W. Burgess and H. A. Scheraga, *Proc. Nat. Acad. Sci. USA*, **72**, 543 (1975); b) J. Snir, R. A. Nemenoff and H. A. Scheraga, *J. Phys. Chem.*, **82**, 2497, 2504, 2513, 2521 (1978).
84) J. Pliva, *Coll. Czech. Chem. Commun.*, **23**, 777 (1958).
85) V. S. Allured, Ch. M. Kelly and C. R. Landis, *J. Am. Chem. Soc.*, **113**, 1 (1991).
86) D. M. Root, C. R. Landis and T. Cleveland, *J. Am. Chem. Soc.*, **115**, 4201 (1993).
87) L. Pauling, *Proc. Nat. Acad. Sci. USA*, **72**, 4200 (1975), *ibid.*, **73**, 274 (1976).
88) T. N. Doman, C. R. Landis and B. Bosnish, *J. Am. Chem. Soc.*, **114**, 7264 (1992).
89) T. K. Brunck and E. Weinhold, *J. Am Chem. Soc.*, **101**, 1700 (1979).
90) L. S. Bartell, *J. Am. Chem. Soc.*, **99**, 3279 (1977).
91) N. L. Allinger, D. Hindman and H. Honig, *J. Am. Chem. Soc.*, **99**, 3282 (1973).

Chapter 3

Equilibrium Structure of Molecules

3.1 Expansion of Potential Functions in Cartesian Displacement Coordinates

The search for equilibrium structures of molecules is probably the most basic purpose of molecular mechanics calculations. The equilibrium structures of a molecule are defined as those structures which give minimum values of the potential function V (Eq.(2-1)). For a molecule with more than three atoms, it is not unusual for the potential function to have more than one minimum, especially when there is at least one degree of freedom of the internal rotation around a single bond.

How to calculate the potential function of a polyatomic molecule for a given nuclear configuration has been described in Chapter 2. To search for a potential minimum in any efficient ways, we need to evaluate not only the potential function itself but also some of its derivatives with respect to a set of independent variables. Minimization of a continuous function is achieved by various iterative methods, some of which utilize only the first derivatives while others require both the first and second derivatives. Obviously, the methods of the former category require shorter time for a single iteration than those of the latter. On the other hand, more cycles are generally necessary to reach the minimum for the former than for the latter. When the initial structure chosen is sufficiently near an equilibrium structure, the most efficient of the iterative methods is the Newton–Raphson method, which belongs to the latter class.

Let the initial Cartesian coordinates of the ith nucleus be denoted by

$$x_{3i-2}^0 = X_i^0, \quad x_{3i-1}^0 = Y_i^0, \quad x_{3i}^0 = Z_i^0 , \tag{3-1}$$

The nuclear displacements from the initial positions are then given in terms of the Cartesian displacement coordinates defined in the form

$$\Delta x_k = x_k - x_k^0 . \tag{3-2}$$

On assuming that the nuclear displacements are small, the potential function (Eq.(2-1)) can be expanded as the Taylor series of Δx_k's around the initial positions of nuclei. Truncating this expansion at the second-order terms, we have

$$\begin{aligned} V(x_1^0 + \Delta x_1, x_2^0 + \Delta x_2, \cdots, x_{3N}^0 + \Delta x_{3N}) &= V(x_1^0, x_2^0, \cdots, x_{3N}^0) \\ &+ \sum_{i=1}^{3N} (\partial V/\partial x_i)_0 \Delta x_i + \tfrac{1}{2} \sum_{i=1}^{3N} \sum_{j=1}^{3N} (\partial^2 V/\partial x_i \partial x_j)_0 \Delta x_i \Delta x_j . \end{aligned} \tag{3-3}$$

The symbol $(\ )_0$ means that the quantity in parentheses is evaluated at the initial position. Differentiation of Eq. (3-3) gives

$$\frac{\partial V}{\partial x_i} = \frac{\partial V}{\partial(\Delta x_i)} = \left(\frac{\partial V}{\partial x_i}\right)_0 + \sum_{j=1}^{3N}\left(\frac{\partial^2 V}{\partial x_i \partial x_j}\right)_0 \Delta x_j \,. \tag{3-4}$$

Equation (3-4) represents a first derivative of the potential function after a slight distortion of the molecule from the initial structure. To simplify the expression of Eq.(3-3), we introduce abbreviated symbols for the first and the second derivatives in the forms

$$F_i^x = (\partial V/\partial x_i)_0 \tag{3-5}$$

and

$$F_{ij}^{xx} = (\partial^2 V/\partial x_i \partial x_j)_0 \,, \tag{3-6}$$

and collect them to form a vector

$$\boldsymbol{F}^x = \begin{bmatrix} F_1^x & F_2^x & \cdots & F_{3N}^x \end{bmatrix}^{\mathrm{T}}, \tag{3-7}$$

and a matrix

$$\boldsymbol{F}^{xx} = \begin{bmatrix} F_{11}^{xx} & F_{12}^{xx} & \cdots & F_{1,3N}^{xx} \\ F_{21}^{xx} & F_{22}^{xx} & \cdots & F_{2,3N}^{xx} \\ \vdots & \vdots & \ddots & \vdots \\ F_{3N,1}^{xx} & F_{3N,2}^{xx} & \cdots & F_{3N,3N}^{xx} \end{bmatrix}, \tag{3-8}$$

respectively. By introducing a vector of the Cartesian displacement coordinates

$$\boldsymbol{\Delta x} = \begin{bmatrix} \Delta x_1 & \Delta x_2 & \cdots & \Delta x_{3N} \end{bmatrix}^{\mathrm{T}}, \tag{3-9}$$

Eq.(3-3) is rewritten simply as

$$\begin{aligned} V &= V^0 + \sum_{i=1}^{3N} F_i^x \Delta x_i + \tfrac{1}{2}\sum_{i=1}^{3N}\sum_{j=1}^{3N} F_{ij}^{xx}\Delta x_i \Delta x_j \\ &= V^0 + \tilde{\boldsymbol{F}}^x \boldsymbol{\Delta x} + \tfrac{1}{2}\tilde{\boldsymbol{\Delta x}}\, \boldsymbol{F}^{xx} \boldsymbol{\Delta x}, \end{aligned} \tag{3-10}$$

where $V_0 = V(x_1{}^0, x_2{}^0, \cdots x_{3N}{}^0)$. If a set of certain displacements of the nuclei from the initial positions, $\boldsymbol{\Delta x}$, leads to an equilibrium structure, the first derivative of the potential function given by Eq. (3-4) should vanish altogether after the nuclear displacements. Then we have

$$F_i^x + \sum_j F_{ij}^{xx} \Delta x_i = 0, \qquad (i = 1\text{–}3N). \tag{3-11}$$

Equation (3-11) takes the form of simultaneous equations with respect to the corrections to the nuclear coordinates required to attain an equilibrium structure, Δx_1, Δx_2, ···, Δx_{3N}. Writing the zero vector in the $3N$-space as $\mathbf{0}$, we obtain a matrix expression of Eq. (3-11) in the form

$$\boldsymbol{F}^x + \boldsymbol{F}^{xx}\boldsymbol{\Delta x} = \mathbf{0}\,. \tag{3-12}$$

One might expect to be able to solve Eq. (3-12) directly, obtaining the vector of the corrections to the coordinates, $\boldsymbol{\Delta x}$, which will be called the correction vector hereafter. This is not true, however, since $\boldsymbol{F}^{xx}$ is a singular matrix which cannot be inverted. This singularity arises from the fact that the elements of $\boldsymbol{F}^{xx}$ are not independent of each other, but are interrelated by five (linear molecules) or six (nonlinear molecules) relations which ascertain the invariance of the potential function under the translation and the rotation of the whole molecule. To obtain the correction vector $\Delta\boldsymbol{x}$, one must eliminate the external variables which do not contribute to the potential function from Eq. (3-11), following the procedure described in the next section.

3.2 Separation of Translation and Rotation

For a general discussion of molecular deformations due to small displacements of nuclei, we must first define the internal coordinates $\Delta R_1, \Delta R_2, \cdots, \Delta R_n$ as the changes of those internal variables which are necessary for describing the nuclear configuration uniquely. Next we expand each internal coordinate in powers of the Cartesian displacements as given in Eq. (2-38), and truncate the series at the linear terms under the assumption of small variables. Then the vector of the internal coordinates, $\Delta\boldsymbol{R}$ in Eq. (2-56), can be expressed in terms of $\Delta\boldsymbol{x}$ defined by Eq. (3-9) in the simple form

$$\Delta\boldsymbol{R}_{\text{in}} = \begin{bmatrix} \Delta R_1 \\ \Delta R_2 \\ \vdots \\ \Delta R_n \end{bmatrix} = \begin{bmatrix} B_{11} & B_{12} & \cdots & B_{1,3N} \\ B_{21} & B_{22} & \cdots & B_{2,3N} \\ \vdots & \vdots & \ddots & \vdots \\ B_{n1} & B_{n1} & \cdots & B_{n,3N} \end{bmatrix} \begin{bmatrix} \Delta x_1 \\ \Delta x_2 \\ \vdots \\ \Delta x_{3N} \end{bmatrix} = \boldsymbol{B}_{\text{in}} \Delta\boldsymbol{x} . \tag{3-13}$$

In Eq. (3-13), the suffix "in" has been attached to $\Delta\boldsymbol{R}$ in the leftmost side in order to distinguish it from the vector of the external variables to be introduced later. The matrix $\boldsymbol{B}_{\text{in}}$ is called the $\boldsymbol{B}$ matrix for the internal coordinates. The external variables which describe the amounts of translations and rotations of the molecule as a whole can be related to the Cartesian displacements as follows.

The displacement of the nucleus i of an N-atomic molecule from its initial position is given by

$$\Delta X_i = X_i - X_i^0, \qquad \Delta Y_i = Y_i - Y_i^0, \qquad \Delta Z_i = Z_i - Z_i^0. \tag{3-14}$$

The displacement vector of the nucleus i is then defined as

$$\Delta\boldsymbol{X}_i = [\Delta X_i \quad \Delta Y_i \quad \Delta Z_i]^{\text{T}}. \tag{3-15}$$

Let m_i be the mass of the nucleus i. Then the initial position of the center of mass of the molecule is given by the coordinates

$$X_g = \frac{\sum_i m_i X_i^0}{M}, \qquad Y_g = \frac{\sum_i m_i Y_i^0}{M}, \qquad Z_g = \frac{\sum_i m_i Z_i^0}{M}, \tag{3-16}$$

where M is the molecular mass given in the form

$$M = \sum_i m_i . \tag{3-17}$$

If we choose the initial position of the center of mass of the molecule at the origin, it holds that

$$\sum_i m_i X_i^0 = 0, \qquad \sum_i m_i Y_i^0 = 0, \qquad \sum_i m_i Z_i^0 = 0. \tag{3-18}$$

The moments of inertia of the molecule around the Cartesian axes are defined as

$$I_X = \sum_i m_i\left(Y_i^{02} + Z_i^{02}\right), \quad I_Y = \sum_i m_i\left(Z_i^{02} + X_i^{02}\right), \quad I_Z = \sum_i m_i\left(X_i^{02} + Y_i^{02}\right). \tag{3-19}$$

The products of inertia are given by

$$I_{XY} = \sum_i m_i X_i Y_i, \quad I_{YZ} = \sum_i m_i Y_i Z_i \quad I_{ZX} = \sum_i m_i Z_i X_i. \tag{3-20}$$

The tensor of inertia of the molecule is given in terms of these quantities in the form

$$\boldsymbol{I} = \begin{bmatrix} I_X & -I_{XY} & -I_{XZ} \\ -I_{XY} & I_Y & -I_{YZ} \\ -I_{XZ} & -I_{YZ} & I_Z \end{bmatrix}. \tag{3-21}$$

Now we look for a matrix $\boldsymbol{L}^P$ which diagonalizes $\boldsymbol{I}$ in the form

$$\boldsymbol{I}^P = \begin{bmatrix} I_X^P & 0 & 0 \\ 0 & I_Y^P & 0 \\ 0 & 0 & I_Z^P \end{bmatrix} = \boldsymbol{L}^P \boldsymbol{I} \tilde{\boldsymbol{L}}^P. \tag{3-22}$$

It is convenient to define the position vector of the nucleus i as the vector of its Cartesian coordinates in the form

$$\boldsymbol{r}_i = [X_i \quad Y_i \quad Z_i]^{\mathrm{T}}. \tag{3-23}$$

The tensor of inertia for the initial configuration can be expressed in terms of the position vectors as

$$\boldsymbol{I} = \sum_i m_i\left(r_i^{0\,2}\boldsymbol{E} - \boldsymbol{r}_i^0 \tilde{\boldsymbol{r}}_i^0\right). \tag{3-24}$$

By transforming the vector $\boldsymbol{r}_i$ with the orthogonal matrix $\boldsymbol{L}^P$, we have

$$\boldsymbol{r}_i^P \equiv \left[X_i^P \quad Y_i^P \quad Z_i^P\right]^{\mathrm{T}} = \boldsymbol{L}^P \boldsymbol{r}_i. \tag{3-25}$$

Then the congruence transformation of Eq. (3-24), followed by the use of the orthogonality of $\boldsymbol{L}^P$, yields

$$\boldsymbol{I}^P = \sum_i m_i\left(\left|\boldsymbol{r}_i^{P0}\right|^2 \boldsymbol{E} - \boldsymbol{r}_i^{P0} \tilde{\boldsymbol{r}}_i^{P0}\right). \tag{3-26}$$

Since $\boldsymbol{I}^P$ is a diagonal tensor according to Eq. (3-22), it follows that

$$\sum_i m_i\, X_i^{P0}\, Y_i^{P0} = \sum_i m_i\, Y_i^{P0}\, Z_i^{P0} = \sum_i m_i\, Z_i^{P0}\, X_i^{P0} = 0. \tag{3-27}$$

The Cartesian coordinate system $(X_i^{P0}, Y_i^{P0}, Z_i^{P0})$ which satisfies Eqs. (3-18) and (3-27) is called the principal system of inertia, and the tensor $\boldsymbol{I}^P$ is called the principal tensor of inertia. Hereafter, the Cartesian coordinate system will be taken as the principal system unless specifically mentioned otherwise, and the superscript P will be omitted for simplicity.

Now we introduce six new variables as the linear combinations of the Cartesian displacement coordinates in the principal system in the form

$$T_X = (1/M)\sum_i m_i \,\Delta X_i \tag{3-28a}$$

$$T_Y = (1/M)\sum_i m_i \,\Delta Y_i \tag{3-28b}$$

$$T_Z = (1/M)\sum_i m_i \,\Delta Z_i \tag{3-28c}$$

$$R_X = (1/I_X)\sum_i m_i \,(Y_i^0 \,\Delta Z_i - Z_i^0 \Delta Y_i) \tag{3-28d}$$

$$R_Y = (1/I_Y)\sum_i m_i \,(Z_i^0 \,\Delta X_i - X_i^0 \Delta Z_i) \tag{3-28e}$$

$$R_Z = (1/I_Z)\sum_i m_i \,(X_i^0 \,\Delta Y_i - Y_i^0 \Delta X_i). \tag{3-28f}$$

If we move the molecule by the amount δT along the positive direction of the X-axis retaining its original orientation, the components of the displacement of the nucleus i is given by

$$\Delta X_i = \delta T,\ \Delta Y_i = 0,\ \Delta Z_i = 0 \tag{3-29}$$

Substituting these values into Eq. (3-28a–f) and using Eqs. (3-17) and (3-18), we have $T_X = \delta T$ and $T_Y = T_Z = R_X = R_Y = R_Z = 0$. Similarly, displacements of the molecule along the Y- and the Z-axes bring forth increase of T_X and T_Y, respectively, by the same amounts as the displacements, leaving the remaining five variables unchanged. It is thus proved that variables T_X, T_Y and T_Z represent the amounts of parallel displacements of the molecule along the X-, the Y- and the Z-axes, respectively, in the same scale as used for the Cartesian coordinates.

To examine the amounts of nuclear displacements on the rotational motion of a molecule, let us place the center of mass of the molecule at the origin, and look at it from the positive direction of the Z-axis as shown in Fig. 3.1. If the whole molecule is rotated around the Z-axis counter-clockwise by $\delta\theta$ radian, the Cartesian displacement coordinates of the nucleus i are given by

$$\left.\begin{aligned} \Delta X_i &= -\,r_i^0 \delta\theta \sin\theta^0 = -\,\delta\theta\, Y_i^0 \\ \Delta Y_i &= r_i^0 \delta\theta \cos\theta^0 = \delta\theta\, X_i^0 \\ \Delta Z_i &= 0 \end{aligned}\right\}. \tag{3-30}$$

curvature not captured
okay for small δθ.

Substituting Eq. (3-30) into Eqs. (3-28a–f) and using Eqs. (3-18) and (3-20), we have $R_Z = \delta\theta$ and $T_X = T_Y = T_Z = R_X = R_Y = 0$. Similarly it is verified that on rotation of the molecule around the X-axis by $\delta\theta$ radian, R_X increases by the same amount, while the other five variables remain unchanged. The molecular rotation around the Y-axis leads to the same increase in R_Y. Thus, the variables R_X, R_Y and R_Z give the amount of rotation of the whole

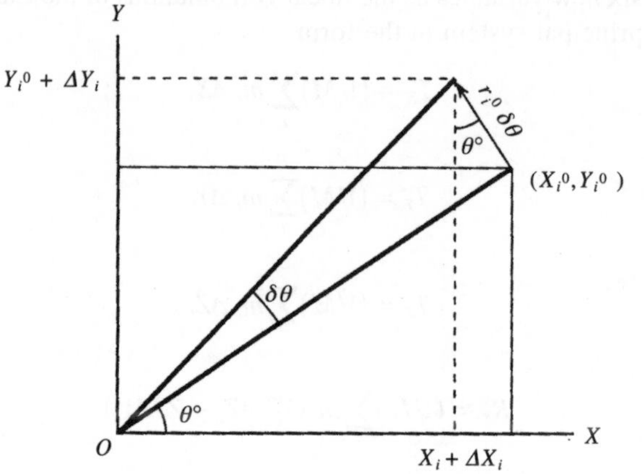

Fig. 3.1 Change in Cartesian coordinates of a nucleus on infinitesimal rotation of the whole molecule.

molecule in radians around the X-, the Y- and the Z-axes, respectively. For linear molecules, we may choose the Z-axis to be parallel with the molecular axis at the equilibrium position. In this case, all the coefficients of ΔX_i and ΔY_i in Eq. (3-28f) vanish, and accordingly the variable R_Z disappears. Hence a given linear molecule should have two degrees of freedom of rotational motion as noted earlier.

Now we introduce a vector of the external variables

$$\Delta \boldsymbol{R}_{\mathrm{ex}} = \begin{bmatrix} T_X & T_Y & T_Z & R_X & R_Y & R_Z \end{bmatrix}^{\mathrm{T}} \tag{3-31}$$

and express Eqs. (3-28a–f) in a compact matrix representation,

$$\Delta \boldsymbol{R}_{\mathrm{ex}} = \boldsymbol{B}_{\mathrm{ex}} \Delta \boldsymbol{x}. \tag{3-32}$$

The matrix of the coefficients in Eq. (3-32), $\boldsymbol{B}_{\mathrm{ex}}$, is derived from Eqs. (3-28a–f) to be given by

$$\boldsymbol{B}_{\mathrm{ex}} = \boldsymbol{M}^{-1} \boldsymbol{D}_{\mathrm{ex}} \boldsymbol{m}, \tag{3-33}$$

where $\boldsymbol{D}_{\mathrm{ex}}$ is a $(6 \times 3N)$ matrix in the form

$$\boldsymbol{D}_{\mathrm{ex}} = \begin{bmatrix} 1 & 0 & 0 & 1 & 0 & 0 & \cdots & 1 & 0 & 0 \\ 0 & 1 & 0 & 0 & 1 & 0 & \cdots & 0 & 1 & 0 \\ 0 & 0 & 1 & 0 & 0 & 1 & \cdots & 0 & 0 & 1 \\ 0 & -Z_1^0 & Y_1^0 & 0 & -Z_2^0 & Y_2^0 & \cdots & 0 & -Z_N^0 & Y_N^0 \\ Z_1^0 & 0 & -X_1^0 & Z_2^0 & 0 & -X_2^0 & \cdots & Z_N^0 & 0 & -X_N^0 \\ -Y_1^0 & X_1^0 & 0 & -Y_2^0 & X_2^0 & 0 & \cdots & -Y_N^0 & X_N^0 & 0 \end{bmatrix} \tag{3-34}$$

and $\boldsymbol{M}$ and $\boldsymbol{m}$ are diagonal matrices given respectively by

$$\boldsymbol{M} = \mathrm{diag}\begin{bmatrix} M & M & M & I_X & I_Y & I_Z \end{bmatrix}$$

and

$$\boldsymbol{m} = \mathrm{diag}\begin{bmatrix} m_1 & m_1 & m_1 & m_2 & m_2 & m_2 & \cdots & m_N & m_N & m_N \end{bmatrix}. \tag{3-35}$$

respectively. Here, diag[$A \;\; B \;\; \cdots$] denotes a diagonal matrix with the elements A, B, $\cdots$.

On taking the number of the internal coordinates to be the same as the internal degrees of freedom of motion, the total number of internal and external variables becomes the same as the degrees of freedom of motion of N particles, $3N$. Then we form a vector $\boldsymbol{\Delta R}$ of the order $3N$ by joining $\boldsymbol{\Delta R}_{\mathrm{in}}$ (Eq. (2-57)) and $\boldsymbol{\Delta R}_{\mathrm{ex}}$ (Eq. 3-31)) to each other and, by introducing a ($3N \times 3N$) matrix $\boldsymbol{B}$, combine Eqs. (3-13) and (3-32) into a single form

$$\boldsymbol{\Delta R} \equiv \begin{bmatrix} \boldsymbol{\Delta R}_{\mathrm{in}} \\ \boldsymbol{\Delta R}_{\mathrm{ex}} \end{bmatrix} = \boldsymbol{B}\,\boldsymbol{\Delta x}, \quad \boldsymbol{B} = \begin{bmatrix} \boldsymbol{B}_{\mathrm{in}} \\ \boldsymbol{B}_{\mathrm{ex}} \end{bmatrix}. \tag{3-36}$$

If all the internal coordinates are taken to be independent of one another, the matrix $\boldsymbol{B}$ has an inverse, which may be written as

$$\boldsymbol{B}^{-1} = \begin{bmatrix} (\boldsymbol{B}^{-1})_{\mathrm{in}} & (\boldsymbol{B}^{-1})_{\mathrm{ex}} \end{bmatrix}. \tag{3-37}$$

this is why I change to cart. first - so I can change back.

The submatrix of $\boldsymbol{B}^{-1}$ corresponding to the external variables, $(\boldsymbol{B}^{-1})_{\mathrm{ex}}$ is given by

$$(\boldsymbol{B}^{-1})_{\mathrm{ex}} = \boldsymbol{m}^{-1}\,\tilde{\boldsymbol{B}}_{\mathrm{ex}}\,\boldsymbol{M} = \tilde{\boldsymbol{D}}_{\mathrm{ex}}\,. \tag{3-38}$$

To prove Eq. (3-38), we premultiply it by Eq. (3-33), and make use of the relation derivable from Eqs. (3-34) and (3-35),

$$\boldsymbol{D}_{\mathrm{ex}}\,\boldsymbol{m}\,\tilde{\boldsymbol{D}}_{\mathrm{ex}} = \boldsymbol{M},$$

obtaining a necessary condition for $(\boldsymbol{B}^{-1})_{\mathrm{ex}}$ in the form

$$\boldsymbol{B}_{\mathrm{ex}}(\boldsymbol{B}^{-1})_{\mathrm{ex}} = \boldsymbol{M}^{-1}\,\boldsymbol{D}_{\mathrm{ex}}\,\boldsymbol{m}\,\tilde{\boldsymbol{D}}_{\mathrm{ex}} = \boldsymbol{E}. \tag{3-39}$$

Note that exchanging the order of multiplication in Eq. (3-39) by using Eqs. (3-33) and (3-38) gives

$$(\boldsymbol{B}^{-1})_{\mathrm{ex}}\,\boldsymbol{B}_{\mathrm{ex}} = \tilde{\boldsymbol{D}}_{\mathrm{ex}}\,\boldsymbol{M}^{-1}\,\boldsymbol{D}_{\mathrm{ex}}\,\boldsymbol{m}. \tag{3-40}$$

The submatrix of $\boldsymbol{B}^{-1}$ corresponding to the internal variables, $(\boldsymbol{B}^{-1})_{\mathrm{in}}$, should satisfy the necessary condition,

$$\boldsymbol{B}_{\mathrm{in}}\,(\boldsymbol{B}^{-1})_{\mathrm{in}} = \boldsymbol{E}. \tag{3-41}$$

To find such a matrix, we introduce the $\boldsymbol{G}$ matrix defined in the form[1)]

$$\boldsymbol{G} = \boldsymbol{B}_{\mathrm{in}}\,\boldsymbol{m}^{-1}\,\tilde{\boldsymbol{B}}_{\mathrm{in}}\,. \tag{3-42}$$

The matrix $(\boldsymbol{B}^{-1})_{\mathrm{in}}$ is then given by[2)]

$$(\boldsymbol{B}^{-1})_{\mathrm{in}} = \boldsymbol{m}^{-1}\,\tilde{\boldsymbol{B}}_{\mathrm{in}}\,\boldsymbol{G}^{-1}. \tag{3-43}$$

Premultiplying Eq. (3-43) by $\boldsymbol{B}_{\mathrm{in}}$ and using Eq. (3-42), we obtain Eq. (3-41). In order for Eq. (3-37) to represent the inverse of $\boldsymbol{B}$ in the satisfactory sense, it should hold that

$$\boldsymbol{B}_{\mathrm{in}}(\boldsymbol{B}^{-1})_{\mathrm{ex}} = \boldsymbol{B}_{\mathrm{in}}\,\tilde{\boldsymbol{D}}_{\mathrm{ex}} = \boldsymbol{0} \tag{3-44}$$

and

$$\begin{aligned} \boldsymbol{B}_{\text{ex}}(\boldsymbol{B}^{-1})_{\text{in}} &= \boldsymbol{M}^{-1}\,\boldsymbol{D}_{\text{ex}}\,\boldsymbol{m}\,\boldsymbol{m}^{-1}\,\tilde{\boldsymbol{B}}_{\text{in}}\,\boldsymbol{G}^{-1} \\ &= \boldsymbol{M}^{-1}\,\boldsymbol{D}_{\text{ex}}\,\tilde{\boldsymbol{B}}_{\text{in}}\,\boldsymbol{G}^{-1} = \boldsymbol{0}, \end{aligned} \tag{3-45}$$

besides Eqs. (3-39) and (3-41). To prove Eq. (3-44), we introduce a (3 × 3) antisymmetric matrix

$$\boldsymbol{D}_i^R = \begin{bmatrix} 0 & -Z_i^0 & Y_i^0 \\ Z_i^0 & 0 & -X_i^0 \\ -Y_i^0 & X_i^0 & 0 \end{bmatrix}, \tag{3-46}$$

and rewrite $\boldsymbol{D}_{\text{ex}}$ in Eq. (3-34) as

$$\boldsymbol{D}_{\text{ex}} = \begin{bmatrix} \boldsymbol{E} & \boldsymbol{E} & \cdots & \boldsymbol{E} \\ \boldsymbol{D}_1^R & \boldsymbol{D}_2^R & \cdots & \boldsymbol{D}_N^R \end{bmatrix}.$$

To simplify the following calculation, the three elements of the $\boldsymbol{B}$ matrix concerning the internal coordinate R_p and the nucleus i are collected into a vector in the form

$$\boldsymbol{s}_{pi} \equiv \begin{bmatrix} s_{pi}^X & s_{pi}^Y & s_{pi}^Z \end{bmatrix}^{\text{T}} = \begin{bmatrix} B_{p,3i-2} & B_{p,3i-1} & B_{p,3i} \end{bmatrix}^{\text{T}}, \tag{3-47}$$

which is well known as the $\boldsymbol{s}$ vector in the theory of molecular vibrations.[1] Premultiplying Eq. (3-47) by Eq. (3-46) and changing the sign, we have

$$-\boldsymbol{D}_i^R\,\boldsymbol{s}_{pi} = \begin{bmatrix} s_{pi}^Y\,Z_i^0 - s_{pi}^Z\,Y_i^0 \\ s_{pi}^Z\,X_i^0 - s_{pi}^X\,Z_i^0 \\ s_{pi}^X\,Y_i^0 - s_{pi}^Y\,X_i^0 \end{bmatrix} = \boldsymbol{s}_{pi} \times \boldsymbol{r}_i^0. \tag{3-48}$$

By expressing the matrix $\boldsymbol{B}_{\text{in}}$ in terms of the $\boldsymbol{s}$ vectors, the matrix product $\boldsymbol{B}_{\text{in}}\,\tilde{\boldsymbol{D}}_{\text{ex}}$ in Eq. (3-44) can be rewritten as

$$\boldsymbol{B}_{\text{in}}\,\tilde{\boldsymbol{D}}_{\text{ex}} = \begin{bmatrix} \tilde{\boldsymbol{s}}_{11} & \tilde{\boldsymbol{s}}_{12} & \cdots & \tilde{\boldsymbol{s}}_{1N} \\ \tilde{\boldsymbol{s}}_{21} & \tilde{\boldsymbol{s}}_{22} & \cdots & \tilde{\boldsymbol{s}}_{2N} \\ \vdots & \vdots & \ddots & \vdots \\ \tilde{\boldsymbol{s}}_{n1} & \tilde{\boldsymbol{s}}_{n2} & \cdots & \tilde{\boldsymbol{s}}_{nN} \end{bmatrix} \begin{bmatrix} \boldsymbol{E} & -\boldsymbol{D}_1^R \\ \boldsymbol{E} & -\boldsymbol{D}_2^R \\ \vdots & \vdots \\ \boldsymbol{E} & -\boldsymbol{D}_N^R \end{bmatrix} = \begin{bmatrix} \sum_i \tilde{\boldsymbol{s}}_{1i} & -\sum_i \tilde{\boldsymbol{s}}_{1i}\,\boldsymbol{D}_i^R \\ \sum_i \tilde{\boldsymbol{s}}_{2i} & -\sum_i \tilde{\boldsymbol{s}}_{2i}\,\boldsymbol{D}_i^R \\ \cdots & \cdots \\ \sum_i \tilde{\boldsymbol{s}}_{ni} & -\sum_i \tilde{\boldsymbol{s}}_{ni}\,\boldsymbol{D}_i^R \end{bmatrix}. \tag{3-49}$$

The internal coordinate ΔR_p can be expressed in terms of the $\boldsymbol{s}$ vectors and the displacement vectors of the relevant nuclei (Eq. (3-15)) in the form

$$\Delta R_p = \tilde{\boldsymbol{s}}_{p1}\,\Delta\boldsymbol{X}_1 + \tilde{\boldsymbol{s}}_{p2}\,\Delta\boldsymbol{X}_2 + \cdots + \tilde{\boldsymbol{s}}_{pN}\,\Delta\boldsymbol{X}_N = \sum_i \tilde{\boldsymbol{s}}_{pi}\,\Delta\boldsymbol{X}_i.$$

On any translation of the molecule as a whole, it holds that

$$\Delta\boldsymbol{X}_1 = \Delta\boldsymbol{X}_2 = \cdots = \Delta\boldsymbol{X}_N \neq \boldsymbol{0},$$

and all the internal coordinates remain unchanged, giving

$$\Delta R_p = 0 = \left(\sum_i \tilde{s}_{pi} \right) \Delta X_1,$$

whence we obtain

$$\sum_i \boldsymbol{s}_{pi} = \mathbf{0}. \tag{3-50}$$

Suppose next that the molecule is rotated around an axis which is directed along the vector

$$\boldsymbol{\delta\Omega} = [\Omega_X \quad \Omega_Y \quad \Omega_Z]^{\mathrm{T}}, \tag{3-51}$$

and let the angle of the rotation be given by

$$\omega = \left(\Omega_X{}^2 + \Omega_Y{}^2 + \Omega_Z{}^2\right)^{1/2}$$

in radians. The displacement of the nucleus i on this rotation is given by

$$\Delta \boldsymbol{X}_i = \begin{bmatrix} \Delta X_i \\ \Delta Y_i \\ \Delta Z_i \end{bmatrix} = - \begin{bmatrix} Y_i^0 \, \Omega_Z - Z_i^0 \, \Omega_Y \\ Z_i^0 \, \Omega_X - X_i^0 \, \Omega_Z \\ X_i^0 \, \Omega_Y - Y_i^0 \, \Omega_X \end{bmatrix} = - \boldsymbol{r}_i^0 \times \boldsymbol{\delta\Omega}, \tag{3-52}$$

according to the definition of a vector product. Since the internal coordinate ΔR_p does not change in this case, either, it holds that

$$\Delta R_p = 0 = - \sum_i \boldsymbol{s}_{pi} \cdot \left(\boldsymbol{r}_i^0 \times \boldsymbol{\delta\Omega}\right) = \boldsymbol{\delta\Omega} \cdot \sum_i \left(\boldsymbol{s}_{pi} \times \boldsymbol{r}_i^0\right). \tag{3-53}$$

Substituting Eq. (3-48) into Eq. (3-53), and noting that $\boldsymbol{\delta\Omega}$ is an arbitrarily chosen vector, we obtain

$$\sum_i \boldsymbol{D}_i^R \, \boldsymbol{s}_{pi} = \mathbf{0} \tag{3-54}$$

From Eqs. (3-50) and (3-54), all the elements of the matrix $\boldsymbol{B}_{\text{in}} \tilde{\boldsymbol{D}}_{\text{ex}}$ in Eq. (3-49) vanish, and Eq. (3-44) is confirmed. Since the transpose of Eq. (3-44), $\boldsymbol{D}_{\text{ex}} \tilde{\boldsymbol{B}}_{\text{in}} = \left(\boldsymbol{B}_{\text{in}} \tilde{\boldsymbol{D}}_{\text{ex}}\right)^{\mathrm{T}}$, is a factor of the matrix product $\boldsymbol{B}_{\text{ex}} \left(\boldsymbol{B}^{-1}\right)_{\text{in}}$ given by Eq. (3-45), the latter should vanish whenever the former vanishes. This completes the proof of Eqs. (3-37), (3-38) and (3-43).

3.3 Transformation between the Internal and the Cartesian Displacement Coordinates

Suppose that the potential energy of an N-atomic molecule is expanded in powers of the internal coordinates at the initial point in the configuration space. If the initial configuration is not an equilibrium structure in contrast to the case of Eq. (2-54), the expansion starts from the linear terms as given by

$$V = V^0(R_1^0, R_2^0, \cdots) + \sum_p F_p \, \Delta R_p + \tfrac{1}{2}\sum_p \sum_q F_{pq} \, \Delta R_p \, \Delta R_q + \cdots , \tag{3-55}$$

where

$$F_p = \left(\partial V / \partial R_p\right)_0 \tag{3-56}$$

and F_{pq} is defined at the initial point in the same form as that in Eq. (2-55a).

The first and the second derivatives of the potential V with respect to the Cartesian displacements of the nuclei are then obtained, by substituting the internal coordinates expanded as Eq. (2-38) into Eq. (3-55), in the forms

$$F_i^x = \left(\frac{\partial V}{\partial x_i}\right)_0 = \sum_p \left(\frac{\partial V}{\partial R_p}\right)_0 \left(\frac{\partial R_p}{\partial x_i}\right)_0 = \sum_p F_p \, B_{pi} \tag{3-57}$$

and

$$\begin{aligned} F_{ij}^{xx} &= \left(\frac{\partial^2 V}{\partial x_i \, \partial x_j}\right)_0 = \sum_p \left(\frac{\partial V}{\partial R_p}\right)_0 \left(\frac{\partial^2 R_p}{\partial x_i \, \partial x_j}\right)_0 + \sum_p \sum_q \left(\frac{\partial^2 V}{\partial R_p \, \partial R_q}\right)_0 \left(\frac{\partial R_p}{\partial x_i}\right)_0 \left(\frac{\partial R_q}{\partial x_j}\right)_0 \\ &= \sum_p F_p \, B_{p,ij} + \sum_p \sum_q F_{pq} \, B_{pi} \, B_{qj} , \end{aligned} \tag{3-58}$$

respectively. As shown in Chapter 2, any type of internal variable used in molecular mechanics can be expressed in terms of the distance vectors between two properly chosen nuclei. The derivatives of the X component of the distance vector connecting the nuclei i and j, X_{ij}, with respect to the X coordinates of i and j are given simply in the form

$$\frac{\partial X_{ij}}{\partial X_i} = \frac{\partial}{\partial X_i}\left(X_j - X_i\right) = -1, \quad \frac{\partial X_{ij}}{\partial X_j} = \frac{\partial}{\partial X_j}\left(X_j - X_i\right) = 1. \tag{3-59}$$

Similar relations hold for the Y and the Z components. Thus the Cartesian derivatives of internal variables can be obtained from the derivatives with respect to the components of these distance vectors. For example, if R_p is a bond length r_{ij}, differentiation of both sides of the well-known relation

$$r_{ij}^{\;2} = \mathbf{r}_{ij} \cdot \mathbf{r}_{ij} = X_{ij}^{\;2} + Y_{ij}^{\;2} + Z_{ij}^{\;2} \tag{3-60}$$

gives

$$\frac{\partial r_{ij}}{\partial X_{ij}} = \frac{X_{ij}}{r_{ij}}, \quad \frac{\partial^2 r_{ij}}{\partial X_{ij}^{\;2}} = \frac{r_{ij}^{\;2} - X_{ij}^{\;2}}{r_{ij}^{\;3}}, \quad \frac{\partial^2 r_{ij}}{\partial X_{ij} \, \partial Y_{ij}} = -\frac{X_{ij} \, Y_{ij}}{r_{ij}^{\;3}}. \tag{3-61}$$

Substitution of Eq. (3-59) into Eq. (3-61) then yields

$$\left.\begin{aligned}
&\frac{\partial r_{ij}}{\partial X_i} = \frac{\partial r_{ij}}{\partial X_{ij}}\frac{\partial X_{ij}}{\partial X_i} = \frac{X_{ij}}{r_{ij}}\frac{\partial X_{ij}}{\partial X_i} = -\frac{X_{ij}}{r_{ij}} \\
&\frac{\partial r_{ij}}{\partial X_j} = \frac{X_{ij}}{r_{ij}}\frac{\partial X_{ij}}{\partial X_j} = \frac{X_{ij}}{r_{ij}} \\
&\frac{\partial^2 r_{ij}}{\partial X_i \partial X_j} = \frac{\partial^2 r_{ij}}{\partial X_{ij}^2}\frac{\partial X_{ij}}{\partial X_i}\frac{\partial X_{ij}}{\partial X_j} = \frac{r_{ij}^2 - X_{ij}^2}{r_{ij}^3} \\
&\frac{\partial^2 r_{ij}}{\partial X_i \partial Y_i} = \frac{\partial^2 r_{ij}}{\partial X_{ij}\partial Y_{ij}}\frac{\partial X_{ij}}{\partial X_i}\frac{\partial Y_{ij}}{\partial Y_i} = -\frac{X_{ij}Y_{ij}}{r_{ij}^3}
\end{aligned}\right\} \tag{3-62}$$

and similar relations for the other derivatives.[3] These derivatives evaluated for the initial structure give the coefficients of the expansion in Eq. (2-38) of the stretching coordinates in powers of the Cartesian displacements of the two nuclei i and j.

When R_p is a variable of an angle type, *i.e.*, the bond angle, the out-of-plane angle, the linear bond angle or the torsion angle, the corresponding angle in Eqs. (2-27), (2-31), (2-32), (2-41), (2-43), (2-48) and (2-49) is expanded as a Taylor series of the Cartesian displacements of the relevant nuclei. The right sides of these equations are all expressed as either R_p = Arcsin f or R_p = Arccos f, where f is a certain combination of products, quotients and square roots of scalar quantities derived from vector calculus on the bond vectors, $\boldsymbol{r}_{ij}$, $\boldsymbol{r}_{ik}$, *etc.*

Let A and B be certain functions of the infinitesimal variables ξ_1, ξ_2, $\cdots$, expanded as the Taylor series in the forms

$$A(\xi_1,\xi_2,\cdots) = A_0 + \sum_i A_i'\xi_i + \tfrac{1}{2}\sum_i\sum_j A_{ij}''\xi_i\xi_j + \cdots \tag{3-63a}$$

and

$$B(\xi_1,\xi_2,\cdots) = B_0 + \sum_i B_i'\xi_i + \tfrac{1}{2}\sum_i\sum_j B_{ij}''\xi_i\xi_j + \cdots, \tag{3-63b}$$

respectively. If C is a function of A and B in the form $C = f(A,B)$, and can be expanded as

$$C(\xi_1,\xi_2,\cdots) = C_0 + \sum_i C_i'\xi_i + \tfrac{1}{2}\sum_i\sum_j C_{ij}''\xi_i\xi_j + \cdots, \tag{3-63c}$$

the coefficients C_i', C_{ij}'', $\cdots$ are expressed in terms of the expansion coefficients of A and B in Eqs. (3-63a,b) as shown in Table 3.1. If the variables ξ_i and ξ_j represent the changes in the components of certain bond vectors from their initial values, and the function A is the scalar product of the bond vectors given by Eq. (2-26), the Taylor series of A terminates at the quadratic terms. Similarly, if A is the scalar triple product in Eq.(2-30), the series terminates at the cubic terms.

Now let us introduce the vector operator defined by

$$\boldsymbol{\nabla}_{ij} = \left[\partial/\partial X_{ij} \quad \partial/\partial Y_{ij} \quad \partial/\partial Z_{ij}\right]^{\mathrm{T}}, \tag{3-64}$$

and expand Eqs. (2-25), (2-26) and (2-30) in powers of the components of the relevant bond vectors. The coefficients of the linear terms in this expansion are given as the components of the vectors

Table 3.1 Expansion coefficients of functions of power series.[a)]

C	C_0	C_i'	C_{ij}''
$A \pm B$	$A_0 \pm B_0$	$A_i' \pm B_i'$	$A_{ij}'' \pm B_{ij}''$
AB	$A_0 B_0$	$A_0 B_i' + B_0 A_i'$	$A_0 B_{ij}'' + B_0 A_{ij}'' + A_i' B_j' + A_j' B_i'$
$\dfrac{A}{B}$	$\dfrac{A_0}{B_0}$	$\dfrac{1}{B_0}(A_i' - C_0 B_i')$	$\dfrac{1}{B_0}(A_{ij}'' - C_0 B_{ij}'' - C_i' B_j' - C_j' B_i')$
$A^{1/2}$	$A_0{}^{1/2}$	$\dfrac{A_i'}{2A_0{}^{1/2}}$	$\dfrac{1}{2A_0{}^{1/2}}\left(A_{ij}'' - \dfrac{A_i' A_j'}{2A_0}\right)$
$\arcsin\dfrac{A}{B}$ [b)]	$\arcsin\dfrac{A_0}{B_0}$	$\dfrac{1}{B_0 c_\alpha}(A_i' - s_\alpha B_i')$	$\dfrac{1}{B_0}\left(\dfrac{A_{ij}''}{s_\alpha} - B_i' C_j' - B_j' C_i'\right) - \dfrac{s_\alpha}{c_\alpha}\left(\dfrac{B_{ij}''}{B_0} - C_i' C_j'\right)$
$\arccos\dfrac{A}{B}$ [b)]	$\arccos\dfrac{A_0}{B_0}$	$\dfrac{-1}{B_0 s_\alpha}(A_i' - c_\alpha B_i')$	$\dfrac{1}{B_0}\left(\dfrac{-A_{ij}''}{c_\alpha} - B_i' C_j' - B_j' C_i'\right) + \dfrac{c_\alpha}{s_\alpha}\left(\dfrac{B_{ij}''}{B_0} - C_i' C_j'\right)$

a) $C_i' = (\partial C/\partial \xi_i)_0$, $C_{ij}'' = (\partial^2 C/\partial \xi_i \partial \xi_j)_0$

b) s_α : $\sin \alpha$, c_α : $\cos \alpha$ (α is the angular variable represented by C)

$$\boldsymbol{\nabla}_{ij}\, r_{ij} = \boldsymbol{r}_{ij}/r_{ij} = \boldsymbol{e}_{ij} \tag{3-65a}$$

$$\boldsymbol{\nabla}_{ji}\left(\boldsymbol{r}_{ji}\cdot\boldsymbol{r}_{jk}\right) = \boldsymbol{r}_{jk}, \quad \boldsymbol{\nabla}_{jk}\left(\boldsymbol{r}_{ji}\cdot\boldsymbol{r}_{jk}\right) = \boldsymbol{r}_{ji} \tag{3-65b}$$

and

$$\begin{aligned} \boldsymbol{\nabla}_{ij}\, v &= \boldsymbol{\nabla}_{ij}\left\{\boldsymbol{r}_{ij}\cdot\left(\boldsymbol{r}_{jk}\times\boldsymbol{r}_{kl}\right)\right\} = \boldsymbol{r}_{jk}\times\boldsymbol{r}_{kl} \\ \boldsymbol{\nabla}_{jk}\, v &= \boldsymbol{\nabla}_{jk}\left\{\boldsymbol{r}_{ij}\cdot\left(\boldsymbol{r}_{jk}\times\boldsymbol{r}_{kl}\right)\right\} = \boldsymbol{r}_{kl}\times\boldsymbol{r}_{ij} \\ \boldsymbol{\nabla}_{kl}\, v &= \boldsymbol{\nabla}_{kl}\left\{\boldsymbol{r}_{ij}\cdot\left(\boldsymbol{r}_{jk}\times\boldsymbol{r}_{kl}\right)\right\} = \boldsymbol{r}_{ij}\times\boldsymbol{r}_{jk}. \end{aligned} \tag{3-66}$$

Similarly, the coefficients of the quadratic terms in the expansion of $\boldsymbol{r}_{ji}\cdot\boldsymbol{r}_{jk}$ and v are the components of the (6×6) matrix,

$$\begin{bmatrix}\boldsymbol{\nabla}_{ji}\\ \boldsymbol{\nabla}_{jk}\end{bmatrix}\begin{bmatrix}\tilde{\boldsymbol{\nabla}}_{ji} & \tilde{\boldsymbol{\nabla}}_{jk}\end{bmatrix}\left(\boldsymbol{r}_{ji}\cdot\boldsymbol{r}_{jk}\right) = \begin{bmatrix}\boldsymbol{0} & \boldsymbol{E}\\ \boldsymbol{E} & \boldsymbol{0}\end{bmatrix} \tag{3-67}$$

and the (9×9) matrix,

$$\begin{bmatrix}\boldsymbol{\nabla}_{ij}\\ \boldsymbol{\nabla}_{jk}\\ \boldsymbol{\nabla}_{kl}\end{bmatrix}\begin{bmatrix}\tilde{\boldsymbol{\nabla}}_{ij} & \tilde{\boldsymbol{\nabla}}_{jk} & \tilde{\boldsymbol{\nabla}}_{kl}\end{bmatrix} v = \begin{bmatrix}\boldsymbol{0} & \boldsymbol{D}_{kl} & -\boldsymbol{D}_{jk}\\ -\boldsymbol{D}_{kl} & \boldsymbol{0} & \boldsymbol{D}_{ij}\\ \boldsymbol{D}_{jk} & -\boldsymbol{D}_{ij} & \boldsymbol{0}\end{bmatrix}, \tag{3-68}$$

respectively. In Eqs. (3-67) and (3-68), $\boldsymbol{0}$ is the (3×3) zero matrix, $\boldsymbol{E}$ is the (3×3) unit matrix and $\boldsymbol{D}_{ij}$ is the (3×3) matrix defined by

$$\boldsymbol{D}_{ij} = \begin{bmatrix} 0 & -Z_{ij} & Y_{ij} \\ Z_{ij} & 0 & -X_{ij} \\ -Y_{ij} & X_{ij} & 0 \end{bmatrix}, \tag{3-69}$$

When the function A is a bond length r_{ij}, the coefficients A_i' and A_{ij}'' are given by Eq. (3-62). The coefficients of the expansion of the other internal variables can also be calculated efficiently by combining the unit operations on two power series listed in Table 3.1 with appropriate scalar and vector products of two vectors whose components are given as power series.

Theoretically, the Hessian matrix $\boldsymbol{F}^{xx}$ with the second derivative F_{ij}^{xx} as the (i, j) element (Eq. (3-8)) should have the same number of vanishing eigenvalues as the degrees of freedom of translation and rotation, *i.e.,* 5 and 6 for linear and nonlinear molecules, respectively. It should be noted, however, that a rotational coordinate is expressed as a linear combination of the Cartesian displacements given in one of Eqs. (3-28d–f). Any nuclear motion along this coordinate is not circular, and necessarily gives rise to a certain distortion of the molecule if infinitesimal quantities in the second-order are not neglected. Such a linearized rotation will be resisted by a restoring force arising from the second-order mixing between the rotational and the internal coordinates if the second derivatives of the internal coordinates with respect to the Cartesian displacements are involved in the calculation. For this reason, the number of vanishing eigenvalues in actual numerical calculations including the coefficients $B_{p,ij}$ in the last formula of Eq. (3-58) is usually smaller than that expected theoretically. Since the condition for the Newton–Raphson method to converge to a minimum is that all the non-vanishing eigenvalues of the Hessian matrix are positive, the convergence condition cannot be decided if any eigenvalues intrinsically vanishing turn out to be apparently positive. The following correction is useful for removing this difficulty.

Construct a new matrix $(\boldsymbol{F}^{xx})_{\text{in}}$ from $\boldsymbol{F}^{xx}$ by using the previously defined matrices $\boldsymbol{B}_{\text{ex}}$ and $(\boldsymbol{B}^{-1})_{\text{ex}}$ in the form [4)]

$$(\boldsymbol{F}^{xx})_{\text{in}} = \left\{\boldsymbol{E} - \tilde{\boldsymbol{B}}_{\text{ex}}\left(\tilde{\boldsymbol{B}}^{-1}\right)_{\text{ex}}\right\}\boldsymbol{F}^{xx}\left\{\boldsymbol{E} - (\boldsymbol{B}^{-1})_{\text{ex}}\boldsymbol{B}_{\text{ex}}\right\}. \tag{3-70}$$

Noting that the product $\boldsymbol{B}^{-1}\,\boldsymbol{B}$ is written in terms of the submatrices for the internal and the external coordinates as

$$\boldsymbol{B}^{-1}\,\boldsymbol{B} = (\boldsymbol{B}^{-1})_{\text{in}}\boldsymbol{B}_{\text{in}} + (\boldsymbol{B}^{-1})_{\text{ex}}\boldsymbol{B}_{\text{ex}} = \boldsymbol{E}, \tag{3-71}$$

we can rewrite Eq. (3-70) in the form

$$(\boldsymbol{F}^{xx})_{\text{in}} = \tilde{\boldsymbol{B}}_{\text{in}}\left(\tilde{\boldsymbol{B}}^{-1}\right)_{\text{in}}\boldsymbol{F}^{xx}(\boldsymbol{B}^{-1})_{\text{in}}\boldsymbol{B}_{\text{in}} = \tilde{\boldsymbol{B}}_{\text{in}}\,\boldsymbol{F}^{RR}\,\boldsymbol{B}_{\text{in}}.$$

Since $\boldsymbol{F}^{RR}$ is the Hessian matrix consisting only of the internal coordinates, the rank of both matrices $\boldsymbol{B}_{\text{in}}$ and $\boldsymbol{F}^{RR}$ equals the number of degrees of freedom of internal motion of the molecule if the internal coordinates used are independent of each other. The number of non-vanishing eigenvalues of $(\boldsymbol{F}^{xx})_{\text{in}}$, that is, the rank of $(\boldsymbol{F}^{xx})_{\text{in}}$, cannot exceed the degrees of freedom of the internal motion of the molecule. The matrix $\boldsymbol{B}_{\text{ex}}$ in Eq. (3-70), which is a $(6 \times 3N)$ matrix for a nonlinear molecule, may be replaced by a $(3 \times 3N)$ matrix for the rotational coordinates R_X, R_Y and R_Z, since the translational coordinates are linearly related to the Cartesian displacements in the rigorous sense, and are not mixed with any internal coordinates. If the Hessian matrix $(\boldsymbol{F}^{xx})_{\text{in}}$ after the correction for the effect of the linearized rotational coordinates is shown to have the same number of positive eigenvalues as the degrees of freedom of internal motion, it is ascertained that the Newton–Raphson method leads to an energy minimum.

The search for any equilibrium structure of a molecule in molecular mechanics calculation is sometimes followed by the normal coordinate analysis for that structure. This calculation can be efficiently performed by transforming the matrix $\boldsymbol{F}^{xx}$ before the diagonalization into a matrix of the form

$$\boldsymbol{F}^{mxx} = \boldsymbol{m}^{-1/2}\boldsymbol{F}^{xx}\boldsymbol{m}^{-1/2}, \tag{3-72}$$

where $\boldsymbol{m}^{-1/2}$ is the inverse of a $(3N \times 3N)$ diagonal matrix $\boldsymbol{m}^{1/2}$ defined by

$$\boldsymbol{m}^{1/2} = \text{diag}\left[m_1^{\ 1/2} \quad m_1^{\ 1/2} \quad m_1^{\ 1/2} \quad m_2^{\ 1/2} \quad m_2^{\ 1/2} \quad \cdots \quad m_N^{\ 1/2}\right]. \tag{3-73}$$

Since $\boldsymbol{F}^{mxx}$ is a symmetric matrix, it can be diagonalized by a congruence transformation with an orthogonal matrix $\boldsymbol{L}^{mx}$ in the form

$$\boldsymbol{\Lambda} = \tilde{\boldsymbol{L}}^{mx}\,\boldsymbol{F}^{mxx}\,\boldsymbol{L}^{mx}, \tag{3-74}$$

where $\boldsymbol{\Lambda}$ is a diagonal matrix consisting of the eigenvalues of $\boldsymbol{F}^{mxx}$, λ_1, λ_2, $\cdots$, arranged on the diagonal in the descending order $\lambda_1 \geq \lambda_2 \geq \cdots \geq \lambda_{3N}$.

The rank of $\boldsymbol{F}^{xx}$ does not change on being multiplied by the $(3N \times 3N)$ non-singular matrix $\boldsymbol{m}^{-1/2}$. Hence the smallest positive eigenvalue of $\boldsymbol{\Lambda}$ is λ_{3N-5} for a linear molecule and λ_{3N-6} for a nonlinear molecule, if $\boldsymbol{\Lambda}$ has no negative eigenvalues, *i.e.*, if $\lambda_{3N} = 0$. Let $\lambda_{\min}$ be this smallest positive eigenvalue, and partition $\boldsymbol{\Lambda}$ into two diagonal submatrices, one consisting of the positive eigenvalues as

$$\boldsymbol{\Lambda}_{\text{in}} = \text{diag}\,[\lambda_1, \lambda_2, \cdots, \lambda_{\min},] \tag{3-75}$$

and the other consisting of the vanishing eigenvalues as

$$\boldsymbol{\Lambda}_{\text{ex}} = \boldsymbol{0}. \tag{3-76}$$

The transformation in Eq. (3-74) is then written as

$$\begin{bmatrix} \boldsymbol{\Lambda}_{\text{in}} & \boldsymbol{0} \\ \boldsymbol{0} & \boldsymbol{\Lambda}_{\text{ex}} \end{bmatrix} = \begin{bmatrix} \boldsymbol{\Lambda}_{\text{in}} & \boldsymbol{0} \\ \boldsymbol{0} & \boldsymbol{0} \end{bmatrix} = \begin{bmatrix} \tilde{\boldsymbol{L}}^{mx}_{\text{in}} \\ \tilde{\boldsymbol{L}}^{mx}_{\text{ex}} \end{bmatrix} \boldsymbol{F}^{mxx}\left[\boldsymbol{L}^{mx}_{\text{in}} \quad \boldsymbol{L}^{mx}_{\text{ex}}\right]. \tag{3-77}$$

The Roman suffixes indicate that $\boldsymbol{\Lambda}_{\text{in}}$ and $\boldsymbol{\Lambda}_{\text{ex}}$ correspond to the internal and the external coordinates, respectively.

It is useful at this point to define a set of new variables called the normal coordinates, Q_i, by the linear transformation from the nuclear Cartesian displacements Δx_j in the form

$$Q_i = \sum_j L^{mx}_{ji}\, m_j^{\ 1/2}\, \Delta x_j. \tag{3-78}$$

The normal coordinates are the most widely used variables for discussing the effect of a small distortion of a given molecule on any physical properties of that molecule.

Let $\boldsymbol{Q}$ be the vector of the normal coordinates Q_1, Q_2, $\cdots$ in the form

$$\boldsymbol{Q} = [Q_1 \quad Q_2 \quad \cdots \quad Q_{3N}]^{\text{T}}.$$

The matrix representation of Eq. (3-78) is then given by

$$\boldsymbol{Q} = \tilde{\boldsymbol{L}}^{mx}\, \boldsymbol{m}^{1/2}\, \Delta\boldsymbol{x}. \tag{3-79}$$

Since $\boldsymbol{L}^{mx}$ is an orthogonal matrix and $\boldsymbol{m}^{1/2}$ is a diagonal matrix, Eq. (3-79) is immediately inverted as given by

$$\Delta\boldsymbol{x} = \boldsymbol{L}^{x}\, \boldsymbol{Q} = \boldsymbol{m}^{-1/2}\, \boldsymbol{L}^{mx}\, \boldsymbol{Q}\,. \tag{3-80}$$

Substituting Eq. (3-80) into Eq. (3-12), premultiplying the result by $\tilde{\boldsymbol{L}}^{mx}\, \boldsymbol{m}^{-1/2}$, and using Eqs. (3-72) and (3-74), we obtain

$$\tilde{\boldsymbol{L}}^{mx}\, \boldsymbol{m}^{-1/2}\left(\boldsymbol{F}^{x} + \boldsymbol{F}^{xx}\, \boldsymbol{m}^{-1/2}\, \boldsymbol{L}^{mx}\, \boldsymbol{Q}\right) = \tilde{\boldsymbol{L}}^{mx}\, \boldsymbol{m}^{-1/2}\, \boldsymbol{F}^{x} + \boldsymbol{\Lambda}\, \boldsymbol{Q} = \boldsymbol{0}. \tag{3-81}$$

Let the number of degrees of freedom of translational and rotational motion (5 for linear and 6 for nonlinear molecules) be denoted by p, and partition the vector $\boldsymbol{Q}$ into two subvectors $\boldsymbol{Q}_{\text{in}}$ and $\boldsymbol{Q}_{\text{ex}}$ in the forms

$$\boldsymbol{Q}_{\text{in}} = \begin{bmatrix} Q_1 & Q_2 & \cdots & Q_{3N-p} \end{bmatrix}^{\text{T}} \tag{3-82a}$$

and

$$\boldsymbol{Q}_{\text{ex}} = \begin{bmatrix} Q_{3N-p+1} & Q_{3N-p+2} & \cdots & Q_{3N} \end{bmatrix}^{\text{T}}. \tag{3-82b}$$

The vectors $\boldsymbol{Q}_{\text{in}}$ and $\boldsymbol{Q}_{\text{ex}}$ correspond to the positive and the vanishing eigenvalues of $\boldsymbol{\Lambda}$, respectively. In terms of these submatrices, the transformation of Eq. (3-79) is written as

$$\boldsymbol{Q} \equiv \begin{bmatrix} \boldsymbol{Q}_{\text{in}} \\ \boldsymbol{Q}_{\text{ex}} \end{bmatrix} = \begin{bmatrix} \tilde{\boldsymbol{L}}^{mx}_{\text{in}} \\ \tilde{\boldsymbol{L}}^{mx}_{\text{ex}} \end{bmatrix} \boldsymbol{m}^{1/2}\, \Delta\boldsymbol{x}, \tag{3-83}$$

and the last equality of Eq. (3-81) is expressed in two parts as

$$\tilde{\boldsymbol{L}}^{mx}_{\text{in}}\, \boldsymbol{m}^{-1/2}\, \boldsymbol{F}^{x} + \boldsymbol{\Lambda}_{\text{in}}\, \boldsymbol{Q}_{\text{in}} = \boldsymbol{0} \tag{3-84a}$$

and

$$\tilde{\boldsymbol{L}}^{mx}_{\text{ex}}\, \boldsymbol{m}^{-1/2}\, \boldsymbol{F}^{x} = \boldsymbol{0}\,. \tag{3-84b}$$

On substituting Eqs. (3-80), (3-72) and (3-74) into Eq. (3-10), the potential function is rewritten in terms of the normal coordinates in the form

$$V = V^{0} + \tilde{\boldsymbol{Q}}\, \tilde{\boldsymbol{L}}^{mx}\, \boldsymbol{m}^{-1/2}\, \boldsymbol{F}^{x} + \tfrac{1}{2} \tilde{\boldsymbol{Q}}\, \boldsymbol{\Lambda}\, \boldsymbol{Q}\,. \tag{3-85}$$

Each of the second and the third terms in the right side of Eq. (3-85) can be divided into two parts corresponding to $\boldsymbol{Q}_{\text{in}}$ and $\boldsymbol{Q}_{\text{ex}}$, of which the part corresponding to $\boldsymbol{Q}_{\text{ex}}$ is found to vanish with the help of Eqs. (3-84b) and (3-76). Hence the potential function can be expressed only in terms of $3N - p$ variables, Q_1–Q_{3N-p}, included in the subvector $\boldsymbol{Q}_{\text{in}}$ as given in the form

$$V = V^{0} + \tilde{\boldsymbol{Q}}_{\text{in}}\, \tilde{\boldsymbol{L}}^{mx}_{\text{in}}\, \boldsymbol{m}^{-1/2}\, \boldsymbol{F}^{x} + \tfrac{1}{2} \tilde{\boldsymbol{Q}}_{\text{in}}\, \boldsymbol{\Lambda}_{\text{in}}\, \boldsymbol{Q}_{\text{in}}\,. \tag{3-86}$$

The condition for the variables $Q_1, \cdots, Q_{3N-p}$ to give the vanishing linear terms in Eq. (3-86) is obtained by solving Eq. (3-84a) in the form

$$\boldsymbol{Q}_{\text{in}} = -\boldsymbol{\Lambda}_{\text{in}}^{-1}\, \tilde{\boldsymbol{L}}^{mx}_{\text{in}}\, \boldsymbol{m}^{-1/2}\, \boldsymbol{F}^{x}. \tag{3-87}$$

Note that the matrix $\boldsymbol{\Lambda}_{\text{in}}$ has no vanishing diagonal elements and can be inverted in contrast to the case of $\boldsymbol{\Lambda}$.

Premultiplying Eq. (3-87) by $\boldsymbol{m}^{-1/2}\,\boldsymbol{L}_{\text{in}}^{mx}$ and using Eq. (3-80) to eliminate $\boldsymbol{Q}_{\text{in}}$, we obtain the solution of Eq. (3-12) as a vector of the Cartesian displacement coordinates in the form

$$\Delta\boldsymbol{x} = -\boldsymbol{m}^{-1/2}\,\boldsymbol{L}_{\text{in}}^{mx}\,\boldsymbol{\Lambda}_{\text{in}}^{-1}\,\tilde{\boldsymbol{L}}_{\text{in}}^{mx}\,\boldsymbol{m}^{-1/2}\,\boldsymbol{F}^{x}. \tag{3-88}$$

If the expansion of the potential function in powers of the Cartesian displacement coordinates terminates rigorously at the quadratic terms, no approximation enters Eq. (3-12) and the components of Eq. (3-88) give the exact corrections to attain the equilibrium condition. In this case, the nuclear coordinates at the potential minimum are obtained, as seen from Eq. (3-2), by adding the components of $\Delta\boldsymbol{x}$ in Eq. (3-88) to the initial coordinates in the form

$$x_k = x_k^0 + \Delta x_k. \tag{3-89}$$

Since the expansion in Eq. (3-3) involves terms with powers higher than the second in actual cases, the correction given by Eq. (3-88) involves more or less error, and the coordinates calculated by Eq. (3-89) correspond only to an approximate equilibrium. In this case, the calculation should be repeated by taking these coordinates as the revised initial values, until the squared sum of the coordinate corrections $|\Delta\boldsymbol{x}|^2$ becomes less than a threshold given beforehand.

If the starting set of nuclear coordinates involves those very far from any equilibrium values for the model potential, the approximation to truncate the expansion at the quadratic terms in Eq. (3-3) is apt to be so poor that the potential energy calculated at the revised coordinates is higher than that before the revision. In this case, Eq. (3-89) is used with a damped correction vector $\Delta\boldsymbol{x}$ obtained by multiplying the vector $\Delta\boldsymbol{x}$ calculated from Eq. (3-88) by a factor less than 1.

Since an array for the $(n \times n)$ Hessian matrix is necessary in the computer minimization of an n variable function by the Newton–Raphson method, the memory size required for searching for a potential minimum of an N-atomic molecule by this method increases in proportion to N^2. On the other hand, the diagonalization of $(n \times n)$ symmetric matrices to obtain the eigenvalues and eigenvectors was done by the Jacobi method in the early stage of the computerization of numerical calculations. In this method, the computer time increases approximately in proportion to $n^3 \ln n$. More efficient methods of computing eigenvalues of matrices based on the tridiagonalization have been developed since 1960.[5,6] At present, the most efficient method based on the combination of the tridiagonalization and the QR decomposition of real symmetric matrices can suppress the computing time to such a low level as to be proportional at most to n^3.[6]

The increase of memory size and computing time in proportion to the square of molecular size is very serious, however, for large molecules such as proteins. Many devices in the algorithm to diminish the powers of proportionality between molecular size and load of computation have been proposed. These are described below.

The most efficient approximation beyond the full matrix Newton–Raphson method to save computer memory is the block diagonalization method. In each step of this method, only the three-dimensional subvector of $\boldsymbol{F}^{x}$ and the (3×3) diagonal block of the Hessian matrix $\boldsymbol{F}^{xx}$ related to a single nucleus, the nucleus i for example, are calculated, and the Newton–Raphson method is applied to the three-dimensional subspace spanned by the coordinates of this nucleus, X_i, Y_i and Z_i. In this way, the optimized position of the nucleus i in the environment formed by the other nuclei fixed at their initial positions is obtained by a few cycles of applying the correction in the form

$$\begin{bmatrix} \Delta X_i \\ \Delta Y_i \\ \Delta Z_i \end{bmatrix} = \begin{bmatrix} \left(\frac{\partial^2 V}{\partial X_i^2}\right)_0 & \left(\frac{\partial^2 V}{\partial X_i \partial Y_i}\right)_0 & \left(\frac{\partial^2 V}{\partial X_i \partial Z_i}\right)_0 \\ \left(\frac{\partial^2 V}{\partial Y_i \partial X_i}\right)_0 & \left(\frac{\partial^2 V}{\partial Y_i^2}\right)_0 & \left(\frac{\partial^2 V}{\partial Y_i \partial Z_i}\right)_0 \\ \left(\frac{\partial^2 V}{\partial Z_i \partial X_i}\right)_0 & \left(\frac{\partial^2 V}{\partial Z_i \partial Y_i}\right)_0 & \left(\frac{\partial^2 V}{\partial Z_i^2}\right)_0 \end{bmatrix}^{-1} \begin{bmatrix} \left(\frac{\partial V}{\partial X_i}\right)_0 \\ \left(\frac{\partial V}{\partial Y_i}\right)_0 \\ \left(\frac{\partial V}{\partial Z_i}\right)_0 \end{bmatrix}. \tag{3-90}$$

This calculation is carried out N times for $i = 1$ through N. If none of the nuclear displacements executed in this cycle exceeds the given threshold, the entire calculation is finished. Otherwise the iteration is repeated again by starting from $i = 1$.[7)] In this method, the memory size need not be increased with the molecular size, and the computer time increases in proportion to the molecular size at most. Furthermore, in cases where the molecular structure is partly established, the method of block diagonalization shows very stable performance since the movement of a single nucleus must be confined in a narrow range in the space where the movements of the other nuclei are fixed at reasonable positions. The method of block diagonalization has been incorporated in MM2.[7)]

Diagonalization of the Hessian matrix can be avoided and corrections to the Cartesian coordinates of the nuclei can be obtained simply by solving simultaneous equations if the external variables describing the rotation and the translation of the molecule are fixed from the beginning. For this purpose, Thomas and Emerson proposed the use of the well-known method of undetermined multipliers by Lagrange.[8)]

If the external variables introduced by Eqs. (3-28a–f) are fixed at 0, an element of Eq. (3-32) is written as

$$(\Delta R_{ex})_i = \sum_j (B_{ex})_{ij}\, \Delta x_j = 0. \tag{3-91}$$

This condition is known as the Eckart condition.[9)] Multiplying Eq. (3-91) by the undetermined multiplier λ_i and adding the result to the function V to be minimized, Eq. (3-10), we obtain

$$V = V^0 + \sum_i F_i^x\, \Delta x_i + \tfrac{1}{2} \sum_i \sum_j F_{ij}^{xx}\, \Delta x_i\, \Delta x_j + \sum_i \lambda_i \sum_j (B_{ex})_{ij}\, \Delta x_j. \tag{3-92}$$

Differentiation of Eq. (3-92) gives

$$\left(\partial^2 V / \partial \lambda_i\, \partial x_j\right)_0 = (B_{ex})_{ij}.$$

The Hessian matrix for the $3N + 6$ variables consisting of Δx_i (i = 1–3N) and λ_i (i =1–6) are then constructed to give the simultaneous equations in the form

$$\begin{bmatrix} \boldsymbol{F}^{xx} & \tilde{\boldsymbol{B}}_{ex} \\ \boldsymbol{B}_{ex} & \boldsymbol{0} \end{bmatrix} \begin{bmatrix} \boldsymbol{\Delta x} \\ \boldsymbol{\lambda} \end{bmatrix} = \begin{bmatrix} -\boldsymbol{F}^x \\ \boldsymbol{0} \end{bmatrix}. \tag{3-93}$$

The undetermined multipliers, which enter Eq. (3-93) in the vector form

$$\boldsymbol{\lambda} = [\lambda_1 \quad \lambda_2 \quad \lambda_3 \quad \lambda_4 \quad \lambda_5 \quad \lambda_6]^T$$

need not be calculated explicitly.

According to Faber and Altona, the requirement of extra memories corresponding to two atoms ($2 \times 3 = 6$) in this method is trivial compared to its advantage, leading in most cases to stable solutions.[10] This method can also be applied not only to fix the translation and rotation, but also to fix a set of specified internal variables one by one or simply the averaged value of these variables, thus imposing an arbitrary restriction on the molecular structure. The fixing of an average of several internal variables is useful for the specification of the angles of internal rotation around single bonds, since there are more than one torsion angles around a single bond in general, and the net angle of internal rotation corresponds to the average of these torsion angles.[11] The change in internal rotation angle as an average of the changes in torsion angles has been used as an internal coordinate in the normal coordinate analysis by Tasumi and Shimanouchi.[12]

3.4 Method of Steepest Descent

Generally, the energy minimization by the Newton–Raphson method cannot be applied when the Hessian matrix has negative eigenvalues. As shown in Eq. (3-11), the Newton–Raphson method is simply a method to search for a point where the gradient vanishes. In order for the direction of the correction vector in Eq. (3-2) to coincide with the direction along which the energy decreases in each cycle of iteration, the eigenvalues of the Hessian matrix must be always positive.

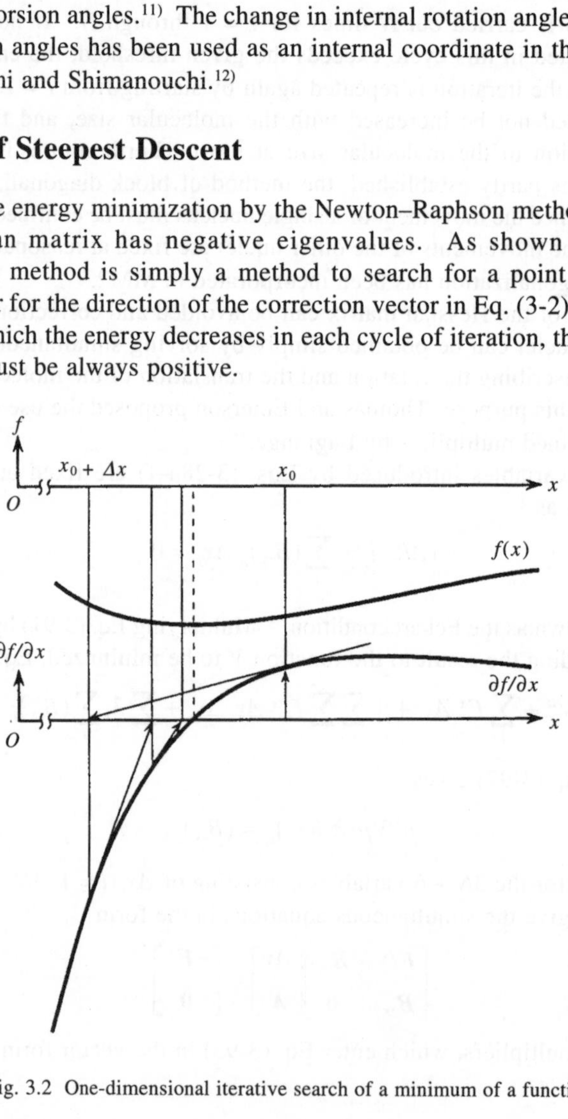

Fig. 3.2 One-dimensional iterative search of a minimum of a function.

In the case of the one-dimensional search shown in Fig. 3.2, the correction is given by

$$\Delta x = -(\partial f/\partial x)_0/(\partial^2 f/\partial x^2)_0.$$

When $(\partial f/\partial x)_0$ is positive, Δx becomes positive in the region where $(\partial^2 f/\partial x^2)_0$ is negative, and the structure changes toward such a direction as to increase the energy. In the case of optimization in a multidimensional configuration space, the energy is expected to decrease irrespective of the signs of eigenvalues of the Hessian matrix, if the configuration point is shifted against the gradient of the potential function f. In this case, the correction vector $\Delta \boldsymbol{x}$ is given in the form

$$\Delta \boldsymbol{x} = -\xi \, \boldsymbol{\nabla} f,$$

where ξ is a constant scale factor. For a given infinitesimal shift of the configuration point, this direction gives the maximum decrease of the potential function, whence the method is called the method of steepest descent.

The choice of the scale factor ξ is very important in the method of steepest descent. If ξ is too small, the number of cycles needed for convergence increases and the efficiency of the calculation is worsened, while a too large ξ may give a shift of the configuration point over the range within which the function f decreases, leading to an increase in the energy. In general, on shifting the configuration point $\boldsymbol{x}$ along a vector given by

$$\boldsymbol{v} = \begin{bmatrix} v_1 & v_2 & \cdots \end{bmatrix}^{\mathrm{T}} \tag{3-94}$$

from the starting point $\boldsymbol{x}_0$, the point at which the energy is minimized can be found as follows.

Suppose that a set of variables x_1, x_2, $\cdots$ satisfies the relation

$$x_i = x_{0i} + \xi \, v_i, \qquad (i = 1, 2, \cdots).$$

Then the problem is reduced to how to find the value of ξ which minimizes the function

$$f\{x_1(\xi), x_2(\xi), \cdots\} = f(\xi).$$

On introducing the position vectors in the configuration space,

$$\boldsymbol{x} = \begin{bmatrix} x_1 & x_2 & \cdots \end{bmatrix}^{\mathrm{T}}, \quad \boldsymbol{x}_0 = \begin{bmatrix} x_{01} & x_{02} & \cdots \end{bmatrix}^{\mathrm{T}}$$

the function $f(x_1, x_2, \cdots)$ may be expressed as

$$f(\boldsymbol{x}) = f(\boldsymbol{x}_0 + \xi \boldsymbol{v}).$$

Expanding $\mathrm{d}f(\boldsymbol{x})/\mathrm{d}\xi$ at the point $\boldsymbol{x}_0$ in powers of ξ up to the linear terms, and putting the result equal to zero, we obtain

$$\frac{d}{d\xi} f(\boldsymbol{x}_0 + \xi \, \boldsymbol{v}) = \left(\frac{df}{d\xi}\right)_0 + \left(\frac{d^2 f}{d\xi^2}\right)_0 \xi = 0. \tag{3-95}$$

From the rule of successive differentiation of multivariable functions, it follows that

$$\left(\frac{df}{d\xi}\right)_0 = \sum_i \left\{\frac{\partial f}{\partial(\xi v_i)}\right\}_0 \frac{d\xi\, v_i}{d\xi} = \sum_i \left(\frac{\partial f}{\partial x_i}\right)_0 v_i \tag{3-96}$$

and

$$\begin{aligned}\left(\frac{d^2 f}{d\xi^2}\right)_0 &= \sum_i \sum_j \left\{\frac{\partial^2 f}{\partial(\xi v_i)\partial(\xi v_j)}\right\}_0 \frac{d\xi\, v_i}{d\xi}\frac{d\xi\, v_j}{d\xi} \\ &= \sum_i \sum_j \left(\frac{\partial^2 f}{\partial x_i \partial x_j}\right)_0 v_i\, v_j .\end{aligned} \tag{3-97}$$

Substituting Eqs. (3-96) and (3-97) into Eq. (3-95) and taking the components of the vector $\boldsymbol{v}$ as

$$v_i = -(\partial f/\partial x_i)_0 \tag{3-98}$$

we obtain the scale factor ξ as

$$\xi = \sum_i \left(\frac{\partial f}{\partial x_i}\right)_0^2 \Bigg/ \sum_i \sum_j \left(\frac{\partial f}{\partial x_i}\right)_0 \left(\frac{\partial^2 f}{\partial x_i \partial x_j}\right)_0 \left(\frac{\partial f}{\partial x_j}\right)_0 . \tag{3-99}$$

The right side of Eq. (3-99) can be expressed compactly by using the matrix representation in the form[13)]

$$\xi = (\tilde{\boldsymbol{v}}\,\boldsymbol{v})/(\tilde{\boldsymbol{v}}\,\boldsymbol{H}\,\boldsymbol{v}) \tag{3-100}$$

where $\boldsymbol{H}$ is the Hessian matrix of the function $f(x_1, x_2, \cdots)$. The calculation of the second derivatives of the function f is necessary for determining the length of the correction vector. A method of minimizing a multivarialbe function without using the derivatives was proposed by Powell.[14)] In this method, the function for an appropriately chosen value ξ_1 is calculated first, and ξ_2 is taken in such a way that $\xi_2 = \xi_1/2$ if $f(0) < f(\xi_1)$ and $\xi_2 = 2\xi_1$ if $f(0) > f(\xi_1)$. Then the minimum is searched by the three point fitting of $f(0)$, $f(\xi_1)$ and $f(\xi_2)$ to a parabola.[14)]

When the potential function is exactly quadratic with respect to the independent variables, the minimum point is reached by the Newton–Raphson method always in a single trial irrespective of the initial position, but not by the method of the steepest descent except for the special case where the vector $\boldsymbol{v}_1$ is exactly directed to the minimum. To illustrate this point, we shall compare the two methods with each other for the minimum search of a simple potential function of two variables:

$$V = \tfrac{1}{2}\left(F_{11} R_1^2 + F_{22} R_2^2\right). \tag{3-101}$$

Let $\boldsymbol{R}_0 = [R_{01} \quad R_{02}]^{\mathrm{T}}$ be the position vector at the starting point. The gradient and the Hessian matrix at this point are given, by differentiating Eq. (3-101), in the form

$$\begin{bmatrix}(\partial V/\partial R_1)_0 \\ (\partial V/\partial R_2)_0\end{bmatrix} = \begin{bmatrix}F_{11} R_{01} \\ F_{22} R_{02}\end{bmatrix} \tag{3-102a}$$

and

$$\begin{bmatrix} (\partial^2 V/\partial R_1^2)_0 & (\partial^2 V/\partial R_1 \partial R_2)_0 \\ (\partial^2 V/\partial R_2 \partial R_1)_0 & (\partial^2 V/\partial R_2^2)_0 \end{bmatrix} = \begin{bmatrix} F_{11} & 0 \\ 0 & F_{22} \end{bmatrix}. \tag{3-102b}$$

From Eqs. (3-102a,b), the correction vector in the Newton–Raphson method is calculated in the form

$$\begin{bmatrix} \Delta R_1 \\ \Delta R_2 \end{bmatrix} = -\begin{bmatrix} F_{11}^{-1} & 0 \\ 0 & F_{22}^{-1} \end{bmatrix} \begin{bmatrix} F_{11} R_{01} \\ F_{22} R_{02} \end{bmatrix} = -\begin{bmatrix} R_{01} \\ R_{02} \end{bmatrix}. \tag{3-103}$$

This vector is minus the position vector $\boldsymbol{R}_0$, indicating that the potential minimum at the origin is reached in the first trial.

The correction vector in the method of steepest descent is given, from Eqs. (3-99) and (3-102a,b), as

$$\begin{bmatrix} \Delta R_1 \\ \Delta R_2 \end{bmatrix} = \xi \begin{bmatrix} F_{11} R_{01} \\ F_{22} R_{02} \end{bmatrix}, \quad \xi = \frac{(F_{11} R_{01})^2 (F_{22} R_{02})^2}{F_{11}^{\ 3} R_{01}^{\ 2} + F_{22}^{\ 3} R_{02}^{\ 2}}. \tag{3-104}$$

The position vector of the corrected point is

$$\begin{aligned} \boldsymbol{R}_1 \equiv \begin{bmatrix} R_{11} \\ R_{12} \end{bmatrix} &= \boldsymbol{R}_0 + \boldsymbol{\Delta R} = \begin{bmatrix} (1-\xi F_{11}) R_{01} \\ (1-\xi F_{22}) R_{02} \end{bmatrix} \\ &= \frac{(F_{22} - F_{11}) R_{01} R_{02}}{F_{11}^{\ 3} R_{01}^{\ 2} + F_{22}^{\ 3} R_{02}^{\ 2}} \begin{bmatrix} F_{22}^{\ 2} R_{02} \\ -F_{11}^{\ 2} R_{01} \end{bmatrix}. \end{aligned} \tag{3-105}$$

This point coincides with the origin if either R_{01} or R_{02} is zero, or $F_{11} = F_{22}$. In Fig. 3.3, the equation of the line of steepest descent passing the starting point is

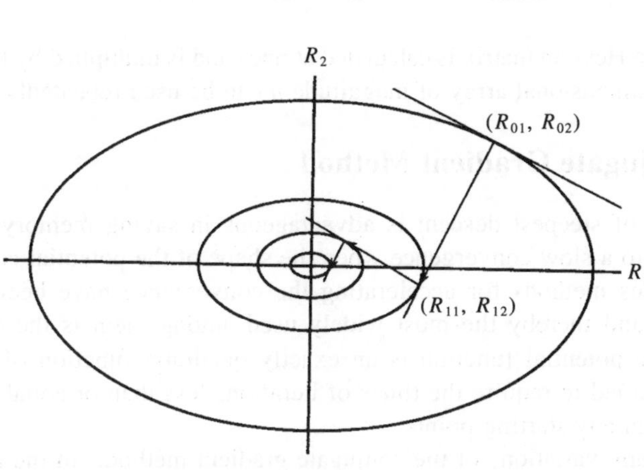

Fig. 3.3 Two-dimensional minimum search by the method of steepest descent.

$$R_2 = \frac{F_{22}\, R_{02}}{F_{11}\, R_{01}}(R_1 - R_{01}) + R_{02}\ . \tag{3-106}$$

The straight line given by Eq. (3-106) contacts a potential contour at the point where the potential V along the line reaches the minimum. The gradient of the tangent of a contour at the point (R_1, R_2) is obtained by regarding R_2 as a function of R_1 and V as a constant in Eq. (3-101), and differentiating both sides of Eq. (3-101) with respect to R_1; thus

$$0 = F_{11}R_1 + \left(\frac{\partial R_2}{\partial R_1}\right)F_{22}R_2, \quad \therefore \quad \left(\frac{\partial R_2}{\partial R_1}\right) = -\frac{F_{11}R_1}{F_{22}R_2}. \tag{3-107}$$

Since a potential contour and the line of Eq. (3-106) contacts each other at the point (R_{11}, R_{12}), the following two conditions hold.
(i) Equation (3-106) is satisfied by $R_1 = R_{11}$ and $R_2 = R_{12}$.
(ii) The right side of Eq. (3-107) in which $R_1 = R_{11}$ and $R_2 = R_{12}$ are substituted is equal to the coefficient of R_1 in Eq. (3-106).
The conditions (i) and (ii) lead to simultaneous equations for the unknowns R_{11} and R_{12} in the forms

$$\left.\begin{aligned} F_{22}\, R_{02}\, R_{11} - F_{11}\, R_{01}\, R_{12} &= (F_{22} - F_{11})R_{01}\, R_{02} \\ F_{11}{}^2\, R_{01}\, R_{11} + F_{22}{}^2\, R_{02}\, R_{12} &= 0 \end{aligned}\right\}. \tag{3-108}$$

The solution of Eq. (3-108) coincides with Eq. (3-105), confirming that the scale factor of Eq. (3-99) gives the minimum point of the potential on the line of steepest descent.

The method of steepest descent has the great advantage that the Hessian matrix need not be inverted so that all the elements of the Hessian matrix are not required to be computed at once. No arrays for a full matrix are necessary in the program. The denominator in the right side of Eq. (3-99) may be calculated as

$$\sum_i \sum_j v_i\, H_{ij}\, v_j = \sum_i v_i \left(\sum_j H_{ij}\, v_j \right) = \tilde{v}(\boldsymbol{H}\,\boldsymbol{v})$$

where a row of the Hessian matrix is calculated at once and is multiplied by the gradient vector $\boldsymbol{v}$, so that a one-dimensional array of magnitude n can be used repeatedly.

3.5 The Conjugate Gradient Method

The method of steepest descent is advantageous in saving memory, but the method sometimes leads to a slow convergence when the shape of the potential contour is far from isotropic. Various methods for accelerating the convergence have been proposed. The most successful and thereby the most widely used among them is the cojugate gradient method.[15] If the potential function is an exactly quadratic function of n variables, this method is ascertained to require the times of iterations less than or equal to n for reaching the minimum from any starting points.

There are many variations of the conjugate gradient method. In the application to the minimum search, the method of Fletcher and Reeves may be regarded as a standard.[15] In this method, the first correction vector $\boldsymbol{d}_1$ is chosen to be $\boldsymbol{v}_1$ in Eq. (3-94) in the method of steepest descent, while the second and later correction vectors are calculated by

$$\boldsymbol{d}_i = -\boldsymbol{v}_i + \frac{\tilde{\boldsymbol{v}}\boldsymbol{H}\boldsymbol{d}_{i-1}}{\tilde{\boldsymbol{d}}_{i-1}\boldsymbol{H}\boldsymbol{d}_{i-1}}\boldsymbol{d}_{i-1} , \tag{3-109}$$

where $\boldsymbol{v}_i$ is the gradient of the function f at the point $\boldsymbol{x}^{(i)}$. If the function $f(x_1, x_2, \cdots, x_n)$ is a quadratic function of x_1, x_2, $\cdots$, x_n in the form

$$f(x_1, x_2, \cdots, x_n) \equiv f(\boldsymbol{x}) = \tfrac{1}{2}\tilde{\boldsymbol{x}}\boldsymbol{H}\boldsymbol{x} + \tilde{\boldsymbol{x}}\boldsymbol{b} ,$$

$\boldsymbol{v}_i$ is given by

$$\boldsymbol{v}_i = \boldsymbol{H}\boldsymbol{x}^{(i)} + \boldsymbol{b}. \tag{3-110}$$

It will be proved later that Eq. (3-109) is equivalent to the expression

$$\boldsymbol{d}_i = -\boldsymbol{v}_i + \frac{\tilde{\boldsymbol{v}}_i\boldsymbol{v}_i}{\tilde{\boldsymbol{v}}_{i-1}\boldsymbol{v}_{i-1}}\boldsymbol{d}_{i-1} = -\boldsymbol{v}_i + \frac{|\boldsymbol{v}_i|^2}{|\boldsymbol{v}_{i-1}|^2}\boldsymbol{d}_{i-1}. \tag{3-111}$$

In the actual numerical calculation, it is advantageous to use Eq. (3-111) which can be determined with less manipulation than Eq. (3-109). The key point of the conjugate gradient method is that the series of correction vectors, $\boldsymbol{d}_1$, $\boldsymbol{d}_2$, $\cdots$, satisfies the condition

$$\tilde{\boldsymbol{d}}_i\boldsymbol{H}\boldsymbol{d}_j = 0, \quad (i > j > 1) \tag{3-112}$$

Any two vectors satisfying Eq. (3-112) are said to be conjugate to each other with respect to $\boldsymbol{H}$. Before proving Eqs. (3-111) and (3-112), let us rewrite Eq. (3-109) in a general form:

$$\boldsymbol{d}_i = -\boldsymbol{v}_i + \gamma_i\boldsymbol{d}_{i-1} . \tag{3-113}$$

In Eq. (3-113), the coefficient γ_i is so chosen that the relation

$$\tilde{\boldsymbol{d}}_i\boldsymbol{H}\boldsymbol{d}_{i-1} = 0 \tag{3-114}$$

holds. Substituting Eq. (3-113) into Eq. (3-114) and solving for γ_i, we obtain

$$\gamma_i = \tilde{\boldsymbol{v}}_i\boldsymbol{H}\boldsymbol{d}_{i-1}/\tilde{\boldsymbol{d}}_{i-1}\boldsymbol{H}\boldsymbol{d}_{i-1} \tag{3-115}$$

which, on substituting back into Eq. (3-113), gives Eq. (3-109), whence $\boldsymbol{d}_i$ calculated by Eq. (3-109) satisfies Eq. (3-114).

When $i = 2$, Eq. (3-111) can easily be derived from Eq. (3-109). The vector of the Cartesian coordinates at the starting point of the $(i + 1)$th iteration is given in the form

$$\boldsymbol{x}^{(i+1)} = \boldsymbol{x}^{(i)} + \lambda_i\boldsymbol{d}_i. \tag{3-116}$$

Let Eq. (3-116) be premultiplied by $\boldsymbol{H}$. The resulting equation is solved for $\boldsymbol{H}\boldsymbol{d}_i$ and use is made of Eq. (3-110) to give

$$\boldsymbol{H}\boldsymbol{d}_i = \boldsymbol{H}(\boldsymbol{x}^{(i+1)} - \boldsymbol{x}^{(i)})/\lambda_i = (\boldsymbol{v}_{i+1} - \boldsymbol{v}_i)/\lambda_i . \tag{3-117}$$

Since the correction vector in the conjugate gradient method is $\boldsymbol{d}_i$ instead of $\boldsymbol{v}_i$ in the method of steepest descent, the condition for the minimum point along the searching direction is given by

$$\frac{d}{d\lambda_i} f(\boldsymbol{x}^{(i)} + \lambda_i \boldsymbol{d}_i) = \left(\frac{df}{d\lambda_i}\right)_0 + \left(\frac{d^2 f}{d\lambda_i^{\,2}}\right)_0 \lambda_i = 0,$$

where

$$\left(\frac{df}{d\lambda_i}\right)_0 = \sum_k \left(\frac{\partial f}{\partial x_k}\right)_0 (d_i)_k$$

and

$$\left(\frac{d^2 f}{d\lambda_i^{\,2}}\right)_0 = \sum_j \sum_k \left(\frac{\partial^2 f}{\partial x_j \partial x_k}\right)_0 (d_i)_j (d_i)_k$$

instead of Eqs. (3-95–97). The step width in the conjugate gradient method is given by

$$\lambda_i = (\tilde{\boldsymbol{v}}_i\, \boldsymbol{d}_i)/(\tilde{\boldsymbol{d}}_i \boldsymbol{H} \boldsymbol{d}_i),$$

which corresponds to Eq. (3-100) in the method of steepest descent.

The line of the $(i-1)$th search contacts a contour curve at the point $\boldsymbol{x}^{(i)}$ where the ith search starts, so that the $(i-1)$th correction vector $\boldsymbol{d}_{i-1}$ is orthogonal to the ith gradient vector $\boldsymbol{v}_i$. Hence it holds that

$$\tilde{\boldsymbol{d}}_{i-1}\, \boldsymbol{v}_i = 0\,. \tag{3-118}$$

Combining Eq. (3-115) with $i = 2$ and Eq. (3-117) with $i = 1$, and making use of the definition of the first correction vector

$$\boldsymbol{d}_1 = -\boldsymbol{v}_1 \tag{3-119}$$

and Eq. (3-114), we obtain

$$\gamma_2 = \frac{\tilde{\boldsymbol{v}}_2\, \boldsymbol{H}\, \boldsymbol{d}_1}{\tilde{\boldsymbol{d}}_1\, \boldsymbol{H}\, \boldsymbol{d}_1} = \frac{\tilde{\boldsymbol{v}}_2(\boldsymbol{v}_2 - \boldsymbol{v}_1)}{\tilde{\boldsymbol{d}}_1(\boldsymbol{v}_2 - \boldsymbol{v}_1)} = \frac{\tilde{\boldsymbol{v}}_2\, \boldsymbol{v}_2}{\tilde{\boldsymbol{v}}_1\, \boldsymbol{v}_1}\,. \tag{3-120}$$

Substituting Eq. (3-120) into Eq. (3-113) leads to Eq. (3-111).

When $i = 2$, Eq. (3-120) is taken to be proved provided that Eq. (3-114) holds. Accordingly, we shall prove Eq. (3-112) for $i \geq 3$ by mathematical induction assuming that the relation obtained by replacing i in Eq. (3-112) with i–1, that is,

$$\tilde{\boldsymbol{d}}_{i-1}\, \boldsymbol{H} \boldsymbol{d}_j = 0, \quad (i-2 \geq j \geq 1) \tag{3-112$'$}$$

holds. The first step of the proof is to show that the orthogonal relation to the ith gradient vector in Eq. (3-118) is satisfied not only by $\boldsymbol{d}_{i-1}$ but also by all the first through the $(i - 2)$th correction vectors as given by

$$\tilde{\boldsymbol{d}}_k\, \boldsymbol{v}_i = 0, \quad (k = 1, 2, \cdots, i-1)\,. \tag{3-121}$$

The equivalence of Eq. (3-111) with Eq. (3-109) for $i \geq 3$ will also be proved in this process. The coordinate vector $\boldsymbol{x}^{(i)}$ in the right side of Eq. (3-110) can be written in terms of $\boldsymbol{x}^{(k)}$ by using Eq. (3-116) repeatedly $(i - k)$ times with decreasing i from $i - 1$ to k. The result is given by

$$\boldsymbol{v}_i = \boldsymbol{H}\boldsymbol{x}^{(i)} + \boldsymbol{b} = \boldsymbol{H}\left(\boldsymbol{x}^{(k)} + \sum_{j=k}^{i-1} \lambda_j \boldsymbol{d}_j\right) + \boldsymbol{b} = \boldsymbol{v}_k + \sum_{j=k}^{i-1} \lambda_j \boldsymbol{H}\boldsymbol{d}_j \, . \tag{3-122}$$

Premultiplying Eq. (3-122) by $\tilde{\boldsymbol{d}}_{k-1}$, we have

$$\tilde{\boldsymbol{d}}_{k-1}\,\boldsymbol{v}_i = \tilde{\boldsymbol{d}}_{k-1}\,\boldsymbol{v}_k + \sum_{j=k}^{i-1} \lambda_j \tilde{\boldsymbol{d}}_{k-1}\,\boldsymbol{H}\boldsymbol{d}_j, \quad (k = 1, 2, \cdots, i-1) \, . \tag{3-123}$$

The first term in the right side of Eq. (3-123) vanishes by virtue of Eq. (3-118). The second term also vanishes if Eq. (3-112′) holds. By combining this result with Eq. (3-118), Eq. (3-121) is proved for $k \le i-1$.

Replacing i in Eq. (3-113) with k and substituting the result into Eq. (121), we have

$$\tilde{\boldsymbol{d}}_k\,\boldsymbol{v}_i = \left(-\tilde{\boldsymbol{v}}_k + \gamma_k \tilde{\boldsymbol{d}}_{k-1}\right)\boldsymbol{v}_i = 0, \quad (1 \le k \le i-1) \, ,$$

where the second term in parentheses vanishes by virtue of Eq. (3-121). It then follows that

$$\tilde{\boldsymbol{v}}_k\,\boldsymbol{v}_i = 0, \quad (1 \le k \le i-1) \, . \tag{3-124}$$

Let us next replace i in Eq. (3-113) with $i-1$ and substitute the result into the denominator of the right side of Eq. (3-115), obtaining

$$\tilde{\boldsymbol{d}}_{i-1}\,\boldsymbol{H}\boldsymbol{d}_{i-1} = -\tilde{\boldsymbol{d}}_{i-1}\,\boldsymbol{H}\,\boldsymbol{v}_{i-1} + \gamma_{i-1}\tilde{\boldsymbol{d}}_{i-1}\,\boldsymbol{H}\boldsymbol{d}_{i-2} \, . \tag{3-125}$$

The second term in the right side of Eq. (3-125) vanishes because of the conjugation between two successive correction vectors indicated by Eq. (3-114). Substituting Eq. (3-125) into Eq. (3-115) and eliminating $\boldsymbol{d}_{i-1}$ and $\boldsymbol{H}$ with the help of Eqs. (3-117) and (3-124), we have

$$\gamma_i = \frac{\tilde{\boldsymbol{d}}_{i-1}\,\boldsymbol{H}\boldsymbol{v}_i}{\tilde{\boldsymbol{d}}_{i-1}\,\boldsymbol{H}\boldsymbol{d}_{i-1}} = -\frac{(\tilde{\boldsymbol{v}}_i - \tilde{\boldsymbol{v}}_{i-1})\boldsymbol{v}_i}{(\tilde{\boldsymbol{v}}_i - \tilde{\boldsymbol{v}}_{i-1})\boldsymbol{v}_{i-1}} = \frac{\tilde{\boldsymbol{v}}_i\,\boldsymbol{v}_i}{\tilde{\boldsymbol{v}}_{i-1}\,\boldsymbol{v}_{i-1}} = \frac{|\boldsymbol{v}_i|^2}{|\boldsymbol{v}_{i-1}|^2} \, ,$$

completing the proof of Eq. (3-111). Finally, Eq. (3-112) can be proved as follows. Rewrite Eq. (3-111) as

$$\frac{\boldsymbol{d}_i}{|\boldsymbol{v}_i|^2} - \frac{\boldsymbol{d}_{i-1}}{|\boldsymbol{v}_{i-1}|^2} = -\frac{\boldsymbol{v}_i}{|\boldsymbol{v}_i|^2}$$

and replace i with $i-1$, $i-2$, …, 2 to obtain

$$\frac{\boldsymbol{d}_{i-1}}{|\boldsymbol{v}_{i-1}|^2} - \frac{\boldsymbol{d}_{i-2}}{|\boldsymbol{v}_{i-2}|^2} = -\frac{\boldsymbol{v}_{i-1}}{|\boldsymbol{v}_{i-1}|^2}$$

$$\cdots \qquad \cdots \qquad \cdots$$

$$\frac{\boldsymbol{d}_2}{|\boldsymbol{v}_2|^2} - \frac{\boldsymbol{d}_1}{|\boldsymbol{v}_1|^2} = -\frac{\boldsymbol{v}_2}{|\boldsymbol{v}_2|^2} \, .$$

Dividing Eq. (3-119) by $|\boldsymbol{v}_1|^2$ gives

$$\frac{\boldsymbol{d}_1}{|\boldsymbol{v}_1|^2} = -\frac{\boldsymbol{v}_1}{|\boldsymbol{v}_1|^2} .$$

Adding up all the above equations and multiplying the sum by $|\boldsymbol{v}_1|^2$, we obtain

$$\boldsymbol{d}_i = -|\boldsymbol{v}_i|^2 \sum_{k=1}^{i} \frac{\boldsymbol{v}_k}{|\boldsymbol{v}_k|^2} . \tag{3-126}$$

The scalar product of Eq. (3-126) with such $\boldsymbol{v}_j$ that $j < i$ is written, with the help of the orthogonality relation in Eq. (3-124), in the form

$$\tilde{\boldsymbol{d}}_i \, \boldsymbol{v}_j = -|\boldsymbol{v}_i|^2 \frac{\tilde{\boldsymbol{v}}_j \, \boldsymbol{v}_j}{|\boldsymbol{v}_j|^2} = -|\boldsymbol{v}_i|^2 . \tag{3-127}$$

On the other hand, replacing i in Eq. (3-122) with $j + 1$ and j and taking the difference, we have

$$\boldsymbol{v}_{j+1} - \boldsymbol{v}_j = \lambda_j \boldsymbol{H} \boldsymbol{d}_j ,$$

which, on being premultiplied by $\tilde{\boldsymbol{d}}_i$, gives

$$\tilde{\boldsymbol{d}}_i \, \boldsymbol{v}_{j+1} - \tilde{\boldsymbol{d}}_i \, \boldsymbol{v}_j = \lambda_j \, \tilde{\boldsymbol{d}}_i \, \boldsymbol{H} \, \boldsymbol{d}_j . \tag{3-128}$$

The left side of Eq. (3-128) vanishes for such j that $1 \le j \le i - 2$ by virtue of Eq. (3-127), so that the right side must vanish too for j in the same range. By combining this result with Eq. (3-114) for the case where $j = i - 1$, Eq. (3-112) is proved for an arbitrary j satisfying the condition $1 \le j < i$.

If a set of n vectors $\boldsymbol{d}_1, \boldsymbol{d}_2, \ldots, \boldsymbol{d}_n$ cannot satisfy the condition

$$\sum_{i=1}^{n} c_i \boldsymbol{d}_i = \boldsymbol{0} \tag{3-129}$$

for any set of constants $c_1, c_2, \ldots, c_n$, except for the trivial one

$$c_1 = c_2 = \cdots = c_n = 0,$$

the vectors $\boldsymbol{d}_1, \boldsymbol{d}_2, \ldots, \boldsymbol{d}_n$ are said to be linearly independent of each other. It will now be proved that any vectors which are conjugate to one another with respect to a positive definite matrix $\boldsymbol{A}$ are linearly independent of one another.

If the vectors $\boldsymbol{d}_1, \boldsymbol{d}_2, \ldots, \boldsymbol{d}_n$ are not linearly independent, there must be such a set of coefficients including a nonzero c_k as to satisfy Eq. (3-129). If $\boldsymbol{d}_1, \boldsymbol{d}_2, \ldots, \boldsymbol{d}_n$ are conjugate to one another with respect to $\boldsymbol{A}$, Eq. (3-129) premultiplied by $\tilde{\boldsymbol{d}}_k \boldsymbol{A}$ must be rewritten, with the help of Eq. (3-112), as

$$0 = \sum_{i=1}^{n} c_i \tilde{\boldsymbol{d}}_k \boldsymbol{A} \boldsymbol{d}_i = c_k \tilde{\boldsymbol{d}}_k \boldsymbol{A} \boldsymbol{d}_k . \tag{3-130}$$

This result contradicts the presumption that $\boldsymbol{A}$ is positive definite ($\tilde{\boldsymbol{d}}_k \boldsymbol{A} \boldsymbol{d}_k > 0$ if $\boldsymbol{d}_k \ne \boldsymbol{0}$) and c_k is not zero. Hence the vectors $\boldsymbol{d}_1, \boldsymbol{d}_2, \ldots, \boldsymbol{d}_n$ must be linearly independent of one another. Since there are no sets of more than n linearly independent vectors in the n-dimensional space, the correction vector $\boldsymbol{d}_{n+1}$ must become a zero vector after n times of trial searches at most.

3.6 Search for Multiminima

Molecules containing a chain of more than three skeletal atoms may have more than one potential minimum in its configuration space. An equilibrium structure at a potential minimum and all the structures from which the equilibrium structure can be formed without any increase of energy are jointly called a conformation. Simple and symmetrical molecules may have more than one equivalent conformations. The number of distinguishable conformations increases very rapidly with molecular size, since the dependence of the former on the latter is expected to be exponential if the atoms behave like non-interacting point masses and there is no steric hindrance among them. It is therefore an important problem to explore efficient methods of picking up really existing conformations from those possible in the absence of steric hindrance.

Consider how to generate different conformations of a molecule systematically. The problem is simple for the case where no ring structures involve single bonds. In this case, one takes up a bond between two skeletal atoms, i–j, and rotates the whole part of the molecule at one side of the bond including the atom j, for example, counting the number of potential minima in the range of the torsion angle between 0 and 2π radians. A new conformation may be generated in this way. The situation is not so simple and we need some device in the case of an endocyclic single bond.

Lipton and Still proposed a systematic method of conformational search applicable to the case involving rings.[16] These authors' method consists of the following three steps.

(1) A hypothetical molecule involving no ring structures is formed by eliminating an appropriately chosen set of the same number of bonds as the rings.

(2) The same systematic search for different conformations as in the acyclic case is applied to the hypothetical acyclic molecule, and the eliminated bonds are restored for each conformation obtained.

(3) The reconstructed ring structure is checked if any bonds or bond angles with too unrealistic values are formed. If the deviations of the restored bond lengths and bond angles from the standard values are considered to be remediable, the structure is taken as a candidate for a new conformation, and is subjected to the potential minimum search. Otherwise, the structure is discarded.

The criterion for selecting a structure as a candidate for a conformation is crucial in this method. Too loose selection may cause useless calculations on inappropriate structures, while too severe criteria may lead to the discarding of a structure that could lead to a realizable conformation. In the test calculation for cycloalkanes by Lipton and Still, the allowable ranges of the bond length and bond angles were taken to be 0.1–0.3 nm and 65–155°, respectively, in the case where the stepwidth of the torsion angle was taken to be 60°.[16]

If a ring contains asymmetric atoms, the process of eliminating and restoring a bond formed by an asymmetric atom may cause inversion of the chirality of that atom. To avoid this confusion, the bonded atom is replaced by a dummy atom on eliminating a bond. After the bond is restored, the dummy atom is confirmed to occupy the original position of the bonded atom and is then replaced by the latter. In this way, Lipton and Still found two and nine new conformers of cyclononane and cyclodecane, respectively.

Gotō and Ōsawa developed an efficient method of generating new conformations of ring structures and called it the corner flap/edge flip method.[17] This method aims at avoiding calculations for trivial conformations by modeling the actual process of conformational transition as faithfully as possible. In this method, two kinds of conformation change, the corner flap and the edge flip, are used. Imagine that five serially bonded nuclei A–B–C–D–E are contained in a ring, *R*, where D and E may be either bonded directly or not, and none of the four bonds, A–B, B–C, C–D or D–E is contained in any ring other than *R*. Condensed rings are therefore not considered here. Let *P* be the plane defined by the nucleus B and D and the midpoint of the line segment AE. Take line BD as the axis of rotation, and rotate the nucleus C and all the exocyclic groups substituted at C as a rigid body around this axis until C reaches its mirror position with respect to plane *P*. The exocyclic groups attached to B and D are then rotated as rigid bodies in such a way that the change of the relevant bond angles is kept to a minimum. This process is called the corner flap.

The change of the bond angle θ_{ABC} on the corner flap is estimated as follows. In the coordinate system shown in Fig. 3.4, the nucleus B is at the origin, and line BD and plane *P* are taken to be the *X*-axis and the *XY*-plane, respectively. Let the projections of A and C on the *XY*-plane be A_P and C_P, respectively, defining angles $\angle ABA_P = \gamma_1$ and $\angle CBC_P = \gamma_2$, and introduce the unit vectors $\boldsymbol{e}_1$ along the bond B–A and $\boldsymbol{e}_2$ along the bond B–C. The *Z* components of $\boldsymbol{e}_1$ and $\boldsymbol{e}_2$ are given by $(\boldsymbol{e}_k)_Z = \sin\gamma_k$, where *k* is 1 or 2. Let the unit vector along the bond B–C after operating the corner flap be denoted by $\boldsymbol{e}'_2$. If the bond angle θ_{ABC} changes to θ'_{ABC} on the corner flap, it follows from the definition of the scalar product that

$$\tilde{e}_1 e_2 = \cos\theta_{ABC} \qquad \tilde{e}_1 e'_2 = \cos\theta'_{ABC}. \tag{3-131}$$

From the definition of the corner flap, the Cartesian components of the vectors $\boldsymbol{e}_2$ and

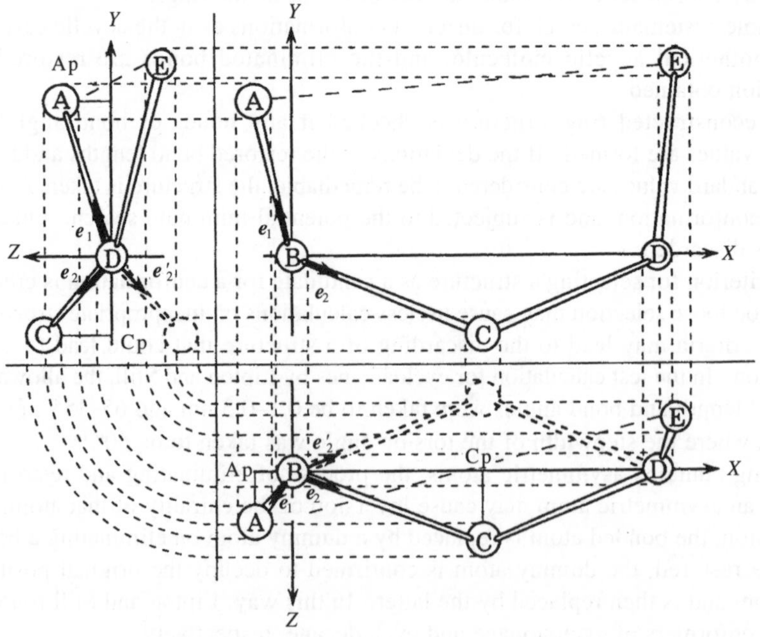

Fig. 3.4 Change in conformation by the corner flap method.

e_2' satisfy the following relations:

$$(e_2')_X = (e_2)_X, \quad (e_2')_Y = (e_2)_Y, \quad (e_2')_Z = -(e_2)_Z = -\sin\gamma_2 \, .$$

Taking then the difference between the scalar products $\tilde{e}_1 e_2$ and $\tilde{e}_1 e_2'$, and using Eq. (3-131), we obtain

$$\tilde{e}_1 e_2 - \tilde{e}_1 e_2' = 2\sin\gamma_1 \sin\gamma_2 = \cos\theta_{ABC} - \cos\theta'_{ABC}.$$

Since the angles γ_1 and γ_2 rarely becomes large simultaneously, the change in the bond angle θ_{ABC} on the corner flap is usually small. The angle of rotation of the bond B–C around the axis along the bond A–B is the dihedral angle between planes ABC and ABC′ and, from the cosine theorem in spherical trigonometry, is given by

$$\tau_1 = \text{Arccos}\frac{\cos 2\gamma_1 - \cos\theta_{ABC}\cos\theta'_{ABC}}{\sin\theta_{ABC}\sin\theta'_{ABC}} \, . \tag{3-132}$$

From the above discussion, it is concluded that the final orientation of the exocyclic groups attached to B is obtained by rotating it first around the axis along the bond A–B by τ_1, and then around the vector $e_1 \times e_2'$ by $(\theta'_{ABC} - \theta_{ABC})/2$. The orientation of the exocyclic groups attached to D is similarly adjusted.

Let the conformations *gauche* and *antigauche* be denoted by G and G′, respectively. If the conformation of the ring skeleton with respect to the bond sequence B–C–D is GG′ or G′G, the nucleus C lies on the plane *P* so that the corner flap cannot be applied. In this case, a new conformation is generated by a procedure called the edge flip as follows.

The edge flip is applied to the structures obtained by replacing the nucleus C in the case of the corner flap with a bond, C–C′, and the nuclei C and C′ are on opposite sides of each other with respect to the plane *P* defined by the three points B, D, and X. To perform the edge flip, the nuclei C and C′ are moved to opposite positions with respect to the plane *P*, and the exocyclic groups attached to B and C are rotated in the same way as the case of the corner flap while those attached to C′ and D are rotated opposite to the corner flap. Obviously, the chirality is conserved during these processes since no bond ruptures are involved.

A very efficient algorithm to systematic search for stable conformations, called the reservoir filling method, has been proposed by Gotō and Ōsawa.[18] In this method, new conformers are generated by successively applying the rotation around the single bonds by ± 120° for the acyclic skeleton and by using cyclically the corner flap/edge flip method for the ring skeleton. The whole procedure consists of a double loop starting from the input structure as the initial structure in the first cycle of the outer loop, $C_{1,1}$. In the ith cycle of the outer loop, all the new conformers formed by applying the conformer-generating operations to different portions of the initial structure $C_{i,1}$ are listed, and each of these conformers is subjected to the energy optimization in a cycle of the inner loop. Let the minimized energy and the equilibrium structure obtained in the jth inner cycle of the ith outer cycle be denoted by $E_{i,j}$ and $C_{i,j}$, respectively.

If $E_{i,j}$ is smaller than a preliminarily given threshold, CSL (chemically significant limit), $C_{i,j}$ is compared with all the structures registered in the CSL list, and if not identified with any structures in the list, added to the CSL list. Furthermore, if $E_{i,j}$ is smaller than another threshold SEL (search limit), set to be a little higher than CSL, $C_{i,j}$ is added to the candidate structure list. After finishing the inner loop, the strucure with the minimum energy among the remaining candidate structures is taken up as the initial structure of the next outer cycle, $C_{i+1,1}$.

By using the reservoir filling method together with the *ab initio* MO method, the number of possible conformers of *n*-alkane $CH_3(CH_2)_nCH_3$ was found to be greater than 3^{n-1}, the number traditionally expected from the theory.[19, 20] The reason for this is that the combination of two successive torsion angles ϕ_i and ϕ_{i+1}, GG′, which has been regarded as a single conformation, actually gives two conformations (ϕ_i = ~ 95°, ϕ_{i+1} = ~ – 65°) and (ϕ_i = ~ 65°, ϕ_{i+1} = ~ – 95°). This result proves the usefulness of combining an accurate *ab initio* calculation for lower alkanes and a wide range search of possible conformers of higher alkanes by molecular mechanics calculation.

References

1) E.B. Wilson, Jr., *J. Chem. Phys.*, **7**, 1047 (1939); *ibid.*, **9**, 76 (1941).
2) B.L. Crawford, Jr. and W. H. Fletcher, *J. Chem. Phys.*, **19**, 141 (1951).
3) M. Pariseau, I. Suzuki and J. Overend, *J. Chem. Phys.*, **42**, 2335 (1965).
4) Y. Miwa and K. Machida, *J. Am. Chem. Soc.*, **110**, 5183 (1988).
5) A.S. Householder, *The Theory of Matrices in Numerical Analysis*, Blaisdell, New York (1964).
6) J.G.F. Francis, *Comput. J.*, **4**, 265, 332 (1961,1962).
7) T. Clark, *A Handbook of Computational Chemistry*, John Wiley and Sons, New York (1985).
8) M. W. Thomas and D. Emerson, *J. Mol. Struct.*, **16**, 473 (1973).
9) C. Eckart, *Phys. Rev.*, **47**, 552 (1935).
10) D.H. Faber and C. Altona, *Comput. Chem.*, **1**, 203 (1977).
11) B. Van de Graaf and J.M.A. Baas, *J. Comput. Chem.*, **5**, 314 (1984).
12) M. Tasumi and T. Shimanouchi, *J. Mol. Spectrosc.*, **11**, 422 (1963).
13) J. Kowalik and M.R. Osborne, *Methods for Unconstrained Optimization Problems*, American Elsevier, New York (1968).
14) M.J.D. Powell, *Comput. J.*, **7**, 155 (1964).
15) R. Fletcher and C.M. Reeves, *Comput. J.*, **7**, 149 (1964)
16) M. Lipton and W.C. Still, *J. Comput. Chem.*, **9**, 343 (1988).
17) H. Gotō and E. Ōsawa, *J. Am. Chem. Soc.*,**111**, 8950 (1989).
18) H. Gotō and E. Ōsawa, *Tetrahedron Lett.*, **33**, 1343 (1992); *idem.*, *J. Chem. Soc., Perkin Trans.*, **2**, 187 (1993).
19) H. Gotō, E. Ōsawa and M. Yamamoto, *Tetrahedron*, **49**, 387 (1993).
20) S. Tsuzuki, L. Schafer, H. Gotō, E.D. Jemmis, H. Hosoya, K. Siam, K. Tanabe and E. Ōsawa, *J. Am. Chem. Soc.*, **113**, 4665 (1993).

Chapter 4

Normal Coordinate Analysis

4.1 Harmonic Approximation

The potential function of a molecule can be expanded as a power series of the Cartesian displacement coordinates around the equilibrium positions of the nuclei, if the molecule does not change its own shape drastically. In this power series, terms higher in order than the second are called anharmonic terms. The harmonic approximation means the approximation in which the anharmonic terms are neglected. A great advantage of the harmonic approximation is that the Schrödinger equation describing the nuclear motions of a non-rotating molecule is separable into the same number of equations as the vibrational degrees of freedom. Each of these equations has a single variable and can be solved analytically to give the vibrational wave functions in closed forms. The separated variables are called the normal coordinates. The wave functions obtained as the solutions of Schrödinger equations under the harmonic approximation are useful as the zeroth-order wave functions in the perturbation theory to take into account the anharmonic terms.

On solving the classical equations of motion under the harmonic approximation, we obtain a solution in which each normal coordinate changes periodically with its own frequency called the normal frequency. The normal frequencies are important not only as the direct source of information regarding the quadratic terms in the Taylor expansion of the potential function, but also as the molecular constants necessary for calculating the thermodynamic functions by the method of statistical mechanics. On the other hand, the normal coordinates are the standard variables to describe changes in physical quantities on nuclear displacement, and are always used in analyses of experiemtal data reflecting the dependence of any molecular properties on the structure. The moments of inertia taken up in Chapter 5 and the dipole moments and the polarizabilities discussed in Chapter 8 are typical examples of such molecular properties.

This chapter presents the theoretical background and the procedure of calculation of the method for deducing the normal coordinates and the normal frequencies from the potential function. Since the automatic generation of the symmetry coordinates by computers is quite elaborate in the programing of normal coordinate analyses, special attention is paid to the treatment of molecular symmetry.

4.2 Normal Vibrations

Suppose that the harmonic approximation is applied to an N-atomic molecule for which an equilibrium structure has been found by any method described in Chapter 3. On placing the origin of the Cartesian displacement coordinates of each nucleus at its equilibrtium position, the linear terms are eliminated from the Taylor series of the potential function, giving

$$V = V^0 + \tfrac{1}{2}\sum_{i,j=1}^{3N} F_{ij}^{xx}\, \Delta x_i\, \Delta x_j = V_0 + \tfrac{1}{2}\Delta\tilde{\boldsymbol{x}}\, \boldsymbol{F}^{xx}\Delta\boldsymbol{x}\ . \tag{4-1}$$

Differentiating Eq. (4-1) with respect to Δx_i (i = 1–3N) and changing the sign, we have the components of the force on the ith nucleus in the form

$$-\partial V/\partial x_i = -\sum_{j=1}^{3N} F_{ij}^{xx}\Delta x_j\ . \tag{4-2}$$

Hereafter, any physical quantities differentiated once and twice with respect to time will be denoted by a dotted and a doubly-dotted symbols, respectively, as illustrated by $(d/dt)a \equiv \dot{a}$ and $(d^2/dt^2)a \equiv \ddot{a}$. Newton's equations of motion for the ith nucleus are then given by

$$\left.\begin{aligned} m_i d^2 x_{3i-2}/dt^2 &\equiv m_i\ddot{x}_{3i-2} = -\partial V/\partial x_{3i-2} \\ m_i d^2 x_{3i-1}/dt^2 &\equiv m_i\ddot{x}_{3i-1} = -\partial V/\partial x_{3i-1} \\ m_i d^2 x_{3i}/dt^2 &\equiv m_i\ddot{x}_{3i} = -\partial V/\partial x_{3i} \end{aligned}\right\}\ . \tag{4-3}$$

Let $\Delta\ddot{\boldsymbol{x}}$ be the vector obtained by differentiating each element of $\Delta\boldsymbol{x}$ twice with respect to t. Noting that

$$\Delta\dot{x}_i = \dot{x}_i - \dot{x}_i^0 = \dot{x}_i$$

from the definition, and substituting Eq. (4-2) into Eq. (4-3), we obtain the matrix representation of the equations of motion in the form

$$\boldsymbol{m}\,\Delta\ddot{\boldsymbol{x}} = -\boldsymbol{F}^{xx}\Delta\boldsymbol{x}\ , \tag{4-4}$$

where $\boldsymbol{m}$ is the vector of atomic masses defined by Eq. (3-35).

Now we introduce a new set of coordinates, called the mass-weighted Cartesian displacement coordinates,[1)] defined by

$$\left.\begin{aligned} (\Delta x_m)_{3i-2} &\equiv (\Delta X_m)_i = m_i^{1/2}\Delta X_i \\ (\Delta x_m)_{3i-1} &\equiv (\Delta Y_m)_i = m_i^{1/2}\Delta Y_i \\ (\Delta x_m)_{3i} &\equiv (\Delta Z_m)_i = m_i^{1/2}\Delta Z_i \end{aligned}\right\}\ . \tag{4-5}$$

The set of mass-weighted Cartesian displacement coordinates of an N-atomic molecule is written in a vector form,

$$\Delta\boldsymbol{x}_m = \left[(\Delta x_m)_1 \quad (\Delta x_m)_2 \quad \cdots \quad (\Delta x_m)_{3N}\right]^{\mathrm{T}}. \tag{4-6}$$

Combining Eqs. (4-6), (3-9) and (3-73), we obtain a compact matrix representation of Eq. (4-5) in the form

$$\Delta\boldsymbol{x}_m = \boldsymbol{m}^{1/2}\,\Delta\boldsymbol{x} \tag{4-7}$$

or in the inverted form

$$\boldsymbol{\Delta x} = \boldsymbol{m}^{-1/2}\, \boldsymbol{\Delta x}_m\,, \tag{4-8}$$

where $\boldsymbol{m}^{-1/2}$ is the inverse of the matrix $\boldsymbol{m}^{1/2}$. The matrix $\boldsymbol{F}^{mxx}$ introduced by Eq. (3-72) in relation to searching for the equilibrium structures is the Hessian matrix of the potential function with respect to the mass-weighted Cartesian displacement coordinates.

Premultiplying Eq. (4-4) by $\boldsymbol{m}^{-1/2}$ and substituting Eqs. (4-8) and (3-72) into the result, we obtain the equations of motion in the mass-weighted Cartesian displacement coordinates in the form

$$\boldsymbol{\Delta \ddot{x}}_m = -\boldsymbol{F}^{mxx}\, \boldsymbol{\Delta x}_m. \tag{4-9}$$

The variables in the simultaneous differential equations (Eq. (4-9)) can be separated from each other if the matrix $\boldsymbol{F}^{mxx}$ is diagonalized. As stated in Chapter 3, the diagonalization is accomplished through the transformation of variables from the mass-weighted Cartesian displacements into the normal coordinates (see Eq. (3-74)). By combining Eqs. (4-8) and (3-80), the required transformation is written simply as

$$\boldsymbol{\Delta x}_m = \boldsymbol{L}^{mx}\, \boldsymbol{Q}. \tag{4-10}$$

Substituting Eq. (4-10) into Eq. (4-9), premultiplying both sides by $\tilde{\boldsymbol{L}}^{mx}$ and using Eq. (3-74), we have

$$\ddot{\boldsymbol{Q}} = -\boldsymbol{\Lambda}\, \boldsymbol{Q}. \tag{4-11}$$

The ith element of Eq. (4-11) gives a second-order ordinary differential equation for a single variable Q_i in the form

$$\ddot{Q}_i \equiv d^2Q_i/dt^2 = -\lambda_i\, Q_i\,. \tag{4-12}$$

The solution of Eq. (4-12), which shows that Q_i is proportional to minus the second derivative of its own, can be written as

$$Q_i = Q_i^0 \cos(2\pi\, c\,\omega_i\, t + \delta_i)\,, \tag{4-13}$$

where c is the speed of light. The function $Q_i(t)$ in the form of Eq. (4-13) satisfies Eq. (4-12) if ω_i is related to λ_i by

$$\lambda_i = 4\pi^2 c^2 \omega_i^{\,2}. \tag{4-14}$$

Equation (4-13) indicates that the ith normal coordinate Q_i changes periodically with the frequency $c\omega_i$. The amplitude Q_i^0 and the phase δ_i, which have been introduced as the integration constants, are determined by the initial condition. Each of the Cartesian displacement coordinates of the kth nucleus as a function of time is given by substituting Eq. (4-13) into the corresponding elements of Eq. (4-10) and then Eq. (4-8). The result is given in the form

$$\Delta x_{3k-l} = m_k^{\;-1/2} \sum_i L^{mx}_{3k-l,i}\, Q_i^0 \cos(2\pi\, c\,\omega_i\, t + \delta_i)\,, \quad (l = 2,\, 1,\, 0). \tag{4-15}$$

If the initial condition is taken to be $Q_j^0 = 0$ and $\delta_j = 0$ for all j ($\neq i$), the solution represents an in-phase motion of all the nuclei vibrating with the frequency $c\omega_i$ and the phase δ_i. This vibration is called a normal vibration of the molecule, and the frequency $c\omega_i$ is called the normal frequency. The quantity ω_i, which is called the wavenumber, indicates the number of nodes in a unit pathlength of the light wave which has the same frequency as the normal mode i. According to Eq. (4-15), any arbitrary motion of the nuclei in a molecule can be described as an overlap of the normal vibrations with appropriately chosen amplitudes and phases. The calculation of the normal frequencies together with the determination of the transformation matrix $\boldsymbol{L}^{mx}$ of a molecule or a molecular system is called normal coordinate analysis.

4.3 The *GF* Matrix Method

To carry out the normal coordinate analysis of a molecule for which an equilibrium structure is known, we can reduce the dimension of the Hessian matrix of the potential energy by using a set of internal coordinates instead of the Cartesian displacement coordinates. The origin of the potential function $V(\Delta R_1, \Delta R_2, \cdots)$ in Eq. (3-55) may be shifted to its minimum without loss of generality. Then we may rewrite $V - V^0$ as V, and expand it in powers of the internal coordinates up to the second-order,

$$V = \tfrac{1}{2}\sum_i \sum_j F_{ij}^{RR}\,\Delta R_i\,\Delta R_j = \tfrac{1}{2}\,\Delta\tilde{\boldsymbol{R}}\,\boldsymbol{F}^{RR}\,\Delta\boldsymbol{R}, \tag{4-16}$$

where $\boldsymbol{F}^{RR}$ is the Hessian matrix for the internal coordinates, known as the $\boldsymbol{F}$ matrix among vibrational spectroscopists. Elements of $\boldsymbol{F}^{RR}$ are the force constants in the GVFF force fields defined by Eq. (2-55a).

Changing the variables $\Delta\boldsymbol{R}_{\text{in}}$ in Eq. (4-16) to $\Delta\boldsymbol{x}$ by the use of Eq. (3-13) and comparing the result with Eq. (4-1), we have

$$V = \tfrac{1}{2}\,\Delta\tilde{\boldsymbol{x}}\,\tilde{\boldsymbol{B}}_{\text{in}}\,\boldsymbol{F}^{RR}\boldsymbol{B}_{\text{in}}\,\Delta\boldsymbol{x} = \tfrac{1}{2}\,\Delta\tilde{\boldsymbol{x}}\,\boldsymbol{F}^{xx}\,\Delta\boldsymbol{x}. \tag{4-17}$$

The matrices $\boldsymbol{F}^{xx}$ and $\boldsymbol{F}^{RR}$ are thus related by Eq. (4-18).

$$\boldsymbol{F}^{xx} = \tilde{\boldsymbol{B}}_{\text{in}}\,\boldsymbol{F}^{RR}\boldsymbol{B}_{\text{in}} \tag{4-18}$$

The transformation of a matrix into another matrix by post- and premultiplying by a matrix and its transpose, respectively, is called a congruence transformation. The congruence transformation is useful for representing the transformation of a Hessian matrix associated with a linear transformation of variables in a quadratic form.

Substituting Eqs. (4-8) and (4-10) into Eq. (4-17), and using Eqs. (3-72) and (3-73), we obtain

$$V = \tfrac{1}{2}\tilde{\boldsymbol{Q}}\boldsymbol{\Lambda}\,\boldsymbol{Q} = \tfrac{1}{2}\tilde{\boldsymbol{Q}}\,\tilde{\boldsymbol{L}}^{mx}\,\boldsymbol{m}^{-1/2}\tilde{\boldsymbol{B}}_{\text{in}}\,\boldsymbol{F}^{RR}\boldsymbol{B}_{\text{in}}\,\boldsymbol{m}^{-1/2}\boldsymbol{L}^{mx}\,\boldsymbol{Q}$$

$$\therefore\quad \boldsymbol{\Lambda} = \tilde{\boldsymbol{L}}^{mx}\,\boldsymbol{m}^{-1/2}\tilde{\boldsymbol{B}}_{\text{in}}\,\boldsymbol{F}^{RR}\boldsymbol{B}_{\text{in}}\,\boldsymbol{m}^{-1/2}\boldsymbol{L}^{mx}. \tag{4-19}$$

Premultiplying Eq. (4-19) by $\boldsymbol{B}_{\text{in}}\,\boldsymbol{m}^{-1/2}\boldsymbol{L}^{mx}$, and using the orthogonality of $\boldsymbol{L}^{mx}$ ($\boldsymbol{L}^{mx}\,\tilde{\boldsymbol{L}}^{mx} = \boldsymbol{E}$), we have

$$\begin{aligned} \boldsymbol{B}_{\text{in}}\,\boldsymbol{m}^{-1/2}\boldsymbol{L}^{mx}\,\boldsymbol{\Lambda} &= \boldsymbol{B}_{\text{in}}\,\boldsymbol{m}^{-1/2}\left(\boldsymbol{L}^{mx}\,\tilde{\boldsymbol{L}}^{mx}\right)\boldsymbol{m}^{-1/2}\tilde{\boldsymbol{B}}_{\text{in}}\,\boldsymbol{F}^{RR}\boldsymbol{B}_{\text{in}}\,\boldsymbol{m}^{-1/2}\boldsymbol{L}^{mx} \\ &= \boldsymbol{B}_{\text{in}}\,\boldsymbol{m}^{-1/2}\,\boldsymbol{m}^{-1/2}\tilde{\boldsymbol{B}}_{\text{in}}\,\boldsymbol{F}^{RR}\boldsymbol{B}_{\text{in}}\,\boldsymbol{m}^{-1/2}\boldsymbol{L}^{mx} \\ &= \boldsymbol{G}^{RR}\,\boldsymbol{F}^{RR}\,\boldsymbol{B}_{\text{in}}\,\boldsymbol{m}^{-1/2}\boldsymbol{L}^{mx}, \end{aligned} \tag{4-20}$$

where $\boldsymbol{G}^{RR}$ is the $\boldsymbol{G}$ matrix introduced in the previous chapter (Eq. (3-42)).[2] The superfix ^{R} is added here to indicate the working coordinate explicitly. By defining a new matrix

$$\boldsymbol{L} = \boldsymbol{B}_{\text{in}}\, \boldsymbol{m}^{-1/2} \boldsymbol{L}^{mx}, \tag{4-21}$$

Eq. (4-20) can be rewritten in the simple form

$$\boldsymbol{L\Lambda} = \boldsymbol{G}^{RR}\, \boldsymbol{F}^{RR}\, \boldsymbol{L}. \tag{4-22}$$

Eliminating $\Delta\boldsymbol{x}$ from Eqs. (3-13) and (3-80), and comparing the result with Eq. (4-21), we obtain

$$\Delta\boldsymbol{R}_{\text{in}} = \boldsymbol{B}_{\text{in}}\, \boldsymbol{m}^{-1/2} \boldsymbol{L}^{mx}\, \boldsymbol{Q} = \boldsymbol{L}\,\boldsymbol{Q}. \tag{4-23}$$

Equation (4-23) shows that $\boldsymbol{L}$ is the matrix of the transformation from the normal coordinates $\boldsymbol{Q}$ to the internal coordinates $\Delta\boldsymbol{R}_{\text{in}}$. If the matrices $\boldsymbol{G}^{RR}$ and $\boldsymbol{F}^{RR}$ are known, it is possible to calculate $\boldsymbol{L}$ and $\boldsymbol{\Lambda}$ as follows, without explicit evaluation of $\boldsymbol{F}^{xx}$ and $\boldsymbol{L}^{mx}$.[3,4]

First, the matrix $\boldsymbol{G}^{RR}$ is diagonalized by an orthogonal transformation to give the matrices of its eigenvalues and eigenvectors denoted by $\boldsymbol{\Lambda}_G$ and $\boldsymbol{L}_G$, respectively. The transformation is written as

$$\tilde{\boldsymbol{L}}_G\, \boldsymbol{G}^{RR}\, \boldsymbol{L}_G = \boldsymbol{\Lambda}_G. \tag{4-24}$$

A new matrix $\boldsymbol{\Lambda}_G{}^{1/2}$ is formed next by replacing the elements $(\Lambda_G)_i$ of the diagonal matrix $\boldsymbol{\Lambda}_G$ with their square roots $(\Lambda_G)_i{}^{1/2}$. Premultiply $\boldsymbol{\Lambda}_G{}^{1/2}$ by $\boldsymbol{L}_G$ to give $\boldsymbol{L}_G^0$ in the form

$$\boldsymbol{L}_G^0 = \boldsymbol{L}_G \boldsymbol{\Lambda}_G{}^{1/2}. \tag{4-25}$$

By noting that $\boldsymbol{L}_G$ is orthogonal, Eq. (4-24) is solved for $\boldsymbol{G}^{RR}$ and Eq. (4-25) is used to yield

$$\boldsymbol{G}^{RR} = \boldsymbol{L}_G\, \boldsymbol{\Lambda}_G\, \tilde{\boldsymbol{L}}_G = \boldsymbol{L}_G\, \boldsymbol{\Lambda}_G{}^{1/2}\, \boldsymbol{\Lambda}_G{}^{1/2}\, \tilde{\boldsymbol{L}}_G = \boldsymbol{L}_G^0\, \tilde{\boldsymbol{L}}_G^0. \tag{4-26}$$

In the diagonal matrix $\boldsymbol{\Lambda}_G$, the eigenvalues of $\boldsymbol{G}^{RR}$ are assumed to be arranged in descending order from upper left to lower right.

If there are N_r relations among N_t internal coordinates, the last N_r eigenvalues of $\boldsymbol{\Lambda}_G$ should vanish and, according to Eq. (4-25), all the elements of $\boldsymbol{L}_G^0$ in the $(N_t - N_r + 1)$th through N_tth columns also vanish. In this case, if the internal coordinates are so chosen as to be able to represent any molecular distortion of an infinitesimal magnitude, the number of non-vanishing eigenvalues, $N_i = N_t - N_r$, should coincide with the internal degrees of freedom of nuclear motion.

Finally, the matrix $\boldsymbol{F}^{RR}$ is premultiplied by $\tilde{\boldsymbol{L}}_G^0$ and postmultiplied by $\boldsymbol{L}_G^0$. The resulting symmetric matrix $\boldsymbol{F}_C$ is diagonalized to give the eigenvalue matrix $\boldsymbol{\Lambda}_C$ and the eigenvector matrix $\boldsymbol{L}_C$. Note that the dimension of $\boldsymbol{L}_G^0$ always makes $\boldsymbol{F}_C$ an $(N_i \times N_i)$ matrix. It follows from the orthogonality of $\boldsymbol{L}_C$ that

$$\boldsymbol{F}_C\, \boldsymbol{L}_C = \left(\tilde{\boldsymbol{L}}_G^0 \boldsymbol{F}^{RR} \boldsymbol{L}_G^0\right) \boldsymbol{L}_C = \boldsymbol{L}_C\, \boldsymbol{\Lambda}_C. \tag{4-27}$$

Premultiplying Eq. (4-27) by $\boldsymbol{L}_G{}^0$ and using Eq. (4-26), we have

$$\boldsymbol{G}^{RR}\, \boldsymbol{F}^{RR}\, \boldsymbol{L}_G^0\, \boldsymbol{L}_C = \boldsymbol{L}_G^0\, \boldsymbol{L}_C\, \boldsymbol{\Lambda}_C. \tag{4-28}$$

Equation (4-28) shows that the matrix $\boldsymbol{G}^{RR}\boldsymbol{F}^{RR}$ is diagonalized to give $\boldsymbol{\Lambda}_C$ by the similarity transformation with the matrix $L_G{}^0 L_C$. Since this sort of diagonalization is unique for a given matrix, Eq. (4-28) should be the same as Eq. (4-22). It follows therefore that $\boldsymbol{\Lambda} = \boldsymbol{\Lambda}_C$ and

$$\boldsymbol{L} = \boldsymbol{L}_G^0 \boldsymbol{L}_C. \tag{4-29}$$

Remembering the orthogonality of $\boldsymbol{L}_C$ and using Eq. (4-29), we may rewrite Eq. (4-26) as

$$\boldsymbol{G}^{RR} = \boldsymbol{L}_G^0 \left(\boldsymbol{L}_C \tilde{\boldsymbol{L}}_C\right) \tilde{\boldsymbol{L}}_G^0 = (\boldsymbol{L}_G^0 \boldsymbol{L}_C)(\boldsymbol{L}_G^0 \boldsymbol{L}_C)^{\mathrm{T}} = \boldsymbol{L} \tilde{\boldsymbol{L}}. \tag{4-30}$$

The method of calculating normal coordinates and normal frequencies by using the $\boldsymbol{G}^{RR}$ and $\boldsymbol{F}^{RR}$ matrices is called the ***GF*** matrix method.[1,2] The minimum number of variables required for applying the ***GF*** matrix method to a nonlinear N-atomic molecule is the same as the internal degrees of freedom of the nuclear motion, $3N - 6$. Each element of the matrix $\boldsymbol{G}^{RR}$ can be expressed in an analytical form not depending on the orientation of the Cartesian coordinate axes but determined only by the types and mutual positions of the relevant internal coordinates, e.g., ΔR_p and ΔR_q for G_{pq}. From the definition of the $\boldsymbol{G}$ matrix in Eq. (3-42), its elements are given as a scalar product of the $\boldsymbol{s}$ vectors (Eq. (3-47)) in the form[1,2]

$$\begin{aligned} G_{pq} &= \sum_i m_i \left(B_{p,3i-2}\, B_{q,3i-2} + B_{p,3i-1}\, B_{q,3i-1} + B_{p,3i}\, B_{q,3i} \right) \\ &= \sum_i m_i\, \boldsymbol{s}_{pi} \cdot \boldsymbol{s}_{qi}. \end{aligned} \tag{4-31}$$

The sum with respect to i in Eq. (4-31) is taken over those nuclei which are used in the definition of both the internal coordinates p and q. To illustrate how to use Eq. (4-31), the elements of the $\boldsymbol{G}$ matrix for a nonlinear molecule j–i–k are calculated as follows.

Let $\boldsymbol{s}(r_{ij})_i$ be the $\boldsymbol{s}$ vector of the bond stretching coordinate Δr_{ij} for the nucleus i. It then follows from Eqs. (3-65a) and (3-59) that

$$\boldsymbol{s}(r_{ij})_i = -\boldsymbol{\nabla}_{ij}\, r_{ij} = -\boldsymbol{e}_{ij}, \tag{4-32}$$

where the operator $\boldsymbol{\nabla}_{ij}$ is defined by Eq. (3-64). Similarly, it holds for the nucleus j that $\boldsymbol{s}(r_{ij})_j = \boldsymbol{e}_{ij}$. The $\boldsymbol{s}$ vector of an angle-bending coordinate $\Delta\theta_{jik}$ is calculated as follows. Let us take the gradients of both sides of the fundamental relation, $\tilde{\boldsymbol{r}}_{ij}\, \boldsymbol{r}_{ik} = r_{ij}\, r_{ik} \cos\theta_{jik}$, and use the well known relation

$$\boldsymbol{\nabla}_{ij} \cos\theta_{jik} = -\sin\theta_{jik}\, \boldsymbol{\nabla}_{ij}\theta_{jik}$$

together with Eqs. (3-65a,b), obtaining

$$\begin{aligned} \boldsymbol{r}_{ik} &= r_{ij}\, r_{ik}\, \boldsymbol{\nabla}_{ij} \cos\theta_{jik} + \left(\boldsymbol{\nabla}_{ij} r_{ij}\right) r_{ik} \cos\theta_{jik} \\ &= -\, r_{ij}\, r_{ik} \sin\theta_{jik}\, \boldsymbol{\nabla}_{ij}\theta_{jik} + \boldsymbol{e}_{ij} r_{ik} \cos\theta_{jik}. \end{aligned}$$

Now we define a unit vector $\boldsymbol{p}_{ij}$, which is perpendicular to the bond i–j in the plane defined by the nuclei j, i and k, and is directed in such a way that $\tilde{\boldsymbol{p}}_{ij}\, \boldsymbol{e}_{ik} < 0$. As seen from Fig. 4.1, the vector $\boldsymbol{p}_{ij}$ is expressed as

$$\boldsymbol{p}_{ij} = \left(-\boldsymbol{e}_{ik} + \boldsymbol{e}_{ij} \cos\theta_{jik}\right) / \sin\theta_{jik}.$$

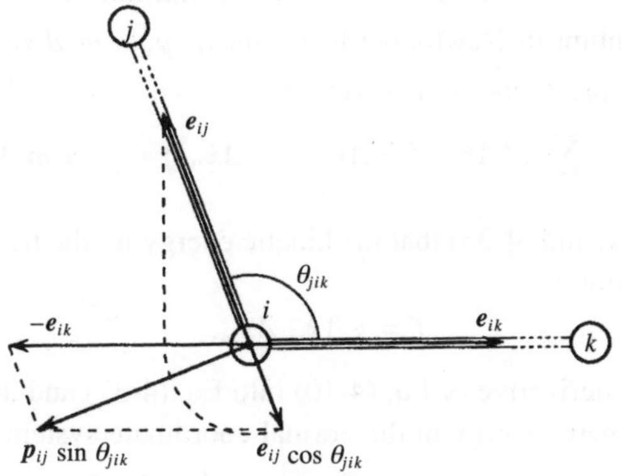

Fig. 4.1 Calculation of *s* vector for a bending coordinate.

Then the *s* vector of the angle bending coordinate $\Delta\theta_{jik}$ for the nucleus j is given by

$$s(\theta_{jik})_j = \boldsymbol{\nabla}_{ij}\theta_{jik} = \frac{-\boldsymbol{e}_{ik} + \boldsymbol{e}_{ij}\cos\theta_{jik}}{r_{ij}\sin\theta_{jik}} = \frac{\boldsymbol{p}_{ij}}{r_{ij}}. \tag{4-33}$$

Substituting Eqs. (4-32) and (4-33) into Eq. (4-31), we obtain the $\boldsymbol{G}$ matrix elements of a water molecule in the forms

$$G(\Delta r_{ij}, \Delta r_{ij}) = (m_i^{-1} + m_j^{-1})\boldsymbol{e}_{ij}\cdot\boldsymbol{e}_{ij} = m_i^{-1} + m_j^{-1} \tag{4-34a}$$

$$G(\Delta r_{ij}, \Delta r_{ik}) = m_i^{-1}\,\boldsymbol{e}_{ij}\cdot\boldsymbol{e}_{ik} = m_i^{-1}\cos\theta_{jik} \tag{4-34b}$$

$$\begin{aligned} G(\Delta\theta_{jik}, \Delta\theta_{jik}) &= \frac{1}{m_i}\left(\frac{\boldsymbol{p}_{ij}}{r_{ij}} + \frac{\boldsymbol{p}_{ik}}{r_{ik}}\right)\cdot\left(\frac{\boldsymbol{p}_{ij}}{r_{ij}} + \frac{\boldsymbol{p}_{ik}}{r_{ik}}\right) + \frac{1}{m_j}\frac{\boldsymbol{p}_{ij}}{r_{ij}}\cdot\frac{\boldsymbol{p}_{ij}}{r_{ij}} + \frac{1}{m_k}\frac{\boldsymbol{p}_{ik}}{r_{ik}}\cdot\frac{\boldsymbol{p}_{ik}}{r_{ik}} \\ &= \frac{1}{m_i}\left(\frac{1}{r_{ij}^2} + \frac{1}{r_{ik}^2} - 2\frac{\cos\theta_{jik}}{r_{ij}r_{ik}}\right) + \frac{1}{m_j}\frac{1}{r_{ij}^2} + \frac{1}{m_k}\frac{1}{r_{ik}^2} \end{aligned} \tag{4-34c}$$

$$\begin{aligned} G(\Delta r_{ij}, \Delta\theta_{jik}) &= -\frac{1}{m_i}\left(\frac{\boldsymbol{p}_{ij}}{r_{ij}} + \frac{\boldsymbol{p}_{ik}}{r_{ik}}\right)\cdot\boldsymbol{e}_{ij} + \frac{1}{m_j}\frac{\boldsymbol{p}_{ij}}{r_{ij}}\cdot\boldsymbol{e}_{ij} \\ &= -\frac{1}{m_i}\frac{\sin\theta_{jik}}{r_{ik}}. \end{aligned} \tag{4-34d}$$

Detailed tables of various types of $\boldsymbol{G}$ matrix elements are found in the literature.[1, 5, 6] Physically, the $\boldsymbol{G}$ matrix is the Hessian matrix of the kinetic energy of the nuclear motion with respect to the momenta conjugate to the internal coordinates. When ξ_i is a coordinate belonging to a certain coordinate system, the momentum conjugate to ξ_i is defined to be

$$(P_\xi)_i = \partial T/\partial\dot{\xi}_i. \tag{4-35}$$

In the case of the Cartesian coordinate system, this definition agrees with the fundamental definition of the momentum in Newtonian mechanics, $p_x = m(dx/dt)$. In the Cartesian coordinate system, the kinetic energy is given by

$$T = \tfrac{1}{2}\sum_i m_i\left(\Delta\dot{x}_{3i-2}{}^2 + \Delta\dot{x}_{3i-1}{}^2 + \Delta\dot{x}_{3i}{}^2\right) = \tfrac{1}{2}\,\Delta\tilde{\dot{\boldsymbol{x}}}\,\boldsymbol{m}\,\Delta\dot{\boldsymbol{x}}. \tag{4-36}$$

It follows from Eqs. (4-8) and (4-36) that the kinetic energy in the mass-weighted Cartesian coordinate system is given by

$$T = \tfrac{1}{2}\,\Delta\tilde{\dot{\boldsymbol{x}}}_m\,\Delta\dot{\boldsymbol{x}}_m. \tag{4-37}$$

By substituting the time derivative of Eq. (4-10) into Eq. (4-37) and using the orthogonality of the matrix $\boldsymbol{L}^{mx}$, the kinetic energy in the normal coordinate system is expressed as

$$T = \tfrac{1}{2}\tilde{\dot{\boldsymbol{Q}}}\dot{\boldsymbol{Q}} = \tfrac{1}{2}\sum_i \dot{Q}_i^2. \tag{4-38}$$

Let the momentum conjugate to the normal coordinate Q_j be denoted by $(P_Q)_j$, and use Eqs. (4-35) and (4-38) to obtain

$$(P_Q)_j \equiv \partial T/\partial\dot{Q}_j = \dot{Q}_j. \tag{4-39}$$

By introducing the momentum vector $\boldsymbol{P}_Q$ defined as

$$\boldsymbol{P}_Q = \left[(P_Q)_1 \quad (P_Q)_2 \quad \cdots\right]^{\mathrm{T}},$$

the kinetic energy can be written in the form

$$T = \tfrac{1}{2}\tilde{\boldsymbol{P}}_Q\,\boldsymbol{P}_Q. \tag{4-40}$$

On the other hand, the momentum conjugate to the internal coordinate R_i is defined by

$$(P_R)_i \equiv \partial T/\partial\dot{R}_i,$$

whence the momentum conjugate to Q_j is expressed as

$$(P_Q)_j \equiv \partial T/\partial\dot{Q}_j = \sum_i\left(\partial T/\partial\dot{R}_i\right)\left(\partial\dot{R}_i/\partial\dot{Q}_j\right) = \sum_i (P_R)_i\,L_{ij}. \tag{4-41}$$

The last equality in Eq. (4-41) results from partially differentiating the relation between the time derivatives of the ith components of the right and the left sides of Eq. (4-23),

$$\Delta\dot{R}_i = \sum_j L_{ij}\,\dot{Q}_j,$$

with respect to $\dot{Q}_j$.

Let the momentum vector consisting of the components $(P_R)_1$, $(P_R)_2$, ⋯ be denoted by $\boldsymbol{P}_R$. Substituting the matrix representation of Eq. (4-41),

$$\boldsymbol{P}_Q = \tilde{\boldsymbol{L}}\,\boldsymbol{P}_R, \tag{4-42}$$

into that of Eq. (4-39), $\boldsymbol{P}_Q = \dot{\boldsymbol{Q}}$, and using Eqs. (4-40), (4-38) and (4-30), we obtain

$$T = \tfrac{1}{2}\tilde{\boldsymbol{P}}_R\,\boldsymbol{L}\tilde{\boldsymbol{L}}\,\boldsymbol{P}_R = \tfrac{1}{2}\tilde{\boldsymbol{P}}_R\,\boldsymbol{G}^{RR}\,\boldsymbol{P}_R. \tag{4-43}$$

Equation (4-43) shows that the components of the matrix $\boldsymbol{G}^{RR}$ are the coefficients of the quadratic terms in the expansion of the kinetic energy in powers of the momenta conjugate to the internal coordinates.

The normal coordinate analysis in terms of internal coordinates is useful for the purpose of discussing a special type of nuclear motion separately from the others. For large molecules such as proteins, the normal coordinate analysis over all the degrees of freedom of the nuclear motion is not easy even using a high-speed computer with a large memory. If only slow and large amplitude motions of the nuclei are of interest, it is useful to diminish the problem size by carrying out the normal coordinate analysis in which only the degrees of freedom of internal rotation around the single bonds are taken into account.

In order to accelerate the search for potential minima and the construction of the $\boldsymbol{F}$ matrix for such calculations, Noguti and Gō derived analytical expressions of the first- and the second- derivatives of the nonbonded internuclear distance r_{ij}' with respect to a given torsion angle.[7)] The elements of the matrix $\boldsymbol{L}$ for low-frequency modes calculated in this way have been shown to be closely correlated to the temperature factors experimentally obtained from X-ray crystallographic analysis.[8, 9)]

4.4 Quantum Mechanics of Harmonic Oscillators

The motion of atomic nuclei in actual molecules is described by quantum mechanics in principle. To apply quantum mechanics to a molecular system, we must express first the potential energy V and the kinetic energy T in a common coordinate system, and then construct the wave equation from the classical Hamiltonian function, which is equivalent to the total energy, $T + V$, for such a conservative system as treated here. The choice of the coordinate system for describing T and V is the key to solving the problem.

Let the potential energy be expanded in powers of the normal coordinates at the equilibrium as given in Eq. (3-86), and let $V - V^0$ be redefined as V. Noting that the nuclear displacements are small, we may drop the terms that are of higher order than quadratic. Furthermore, since the linear terms vanish at an equilibrium, the potential energy is written simply in the form

$$V = \tfrac{1}{2}\tilde{\boldsymbol{Q}}\boldsymbol{\Lambda}\boldsymbol{Q} = \tfrac{1}{2}\sum_i \lambda_i Q_i^2. \tag{4-44}$$

Since the normal coordinates are transformed into the Cartesian displacement coordinates by a linear transformation, Eq. (3-80), the dropping of terms higher than quadratic in the normal coordinates corresponds to the harmonic approximation.

Adding the kinetic energy in Eq. (4-38) to Eq. (4-44), and substituting Eq. (4-14) to eliminate λ_i , we obtain

$$H = T + V = \tfrac{1}{2}\sum_i \left(\dot{Q}_i^2 + \lambda_i Q_i^2\right) = \tfrac{1}{2}\sum_i \left(\dot{Q}_i^2 + 4\pi^2 c^2 \omega_i^2 Q_i^2\right). \tag{4-45}$$

The coefficient of the linear transformation connecting a normal coordinate Q_i to a mass-weighted Cartesian displacement coordinate $(\Delta x_m)_j$ is an element of the orthogonal matrix $\boldsymbol{L}^{mx}$, and should thus be dimensionless. Accordingly, the normal coordinate Q_i has the same dimension as the mass-weighted Cartesian displacement coordinates, that is,

$$[Q_i] = [\text{mass}^{1/2} \times \text{distance}]. \tag{4-46}$$

The physical meaning of the dimension involving a square root is not so intuitive. In order for the nuclear motion to be described by any variables with a clearcut dimension, we introduce a scale factor in the form

$$\alpha_i = 2\pi (c\,\omega_i/h)^{1/2} \tag{4-47}$$

and define a new variable called the dimensionless normal coordinate, q_i,[10] which is linearly related to the normal coordinate Q_i by

$$q_i = \alpha_i Q_i = (4\pi^2 c\,\omega_i/h)^{1/2} Q_i. \tag{4-48}$$

That q_i is dimensionless is confirmed by comparing Eq. (4-46) with the dimension of α_i derived from Eq. (4-47) in the form

$$[\alpha_i] = \left[\frac{\text{speed} \times \text{wavenumber}}{\text{energy} \times \text{time}}\right]^{1/2}$$

$$= \left[\frac{\text{time}^{-1}}{(\text{mass} \times \text{distance}^2 \times \text{time}^{-2}) \times \text{time}}\right]^{1/2}.$$

Let the momentum conjugate to the dimensionless normal coordinate q_i be denoted by p_i. Then it follows from the definition that

$$\begin{aligned} p_i &= \partial T/\partial \dot{q}_i = (\partial T/\partial \dot{Q}_i)(\partial \dot{Q}_i/\partial \dot{q}_i) \\ &= \dot{Q}_i/\alpha_i = (h/4\pi^2 c\,\omega_i)\dot{Q}_i. \end{aligned} \tag{4-49}$$

On substituting Eqs. (4-48) and (4-49) into Eq. (4-45), the Hamiltonian is expressed in terms of the dimensionless normal coordinates and their conjugate momenta in the form

$$H = \tfrac{1}{2}\sum_i \left(\frac{4\pi^2 c\,\omega_i}{h} p_i^{\,2} + \frac{4\pi^2 c^2 \omega_i^{\,2} h}{4\pi^2 c\,\omega_i} q_i^{\,2}\right) = \tfrac{1}{2} hc \sum_i \omega_i \left(\frac{p_i^{\,2}}{\hbar^2} + q_i^{\,2}\right), \tag{4-50}$$

where $\hbar = h/2\pi$.

Only a single variable q_i and its conjugate momentum p_i appear in each term in the sums in Eq. (4-50); hence we can solve it term by term as a single variable wave equation. The total wave function ψ is then given as the product of individual wave functions ψ_i and the total energy E as the sum of individual energies E_i. Following the standard procedure of quantum mechanics, we take out the term containing q_i in Eq. (4-50) and replace the momentum p_i in it with the differential operator $-i\hbar\,(d/d\,q_i)$ to obtain the wave equation in the form

$$hc\,\omega_i \left(-\frac{d^2\psi_i}{d\,q_i^{\,2}} + q_i^{\,2}\psi_i\right) = 2\,E_i\,\psi_i$$

or

$$\frac{d^2\psi_i}{d q_i^2} - \left(q_i^2 - \frac{2E}{hc\omega_i} \right)\psi_i = 0 . \tag{4-51}$$

The physically acceptable solution of Eq. (4-51), $\psi_i(q_i)$, should be such that the integral of its square over the whole variable range $(-\infty, +\infty)$ of q_i is unity, that is, it should hold that

$$\int_{-\infty}^{+\infty} |\psi_i(q_i)|^2 \, d q_i = 1. \tag{4-52}$$

As shown in the following, the product of a Gaussian distribution function and a polynomial of the nth order in the form

$$\psi_i(q_i;n) = N_n \exp(-a q_i^2)(b_n q_i^n + b_{n-1} q_i^{n-1} + \cdots) \tag{4-53}$$

with a suitably chosen factor N_n satisfies Eq. (4-52).

Let the nth order polynomial in Eq. (4-53) be denoted as

$$H_n(y) = \sum_{k=0}^{n} b_k y^k, \qquad (b_n \neq 0) \tag{4-54}$$

and differentiate Eq. (4-53) once and twice with respect to q_i, obtaining

$$\psi_i'(q_i;n) = N_n\{-2 a q_i H_n(q_i) + H_n'(q_i)\}\exp(-a q_i^2)$$

and

$$\psi_i''(q_i;n) = N_n\{(4a^2 q_i^2 - 2a)H_n(q_i) - 4 a q_i H_n'(q_i) + H_n''(q_i)\}\exp(-a q_i^2),$$

respectively. Substituting Eq. (4-53) together with these derivatives into Eq. (4-51), we have

$$\left\{(4a^2 - 1)q_i^2 - 2a + \frac{2E_i}{hc\omega_i}\right\}H_n(q_i) - 4 a q_i H_n'(q_i) + H_n''(q_i) = 0. \tag{4-55}$$

In order for Eq. (4-55) to hold identically, the coefficient of every power of q_i in the left side should vanish individually. On putting the coefficient of the highest power term $(4a^2 - 1)b_n q_i^{n+2}$ to zero, the parameter a, which must be positive in the Gaussian distribution function in Eq. (4-53), is determined as $a = 1/2$. This value is then substituted, together with $H_n'(q_i)$ calculated from Eq. (4-54), into Eq. (4-55), and the coefficient of q_i^n is put equal to zero; thus

$$\left(-1 + \frac{2E_i}{hc\omega_i} - 2n\right)b_n = 0, \quad \therefore \quad E_i = hc\omega_i(n + \tfrac{1}{2}). \tag{4-56}$$

On replacing q_i with x, putting $a = 1/2$ and using Eq. (4-56), Eq. (4-55) is rewritten in the form

$$H_n''(x) - 2x H_n'(x) + 2n H_n(x) = 0. \tag{4-57}$$

This equation is known as the Hermite equation, the solution of which is a polynomial in the nth power expressible in the form

$$H_n(x) = (-1)^n \exp(x^2)\{d^n \exp(-x^2)/d x^n\} \tag{4-58a}$$

or

$$H_n(x) = \sum_{m=0}^{[n/2]} (-1)^m \frac{n!}{m!(n-2m)!}(2x)^{n-2m}, \tag{4-58b}$$

where $[n/2]$ denotes the largest integer not exceeding $n/2$. The polynomial defined by Eq. (4-58a) or (4-58b) is called the Hermite polynomial of order n. The Hermite polynomials of orders 0, 1 and 2 are given from either Eq. (4-58a) or (4-58b) as

$$H_0(x) = 1, \quad H_1(x) = 2x, \quad H_2(x) = 4x^2 - 2.$$

That Eq. (4-57) is satisfied by Eqs. (4-58a,b) is proved as follows. On abbreviating the function $\exp(-x^2)$ as $f(x)$, Eq. (4-58a) is rewritten in the form

$$f(x) H_n(x) = (-1)^n (d^n/d x^n) f(x). \tag{4-59}$$

Replacing n in Eq. (4-59) with $n + 1$, we can differentiate the right side as given by

$$\begin{aligned} f(x) H_{n+1}(x) &= (-1)^{n+1}(d^{n+1}/d x^{n+1}) f(x) = (-1)^{n+1}(d^n/d x^n) f'(x) \\ &= (-1)^{n+1}(d^n/d x^n)\{-2x f(x)\} = (-1)^n (d^n/d x^n)\{2x f(x)\} \\ &= (-1)^n\{2x(d^n/d x^n) f(x) + 2n(d^{n-1}/d x^{n-1}) f(x)\}. \end{aligned} \tag{4-60}$$

To derive the last equality in Eq. (4-60), use is made of the Leibnitz formula for the higher derivatives of a product,

$$(f g)^{(n)} = f^{(n)} g + n f^{(n-1)} g' + \frac{n(n-1)}{2!} f^{(n-2)} g'' + \cdots,$$

applied to the case $g = 2x$. Substituting Eq. (4-59) into the last formula in Eq. (4-60) and dividing by $f(x)$, we obtain a useful recursion formula for the Hermite polynomial

$$H_{n+1}(x) = 2x H_n(x) - 2n H_{n-1}(x). \tag{4-61}$$

On the other hand, differentiating both sides of Eq. (4-59) once with respect to x, we have

$$\{-2x H_n(x) + H_n'(x)\} f(x) = (-1)^n (d^{n+1}/d x^{n+1}) f(x) = -f(x) H_{n+1}(x)$$

or

$$H_{n+1}(x) = 2x H_n(x) - H_n'(x). \tag{4-62}$$

Combining Eq. (4-61) and (4-62) leads to

$$H_n'(x) = 2n H_{n-1}(x). \tag{4-63}$$

Differentiating Eq. (4-63), we have

$$H_n''(x) = 2n H_{n-1}'(x). \tag{4-64}$$

Now the recursion formula (Eq. (4-61)) is differentiated to give

$$H_{n+1}'(x) = 2 H_n(x) + 2x H_n'(x) - 2n H_{n-1}'(x),$$

which leads to the Hermite equation Eq. (4-57) on unifying the suffixes to n with the help of Eq. (4-63), where n has been replaced by $n + 1$, and Eq. (4-64). It then follows that Eq. (4-58a) is a solution of Eq. (4-57).

From Eq. (4-53), the form of a complete wave function is shown to be

$$\psi_i(q_i;n) = N_n\{f(q_i)\}^{1/2} H_n(q_i) = N_n \exp(-q_i^2/2)H_n(q_i). \tag{4-65}$$

The normalization constant N_n is so determined to be

$$N_n = \left(2^n\, n!\sqrt{\pi}\right)^{-1/2} \tag{4-66}$$

that the function (Eq. (4-65)) satisfies the condition of the orthonormality :

$$\int_{-\infty}^{+\infty} \psi^*(x;n)\,\psi(x;m)\,dx = \delta_{nm}. \tag{4-67}$$

To confirm Eq. (4-67), substitute Eq. (4-65) into the left side, divide by N_nN_m, substitute Eq. (4-59) and integrate by parts repeatedly. Noting that the integrated term is a product of a polynomial and a Gaussian distribution function vanishing at $x = \pm\infty$, we have

$$\int_{-\infty}^{+\infty} f(x)\,H_n(x)\,H_m(x)\,dx = (-1)^n \int_{-\infty}^{+\infty} \frac{d^n f(x)}{dx^n} H_m(x)\,dx$$

$$= \left[(-1)^n \frac{d^{n-1}f(x)}{dx^{n-1}} H_m(x)\right]_{-\infty}^{+\infty} + (-1)^{n-1} \int_{-\infty}^{+\infty} \frac{d^{n-1}f(x)}{dx^{n-1}} \frac{d\,H_m(x)}{dx}\,dx$$

$$= (-1)^{n-2} \int_{-\infty}^{+\infty} \frac{d^{n-2}f(x)}{dx^{n-2}} \frac{d^2\,H_m(x)}{dx^2}\,dx = \cdots$$

$$= \int_{-\infty}^{+\infty} f(x) \frac{d^n\,H_m(x)}{dx^n}\,dx = 2^n n!\sqrt{\pi}\,\delta_{nm}.$$

Multiplying the last formula by N_n^2 and using Eq. (4-66), we obtain Eq. (4-67). Important formulae of the definite integrals[11)] involving the wave functions of the harmonic oscillator are given in Appendix 2.

A wave function for the internal motion of all the nuclei is given as the product of the one-dimensional wave functions for individual normal modes in the form

$$\psi(Q_1,Q_2,\cdots,Q_n;n_1,n_2,\cdots,n_n) = \psi(Q_1;n_1)\,\psi(Q_2;n_2)\cdots\psi(Q_n;n_n), \tag{4-68}$$

and the corresponding energy level is given as the sum

$$E_n = hc\sum_i \omega_i(v_i + 1/2), \tag{4-69}$$

where v_i is a non-negative integer called the vibrational quantum number for the normal coordinate Q_i.

According to Eq. (4-69), equal energy levels with different combinations of vibrational quantum numbers appear if there are different normal modes having equal normal frequencies. If there are d_i normal frequencies such that

$$\omega_i = \omega_{i+1} = \cdots = \omega_{i+d_i-1},$$

the portion related to these frequencies in the sum in Eq. (4-69) can be rewritten in the form

$$\omega_i\{v_i + v_{i+1} + \cdots + v_{i+d_i-1} + (d_i/2)\}. \tag{4-70}$$

Within the framework of harmonic approximation, the vibrational energy levels depend solely on the sum of vibrational quantum numbers in the braces in Eq. (4-70). In this case, we cannot distinguish between different combinations of individual quantum numbers, for example, between the two states ($v_i = 1$, $v_{i+1} = 0$) and ($v_i = 0$, $v_{i+1} = 1$). To further clarify the situation, the vibrational quantum numbers are so renumbered that the sum of d_i quantum numbers in braces in Eq. (4-70) is given the number v_i, and the total energy of a given vibrational level of the molecule is written in the form

$$E_n = hc\sum_i \omega_i(v_i + d_i/2). \tag{4-71}$$

The number of the equal vibrational frequencies d_i is called the degeneracy of the normal mode i. The normal modes with $d_i = 1$ are called the non-degenerate normal modes, while those with $d_i > 1$ are called the degenerate normal modes. Note that the upper limit of the sum in Eq. (4-71) for the molecules having degenerate normal modes is less than the number of degrees of freedom of the internal motion.

The degeneracy of a given normal mode is determined by the symmetry of the molecule and the type of the normal mode. Degenerate normal modes are classified into the doubly, the triply, ··· degenerate normal modes according to $d_i = 2, 3, \cdots$. As the symmetry of the molecule is raised, the more highly degenerate normal modes appear. For example, fullerene, C_{60}, has fivefold degenerate normal modes. The relation between the symmetry of molecules and the degeneracy is discussed in detail in section 4.6.

Substituting $v_i = n$ and $v_i = n + 1$ into Eq. (4-71) and taking the difference, we obtain $hc\omega_i$. Thus the vibrational levels associated with the normal coordinate i are spaced equidistantly at the interval of $hc\omega_i$. The lowest vibrational level is given, by substituting all v_i with 0, in the form

$$E^0 = hc\sum_i \omega_i\, d_i/2. \tag{4-72}$$

This level is called the vibrational ground level. Since no energy level lower than the vibrational ground level exists, the molecule even at 0 K has an energy higher than the potential minimum by E^0. The energyE^0 is called the zero point energy.

4.5 Use of Molecular Symmetry

Knowledge of the symmetry properties of normal vibrations is indispensable for assigning normal frequencies correctly based on spectroscopic data. Many simple polyatomic molecules have more than one equivalent nuclei. The equivalent nuclei can be exchanged with one another by carrying out appropriate linear transformations of their Cartesian coordinates without any changes in the apparent feature of the molecule. Let the position vectors of all the nuclei in a molecule be premultiplied by a common (3 × 3) orthogonal matrix. If the Cartesian coordinates of every nucleus generated by such a multiplication are the same as the Cartesian coordinates of any one of its equivalent nuclei, including itself, before the multiplication, the transformation of the nuclear coordinates by this multiplication is called a symmetry operation. The common (3 × 3) matrix may be a unit matrix. The operation in this case is called the identity operation, which leaves the coordinates of every nucleus unchanged. This operation, which actually is to do nothing, is a very important operation in discussing molecular symmetry. Including the identity operation, every molecule has at least one symmetry operation. Mathematically, a symmetry operation is represented by any one of a rotation, a reflection and an inversion, or a combination of the first two (called a rotoreflection). If a molecule has a set of two or more equivalent atoms, it must also have at least one set of the same number of equivalent internal coordinates, and correspondingly, a set of equivalent elements appears in each of the $\boldsymbol{G}$ and $\boldsymbol{F}$ matrices.

For example, let us take the internal coordinates of a water molecule as shown in Fig. 4.2, and define the vector of these coordinates as

$$\Delta\boldsymbol{R} = \left[\Delta r_{12} \quad \Delta r_{13} \quad \Delta\theta_{213}\right]^{\mathrm{T}}. \tag{4-73}$$

From the equivalence between Δr_{12} and Δr_{13}, the $\boldsymbol{G}$ matrix must have the form

$$\boldsymbol{G}^{RR} = \begin{bmatrix} G_{rr} & g_{rr} & g_{r\theta} \\ g_{rr} & G_{rr} & g_{r\theta} \\ g_{r\theta} & g_{r\theta} & G_{\theta\theta} \end{bmatrix}. \tag{4-74}$$

The elements of $\boldsymbol{G}^{RR}$ are calculated from Eq. (4-34a–d) and Fig. 4.2 as follows

$$\left.\begin{aligned} G_{rr} &= m_{\mathrm{O}}^{-1} + m_{\mathrm{H}}^{-1}, \qquad g_{rr} = m_{\mathrm{O}}^{-1}\cos\theta_{\mathrm{HOH}} \\ g_{r\theta} &= -r_{\mathrm{OH}}^{-1}\, m_{\mathrm{O}}^{-1}\sin\theta_{\mathrm{HOH}} \\ G_{\theta\theta} &= 2\, r_{\mathrm{OH}}^{-2}\left\{m_{\mathrm{H}}^{-1} + m_{\mathrm{O}}^{-1}(1-\cos\theta_{\mathrm{HOH}})\right\} \end{aligned}\right\}.$$

On the other hand, the GVFF type potential function is given by

$$V = \tfrac{1}{2}F_{rr}\left(\Delta r_{12}^{\,2} + \Delta r_{13}^{\,2}\right) + f_{rr}\Delta r_{12}\Delta r_{13} + f_{r\theta}\left(\Delta r_{12} + \Delta r_{13}\right)\Delta\theta_{213} + \tfrac{1}{2}F_{\theta\theta}\Delta\theta_{213}^{\,2},$$

leading to the $\boldsymbol{F}$ matrix in the form

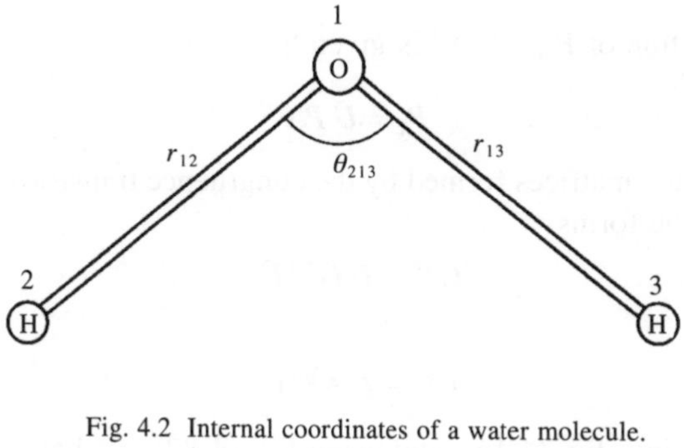

Fig. 4.2 Internal coordinates of a water molecule.

$$\boldsymbol{F}^{RR} = \begin{bmatrix} F_{rr} & f_{rr} & f_{r\theta} \\ f_{rr} & F_{rr} & f_{r\theta} \\ f_{r\theta} & f_{r\theta} & F_{\theta\theta} \end{bmatrix}, \tag{4-75}$$

which has the same pattern of arrangement of equivalent elements as $\boldsymbol{G}^{RR}$.

Generally, we can save computing time and memory considerably in such a case by using a set of variables derived from the internal coordinates, ΔR_1, ΔR_2, $\cdots$,ΔR_n, through a linear transformation in the form

$$\Delta S_i = \sum_j U_{ij}\, \Delta R_j, \tag{4-76}$$

The new variables ΔS_1, ΔS_2, $\cdots$,ΔS_n are called the internal symmetry coordinates. In most cases, the transformation coefficient U_{ij} is so chosen as to be the (i, j) element of an orthogonal matrix $\boldsymbol{U}$. Defining the vector of the internal symmetry coordinates as

$$\Delta\boldsymbol{S} = [\Delta S_1 \quad \Delta S_2 \quad \cdots \quad \Delta S_n]^{\mathrm{T}},$$

we have the matrix representation of Eq. (4-76) in the form

$$\Delta\boldsymbol{S} = \boldsymbol{U}\, \Delta\boldsymbol{R}. \tag{4-77}$$

It then follows from the orthogonality of $\boldsymbol{U}$ that

$$\Delta\boldsymbol{R} = \tilde{\boldsymbol{U}}\, \Delta\boldsymbol{S}. \tag{4-78}$$

Let $(P_S)_i$ be the momentum conjugate to the coordinate ΔS_i, and $\boldsymbol{P}_S$ be the vector containing $(P_S)_i$ as the ith element. Similarly, the momentum conjugate to the internal coordinate ΔR_j and the vector formed therefrom is denoted by $(P_R)_j$ and $\boldsymbol{P}_R$, respectively. By partial differentiation, we have

$$(P_S)_i \equiv \partial T/\partial S_i = \sum_j (\partial T/\partial R_j)(\partial R_j/\partial S_i) = \sum_j U_{ij}(P_R)_j$$

or, in the matrix representation,

$$\boldsymbol{P}_S = \boldsymbol{U}\, \boldsymbol{P}_R. \tag{4-79}$$

The inverse transformation of Eq. (4-79) is given by

$$\boldsymbol{P}_R = \tilde{\boldsymbol{U}}\, \boldsymbol{P}_S. \tag{4-80}$$

Let $\boldsymbol{G}^{SS}$ and $\boldsymbol{F}^{SS}$ be the matrices formed by the congruence transformation of $\boldsymbol{G}^{RR}$ and $\boldsymbol{F}^{RR}$ by $\boldsymbol{U}$, respectively, in the forms

$$\boldsymbol{G}^{SS} = \boldsymbol{U}\, \boldsymbol{G}^{RR}\, \tilde{\boldsymbol{U}} \tag{4-81}$$

and

$$\boldsymbol{F}^{SS} = \boldsymbol{U}\, \boldsymbol{F}^{RR}\, \tilde{\boldsymbol{U}}. \tag{4-82}$$

Substituting Eq. (4-78) into Eq. (4-16), and using Eq. (4-82), we have

$$V = \tfrac{1}{2}\, \Delta\tilde{\boldsymbol{S}}\, \boldsymbol{U}\, \boldsymbol{F}^{RR}\, \tilde{\boldsymbol{U}}\, \Delta\boldsymbol{S} = \tfrac{1}{2}\, \Delta\tilde{\boldsymbol{S}}\, \boldsymbol{F}^{SS}\, \Delta\boldsymbol{S}. \tag{4-83}$$

Similarly, substituting Eqs. (4-80) and (4-81) into Eq. (4-43), we have

$$T = \tfrac{1}{2}\tilde{\boldsymbol{P}}_S \boldsymbol{U} \boldsymbol{G}^{RR} \tilde{\boldsymbol{U}} \boldsymbol{P}_S = \tfrac{1}{2}\tilde{\boldsymbol{P}}_S \boldsymbol{G}^{SS} \boldsymbol{P}_S. \tag{4-84}$$

Equations (4-83) and (4-84) show that the transformation from $\boldsymbol{G}^{RR}$ and $\boldsymbol{F}^{RR}$ to $\boldsymbol{G}^{SS}$ and $\boldsymbol{F}^{SS}$ means the transformation of the coordinate system from the internal coordinates to the internal symmetry coordinates.

Forming the product of Eq. (4-81) with Eq. (4-82), and making use of the orthogonality of $\boldsymbol{U}$, *i.e.*, $\tilde{\boldsymbol{U}} = \boldsymbol{U}^{-1}$, we have

$$\boldsymbol{G}^{SS}\boldsymbol{F}^{SS} = \boldsymbol{U}\boldsymbol{G}^{RR}\tilde{\boldsymbol{U}}\boldsymbol{U}\boldsymbol{F}^{RR}\tilde{\boldsymbol{U}} = \boldsymbol{U}\boldsymbol{G}^{RR}\boldsymbol{F}^{RR}\boldsymbol{U}^{-1}. \tag{4-85}$$

Since the eigenvalues of a matrix are kept invariant on a similarity transformation, the eigenvalues of $\boldsymbol{G}^{RR}\boldsymbol{F}^{RR}$ can be obtained by diagonalizing $\boldsymbol{G}^{SS}\boldsymbol{F}^{SS}$ instead of $\boldsymbol{G}^{RR}\boldsymbol{F}^{RR}$ itself. In this case, if the matrix $\boldsymbol{U}$ has been chosen appropriately, each of the matrices $\boldsymbol{G}^{SS}$ and $\boldsymbol{F}^{SS}$ can be expressed in the form of a direct sum of matrices of smaller sizes on the diagonal. For example, in the case of the water molecule in Fig. 4.2, by choosing the matrix $\boldsymbol{U}$ in the form

$$\boldsymbol{U} = \begin{bmatrix} 2^{-1/2} & 2^{-1/2} & 0 \\ 0 & 0 & 1 \\ 2^{-1/2} & -2^{-1/2} & 0 \end{bmatrix}, \tag{4-86}$$

the matrix $\boldsymbol{G}^{RR}$ is transformed into $\boldsymbol{G}^{SS}$, which consists of two- and one-dimensional diagonal blocks in the form

$$\boldsymbol{G}^{SS} = \begin{bmatrix} G_{rr} + g_{rr} & 2^{1/2} g_{r\theta} & 0 \\ 2^{1/2} g_{r\theta} & G_{\theta\theta} & 0 \\ 0 & 0 & G_{rr} - g_{rr} \end{bmatrix}. \tag{4-87}$$

The matrix $\boldsymbol{F}^{SS}$ is factored in the same way, so that the calculation from Eq. (4-26) to Eq (4-30) may be performed for each pair of submatrices of $\boldsymbol{G}^{SS}$ and $\boldsymbol{F}^{SS}$.

The transformation given by Eq. (4-77) can also be used to generate a maximum number of independent coordinates from a set of mutually dependent internal coordinates. If the relation

$$f(R_1, R_2, \cdots) = 0$$

holds among the internal variables R_1, R_2, $\cdots$, the corresponding internal coordinates, ΔR_1, ΔR_2, $\cdots$, satisfy the relation derived from the expansion of $f(R_1, R_2, \cdots)$ in the form

$$\Delta f = \sum_i \left(\frac{\partial f}{\partial R_i}\right)_0 \Delta R_i + \tfrac{1}{2}\sum_i\sum_j \left(\frac{\partial^2 f}{\partial R_i \partial R_j}\right)_0 \Delta R_i \Delta R_j + \cdots = 0. \tag{4-88}$$

The quantity obtained by terminating the expansion in the right side of Eq. (4-88) at the linear terms and normalizing the coefficients in the form

$$\Delta R_r = \sum_i f_i \Delta R_i, \quad \left[f_i = (\partial f/\partial R_i)_0 \Big/ \left\{ \sum_j (\partial f/\partial R_j)_0^{\,2} \right\} \right] \tag{4-89}$$

is called the redundant coordinate formed from the internal coordinates ΔR_1, ΔR_2, $\cdots$ based on the relation in Eq. (4-88). A redundant coordinate is identically zero within the framework of the first order approximation. Accordingly, substituting the ith row of Eq. (3-13) into Eq. (4-89), we have

$$\Delta R_r = \sum_i f_i \sum_j B_{ij}\, \Delta x_j = \sum_j \left(\sum_i f_i\, B_{ij} \right) \Delta x_j = 0. \tag{4-90}$$

In order for the redundant coordinate in Eq. (4-90) to vanish for any arbitrary choice of Δx_j's, the coefficient of the individual Δx_j should vanish, *i.e.,* it should hold that

$$\sum_i f_i\, B_{ij} = 0. \tag{4-91}$$

Suppose that a component ΔS_p of the vector $\boldsymbol{\Delta S}$, which is constructed from a set of mutually dependent internal coordinates, is chosen to be the redundant coordinate ΔR_r with the coefficients $U_{pi} = f_i$. It then follows from Eqs. (4-81), (3-42) and (4-91) that the elements in the pth row of the matrix $\boldsymbol{G}^{SS}$ vanish as shown by

$$G_{pq}^{SS} = \sum_i \sum_j U_{pi} U_{qj} G_{ij}^{RR} = \sum_i \sum_j U_{pi} U_{qj} \sum_k B_{ik} B_{jk} m_k^{-1}$$

$$= \sum_k \left(\sum_i U_{pi} B_{ik} \right) \left(\sum_j U_{qj} B_{jk} \right) m_k^{-1} = 0.$$

The elements in the pth column vanish in the same way. Thus the dimension of the matrix $\boldsymbol{G}^{SS}$ is made smaller than that of $\boldsymbol{G}^{RR}$ by one. The elements of the matrix $\boldsymbol{F}^{SS}$ in the rows and columns corresponding to any redundant coordinates may have certain finite values when $\boldsymbol{F}^{SS}$ is formed as the Hessian matrix of a model potential function. In this case the calculation is performed simply by dropping the columns and rows of $\boldsymbol{F}^{SS}$ corresponding to the redundant coordinates.

4.6 Molecular Point Groups

Generally, the internal symmetry coordinates of a molecule are constructed according to the prescription based on the theory of molecular point groups. The collection of all the symmetry operations of a molecule including the identity operation forms a group. A molecule usually has many types of similar physical quantities representing certain properties or local structures. They are, for example, a set of components of a vector or a tensor quantity such as the dipole moment or polarizability, a set of the coordinates of equivalent atoms or a set of equivalent internal variables such as bond lengths, bond angles and torsion angles. If the members of such a set of similar quantities undergo a linear transformation among themselves on a symmetry operation of the molecule, the matrix of the linear transformation is called the representation, and the set of similar quantities is called the basis of the representation. A set of equivalent internal coordinates, which are the variations of the structure parameters, may also be used as the basis of the representations.

Suppose that the vectors of two bases $\boldsymbol{R}$ and $\boldsymbol{S}$, the internal coordinates and the internal symmetry coordinates, for example, are related to each other by a linear transformation in the form

$$\boldsymbol{S} = \boldsymbol{U}\boldsymbol{R}. \tag{4-92}$$

Now the basis vectors $\boldsymbol{R}$ and $\boldsymbol{S}$ are assumed to be transformed into $\boldsymbol{R}'$ and $\boldsymbol{S}'$, respectively, by the operation A. On writing the representations of operation A on the bases $\boldsymbol{R}$ and $\boldsymbol{S}$ as $\boldsymbol{D}_{AR}$ and $\boldsymbol{D}_{AS}$, respectively, the basis vectors after operation A are given in terms of those before the operation in the forms

$$\boldsymbol{R}' = \boldsymbol{D}_{AR}\,\boldsymbol{R} \tag{4-93a}$$

and

$$\boldsymbol{S}' = \boldsymbol{D}_{AS}\,\boldsymbol{S}. \tag{4-93b}$$

Since the transformation from $\boldsymbol{R}'$ to $\boldsymbol{S}'$ should be in the same form as that from $\boldsymbol{R}$ to $\boldsymbol{S}$, it holds that

$$\boldsymbol{S}' = \boldsymbol{U}\boldsymbol{R}'. \tag{4-94}$$

Substituting Eqs. (4-93a,b) into Eq. (4-94) and using Eq. (4-92), we have $\boldsymbol{D}_{AS}\boldsymbol{U}\boldsymbol{R} = \boldsymbol{U}\boldsymbol{D}_{AR}\,\boldsymbol{R}$, so that the relation

$$\boldsymbol{D}_{AS} = \boldsymbol{U}\boldsymbol{D}_{AR} \tag{4-95}$$

should hold. Postmultiplying Eq. (4-95) by $\boldsymbol{U}^{-1}$ shows that the two representations $\boldsymbol{D}_{AS}$ and $\boldsymbol{D}_{AR}$ are related to each other by the similarity transformation in the form

$$\boldsymbol{D}_{AS} = \boldsymbol{U}\boldsymbol{D}_{AR}\boldsymbol{U}^{-1}. \tag{4-96}$$

Once a basis $\boldsymbol{R}$ is given, it is sometimes possible to choose such a matrix $\boldsymbol{U}$ for generating a new basis $\boldsymbol{S}$ with Eq. (4-92) that the new representation $\boldsymbol{D}_{AS}$ constructed for each operation according to Eq. (4-96) can be expressed as a direct sum of smaller matrices than $\boldsymbol{D}_{AR}$. This process is called a reduction of the representation $\boldsymbol{D}_{AR}$, and a representation for which any reduction is possible is called a reducible representation. On the other hand, a representation which cannot be reduced further is called an irreducible representation. When a reducible representation is reduced repeatedly until no further reduction is possible, the resulting representation should be the direct sum of a certain number of irreducible representations. Obviously, all one-dimensional representations are irreducible. The irreducible representation is an indispensable concept in the theory of molecular structure. It will be shown in the following that a normal coordinate can be used as the basis of an irreducible representation of the molecule.

A complete set of normal coordinates receives a transformation by a symmetry operation. Let the normal coordinates after a symmetry operation be denoted in the vector form by

$$\boldsymbol{Q}' = [Q'_1 \quad Q'_2 \quad \cdots\ ]^{\mathrm{T}}.$$

Since both the kinetic energy T and the potential energy V of the molecule are left invariant on any symmetry operation, it should hold that

$$T = \tfrac{1}{2}\sum_i \dot{Q}_i^2 = \tfrac{1}{2}\sum_i \dot{Q}_i'^2 \tag{4-97}$$

and

$$V = \tfrac{1}{2}\sum_i \lambda_i Q_i^2 = \tfrac{1}{2}\sum_i \lambda_i \dot{Q}_i'^2. \tag{4-98}$$

Let the transformation from Q_1, Q_2, $\cdots$ to Q'_1, Q'_2, $\cdots$ be denoted by the vector notation in the form

$$Q' = D_{AQ}\, Q. \tag{4-99}$$

In order for Eq. (4-97) to be satisfied, it should hold that

$$2T = \tilde{\dot{Q}}'\dot{Q}' = \tilde{\dot{Q}}\tilde{D}_{AQ} D_{AQ}\dot{Q} = \tilde{\dot{Q}}\dot{Q}$$

whence we must have

$$\tilde{D}_{AQ} D_{AQ} = E, \quad \therefore \quad \tilde{D}_{AQ} = D_{AQ}^{-1}.$$

That is, the transformation matrix D_{AQ} must be an orthogonal matrix. An additional condition is necessary in order for D_{AQ} to satisfy Eq. (4-98). In the case where the molecule has no degeneracy and the λ_i's are all different from one another, the relation

$$Q'^2_i = Q_i^2 \tag{4-100}$$

must hold for each i so that the potential function V satisfies Eq. (4-98) irrespective of the values of individual normal coordinates. The normal coordinates, which are generated from the Cartesian displacements of nuclei by a linear transformation with real coefficients, are real variables. Then Eq. (4-100) requires that

$$Q'_i = \pm Q_i. \tag{4-101}$$

According to Eq. (4-101), a non-degenerate normal coordinate is either left invariant or changes only in the sign on any symmetry operation of the molecule. On regarding Q_i and Q'_i as one-dimensional vectors and comparing Eq. (4-101) with Eq. (4-99), the associated irreducible representation D_{AQ} is written as $D_{AQ} = [\pm 1]$. This result means that a non-degenerate normal coordinate is the basis of a one-dimensional irreducible representation with the members [1] and/or [–1].

For any point group involving neither rotation axes nor rotation–reflection axes of more than twofold, a repetition of any symmetry operation twice leads necessarily to the identity operation. Correspondingly, each matrix of the representation multiplied by itself must be a unit matrix. This condition is satisfied by both one-dimensional matrices [1] and [–1].

Suppose that a reducible representation D_{AR} is reduced to the direct sum of a few one-dimensional irreducible representations D_{AS} by a similarity transformation in the form

$$D_{AS} = U\, D_{AR}\, U^{-1}.$$

Since D_{AS} is diagonal by the presumption, it holds obviously that $D_{AS} = \tilde{D}_{AS}$. The orthogonality of the irreducible representation requires also that $\tilde{D}_{AS} = D_{AS}^{-1}$. Accordingly, we have

$$D_{AS}\tilde{D}_{AS} = D_{AS}^2 = (U\, D_{AR}\, U^{-1})(U\, D_{AR}\, U^{-1}) = U\, D_{AR}^2\, U^{-1} = E$$

$$\therefore \quad D_{AR}^2 = E \quad \text{or} \quad D_{AR}^{-1} = \tilde{D}_{AR} = D_{AR}.$$

Thus the matrix D_{AR} must be symmetric. Conversely, if D_{AR} is not symmetric, D_{AS} cannot be symmetric either.

In cases involving degenerate vibrations, the condition required for Eq. (4-98) to hold is somewhat relaxed. For example, if

$$\lambda_j = \lambda_{j+1} = \cdots = \lambda_{j+k-1}, \tag{4-102}$$

the part of the sum corresponding to $i = j$ through $i = j + k - 1$ in Eq. (4-98) can be written as

$$\sum_{i=j}^{j+k-1} \lambda_i Q_i^2 = \lambda_j \sum_{i=j}^{j+k-1} Q_i^2.$$

Accordingly we need a single condition,

$$\sum_{i=j}^{j+k-1} Q_i^2 = \sum_{i=j}^{j+k-1} Q_i'^2, \tag{4-103}$$

instead of k conditions in the form of Eq. (4-100) for $i = j$ through $i = j + k - 1$. If there are k eigenvalues equal to one another as given in Eq. (4-102), the corresponding normal frequencies ω_j through ω_{j+k-1} must also be equal to one another. The k normal coordinates $Q_j \cdots Q_{j+k-1}$ then form a basis of the k-dimensional irreducible representation of the point group of the molecule.

A polyatomic molecule having a single n-fold axis ($n \geq 3$) must have at least one set of n equivalent internal coordinates around the n-fold axis. That such a molecule has doubly degenerate vibrations can be verified as follows. Let a set of n equivalent internal coordinates around the n-fold axis be given in the vector form as

$$\boldsymbol{\Delta R} = [\Delta R_1 \quad \Delta R_2 \quad \cdots \quad \Delta R_n]^{\mathrm{T}}.$$

The rotation of the molecule around the n-fold axis by an angle of $\phi = 2\pi/n$ causes the transformation of this set in the form

$$\begin{bmatrix} \Delta R_1' \\ \Delta R_2' \\ \Delta R_3' \\ \vdots \\ \Delta R_n' \end{bmatrix} = \begin{bmatrix} \Delta R_2 \\ \Delta R_3 \\ \vdots \\ \Delta R_n \\ \Delta R_1 \end{bmatrix} = \begin{bmatrix} 0 & 1 & 0 & \cdots & 0 \\ 0 & 0 & 1 & \cdots & 0 \\ \vdots & \vdots & \vdots & \ddots & \vdots \\ 0 & 0 & 0 & \cdots & 1 \\ 1 & 0 & 0 & \cdots & 0 \end{bmatrix} \begin{bmatrix} \Delta R_1 \\ \Delta R_2 \\ \Delta R_3 \\ \vdots \\ \Delta R_n \end{bmatrix}. \tag{4-104}$$

Let the $(n \times n)$ representation in Eq. (4-104) be denoted by $\boldsymbol{C}_n$. The representations for the rotations around the n-fold axis by the angles 2ϕ, 3ϕ, $\cdots$, $n\phi$ $(= 2\pi)$ are then obtained from successive multiplications of matrices as $\boldsymbol{C}_n^2$, $\boldsymbol{C}_n^3$, $\cdots$, $\boldsymbol{C}_n^n$ $(= \boldsymbol{E})$, respectively. These representations except for $\boldsymbol{C}_n^{n/2}$, which appears only if n is even, and $\boldsymbol{C}_n^n$ $(= \boldsymbol{E})$ are not symmetric, and therefore cannot be reduced into any diagonal and thereby symmetric matrices which are direct sums of one-dimensional irreducible representations.

On the other hand, there can be a two-dimensional real orthogonal matrix the nth power of which is the two-dimensional unit matrix. Generally, there are two types of two-dimensional real orthogonal matrices expressible in the forms

$$\boldsymbol{D}_\theta = \begin{bmatrix} \cos\theta & \sin\theta \\ -\sin\theta & \cos\theta \end{bmatrix}, \quad \boldsymbol{D}_\theta' = \begin{bmatrix} \cos\theta & \sin\theta \\ \sin\theta & -\cos\theta \end{bmatrix}. \tag{4-105}$$

Of these two matrices, $\boldsymbol{D}_\theta'$ is symmetric and it holds that $\boldsymbol{D}_\theta'^{\,2} = \boldsymbol{E}$ irrespective of the value of θ. Accordingly, the representation $\boldsymbol{D}_\theta'$ cannot be generated by any reduction of non-symmetric reducible representations. On the other hand, $\boldsymbol{D}_\theta$ can be so chosen that the relation $\boldsymbol{D}_\theta^n = \boldsymbol{E}$ holds by taking θ as

$$\phi = 2\pi/n, \quad \theta = k\phi = 2k\pi/n \tag{4-106}$$

and thus satisfies the condition required for a representation of C_n. In fact, the reduction from $\boldsymbol{C}_n$ to $\boldsymbol{D}_{k\phi}$ can be accomplished by introducing a pair of new variables ΔS_{ka} and ΔS_{kb} as the linear combinations of ΔR_1, ΔR_2, $\cdots$, ΔR_n in the forms

$$\Delta S_{ka} = \sum_{i=1}^{n} a_{ki}\, \Delta R_i, \quad \Delta S_{kb} = \sum_{i=1}^{n} b_{ki}\, \Delta R_i, \tag{4-107}$$

where the coefficients a_{ki} and b_{ki} are chosen to be

$$a_{ki} = N_k \cos k(i-1)\phi = N_k \cos 2\pi\, k(i-1)/n \tag{4-108a}$$

$$b_{ki} = N_k \sin k(i-1)\phi = N_k \sin 2\pi\, k(i-1)/n. \tag{4-108b}$$

The normalization coefficient N_k is so determined as to make the product $\boldsymbol{U}\tilde{\boldsymbol{U}}$ a unit matrix. Expressing the bases $(\Delta S_{ka}, \Delta S_{kb})$ in the vector form

$$\boldsymbol{\Delta S}_k = [\Delta S_{ka} \quad \Delta S_{kb}]^{\mathrm{T}} \tag{4-109a}$$

and introducing the $(2 \times n)$ matrix of the coefficients (a_{ki}, b_{ki})

$$\boldsymbol{U}_k = \begin{bmatrix} a_{k1} & a_{k2} & \cdots & a_{kn} \\ b_{k1} & b_{k2} & \cdots & b_{kn} \end{bmatrix}, \tag{4-109b}$$

we can rewrite Eq. (4-107) as

$$\boldsymbol{\Delta S}_k = \boldsymbol{U}_k \boldsymbol{\Delta R} \tag{4-110}$$

If the basis $\boldsymbol{\Delta S}_k$ is transformed into $\boldsymbol{\Delta S}_k' = [\Delta S_{ka}' \quad \Delta S_{kb}']^{\mathrm{T}}$ by the operation C_n, it holds that

$$\boldsymbol{\Delta S}_k' = \boldsymbol{D}_{k\phi}\, \boldsymbol{\Delta S}_k = \boldsymbol{D}_{k\phi}\, \boldsymbol{U}_k \boldsymbol{\Delta R}. \tag{4-111}$$

On the other hand, since $\boldsymbol{\Delta R}'$ is transformed into $\boldsymbol{\Delta S}_k'$ by the same relation as Eq. (4-110), and $\boldsymbol{\Delta R}'$ and $\boldsymbol{\Delta R}$ are related to each other by Eq. (4-104), we have

$$\boldsymbol{\Delta S}_k' = \boldsymbol{U}_k \boldsymbol{\Delta R}' = \boldsymbol{U}_k\, \boldsymbol{C}_n \boldsymbol{\Delta R}. \tag{4-112}$$

Since the transformation given by Eq. (4-111) is the same as that given by Eq. (4-112), the matrices of the coefficients in the right sides must be equal to each other. Hence it must hold that

$$\boldsymbol{D}_{k\phi}\, \boldsymbol{U}_k = \boldsymbol{U}_k\, \boldsymbol{C}_n. \tag{4-113}$$

In fact, by substituting Eqs. (4-108a,b) into the elements in the $(p+1)$th column $(0 < p < n)$ of Eq. (4-113) and using the addition theorems of sines and cosines, it is shown that

$$a_{k,p+1} \cos k\phi + b_{k,p+1} \sin k\phi = N_k \cos(p-1)k\phi = a_{kp}$$

$$-a_{k,p+1}\sin k\phi + b_{k,p+1}\cos k\phi = N_k \sin(p-1)k\phi = b_{kp}\ ,$$

which confirms Eq. (4-113).

The number of distinguishable coordinate pairs (ΔS_{ka}, ΔS_{kb}) possible for a given n is limited. If $k > n$, obviously the same coordinate pair is obtained as in the case where $k' = \mathrm{mod}(k, n) < n$ is taken. If $n > k > n/2$, putting $k' = n - k$ gives

$$\cos(n-k)p\phi = \cos kp\phi, \quad \therefore \quad \Delta S_{k'a} = \Delta S_{ka}$$

$$\sin(n-k)p\phi = -\sin kp\phi, \quad \therefore \quad \Delta S_{k'b} = -\Delta S_{kb}$$

so that $\Delta S_{k'b}$ and ΔS_{kb} represent the same variable as each other except for the sign. The case $k = n$, which is the same as the case $k = 0$, gives

$$a_{0p} = \cos 2p\pi = 1, \quad b_{0p} = \sin 2p\pi = 0$$

for all p, so that the coordinate ΔS_{nb} vanishes identically in this case. Similarly, in the case $k = n/2$ for an even n, we have

$$a_{n/2,p} = \cos p\pi = (-1)^p, \quad b_{n/2,p} = \sin p\pi = 0.$$

In either case, the coordinate ΔS_{kp} vanishes and the remaining ΔS_{ka} forms the basis of a one-dimensional irreducible representation. Thus, it is meaningless to consider k greater than or equal to $n/2$, so that the number of possible coordinate pairs must be the maximum integer m not greater than $(n-1)/2$.

The numbers of the one- and two-dimensional irreducible representations formed by the above reduction are as follows. When n is even, we have two one-dimensional representations and $(n-2)/2$ two-dimensional representations. The number of basis coordinates is therefore $1 \times 2 + 2 \times (n-2)/2 = n$. When n is odd, we have a one-dimensional representation and $(n-1)/2$ two-dimensional representations. The number of basis coordinates is $1 \times 1 + 2 \times (n-1)/2 = n$, again. In either case, the number of coordinates to be used as the basis of the irreducible representations coincides with the number of the original internal coordinates.

Generally, the variables formed as such linear combinations of internal coordinates as to be usable as the basis of the irreducible representations of the molecule are called the internal symmetry coordinates. Let $\Delta\boldsymbol{S}$ be the vector of n internal symmetry coordinates generated through the reduction of the reducible representation formed by using the internal coordinates ΔR_1, ΔR_2, $\cdots$, ΔR_n. The n-dimensional vector $\Delta\boldsymbol{S}$ can be expressed as the collection of the vectors $\Delta\boldsymbol{S}_k$ in Eq. (4-110) in the form

$$\Delta\boldsymbol{S} = \begin{bmatrix}\Delta S_0 & \Delta\tilde{\boldsymbol{S}}_1 & \Delta\tilde{\boldsymbol{S}}_2 & \cdots & \Delta\tilde{\boldsymbol{S}}_m & \Delta S_{n/2}\end{bmatrix}^{\mathrm{T}} \quad : \quad m = (n-2)/2$$

if n is even, and

$$\Delta\boldsymbol{S} = \begin{bmatrix}\Delta S_0 & \Delta\tilde{\boldsymbol{S}}_1 & \Delta\tilde{\boldsymbol{S}}_2 & \cdots & \Delta\tilde{\boldsymbol{S}}_m\end{bmatrix}^{\mathrm{T}} \quad : \quad m = (n-1)/2$$

if n is odd.

Then the transformation from $\Delta\boldsymbol{R}$ to $\Delta\boldsymbol{S}$ is written in the matrix notation as

$$\Delta\boldsymbol{S} = \boldsymbol{U}\Delta\boldsymbol{R}.$$

The matrix $\boldsymbol{U}$ is given by referring to Eq. (4-109) as

$$\boldsymbol{U} = \begin{bmatrix} \boldsymbol{U}_0 & \tilde{\boldsymbol{U}}_1 & \tilde{\boldsymbol{U}}_2 & \cdots & \tilde{\boldsymbol{U}}_L \end{bmatrix}^{\mathrm{T}}, \tag{4-114}$$

where

$$L = n/2 \quad (n : \text{even}), \qquad L = (n-1)/2 \quad (n : \text{odd})$$

The orthogonality of the transformation matrix $\boldsymbol{U}$ is confirmed by postmultiplying $\boldsymbol{U}$ by its transpose and calculating each block $\boldsymbol{U}_j \tilde{\boldsymbol{U}}_k$ of the product

$$\boldsymbol{U}\tilde{\boldsymbol{U}} = \begin{bmatrix} \boldsymbol{U}_0\tilde{\boldsymbol{U}}_0 & \boldsymbol{U}_0\tilde{\boldsymbol{U}}_1 & \boldsymbol{U}_0\tilde{\boldsymbol{U}}_2 & \cdots & \boldsymbol{U}_0\tilde{\boldsymbol{U}}_L \\ \boldsymbol{U}_1\tilde{\boldsymbol{U}}_0 & \boldsymbol{U}_1\tilde{\boldsymbol{U}}_1 & \boldsymbol{U}_1\tilde{\boldsymbol{U}}_2 & \cdots & \boldsymbol{U}_1\tilde{\boldsymbol{U}}_L \\ \boldsymbol{U}_2\tilde{\boldsymbol{U}}_0 & \boldsymbol{U}_2\tilde{\boldsymbol{U}}_1 & \boldsymbol{U}_2\tilde{\boldsymbol{U}}_2 & \cdots & \boldsymbol{U}_2\tilde{\boldsymbol{U}}_L \\ \vdots & \vdots & \vdots & \ddots & \vdots \\ \boldsymbol{U}_L\tilde{\boldsymbol{U}}_0 & \boldsymbol{U}_L\tilde{\boldsymbol{U}}_1 & \boldsymbol{U}_L\tilde{\boldsymbol{U}}_2 & \cdots & \boldsymbol{U}_L\tilde{\boldsymbol{U}}_L \end{bmatrix}. \tag{4-115}$$

Suppose a regular $n$gon inscribed by a unit circle centered at the origin of a Cartesian coordinate system, and let one of its vertices be at the point (1,0) on the X-axis as shown in Fig. 4.3. The sum of n unit vectors directed from the center to the n vertices is obviously a $\mathbf{0}$ vector. The X- and the Y-components of this sum are expressed, by using such an integer K that

$$1 \leq K \leq n-1, \tag{4-116}$$

in the forms

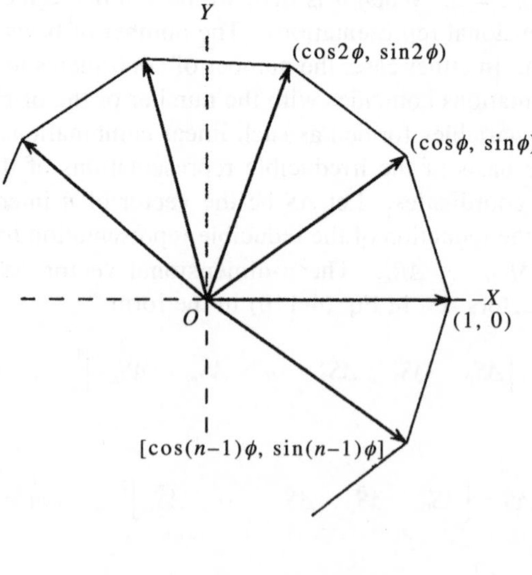

Fig. 4.3 Radial vectors pointing to the vertices of a regular polygon inscribed in the unit circle centered at the origin.

$$\sum_{p=0}^{n-1} \cos pK\phi = 0 \quad \text{and} \quad \sum_{p=0}^{n-1} \sin pK\phi = 0, \tag{4-117}$$

respectively. Now let the integer K be rewritten as

$$K = |j \pm k| \tag{4-118}$$

and substitute it into Eq. (4-117). Then we have, with the help of the addition theorems for sines

$$\sum_{p=0}^{n-1} \sin p(j \pm k)\phi = \sum_{p=0}^{n-1} \sin pj\phi \cos pk\phi \pm \sum_{p=0}^{n-1} \cos pj\phi \sin pk\phi = 0$$

and for cosines

$$\sum_{p=0}^{n-1} \cos p(j+k)\phi = \sum_{p=0}^{n-1} \cos pj\phi \cos pk\phi - \sum_{p=0}^{n-1} \sin pj\phi \sin pk\phi = 0$$

$$\sum_{p=0}^{n-1} \cos p(j-k)\phi = \sum_{p=0}^{n-1} \cos pj\phi \cos pk\phi + \sum_{p=0}^{n-1} \sin pj\phi \sin pk\phi = \delta_{jk}\, n.$$

By taking the sums and the differences of these formulae, it follows that

$$\sum_{p=0}^{n-1} \sin pj\phi \cos pk\phi = 0, \quad \sum_{p=0}^{n-1} \cos pj\phi \sin pk\phi = 0 \tag{4-119}$$

$$\sum_{p=0}^{n-1} \cos pj\phi \cos pk\phi = \delta_{jk} \frac{n}{2}, \quad \sum_{p=0}^{n-1} \sin pj\phi \sin pk\phi = \delta_{jk} \frac{n}{2}. \tag{4-120}$$

Since j and k satisfy the conditions $1 \le j, k \le (n-1)/2$, K satisfies Eq. (4-116). Accordingly, by taking the normalization coefficient $N_k = (2/n)^{1/2}$, we have

$$\boldsymbol{U}_j \tilde{\boldsymbol{U}}_k = \begin{bmatrix} \sum_{p=0}^{n-1} \cos pj\phi \cos pk\phi & \sum_{p=0}^{n-1} \cos pj\phi \sin pk\phi \\ \sum_{p=0}^{n-1} \sin pj\phi \cos pk\phi & \sum_{p=0}^{n-1} \sin pj\phi \sin pk\phi \end{bmatrix} = \delta_{jk}\boldsymbol{E}. \tag{4-121}$$

In the cases where j and/or k are 0 or n(even)/2, the matrix $\boldsymbol{U}_j \tilde{\boldsymbol{U}}_k$ is expressed by dropping the columns or the rows from Eq. (4-121). Note, however, that we must put $N_0 = N_{n/2} = n^{-1/2}$ in order for the conditions

$$\boldsymbol{U}_0 \tilde{\boldsymbol{U}}_0 = \boldsymbol{E}, \quad \boldsymbol{U}_{n/2} \tilde{\boldsymbol{U}}_{n/2} = \boldsymbol{E}$$

to hold. Premultiplying both sides of Eq. (4-113) by $\tilde{\boldsymbol{U}}_{k'}$ and making use of Eq. (4-121), we have

$$\boldsymbol{D}_{k\phi} \boldsymbol{U}_k \tilde{\boldsymbol{U}}_{k'} = \boldsymbol{U}_k \boldsymbol{C}_n \tilde{\boldsymbol{U}}_{k'} = \delta_{kk'} \boldsymbol{D}_{k\phi}. \tag{4-122}$$

On the similarity transformation of the reducible representation $\boldsymbol{C}_n$ by the orthogonal matrix $\boldsymbol{U}$, all the off-diagonal submatrices vanish according to Eq. (4-122), and the irreducible representations are obtained as shown by

$$\boldsymbol{U}\boldsymbol{C}_n\tilde{\boldsymbol{U}} = \begin{bmatrix} \boldsymbol{U}_0\boldsymbol{C}_n\tilde{\boldsymbol{U}}_0 & \boldsymbol{U}_0\boldsymbol{C}_n\tilde{\boldsymbol{U}}_1 & \boldsymbol{U}_0\boldsymbol{C}_n\tilde{\boldsymbol{U}}_2 & \cdots & \boldsymbol{U}_0\boldsymbol{C}_n\tilde{\boldsymbol{U}}_L \\ \boldsymbol{U}_1\boldsymbol{C}_n\tilde{\boldsymbol{U}}_0 & \boldsymbol{U}_1\boldsymbol{C}_n\tilde{\boldsymbol{U}}_1 & \boldsymbol{U}_1\boldsymbol{C}_n\tilde{\boldsymbol{U}}_2 & \cdots & \boldsymbol{U}_1\boldsymbol{C}_n\tilde{\boldsymbol{U}}_L \\ \boldsymbol{U}_2\boldsymbol{C}_n\tilde{\boldsymbol{U}}_0 & \boldsymbol{U}_2\boldsymbol{C}_n\tilde{\boldsymbol{U}}_1 & \boldsymbol{U}_2\boldsymbol{C}_n\tilde{\boldsymbol{U}}_2 & \cdots & \boldsymbol{U}_2\boldsymbol{C}_n\tilde{\boldsymbol{U}}_L \\ \vdots & \vdots & \vdots & \ddots & \vdots \\ \boldsymbol{U}_L\boldsymbol{C}_n\tilde{\boldsymbol{U}}_0 & \boldsymbol{U}_L\boldsymbol{C}_n\tilde{\boldsymbol{U}}_1 & \boldsymbol{U}_L\boldsymbol{C}_n\tilde{\boldsymbol{U}}_2 & \cdots & \boldsymbol{U}_L\boldsymbol{C}_n\tilde{\boldsymbol{U}}_L \end{bmatrix}$$

$$= \begin{bmatrix} \boldsymbol{D}_{0\phi} & \boldsymbol{0} & \boldsymbol{0} & \cdots & \boldsymbol{0} \\ \boldsymbol{0} & \boldsymbol{D}_{1\phi} & \boldsymbol{0} & \cdots & \boldsymbol{0} \\ \boldsymbol{0} & \boldsymbol{0} & \boldsymbol{D}_{2\phi} & \cdots & \boldsymbol{0} \\ \vdots & \vdots & \vdots & \ddots & \vdots \\ \boldsymbol{0} & \boldsymbol{0} & \boldsymbol{0} & \cdots & \boldsymbol{D}_{L\phi} \end{bmatrix} \equiv \boldsymbol{D}_{0\phi} \dotplus \boldsymbol{D}_{1\phi} \dotplus \cdots \dotplus \boldsymbol{D}_{L\phi}, \tag{4-123}$$

where the symbol $\dotplus$ indicates the direct sum of matrices. For $\boldsymbol{C}_n{}^p$, it follows from

$$\boldsymbol{U}\boldsymbol{C}_n{}^p\tilde{\boldsymbol{U}} = (\boldsymbol{U}\boldsymbol{C}_n\tilde{\boldsymbol{U}})(\boldsymbol{U}\boldsymbol{C}_n\tilde{\boldsymbol{U}})\cdots(\boldsymbol{U}\boldsymbol{C}_n\tilde{\boldsymbol{U}}) = (\boldsymbol{U}\boldsymbol{C}_n\tilde{\boldsymbol{U}})^p$$

that

$$\boldsymbol{U}\boldsymbol{C}_n{}^p\tilde{\boldsymbol{U}} = \boldsymbol{D}_{0\phi}{}^p \dotplus \boldsymbol{D}_{1\phi}{}^p \dotplus \boldsymbol{D}_{2\phi}{}^p \dotplus \cdots \dotplus \boldsymbol{D}_{L\phi}{}^p.$$

For $\boldsymbol{D}_{k\phi}{}^p$, the relation $\tilde{\boldsymbol{D}}_{k\phi}{}^p = \boldsymbol{D}_{k\phi}{}^{n-p}$ follows from the definition in Eq. (4-105), whence we obtain

$$\boldsymbol{D}_{k\phi}{}^p + \boldsymbol{D}_{k\phi}{}^{n-p} = \begin{bmatrix} \cos pk\phi & \sin pk\phi \\ -\sin pk\phi & \cos pk\phi \end{bmatrix} + \begin{bmatrix} \cos pk\phi & -\sin pk\phi \\ \sin pk\phi & \cos pk\phi \end{bmatrix} = \begin{bmatrix} 2\cos pk\phi & 0 \\ 0 & 2\cos pk\phi \end{bmatrix}. \tag{4-124}$$

The $\boldsymbol{F}$ and $\boldsymbol{G}$ matrices of a molecule having an n-fold axis contain $(n \times n)$ submatrices of the form

$$A_0\boldsymbol{E} + \sum_{p=1}^{n-1} A_p\boldsymbol{C}_n{}^p. \tag{4-125}$$

If the molecule has n planes of symmetry containing the n-fold axis and/or n twofold axes perpendicular to the n-fold axis, we can find a symmetry operation by which a pair of internal coordinates associated with the coefficient A_n is exchanged with a coordinate pair associated with A_{n-p}. Then the coefficients A_n and A_{n-p} must be equal to each other in order for the kinetic and the potential energies to be invariant under the symmetry operation. In this case, Eq. (4-125) may be rewritten in the form

$$A_0\boldsymbol{E} + \sum_{p=1}^{n/p} A_p(\boldsymbol{C}_n{}^p + \boldsymbol{C}_n{}^{n-p}). \tag{4-126}$$

According to Eq. (4-124), such a submatrix is diagonalized by the similarity transformation on changing the variables from $\Delta\boldsymbol{R}$ to $\Delta\boldsymbol{S}$. Of the submatrices after the transformation, $\boldsymbol{D}_{0\phi}$ and $\boldsymbol{D}_{n/2,\phi}$ are one-dimensional, the latter appearing only if n is even. All the other submatrices are two-dimensional and consist of two equal diagonal elements, which give rise to doubly degenerate eigenvalues. The point groups D_n, D_{nd}, D_{nh}, and C_{nv} contain n elements derived from an n-fold axis of rotation. In addition, the first three point groups have elements derived from the two-fold axis C'_2 and/or C''_2, which are perpendicular to the n-fold axis, while the last has elements derived from symmetry planes σ_v and/or σ_d, which contain the n-fold axis. Accordingly, all the $(n \times n)$ submatrices in the $\boldsymbol{F}$ and the $\boldsymbol{G}$ matrices of the molecules belonging to these point groups can be written in the form of Eq. (4-126) and, on classifying the internal symmetry coordinates into S_a and S_b, the cross submatrices between the two classes vanish completely.

On the other hand, for the point groups C_n and C_{nh} ($n \geq 3$), the latter having a plane of symmetry σ_h perpendicular to the n-fold axis, the $\boldsymbol{F}$ and $\boldsymbol{G}$ matrices for the internal coordinates contain certain submatrices only expressible by Eq. (4-125). In this case, the similarity transformation by $\boldsymbol{U}_k$ leaves the off-diagonal terms of the two-dimensional irreducible representation $\boldsymbol{D}_{k\phi}{}^p$ uneliminated. Since it hold from Eq. (4-105) that

$$\left(D_{k\phi}{}^{p}\right)_{12} = \sin pk\phi = -\left(D_{k\phi}{}^{p}\right)_{21},$$

the transformed matrices $\boldsymbol{G}^S$ and $\boldsymbol{F}^S$ contains those off-diagonal blocks between the coordinate sets ΔS_a and ΔS_b which satisfy the relations $\boldsymbol{G}^S_{ab} = -\boldsymbol{G}^S_{ba}$ and $\boldsymbol{F}^S_{ab} = -\boldsymbol{F}^S_{ba}$, respectively. Hence the transformed matrices are written in the forms

$$\boldsymbol{G}^S = \begin{bmatrix} \boldsymbol{G}_1^S & \tilde{\boldsymbol{G}}_2^S \\ \boldsymbol{G}_2^S & \boldsymbol{G}_1^S \end{bmatrix} = \begin{bmatrix} \boldsymbol{G}_1^S & -\boldsymbol{G}_2^S \\ \boldsymbol{G}_2^S & \boldsymbol{G}_1^S \end{bmatrix} \tag{4-127a}$$

and

$$\boldsymbol{F}^S = \begin{bmatrix} \boldsymbol{F}_1^S & \tilde{\boldsymbol{F}}_2^S \\ \boldsymbol{F}_2^S & \boldsymbol{F}_1^S \end{bmatrix} = \begin{bmatrix} \boldsymbol{F}_1^S & -\boldsymbol{F}_2^S \\ \boldsymbol{F}_2^S & \boldsymbol{F}_1^S \end{bmatrix}. \tag{4-127b}$$

Diagonalizing a real symmetric matrix $\boldsymbol{G}^S$ in Eq. (4-127a) is equivalent to diagonalizing a complex Hermitian matrix $\boldsymbol{G}_1{}^S + i\boldsymbol{G}_2{}^S$ by a unitary transformation in the form

$$\left(\tilde{\boldsymbol{L}}^S_{G1} - i\,\tilde{\boldsymbol{L}}^S_{G2}\right)\left(\boldsymbol{G}^S_1 + i\,\boldsymbol{G}^S_2\right)\left(\boldsymbol{L}^S_{G1} + i\,\boldsymbol{L}^S_{G2}\right) = \boldsymbol{\Lambda}^S_G. \tag{4-128}$$

A Hermitian matrix is defined as a complex matrix which is equal to its conjugate transpose. That the matrix $\boldsymbol{G}_1{}^S + i\boldsymbol{G}_2{}^S$ is Hermitian is confirmed by taking its conjugate transpose and noting that $\boldsymbol{G}_1{}^S$ is real symmetric and $\boldsymbol{G}_2{}^S$ is real antisymmetric; that is,

$$\tilde{\boldsymbol{G}}^{S*}_1 - i\,\tilde{\boldsymbol{G}}^{S*}_2 = \tilde{\boldsymbol{G}}^S_1 - i\,\tilde{\boldsymbol{G}}^S_2 = \boldsymbol{G}^S_1 + i\,\boldsymbol{G}^S_2.$$

where * denotes the conjugate. The unitary transformation is the similarity transformation with a unitary matrix which is defined to be equal to the inverse of its conjugate transpose. Expanding the left side of Eq. (4-128), we have

$$\tilde{\boldsymbol{L}}^S_{G1}\boldsymbol{G}^S_1\boldsymbol{L}^S_{G1} + \tilde{\boldsymbol{L}}^S_{G2}\boldsymbol{G}^S_1\boldsymbol{L}^S_{G2} + \tilde{\boldsymbol{L}}^S_{G2}\boldsymbol{G}^S_2\boldsymbol{L}^S_{G1} - \tilde{\boldsymbol{L}}^S_{G1}\boldsymbol{G}^S_2\boldsymbol{L}^S_{G2}$$

$$+ i\left(\tilde{L}_{G1}^{S} G_1^S L_{G2}^{S} - \tilde{L}_{G2}^{S} G_1^S L_{G1}^{S} + \tilde{L}_{G1}^{S} G_2^S L_{G1}^{S} + \tilde{L}_{G2}^{S} G_2^S L_{G2}^{S}\right) = \Lambda_G^S . \qquad (4\text{-}129)$$

The real and imaginary parts of the left side of Eq. (4-129) are verified to be symmetric and antisymmetric, respectively, by taking the transpose of each term. Since the diagonal elements of an antisymmetric matrix are zero, the diagonal matrix in the right side of Eq. (4-129), Λ_G^S, must be real, as expected from the well known theorem stating that a Hermitian matrix has real eigenvalues. It then follows that

$$\tilde{L}_{G1}^{S} G_1^S L_{G1}^{S} + \tilde{L}_{G2}^{S} G_1^S L_{G2}^{S} + \tilde{L}_{G2}^{S} G_2^S L_{G1}^{S} - \tilde{L}_{G1}^{S} G_2^S L_{G2}^{S} = \Lambda_G^S \qquad (4\text{-}130a)$$

$$\tilde{L}_{G1}^{S} G_1^S L_{G2}^{S} - \tilde{L}_{G2}^{S} G_1^S L_{G1}^{S} + \tilde{L}_{G1}^{S} G_2^S L_{G1}^{S} + \tilde{L}_{G2}^{S} G_2^S L_{G2}^{S} = \mathbf{0}. \qquad (4\text{-}130b)$$

If Eq. (4-129) holds, the orthogonal transformation to diagonalize the real symmetric matrix G^S in Eq. (4-127a) is given by

$$\tilde{L}_G^S G^S L_G^S = \begin{bmatrix} \tilde{L}_{G1}^S & \tilde{L}_{G2}^S \\ -\tilde{L}_{G2}^S & \tilde{L}_{G1}^S \end{bmatrix} \begin{bmatrix} G_1^S & -G_2^S \\ G_2^S & G_1^S \end{bmatrix} \begin{bmatrix} L_{G1}^S & -L_{G2}^S \\ L_{G2}^S & L_{G1}^S \end{bmatrix} = \begin{bmatrix} \Lambda_G^S & \mathbf{0} \\ \mathbf{0} & \Lambda_G^S \end{bmatrix}. \qquad (4\text{-}131)$$

The last equality in Eq. (4-131) is confirmed by performing the matrix multiplication and comparing the result with Eqs. (4-130a,b). Once the matrix G^S is diagonalized, it is straightforward from analogy with the case of Eq. (4-25) to construct the matrix

$$L_G^{S0} = \begin{bmatrix} L_{G1}^S \Lambda_G^{S\,1/2} & -L_{G2}^S \Lambda_G^{S\,1/2} \\ L_{G2}^S \Lambda_G^{S\,1/2} & L_{G1}^S \Lambda_G^{S\,1/2} \end{bmatrix}$$

and subject the matrix F^S to the congruence transformation

$$\tilde{L}_G^{S0} F^S L_G^{S0} = \tilde{L}_G^{S0} \begin{bmatrix} F_1^S & -F_2^S \\ F_2^S & F_1^S \end{bmatrix} L_G^{S0} = \begin{bmatrix} F_{C1}^S & -F_{C2}^S \\ F_{C2}^S & F_{C1}^S \end{bmatrix} = F_C^S. \qquad (4\text{-}132)$$

The matrix F_C^S thus obtained has the same structure as G^S, and can be diagonalized to give doubly degenerated eigenvalues in analogy with the case of Eq. (4-131).[12] Diagonalization of a Hermitian matrix with a unitary transformation can be performed by a successive use of tridiagonalization and QR decomposition slightly modified from the case of a real symmetric matrix.

Simple molecules with relatively low molecular weights rarely belong to point groups C_n with $n \geq 3$, while many chain polymers having one-dimensional translational symmetry have only the $n(\geq 3)$-fold helix axis as symmetry elements. Diagonalization of matrices of the type in Eqs. (4-127a,b) is required also in the analysis of thermal motion of nuclei in crystals. The details of this procedure are given in Chapter 9 in connection with the calculation of temperature factors used in X-ray crystallography. If complex numbers are allowed as elements of irreducible representations, a pair of one-dimensional irreducible representations $e^{imk\phi}$ and $e^{-imk\phi}$ ($\phi = 2\pi/n$) for the operation C_n^m can be formed from the complex internal symmetry coordinates

$$\Delta S_{ka} = \sum_p e^{2ipk\pi/n} \Delta R_p, \quad \Delta S_{kb} = \sum_p e^{-2ipk\pi/n} \Delta R_p \qquad (4\text{-}133)$$

as the basis.[13] For the point groups D_n, D_{nd}, D_{nh}, and C_{nv}, there are some symmetry operations which exchange ΔS_{ka} and ΔS_{kb} in Eq. (4-133) with each other. In these cases, ΔS_{ka} and ΔS_{kb} each by itself cannot be a basis of the irreducible representation as a single variable, necessitating two-dimensional irreducible representations.

4.7 Characters of Irreducible Representations

The sum of the diagonal elements of a representation matrix is called the character of the representation. The character is invariant on the similarity transformation. Let a matrix $\boldsymbol{P}$ be subjected to a similarity transformation by another matrix $\boldsymbol{X}$ to give a new matrix $\boldsymbol{Q} = \boldsymbol{X}\,\boldsymbol{P}\,\boldsymbol{X}^{-1}$. Then we have

$$\sum_i Q_{ii} = \sum_i \left\{ \sum_j \sum_k X_{ij} P_{jk} (X^{-1})_{ki} \right\} = \sum_j \sum_k P_{jk} \left\{ \sum_i X_{ij} (X^{-1})_{ki} \right\}$$
$$= \sum_j \sum_k P_{jk} \delta_{jk} = \sum_j P_{jj},$$

which proves the statement. If any of three symmetry operations A, B and X satisfy the relation

$$AX = XB, \qquad \therefore \quad A = XBX^{-1}.$$

then operations A and B are said to be conjugate to each other. If it holds also that $B = YCY^{-1}$, we have

$$A = X(YCY^{-1})\,X^{-1} = (XY)\,C\,(XY)^{-1}.$$

That is, if A and B are conjugate to each other and B and C are conjugate to each other, then A and C are conjugate to each other. A set of all the mutually conjugate elements of a group is called a class. Since the character of a representation is not changed by any similarity transformation, the representations of symmetry operations belonging to a common class have the same character.

For each of the point groups generated from an n-fold axis of rotation and any of the symmetry elements C_2', C_2'', σ_v, σ_d, σ_h and i (the center of symmetry), all symmetry operations are given as the products of C_n^p (p = 1–n) and the operations of some of point groups C_2, C_s and C_i. Here, C_s is the point group consisting of the identity operation E and the reflection at a plane of symmetry, and C_i is the point group consisting of E and the inversion at a center of symmetry. Generally, if each operation of a point group G_A is given as the product of operations taken one by one from point groups G_B, G_C, $\cdots$, group G_A is called the direct product group of G_B, G_C, $\cdots$, and is represented as $G_A = G_B \times G_C \times \cdots$, while groups G_B, G_C, $\cdots$, are called the factor group of G_A. Let the operation of the direct product group $G_A = G_B \times G_C$ formed as the product of the operation a_B taken from G_B and the operation c_C taken from G_C be denoted by $(a_B \times c_C)_A$. Suppose further that the product of operations a_B and b_B is p_B, *i.e.*, $p_B = a_B b_B$ in group G_B, and similarly, operations q_C, c_C and d_C in group G_C satisfy the relation $q_C = c_C d_C$. Then the operation in the direct product group G_A formed as the product of p_B and q_C is represented as

$$(p_B \times q_C)_A = (a_B \times c_C)_A (b_B \times d_C)_A \qquad (4\text{-}134)$$

If the $(n_B \times n_B)$ matrix $\boldsymbol{B}^a$ is a representation of the operation a_B of the point group G_B

Table 4.1 Character tables in terms of real symmetry coordinates.

C_2	D_2	C_{2v}	I	C_{2Z}	σ_{ZX}	σ_{ZY} [a]
		D_2	I	C_{2Z}	C_{2Y}	C_{2X}
		C_2	I	C_{2Z}		
A [a]	A	A_1	1	1	1	1
	B_1	A_2	1	1	−1	−1
B	B_2	B_1	1	−1	1	−1
	B_3	B_2	1	−1	−1	1

C_3	C_{3v} / D_3	I	$2C_3$	$3\sigma_v$ / $3C_2$
A	A_1	1	1	1
	A_2	1	1	−1
E	E	2	−1	0

C_4	C_{4v}	I	$2C_4$	C_2	$2\sigma_v$	$2\sigma_d$
	D_4 [b]	I	$2C_4$	C_2	$2C_2$	$2C_2$'
	C_4	I	$2C_4$	C_2		
A	A_1	1	1	1	1	1
	A_2	1	1	1	−1	−1
B	B_1	1	−1	1	1	−1
	B_2	1	−1	1	−1	1
E	E	2	0	−2	0	0

C_5	C_{5v}	I	$2C_5$	$2C_5^2$	$5\sigma_v$
	D_5	I	$2C_5$	$2C_5^2$	$5C_2$
	C_5	I	$2C_5$	$2C_5^2$	
A	A_1	1	1	1	1
	A_2	1	1	1	−1
E	E_1	2	$2\cos\theta$	$2\cos 2\theta$	0
	E_2	2	$2\cos 2\theta$	$2\cos 4\theta$	0

$\theta = 2\pi/5$

C_6	C_{6v}	I	$2C_6$	$2C_3$	C_2	$3\sigma_v$	$3\sigma_d$
	D_6	I	$2C_6$	$2C_3$	C_2	$3C_2$	$3C_2$'
	C_6	I	$2C_6$	$2C_3$	C_2		
A	A_1	1	1	1	1	1	1
	A_2	1	1	1	1	−1	−1
B	B_1	1	−1	1	−1	1	−1
	B_2	1	−1	1	−1	−1	1
E_1	E_1	2	1	−1	−2	0	0
E_2	E_2	2	−1	−1	2	0	0

C_{nh} (D_{nh}, if C is replaced by D) [c]

n_{odd}	n_{even}	$R(C_n)$	$R(C_n) \times \sigma_h$ / $R(C_n) \times i$
$\Gamma(C_n)'$	$\Gamma(C_n)_g$	$1\times \chi(C_n)$	$1\times \chi(C_n)$
$\Gamma(C_n)''$	$\Gamma(C_n)_u$	$1\times \chi(C_n)$	$-1\times \chi(C_n)$

$D_n(n_{odd}) \times C_i = D_{nd}$

[a] The column and row headings show the irreducible representations and the symmetry operations, respectively, of the point groups in the upper left corner.
[b] Replacing the operations C_4 and C''_2 with S_4 and σ_d, respectively, gives the character table for D_{2d}.
[c] R(X), Γ(X) and χ(X) are the operations, the irreducible representations and the characters for the point group X.

and the $(n_C \times n_C)$ matrix $\boldsymbol{C}^c$ is a representation of the operation c_C of the point group G_C, the direct product of $\boldsymbol{B}^a$ and $\boldsymbol{C}^c$ defined as the $(n_B n_C \times n_B n_C)$ matrix in the form

$$\boldsymbol{A}^p = \boldsymbol{A}^{a\times c} = \boldsymbol{B}^a \times \boldsymbol{C}^c = \begin{bmatrix} B^a_{11}\boldsymbol{C}^c & B^a_{12}\boldsymbol{C}^c & \cdots & B^a_{1n}\boldsymbol{C}^c \\ B^a_{21}\boldsymbol{C}^c & B^a_{22}\boldsymbol{C}^c & \cdots & B^a_{2n}\boldsymbol{C}^c \\ \vdots & \vdots & \ddots & \vdots \\ B^a_{n1}\boldsymbol{C}^c & B^a_{n2}\boldsymbol{C}^c & \cdots & B^a_{nn}\boldsymbol{C}^c \end{bmatrix}$$

is a representation of the operation $p_A = a_B \times c_C$ of the direct product group $G_A = G_B \times G_C$. Then the relation among the operations in Eq. (4-134) is rewritten as a relation among the matrices in terms of the matrix product in the form

$$\boldsymbol{A}^{p\times q} = \boldsymbol{B}^p \times \boldsymbol{C}^q = (\boldsymbol{B}^a \times \boldsymbol{C}^c)(\boldsymbol{B}^b \times \boldsymbol{C}^d)$$

$$= \begin{bmatrix} B^a{}_{11}\boldsymbol{C}^c & B^a{}_{12}\boldsymbol{C}^c & \cdots & B^a{}_{1n}\boldsymbol{C}^c \\ B^a{}_{21}\boldsymbol{C}^c & B^a{}_{22}\boldsymbol{C}^c & \cdots & B^a{}_{2n}\boldsymbol{C}^c \\ \vdots & \vdots & \ddots & \vdots \\ B^a{}_{n1}\boldsymbol{C}^c & B^a{}_{n2}\boldsymbol{C}^c & \cdots & B^a{}_{nn}\boldsymbol{C}^c \end{bmatrix} \begin{bmatrix} B^b{}_{11}\boldsymbol{C}^d & B^b{}_{12}\boldsymbol{C}^d & \cdots & B^b{}_{1n}\boldsymbol{C}^d \\ B^b{}_{21}\boldsymbol{C}^d & B^b{}_{22}\boldsymbol{C}^d & \cdots & B^b{}_{2n}\boldsymbol{C}^d \\ \vdots & \vdots & \ddots & \vdots \\ B^b{}_{n1}\boldsymbol{C}^d & B^b{}_{n2}\boldsymbol{C}^d & \cdots & B^b{}_{nn}\boldsymbol{C}^d \end{bmatrix}$$

$$= \boldsymbol{B}^a\boldsymbol{B}^b \times \boldsymbol{C}^c\boldsymbol{C}^d$$

$$= \begin{bmatrix} (B^aB^b)_{11}\boldsymbol{C}^c\boldsymbol{C}^d & (B^aB^b)_{12}\boldsymbol{C}^c\boldsymbol{C}^d & \cdots & (B^aB^b)_{1n}\boldsymbol{C}^c\boldsymbol{C}^d \\ (B^aB^b)_{21}\boldsymbol{C}^c\boldsymbol{C}^d & (B^aB^b)_{22}\boldsymbol{C}^c\boldsymbol{C}^d & \cdots & (B^aB^b)_{2n}\boldsymbol{C}^c\boldsymbol{C}^d \\ \vdots & \vdots & \ddots & \vdots \\ (B^aB^b)_{n1}\boldsymbol{C}^c\boldsymbol{C}^d & (B^aB^b)_{2n}\boldsymbol{C}^c\boldsymbol{C}^d & \cdots & (B^aB^b)_{nn}\boldsymbol{C}^c\boldsymbol{C}^d \end{bmatrix}.$$

It follows from the definition that the trace of the direct product of matrices satisfies the relation

$$\mathrm{Tr}(\boldsymbol{B}\times\boldsymbol{C})^{a\times c} = \sum_i\sum_j B^a_{ii}C^c_{jj} = \sum_i B^a_{ii}\sum_j C^c_{jj} = \mathrm{Tr}(\boldsymbol{B}^a)\mathrm{Tr}(\boldsymbol{C}^c).$$

Accordingly, by preparing the list of characters of selected point groups of lower orders in the computer program, the characters of their direct product groups can be generated by repeating the multiplication of the characters of the factor groups. A table of the characters of all the irreducible representations for all the operations of a given group is called the character table. Once the character table of a point group is obtained, the internal symmetry coordinates and the symmetry Cartesian displacement coordinates are generated by the procedure described in the following section. Table 4.1 shows the character tables based on the real irreducible representation of cyclic groups C_n with n less than 7 and their direct product groups with C_2, C_i and C_s. More detailed character tables of point groups encountered in molecular structures are given in the literature.[14)]

4.8 Computer Program to Generate Symmetry Coordinates

The procedure of generating the internal symmetry coordinates ΔS_{ka} and ΔS_{kb} from the set of equivalent internal coordinates transformed cyclically by the rotations C_n, C_n^2, $\cdots$ is extended to sets of equivalent coordinates generated by all symmetry operations in general. Suppose that a symmetry coordinate ΔS^J_{ka}, which forms the basis of an irreducible representation of the order d_k, is generated from an equivalent set of internal coordinates ΔR^J_1, ΔR^J_2, $\cdots$, ΔR^J_{Nj}, in the form

$$\Delta S^J_{ka} = \sum_{i=1}^{N_j} (U_k)_{ai} \Delta R^J_i \qquad (a = 1\text{–}d_k). \tag{4-135}$$

In this case the coefficient $(U_k)_{1i}$ can be expressed in terms of the characters in a convenient form

$$(U_k)_{1i} = N_k' \sum_{P(1\to i)} \chi_{kP} P^{(i)}, \tag{4-136}$$

where χ_{kP} is the character of the irreducible representation Γ_k for the operation P. The sum is taken over the operations which transfer ΔR^J_1 into $P^{(i)}\Delta R^J_i$, whereby $P^{(i)}$ is $\pm$ 1. The coefficient N_k' is the normalization constant to be so taken that the sum of the squares of the right sides becomes unity.

Although a general proof of Eq. (4-136) requires a lengthy consideration,[15] it is easily confirmed that Eq. (4-136) holds for the case of the equivalent set of coordinates formed by an n-fold axis of rotation. Among the operations C_n, C_n^2, $\cdots$, that which transfers ΔR^J_1 into ΔR^J_i is C_n^{n-i+1}. Let us replace $\boldsymbol{C}_n$ and $\boldsymbol{D}_{k\phi}$ in Eq. (4-113) by $\boldsymbol{C}_n^{n-i+1}$ and $\boldsymbol{D}_{k,(n-i+1)\phi}$, respectively, and take the (1,1) elements of the left and right sides. Then noting that $(C_n^{n-i+1})_{j1} = \delta_{ji}$, we have

$$\sum_{a=1}^{2} \left\{D_{k,(n-i+1)\phi}\right\}_{1a} (U_k)_{a1} = \sum_{j=1}^{n} (U_k)_{1a} \left(C_n^{\,n-i+1}\right)_{j1} = (U_k)_{1i}. \tag{4-137}$$

From Eqs. (4-108a,b), the elements in the first column of the transformation matrix $\boldsymbol{U}_k$ are given by

$$(U_k)_{11} = N_k, \quad (U_k)_{21} = 0. \tag{4-138}$$

By using Eqs. (4-105) and (4-106), the character of C_n^{n-i+1} is calculated as

$$2\left\{D_{k,(n-i+1)\phi}\right\}_{11} = 2\cos\frac{2\pi k(n-i+1)}{n} = 2\cos\frac{2\pi k(i-1)}{n}, \tag{4-139}$$

which coincides with Eq. (4-108a) except for a constant coefficient. Accordingly, by using Eq. (4-139) as χ_{kP} and substituting together with Eq. (4-138) into Eq. (4-137), we obtain the expression in which the sum over P is omitted by putting $P^{(i)} = 1$ in Eq. (4-136). In a direct product group formed from a cyclic group multiplied by any of C_2, C_s and C_i, more than one operations may transfer ΔR^J_1 into ΔR^J_i. The summation in Eq. (4-136) is introduced to take account of all these operations. The factor $P^{(i)}$ is necessary for the linear bending, the out-of-plane bending and torsional coordinates which may change the sign when subjected to certain symmetry operations. If Eq. (4-136) is applied to the case where no internal symmetry coordinates belonging to Γ_k are generated from the equivalent set of internal coordinates, ΔR^J_1, ΔR^J_2, $\cdots$, all coefficients of ΔR^J_i vanish because the contributions from different operations cancel out one another. By utilizing this result, the internal symmetry coordinates belonging to each irreducible representation can be generated automatically as described below.

(1) The local symmetry coordinates are formed by taking account of the local symmetry of the molecule. A local symmetry coordinate is a linear combination of certain internal coordinates within a group. A set of local symmetry coordinates forms a basis of the irreducible representation of the point group to which the group must belong if isolated. Possible redundancies with respect to the bond angles around a tetrahedral structure are eliminated in this step. The most frequently used local symmetry coordinates are those for

Table 4.2 Local symmetry coordinates.

No.	Name[a)]	N[b)]	Linear combination of internal coordinates
CH_3X			
ΔS_1	CH_3 sym. str.	$(1/\sqrt{3})$	$(\Delta r_{12} + \Delta r_{13} + \Delta r_{14})$
ΔS_2	CH_3 asym.str.	$(1/\sqrt{6})$	$(2\Delta r_{12} - \Delta r_{13} - \Delta r_{14})$
ΔS_3	CH_3 asym. str.	$(1/\sqrt{2})$	$(\Delta r_{13} - \Delta r_{14})$
ΔS_4	CH_3 sym. def.	$(1/\sqrt{6})$	$(\Delta\theta_{314} + \Delta\theta_{213} + \Delta\theta_{214} - \Delta\theta_{215} - \Delta\theta_{315} - \Delta\theta_{415})$
ΔS_5	CH_3 asym. def.	$(1/\sqrt{6})$	$(2\Delta\theta_{314} - \Delta\theta_{213} - \Delta\theta_{214})$
ΔS_6	CH_3 rocking	$(1/\sqrt{6})$	$(2\Delta\theta_{215} - \Delta\theta_{315} - \Delta\theta_{415})$
ΔS_7	CH_3 asym. def.	$(1/\sqrt{2})$	$(\Delta\theta_{213} - \Delta\theta_{214})$
ΔS_8	CH_3 rocking	$(1/\sqrt{2})$	$(\Delta\theta_{315} - \Delta\theta_{415})$
ΔS_{red}		$(1/\sqrt{6})$	$(\Delta\theta_{314} + \Delta\theta_{213} + \Delta\theta_{214} + \Delta\theta_{215} + \Delta\theta_{315} + \Delta\theta_{415})$
CH_2XY			
ΔS_1	CH_2 sym. str.	$(1/\sqrt{2})$	$(\Delta r_{12} + \Delta r_{13})$
ΔS_2	CH_2 antisym. str.	$(1/\sqrt{2})$	$(\Delta r_{12} - \Delta r_{13})$
ΔS_3	CH_2 bending	$(1/\sqrt{20})$	$(4\Delta\theta_{213} - \Delta\theta_{214} - \Delta\theta_{215} - \Delta\theta_{314} - \Delta\theta_{315})$
ΔS_4	CXY def.	$(1/\sqrt{30})$	$(-\Delta\theta_{213} - \Delta\theta_{214} - \Delta\theta_{215} + 5\Delta\theta_{415} - \Delta\theta_{314} - \Delta\theta_{315})$
ΔS_5	CH_2 wagging	(1/2)	$(\Delta\theta_{214} - \Delta\theta_{215} + \Delta\theta_{314} - \Delta\theta_{315})$
ΔS_6	CH_2 rocking	(1/2)	$(\Delta\theta_{214} + \Delta\theta_{215} - \Delta\theta_{314} - \Delta\theta_{315})$
ΔS_7	CH_2 twisting	(1/2)	$(\Delta\theta_{214} - \Delta\theta_{215} - \Delta\theta_{314} + \Delta\theta_{315})$
ΔS_{red}		$(1/\sqrt{6})$	$(\Delta\theta_{213} + \Delta\theta_{214} + \Delta\theta_{215} + \Delta\theta_{415} + \Delta\theta_{314} + \Delta\theta_{315})$
$CH_2{=}X$			
ΔS_1	CH_2 sym. str.	$(1/\sqrt{2})$	$(\Delta r_{12} + \Delta r_{13})$
ΔS_2	CH_2 antisym. str.	$(1/\sqrt{2})$	$(\Delta r_{12} - \Delta r_{13})$
ΔS_3	CH_2 bending	$(1/\sqrt{6})$	$(2\Delta\theta_{213} - \Delta\theta_{214} - \Delta\theta_{314})$
ΔS_4	CH_2 rocking	$(1/\sqrt{2})$	$(\Delta\theta_{214} - \Delta\theta_{314})$
ΔS_{red}		$(1/\sqrt{3})$	$(\Delta\theta_{213} + \Delta\theta_{214} + \Delta\theta_{314})$

[a)] abbreviations : sym. = symmetric, asym. = asymmetric, str. = stretching, def. = deformation, antisym. = antisymmetric.

[b)] Normalization coefficient

methyl and methylene groups of hydrocarbon residues listed in Table 4.2, where the valence angles for CH_3X and CH_2XY structures are assumed to be tetrahedral. The local symmetry coordinates for the other groups having the same topology as the methyl or the methylene group are defined in analogous ways. The force constants in terms of local symmetry coordinates have been extensively used in the normal coordinate analyses of organic molecules.[16)]

(2) The symmetry operations of the molecule are listed up, the point group is determined and the character table is constructed.

(3) The matrices of transformation of the nuclear Cartesian coordinates associated with all the symmetry operations of the point group are calculated, and the multiplication table of the point group is deduced from these matrices. The local symmetry coordinates are classified into equivalent sets in which the first member lies on a specified symmetry element.

(4) In a triple loop in which the inner loop runs over the symmetry operations, the middle over the equivalent coordinates and the outer over the irreducible representations, each term of the right side of Eq. (4-136) is calculated and stored in the array for the coefficients $(U_k)_{1i}$. On getting out of the inner loop, whether all the members of the coefficient array are vanishing is checked. If there are any non-vanishing $(U_k)_{1i}$, the equivalent set of the local symmetry coordinates specified by the running index of the middle loop is registered as giving a member of the internal symmetry coordinates of the irreducible representation specified by the running index of the outer loop.

By following a similar procedure, the symmetry Cartesian displacement coordinates can be constructed from the nuclear Cartesian displacements. For this purpose, the sets of equivalent nuclei are taken up in place of the sets of equivalent local symmetry coordinates in the case of internal symmetry coordinates. The symmetry Cartesian displacements are generated from the nuclear displacement vector ΔX_i^J according to the equation corresponding to Eqs. (4-135) and (4-136) in the form

$$\Delta S(X_k^J) = N_k^J \sum_i \left(\sum_{P(1\rightarrow i)} \chi_{kP} \, \boldsymbol{R}_P^X \right) \Delta X_i^J, \tag{4-140}$$

where $\boldsymbol{R}_P^X$ is the (3×3) matrix transforming the coordinates according to the operation P, and the sum in parentheses is taken over the operations transferring the nucleus 1 in the Jth equivalent set into the nucleus i in the same set.

Since the sum over such operations and then over all the nuclei i belonging to the Jth equivalent set covers all the operations in the point group, Eq. (4-140) can be rewritten as

$$\Delta S(X_k^J) = N_k^J \sum_P \chi_{kP} \, \boldsymbol{R}_P^X (O^P \Delta X_1^J), \tag{4-141}$$

where $O^P \Delta X_1^J$ is the displacement vector of the nucleus formed by the operation P from the nucleus 1 of the Jth set. Non-vanishing components of the vector $\Delta S(X_k^J)$, if any, represent the symmetry Cartesian displacement coordinates in the irreducible representation Γ_k generated from the Jth equivalent set. In actual calculations, the components of the vector in parentheses in Eq. (4-140) are stored as the coefficients of the terms involving ΔX_i^J. After the summation, the coefficients of all the terms are compared to a given threshold for each of the three components, and the components having any non-vanishing term are registered as the symmetry Cartesian displacement coordinates in each cycle of the loop over the irreducible representations.

The number of symmetry Cartesian displacement coordinates registered in a given irreducible representation Γ_k, N_k^X, is the same as the degrees of the freedom of the nuclear motion in the irreducible representation Γ_k. Accordingly, if the translational and the rotational degrees of freedom belonging to Γ_k are denoted by N_k^t and N_k^r, respectively, the vibrational degrees of freedom belonging to Γ_k are given by

$$N_k^v = N_k^X - N_k^t - N_k^r .$$

The degrees of freedom N_k^t and N_k^r are calculated by substituting the characters χ_P^t and χ_P^r of the representations D_P^t and D_P^r, which are formed from the translational and rotational coordinates defined by Eq. (3-28a–f) as the bases, respectively, into the expressions

$$N_k^t = \sum_P \chi_{kP}\, \chi_P^t \tag{4-142a}$$

and

$$N_k^r = \sum_P \chi_{kP}\, \chi_P^r . \tag{4-142b}$$

The components associated with the three axes of the translational and rotational coordinates are known from a modification of Eqs. (4-142a,b) in which the characters χ_P^t and χ_P^r are replaced by the individual diagonal terms of the corresponding representations, $(D_P^t)_{AA}$ and $(D_P^r)_{AA}$ $(A = X, Y, Z)$, respectively. The matrix $\boldsymbol{D}_P^t$ is obviously the matrix of transformation $\boldsymbol{R}_P^X$ itself, while the matrix $\boldsymbol{D}_P^r$ is obtained as follows.

Let the transformation of the Cartesian coordinates of nuclei by a symmetry operation be written as

$$\boldsymbol{X}_i' = \begin{bmatrix} X_i' \\ Y_i' \\ Z_i' \end{bmatrix} = \boldsymbol{D}_P^t\, \boldsymbol{X}_i = \begin{bmatrix} C_{Xx} & C_{Xy} & C_{Xz} \\ C_{Yx} & C_{Yy} & C_{Yz} \\ C_{Zx} & C_{Zy} & C_{Zz} \end{bmatrix} \begin{bmatrix} X_i \\ Y_i \\ Z_i \end{bmatrix} . \tag{4-143}$$

The rotational coordinate around the X-axis after the symmetry operation is given, according to Eq. (3-28d), in the form

$$R_X' = I_X^{-1} \sum_i m_i (Y_i'^0 \Delta Z_i' - Z_i'^0 \Delta Y_i'). \tag{4-144}$$

Rewriting $Y_i'^0$, $\Delta Z_i'$, *etc.* in Eq. (4-144) in terms of the coordinates before the transformation according to Eq. (4-143), and using Eqs. (3-28d–f), we obtain, after rearrangement,

$$R_X' = I_X^{-1} \sum_i m_i \{ -(C_{Zx} X_i^0 + C_{Zy} Y_i^0 + C_{Zz} Z_i^0)(C_{Yx} \Delta X_i + C_{Yy} \Delta Y_i + C_{Yz} \Delta Z_i)$$
$$+ (C_{Yx} X_i^0 + C_{Yy} Y_i^0 + C_{Yz} Z_i^0)(C_{Zx} \Delta X_i + C_{Zy} \Delta Y_i + C_{Zz} \Delta Z_i)$$

$$= (C_{Zz}C_{Yy} - C_{Zy}C_{Yz})R_X + (C_{Yz}C_{Zx} - C_{Yx}C_{Zz})R_Y + (C_{Yx}C_{Zy} - C_{Yy}C_{Zx})R_Z.$$

The rotational coordinates around the Y' and the Z' axes are similarly expressed in terms of the rotational coordinates before the transformation. Collecting the three rotational coordinates into a matrix representation gives

$$\begin{bmatrix} R'_X \\ R'_Y \\ R'_Z \end{bmatrix} = \begin{bmatrix} C_{Zz}C_{Yy} - C_{Zy}C_{Yz} & C_{Zx}C_{Yz} - C_{Zz}C_{Yx} & C_{Zy}C_{Yx} - C_{Zx}C_{Yy} \\ C_{Xz}C_{Zy} - C_{Xy}C_{Zz} & C_{Xx}C_{Zz} - C_{Xz}C_{Zx} & C_{Xy}C_{Zx} - C_{Xx}C_{Zy} \\ C_{Yz}C_{Xy} - C_{Yy}C_{Xz} & C_{Yx}C_{Xz} - C_{Yz}C_{Xx} & C_{Yy}C_{Xx} - C_{Yx}C_{Xy} \end{bmatrix} \begin{bmatrix} R_X \\ R_Y \\ R_Z \end{bmatrix}. \quad (4\text{-}145)$$

By subtracting the number of vibrational degrees of freedom N_k^v from the number of internal symmetry coordinates in each irreducible representation Γ_k, the number of redundant coordinates in Γ_k is obtained. Such a redundant coordinate arises mainly from the ring formation. On diagonalizing the matrix $\boldsymbol{G}^{SS}$, non-vanishing elements in the column of the eigenvector matrix $\boldsymbol{L}_G$ corresponding to the vanishing eigenvalues indicates which internal coordinates are involved in the redundancy. A submatrix of $\boldsymbol{G}^{SS}$ is then constructed from the columns and rows corresponding to these coordinates, and is diagonalized again to give vanishing eigenvalues. The corresponding eigenvectors give the linear combinations of internal coordinates which represent the redundant coordinates exactly.

Once the number and the form of independent internal symmetry coordinates in each irreducible representation are determined in this way, the Hessian matrix for the mass-weighted Cartesian displacement coordinates is diagonalized. If the equilibrium structure is not yet obtained, the diagonalization is carried out only for the totally symmetric irreducible representation, and it is checked if the number of positive eigenvalues coincides with the totally symmetric vibrational degrees of freedom. The potential minimum is searched by the Newton–Raphson method if the number of positive eigenvalues coincides with N_k^v, and by some other method if not. At the final stage of the minimum search, the conditions for the use of the Newton–Raphson method must be fulfilled in order for the potential function to give a reasonable equilibrium structure.

To specify the nature of the vibrational form of each normal mode, it is convenient to use the potential energy distribution, P.E.D., which shows the distribution of the potential energy stored in various parts of the molecule during the normal vibration in average.[17] Substituting Eq. (4-21) into Eq. (4-19) and taking the ith diagonal elements of both sides, we obtain

$$\Lambda_{ii} = \sum_j \sum_k L_{ji} F_{jk}^{SS} L_{ki}. \quad (4\text{-}146)$$

On neglecting the terms with $k \neq j$ in Eq. (4-146), the ratio of the potential energy stored in the jth internal symmetry coordinate in the course of the ith normal vibration is given by

$$(\text{PED})_{ji}(\%) = 100 \times F_{jj}^{SS} L_{ji}^{\ 2} / \Lambda_{ii}.$$

The contribution of ΔS_j to the normal coordinate Q_i may be estimated from the magnitude of the element of the $\boldsymbol{L}$ matrix, L_{ji}. However, since the amplitude of such a light nucleus as hydrogen is large compared to that of heavy nuclei, there may be cases in which the energetic contribution is small even if L_{ji} is large and the judgment is obscured.

An example of the output of the normal coordinate analysis by the program RISE, for

```
                 FORMIC  ACID

CARTESIAN COORDINATES AND ATOMIC CHARGES

NO.           X             Y             Z            FINAL      INITIAL

 1  H     0.000000    -1.502580     0.082859        0.0832      0.0852
 2  C     0.000000    -0.407411     0.105097        0.3042      0.2866
 3  O     0.000000     0.246952     1.115855       -0.3604     -0.3506
 4  O     0.000000     0.085247    -1.133805       -0.3893     -0.3797
 5  H     0.000000     1.081276    -1.049344        0.3623      0.3585

LIST OF INTERNAL COORDINATES(STRUCTURE PARAMETERS)

 NO.           DEFINITION                       CODE      R(FINAL)  R(INIT)  R(STD)

  1   1    H(  1)- C(  2)                         1  STR    1.0954   1.0840   1.0850
  2   2    C(  2)= O(  3)                         1  STR    1.2041   1.2150   1.1900
  3   3    C(  2)- O(  4)                         1  STR    1.3330   1.4310   1.3300
  4   4    O(  4)- H(  5)                         1  STR    0.9996   0.9730   1.0000
  5   5    H(  1)- C(  2)= O(  3)                 2  BEND   124.07   120.00   125.50
  6   6    H(  1)- C(  2)- O(  4)                 2  BEND   110.53   120.00   110.00
  7   7    O(  3)= C(  2)- O(  4)                 2  BEND   125.40   120.00   124.00
  8   8    O(  3)= C(  2)( O(  4), H(  1)) 3  OP B     0.00     0.00
  9   9    C(  2)- O(  4)- H(  5)                 2  BEND   106.84   109.47   102.50
 10  10    H(  1)- C(  2)- O(  4)- H(  5)  4  TOR    180.00   180.00
 11   0    O(  3)= C(  2)- O(  4)- H(  5)  4  TOR      0.00     0.00

LIST OF LOCAL SYMMETRY COORDINATES

NO.    NAME                    DEFINITION

 1  C-H STR       1  1.00000
 2  O=C STR       2  1.00000
 3  CHO BEND      5  0.81650   6 -0.40825   7 -0.40825
 4  CHO ROCK      6  0.70711   7 -0.70711
 5  CHO OPL BEND  8  1.00000
 6  O-H STR       4  1.00000
 7  O-C STR       3  1.00000
 8  HOC DEF       9  1.00000
 9  O-C TOR      10  1.00000


SPECIES NO.   1(A' )

LIST OF INTERNAL SYMMETRY COORDINATES

NO.    NAME                    DEFINITION
 1  O-C STR       7  1.00000
 2  C-H STR       1  1.00000
 3  O=C STR       2  1.00000
 4  CHO BEND      3  1.00000
 5  CHO ROCK      4  1.00000
 6  O-H STR       6  1.00000
 7  HOC DEF       8  1.00000

FREQUENCIES AND INTENSITIES

 1   3583.8 (CM-1)    PED  -99(  6)

 2   2956.3 (CM-1)    PED  -98(  2)

 3   1777.4 (CM-1)    PED  -81(  3)    7(  1)  -5(  7)  -4(  5)

 4   1415.8 (CM-1)    PED   65(  4) -14(  5)  12(  7)  -4(  3)   3(  1)
```

Fig. 4.4 Output of a normal coordinate analysis of formic acid monomer by RISE.

```
5   1239.1 (CM-1)    PED    37(  7)  33(  5) -14(  4)  13(  1)  -1(  3)

6   1115.7 (CM-1)    PED   -58(  1)  23(  7) -11(  4)  -4(  3)

7    633.9 (CM-1)    PED    59(  5) -15(  7)  12(  4)  -7(  1)  -3(  3)

SPECIES NO.  2(A" )

LIST OF INTERNAL SYMMETRY COORDINATES

NO.    NAME                DEFINITION

 8  O-C TOR          9  1.00000
 9  CHO OPL BEND     5  1.00000

FREQUENCIES AND INTENSITIES

 1   1027.4 (CM-1)    PED   -96(  9)    3(  8)

 2    642.8 (CM-1)    PED   -95(  8)   -4(  9)
```

Fig. 4.4 Continued.

formic acid monomer,[18] is shown in Fig. 4.4. In the list of the internal coordinates, R(FINAL), R(INT) and R(STD) are the final values (equilibrium value, R^e), the initial values and the standard values (R^0) of the corresponding internal variables, respectively. The local symmetry coordinates are defined first as linear combinations of the internal coordinates. Then the internal symmetry cordinates are defined as linear combinations of the local symmetry coordinates, and the numbers and values of the coefficients are listed. The PED's are listed in the form of %(j), where j is the index of the internal symmetry coordinates, following the wavenumber ω_i for each normal coordinate Q_i. The symbol % is preceded by a minus sign if the corresponding $\boldsymbol{L}$ matrix element L_{ji} is negative.

References

1) E.B. Wilson, Jr., J.C. Decius and P.C. Cross, *Molecular Vibrations*, McGraw Hill, New York (1955).
2) E.B. Wilson, Jr., *J. Chem. Phys.*, **7**, 1047 (1939).
3) W.J. Taylor, *J. Chem. Phys.*, **18**, 1301 (1950).
4) T. Miyazawa, *J. Chem. Phys.*, **29**, 246 (1958).
5) J.C. Decius, *J. Chem. Phys.*, **16**, 1025 (1949); Ref. 1, p. 303.
6) S.M. Ferigle and A.G. Meister, *J. Chem. Phys.*, **19**, 982 (1951).
7) T. Noguti and N. Gō, *J. Phys. Soc. Jpn.*, **52**, 3283, 3685 (1983).
8) A. Kidera and N. Gō, *Proc. Nat. Acad. Sci.*, **87**, 3718 (1990); *idem.*, *J. Mol. Biol.*, **225**, 457 (1992).
9) A. Kidera, K. Inaka, M. Matsushima and N. Gō, *J. Mol. Biol.*, **225**, 477 (1992).
10) B.T. Darling and D.M. Dennison, *Phys. Rev.*, **57**, 128 (1950).
11) Ref. 1, p. 289.
12) T. Miyazawa, Y. Ideguchi and K. Fukushima, *J. Chem. Phys.*, **38**, 2709 (1963).
13) P. Higgs, *Proc. Roy. Soc.* (London), **A220**, 472 (1953).
14) Ref. 1, p. 323.
15) Ref. 1, p. 341.
16) T. Shimanouchi, H. Matsuura, Y. Ogawa and I. Harada, *J. Phys. Chem. Ref. Data*, **7**, 1323 (1977).
17) Y. Morino and K. Kuchitsu, *J. Chem. Phys.*, **20**, 1809 (1953).
18) I. Yokoyama, Y. Miwa and K. Machida, *J. Am. Chem. Soc.*, **113**, 6458 (1991).

Chapter 5

Rotations and Anharmonic Vibrations of Molecules

5.1 Anharmonicity of Molecular Potential Functions

If the molecular vibrations are rigorously harmonic, the average positions of atomic nuclei should coincide with the equilibrium positions, but the nuclei in actual molecules under the influence of an anharmonic potential function deviate appreciably from the equilibrium positions in average. As a result, the molecular structure determined from the electron diffraction or the microwave experiments is systematically different from the true equilibrium structure corresponding to a potential minimum. These differences often exceed experimental accuracy. We should keep this point in mind when comparing the equilibrium structure obtained from the molecular mechanics calculation with experimental data. The effect of anharmonic terms in the potential function on nuclear motion becomes more and more significant as the molecular distortion increases, and eventually governs the direction and ease of chemical reactions.

In this chapter, the theoretical formulae connecting the potential function and the molecular constants obtained from diffraction and spectroscopic measurements are derived as easily as possible, and various programing techniques are shown together with some examples of calculated results. The energy levels estimated from recent high-resolution experiments are correct within an accuracy of 10^{-3}–10^{-4} cm^{-1}. The use of such experimental data may contribute to improving the model potential function in molecular mechanics in the future.

5.2 Nuclear Kinetic Energies of Molecules

In Chapter 3, the principal system of inertia has been chosen as the coordinate system for describing the internal distortion of a molecule. In order to express the total kinetic energy of all the nuclei constituting the molecule, we must take into account the translation and rotation of the molecule as a whole in terms of a space-fixed coordinate system.[1] Let the position vector of the origin of the molecule-fixed system in the space-fixed system be denoted by

$$\boldsymbol{R} = \begin{bmatrix} X_g & Y_g & Z_g \end{bmatrix}^{\mathrm{T}} . \tag{5-1}$$

Then the velocity of translation of the molecule is given by

$$\dot{\boldsymbol{R}} = \partial \boldsymbol{R}/\partial t = \begin{bmatrix} \partial X_g/\partial t & \partial Y_g/\partial t & \partial Z_g/\partial t \end{bmatrix}^{\mathrm{T}} .$$

The position vector of the atomic nucleus i in the molecule-fixed coordinate system is denoted by

$$\boldsymbol{r}_i = [X_i \quad Y_i \quad Z_i]^{\mathrm{T}}. \tag{5-2}$$

Suppose that the molecule-fixed system rotates with the angular velocity $\boldsymbol{\omega}$ in the space-fixed system, and let $d\boldsymbol{r}_i/dt$ be denoted by $\dot{\boldsymbol{r}}_i$. The angular velocity is defined as a vector in the form

$$\boldsymbol{\omega} = \delta\dot{\boldsymbol{\Omega}} = \left[\dot{\Omega}_X \quad \dot{\Omega}_Y \quad \dot{\Omega}_Z\right]^{\mathrm{T}},$$

which is obtained by taking the time derivatives of each component of the vector of rotational angles $\boldsymbol{\delta\Omega}$ introduced as given by Eq. (3-51). Since

$$d\Omega_X/\dot{\Omega}_X = d\Omega_Y/\dot{\Omega}_Y = d\Omega_Z/\dot{\Omega}_Z = dt,$$

the direction of the vector $\boldsymbol{\omega}$ coincides with the positive direction of the axis of rotation. Furthermore, it is seen from Eq. (3-51) that the magnitude of $\boldsymbol{\omega}$ is the angle of rotation per unit time, that is,

$$|\boldsymbol{\omega}| = d\Omega/dt. \tag{5-3}$$

Then the velocity of the nucleus i in the space-fixed system is given, as seen from Fig. 5.1, by

$$\dot{\boldsymbol{r}}_i^s = \dot{\boldsymbol{R}} + \boldsymbol{\omega} \times \boldsymbol{r}_i + \dot{\boldsymbol{r}}_i. \tag{5-4}$$

Multiplying the square of Eq. (5-4) by the mass of the nucleus i, m_i, and summing over i, we have

$$\begin{aligned} T = \tfrac{1}{2}\Bigg\{ & |\dot{\boldsymbol{R}}|^2 \sum_{i=1}^{N} m_i + \sum_{i=1}^{N} m_i |\boldsymbol{\omega} \times \boldsymbol{r}_i|^2 + \sum_{i=1}^{N} m_i |\dot{\boldsymbol{r}}_i|^2 + 2\dot{\boldsymbol{R}} \cdot \sum_{i=1}^{N} m_i \dot{\boldsymbol{r}}_i \\ & + 2\dot{\boldsymbol{R}} \cdot \sum_{i=1}^{N} m_i (\boldsymbol{\omega} \times \boldsymbol{r}_i) + 2\sum_{i=1}^{N} m_i \dot{\boldsymbol{r}}_i \cdot (\boldsymbol{\omega} \times \boldsymbol{r}_i) \Bigg\}. \end{aligned} \tag{5-5}$$

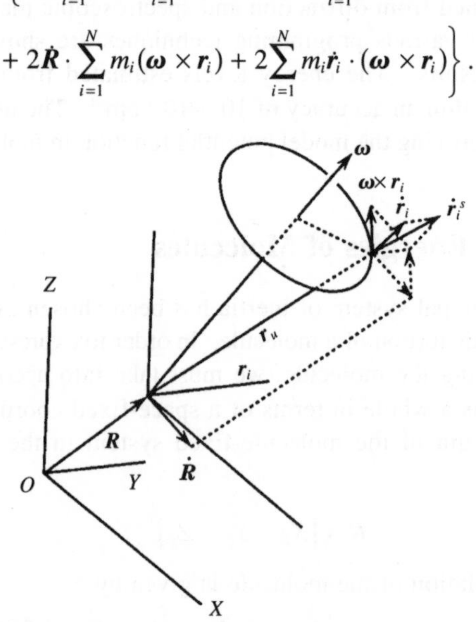

Fig. 5.1 Displacement of nucleus i in a space-fixed coordinate system.

Since the center of mass is the origin of the molecule-fixed system, we have

$$\sum_n m_n \boldsymbol{r}_n = \boldsymbol{0}, \qquad \therefore \quad \sum_n m_n \dot{\boldsymbol{r}}_n = \boldsymbol{0},$$

whence the fourth term in the braces in Eq. (5-5) disappears. The fifth term disappears, too, since it is rewritten as

$$2\dot{\boldsymbol{R}} \cdot \sum_n m_n(\boldsymbol{\omega} \times \boldsymbol{r}_n) = 2\left(\sum_n m_n \boldsymbol{r}_n\right) \cdot (\dot{\boldsymbol{R}} \times \boldsymbol{\omega}) = \boldsymbol{0} \cdot (\dot{\boldsymbol{R}} \times \omega). \tag{5-6}$$

Thus, Eq. (5-5) is now simplified as

$$T = \tfrac{1}{2}\left\{ M|\dot{\boldsymbol{R}}|^2 + \sum_{n=1}^{N} m_n |\boldsymbol{\omega} \times \boldsymbol{r}_n|^2 + \sum_{n=1}^{N} m_n |\dot{\boldsymbol{r}}_n|^2 + 2\sum_{n=1}^{N} m_n \dot{\boldsymbol{r}}_n \cdot (\boldsymbol{\omega} \times \boldsymbol{r}_n) \right\}. \tag{5-7}$$

The position vector of the nucleus n in the molecule-fixed system is the sum of its position vector for the equilibrium structure and the displacement vector as shown by

$$\boldsymbol{r}_n = \boldsymbol{r}_n^0 + \Delta\boldsymbol{r}_n. \tag{5-8}$$

If the axes of the molecule-fixed system are taken to be the principal axes of inertia, it holds that

$$\sum_{n=1}^{N} m_n(\boldsymbol{\omega} \times \boldsymbol{r}_n^0) = \boldsymbol{\omega} \times \sum_{n=1}^{N} m_n \boldsymbol{r}_n^0 = \boldsymbol{\omega} \times \boldsymbol{0} = \boldsymbol{0}.$$

From the property of a scalar triple product of three vectors, it follows that

$$\dot{\boldsymbol{r}}_n \cdot (\boldsymbol{\omega} \times \Delta\boldsymbol{r}_n) = \boldsymbol{\omega} \cdot (\Delta\boldsymbol{r}_n \times \dot{\boldsymbol{r}}_n). \tag{5-9}$$

Hence the kinetic energy in Eq. (5-7) is rewritten in the form

$$T = \tfrac{1}{2}\left\{ M|\dot{\boldsymbol{R}}|^2 + \sum_n m_n |\omega \times \boldsymbol{r}_n|^2 + \sum_n m_n |\dot{\boldsymbol{r}}_n|^2 + 2\boldsymbol{\omega} \cdot \sum_n m_n(\Delta\boldsymbol{r}_n \times \dot{\boldsymbol{r}}_n) \right\}. \tag{5-10}$$

The first, second and third terms in braces in Eq. (5-10) represent the translational, rotational and vibrational energies, respectively. The fourth term represents the interaction energy between the rotation and vibration. This term causes the inseparability of variables between the molecular rotation as a whole and the nuclear vibrations in describing the kinetic energy.

For the purpose of discussing the nuclear motion of a molecule in terms of quantum mechanics, the normal coordinates are more convenient than the Cartesian displacement coordinates (Cartesian displacements). As seen from Eqs. (4-36) and (4-38), the time derivatives of the normal coordinates and the Cartesian displacements satisfy the following relation:

$$\sum_n m_n\left(\Delta\dot{X}_n^2 + \Delta\dot{Y}_n^2 + \Delta\dot{Z}_n^2\right) = \sum_n m_n |\dot{\boldsymbol{r}}_n|^2 = \sum_i \dot{Q}_i^2. \tag{5-11}$$

The transformation between the Cartesian displacements and the normal coordinates is given by Eq. (3-80). Let us now rewrite the elements of the matrix $\boldsymbol{L}^x$ for the nth nucleus in the forms in which the coordinate axes are explicitly given as

$$L^x_{3n-2,i} = L^X_{ni}, \qquad L^x_{3n-1,i} = L^Y_{ni}, \qquad L^x_{3n,i} = L^Z_{ni}.$$

By collecting these elements into a (3 × 1) vector

$$\boldsymbol{l}^X_{ni} = [L^X_{ni} \quad L^Y_{ni} \quad L^Z_{ni}]^{\mathrm{T}}, \tag{5-12}$$

the portion of the transformation in Eq. (3-80) related to the nucleus n is written as

$$\Delta\boldsymbol{r}_n = \sum_i \boldsymbol{l}^X_{ni} Q_i. \tag{5-13}$$

The vector product of $\Delta\boldsymbol{r}_n$ in Eq. (5-13) and its time derivative, $\Delta\dot{\boldsymbol{r}}_n$, is given by

$$\Delta\boldsymbol{r}_n \times \dot{\boldsymbol{r}}_n = \left(\sum_i \boldsymbol{l}^X_{ni} Q_i\right) \times \left(\sum_j \boldsymbol{l}^X_{nj} \dot{Q}_j\right) = \sum_i \sum_j \left(\boldsymbol{l}^X_{ni} \times \boldsymbol{l}^X_{nj}\right) Q_i \dot{Q}_j. \tag{5-14}$$

Multiplying Eq. (5-14) by m_n and summing over n, we obtain one of the vectors appearing in the fourth term in braces in Eq. (5-10). Let this vector be denoted by

$$\sum_n m_n (\Delta\boldsymbol{r}_n \times \dot{\boldsymbol{r}}_n) = \sum_i \sum_j \boldsymbol{\zeta}_{ij} Q_i \dot{Q}_j \tag{5-15}$$

where

$$\boldsymbol{\zeta}_{ij} \equiv \begin{bmatrix} \zeta^X_{ij} \\ \zeta^Y_{ij} \\ \zeta^Z_{ij} \end{bmatrix} = \sum_n m_n \left(\boldsymbol{l}^X_{ni} \times \boldsymbol{l}^X_{nj}\right) = \begin{bmatrix} \sum_n m_n \left(L^Y_{ni} L^Z_{nj} - L^Z_{ni} L^Y_{nj}\right) \\ \sum_n m_n \left(L^Z_{ni} L^X_{nj} - L^X_{ni} L^Z_{nj}\right) \\ \sum_n m_n \left(L^X_{ni} L^Y_{nj} - L^Y_{ni} L^X_{nj}\right) \end{bmatrix}. \tag{5-16}$$

The components of the vector $\boldsymbol{\zeta}_{ij}$ defined by Eq. (5-16) are called the Coriolis coupling constants through which the normal coordinates i and j interact with each other.

The vector product $(\boldsymbol{\omega} \times \boldsymbol{r}_n)$ can be written in terms of the matrix $\boldsymbol{D}^R_n$ defined by Eq.(3-46) as

$$\begin{bmatrix} (\boldsymbol{\omega} \times \boldsymbol{r}_n)_X \\ (\boldsymbol{\omega} \times \boldsymbol{r}_n)_Y \\ (\boldsymbol{\omega} \times \boldsymbol{r}_n)_Z \end{bmatrix} = \begin{bmatrix} 0 & -Z^0_n & Y^0_n \\ Z^0_n & 0 & -X^0_n \\ -Y^0_n & X^0_n & 0 \end{bmatrix} \begin{bmatrix} \omega_X \\ \omega_Y \\ \omega_Z \end{bmatrix} = \boldsymbol{D}^R_n \boldsymbol{\omega}. \tag{5-17}$$

Substitution of Eq. (5-17) into the second term in braces in Eq.(5-10) leads to

$$\sum_n m_n |\boldsymbol{\omega} \times \boldsymbol{r}_n|^2 = \tilde{\boldsymbol{\omega}} \left(\sum_n m_n \tilde{\boldsymbol{D}}^R_n \boldsymbol{D}^R_n\right) \boldsymbol{\omega} = \tilde{\boldsymbol{\omega}}\, \boldsymbol{I}\, \boldsymbol{\omega}, \tag{5-18}$$

where $\boldsymbol{I}$ is the principal tensor of inertia defined by Eq. (3-26).

Substituting Eqs. (5-11), (5-15) and (5-18) into Eq. (5-10), we obtain

$$T = \tfrac{1}{2}\left\{ M\dot{\boldsymbol{R}}^2 + \tilde{\boldsymbol{\omega}}\, \boldsymbol{I} \boldsymbol{\omega} + \sum_i \dot{Q}_i^2 + 2\tilde{\boldsymbol{\omega}} \sum_i \sum_j \boldsymbol{\zeta}_{ij} Q_i \dot{Q}_j \right\}, \tag{5-19}$$

where the scalar product of two vectors $\boldsymbol{\omega} \cdot \boldsymbol{\zeta}_{ij}$ has been expressed as the product of (1×3) matrix $\tilde{\boldsymbol{\omega}}$ and (3×1) matrix $\boldsymbol{\zeta}_{ij}$.

The tensor of inertia undergoes an infinitesimal change on any infinitesimal displacement of atomic nuclei. Such a change in tensor of inertia may be expressed in terms of the normal coordinates as follows.[2] Substituting Eqs. (5-8) and (5-13) into Eq. (3-26) in which the superscript $P0$ has been dropped, and rearranging the terms in ascending order in powers of Q_i, we have

$$\begin{aligned} \boldsymbol{I} &= \sum_n m_n \left\{ \left(\tilde{\boldsymbol{r}}_n^0 + \sum_i \tilde{\boldsymbol{l}}_{ni}^X Q_i \right) \left(\boldsymbol{r}_n^0 + \sum_j \boldsymbol{l}_{nj}^X Q_j \right) \boldsymbol{E} - \left(\boldsymbol{r}_n^0 + \sum_i \boldsymbol{l}_{ni}^X Q_i \right) \left(\tilde{\boldsymbol{r}}_n^0 + \sum_j \tilde{\boldsymbol{l}}_{nj}^X Q_j \right) \right\} \\ &= \boldsymbol{I}^e + \sum_n m_n \left[\left\{ 2\tilde{\boldsymbol{r}}_n^0 \left(\sum_i \boldsymbol{l}_{ni}^X Q_i \right) + \sum_i \sum_j \left(\tilde{\boldsymbol{l}}_{ni}^X \boldsymbol{l}_{nj}^X \right) Q_i Q_j \right\} \boldsymbol{E} \right. \\ &\qquad \left. - \left\{ \boldsymbol{r}_n^0 \left(\sum_i Q_i \tilde{\boldsymbol{l}}_{ni}^X \right) + \left(\sum_i \boldsymbol{l}_{ni}^X Q_i \right) \tilde{\boldsymbol{r}}_n^0 + \sum_i \sum_j \left(\boldsymbol{l}_{ni}^X \tilde{\boldsymbol{l}}_{nj}^X \right) Q_i Q_j \right\} \right] \\ &= \boldsymbol{I}^e + \sum_i \boldsymbol{a}_i Q_i + \sum_i \sum_j \boldsymbol{A}_{ij} Q_i Q_j . \end{aligned} \tag{5-20}$$

The (3 × 3) matrices of the coefficients of the linear and the quadratic terms in the last formula of Eq. (5-20) are given by

$$\boldsymbol{a}_i = \sum_n m_n \left\{ (2\tilde{\boldsymbol{r}}_n^0 \boldsymbol{l}_{ni}^X)\boldsymbol{E} - \left(\boldsymbol{r}_n^0 \tilde{\boldsymbol{l}}_{ni}^X + \boldsymbol{l}_{ni}^X \tilde{\boldsymbol{r}}_n^0 \right) \right\}$$

and

$$\boldsymbol{A}_{ij} = \sum_n m_n \left\{ \left(\tilde{\boldsymbol{l}}_{ni}^X \boldsymbol{l}_{nj}^X \right)\boldsymbol{E} - \boldsymbol{l}_{ni}^X \tilde{\boldsymbol{l}}_{nj}^X \right\}$$

respectively. The elements of $\boldsymbol{a}_i$ are expressed, for example, as

$$\left. \begin{aligned} a_i^{XX} &= 2\sum_n m_n (Y_n^0 L_{ni}^Y + Z_n^0 L_{ni}^Z) \\ a_i^{XY} &= -\sum_n m_n (X_n^0 L_{ni}^Y + Y_n^0 L_{ni}^X) \end{aligned} \right\}, \tag{5-21}$$

which are obtained by subtracting

$$\sum_n m_n \boldsymbol{r}_n^0 \tilde{\boldsymbol{l}}_{ni}^X = \sum_n m_n \begin{bmatrix} X_n^0 L_{ni}^X & X_n^0 L_{ni}^Y & X_n^0 L_{ni}^Z \\ Y_n^0 L_{ni}^X & Y_n^0 L_{ni}^Y & Y_n^0 L_{ni}^Z \\ Z_n^0 L_{ni}^X & Z_n^0 L_{ni}^Y & Z_n^0 L_{ni}^Z \end{bmatrix}$$

and its transpose from

$$2\left(\sum_n m_n \tilde{\boldsymbol{r}}_n^0 \boldsymbol{l}_{ni}^X \right)\boldsymbol{E} = 2\left\{ \sum_n m_n (X_n^0 L_{ni}^X + Y_n^0 L_{ni}^Y + Z_n^0 L_{ni}^Z) \right\} \boldsymbol{E}.$$

Similarly, the elements of $\boldsymbol{A}_{ij}$ are expressed, for example, as

$$A_{ij}^{XX} = \sum_n m_n \left(L_{ni}^Y L_{nj}^Z + L_{ni}^Z L_{nj}^Y \right)$$

$$A_{ij}^{XY} = -\sum_n m_n L_{ni}^X L_{nj}^Y .$$

Substitution of Eq. (5-20) into Eq. (5-19) yields

$$T = \tfrac{1}{2}\left\{ M\dot{\boldsymbol{R}}^2 + \tilde{\boldsymbol{\omega}}\left(\boldsymbol{I}^e + \sum_i \boldsymbol{a}_i Q_i + \sum_i \sum_j \boldsymbol{A}_{ij} Q_i Q_j \right)\boldsymbol{\omega} + \sum_i \dot{Q}_i^2 + 2\tilde{\boldsymbol{\omega}} \sum_i \sum_j \boldsymbol{\zeta}_{ij} Q_i \dot{Q}_j \right\}. \tag{5-22}$$

Equation (5-22) represents the kinetic energy of an isolated molecule in terms of the velocity of translation, the angular velocity around its center of gravity, the normal coordinates and the time derivatives of the normal coordinates. The first term in braces in Eq. (5-22) will be omitted hereafter, since this term denotes the contribution from the translational motion and is not related to the potential energy.

In order to cast Eq. (5-22) into the quantum mechanical formulation, we must use the angular momentum defined by

$$\boldsymbol{P} = \left[\partial T/\partial \omega_x \quad \partial T/\partial \omega_y \quad \partial T/\partial \omega_z \right]^{\mathrm{T}} \tag{5-23}$$

and the momentum conjugate to each normal coordinate,

$$\bar{p}_j = \partial T/\partial \dot{Q}_j \tag{5-24}$$

instead of the angular velocity $\boldsymbol{\omega}$ and the time derivative of each normal coordinate Q_j, respectively, as the independent variables to describe the kinetic energy. Differentiating Eq. (5-22) with respect to the components of $\boldsymbol{\omega}$ and substituting the results into Eq. (5-23), we have

$$\boldsymbol{P} = \left(\boldsymbol{I}^e + \sum_i \boldsymbol{a}_i Q_i + \sum_i \sum_j \boldsymbol{A}_{ij} Q_i Q_j \right)\boldsymbol{\omega} + \sum_i \sum_j \boldsymbol{\zeta}_{ij} Q_i \dot{Q}_j . \tag{5-25}$$

Similarly, it follows from Eqs. (5-22) and (5-24) that

$$\bar{p}_j = \dot{Q}_j + \tilde{\boldsymbol{\omega}} \sum_i \boldsymbol{\zeta}_{ij} Q_i$$

or

$$\dot{Q}_j = \bar{p}_j - \tilde{\boldsymbol{\omega}} \sum_i \boldsymbol{\zeta}_{ij} Q_i . \tag{5-26}$$

The variable $\dot{Q}_j$ in the last term of Eq. (5-25) can be eliminated by using Eq. (5-26) in which the suffix i in the right side is replaced by k. The result is

$$\boldsymbol{P} = \left(\boldsymbol{I}^e + \sum_i \boldsymbol{a}_i Q_i + \sum_i \sum_j \boldsymbol{A}_{ij} Q_i Q_j \right)\boldsymbol{\omega} + \sum_i \sum_j \boldsymbol{\zeta}_{ij} Q_i \left(\bar{p}_j - \tilde{\boldsymbol{\omega}} \sum_k \boldsymbol{\zeta}_{kj} Q_k \right). \tag{5-27}$$

By collecting the terms not including $\boldsymbol{\omega}$ into a vector

$$\boldsymbol{p} = \sum_i \sum_j \boldsymbol{\zeta}_{ij} Q_i \bar{p}_j \tag{5-28}$$

and using the relation $\tilde{\boldsymbol{\omega}} \boldsymbol{\zeta}_{ij} = \tilde{\boldsymbol{\zeta}}_{ij} \boldsymbol{\omega}$, Eq. (5-27) can be rewritten as

$$\boldsymbol{P} - \boldsymbol{p} = \left(\boldsymbol{I}^e + \sum_i \boldsymbol{a}_i Q_i + \sum_i \sum_j \boldsymbol{A}_{ij} Q_i Q_j - \sum_i \sum_j \boldsymbol{\zeta}_{ij} \sum_k \tilde{\boldsymbol{\zeta}}_{kj} Q_i Q_k \right) \boldsymbol{\omega} \tag{5-29}$$

Exchanging the suffixes j and k in the fourth term in parentheses in Eq. (5-29) as

$$\sum_i \sum_k \boldsymbol{\zeta}_{ik} \left(\sum_j \tilde{\boldsymbol{\zeta}}_{jk} Q_i Q_j \right) = \sum_i \sum_j \left(\sum_k \boldsymbol{\zeta}_{ik} \tilde{\boldsymbol{\zeta}}_{jk} \right) Q_i Q_j,$$

and introducing an abbreviated notation

$$\boldsymbol{A}'_{ij} = \boldsymbol{A}_{ij} - \sum_k \boldsymbol{\zeta}_{ik} \tilde{\boldsymbol{\zeta}}_{jk},$$

we obtain

$$\boldsymbol{P} - \boldsymbol{p} = \left(\boldsymbol{I}^e + \sum_i \boldsymbol{a}_i Q_i + \sum_i \sum_j \boldsymbol{A}'_{ij} Q_i Q_j \right) \boldsymbol{\omega}$$

or simply

$$\boldsymbol{P} - \boldsymbol{p} = \boldsymbol{I}' \boldsymbol{\omega} \tag{5-30}$$

where

$$\boldsymbol{I}' = \boldsymbol{I}^e + \sum_i \boldsymbol{a}_i Q_i + \sum_i \sum_j \boldsymbol{A}'_{ij} Q_i Q_j = \boldsymbol{I} - \sum_i \sum_j \left(\sum_k \boldsymbol{\zeta}_{ik} \tilde{\boldsymbol{\zeta}}_{jk} \right) Q_i Q_j. \tag{5-31}$$

According to Eq. (5-30), the total angular momentum minus $\boldsymbol{p}$ is proportional to the angular velocity $\boldsymbol{\omega}$ of the molecular rotation. The quantity $\boldsymbol{p}$ represents the angular momentum associated with the intramolecular vibrations of the molecule, and is called the internal angular momentum.

We may write the inverse transformation of Eq. (5-30) as

$$\boldsymbol{\omega} = \boldsymbol{I}'^{-1}(\boldsymbol{P} - \boldsymbol{p}) = \boldsymbol{\mu}\,(\boldsymbol{P} - \boldsymbol{p}). \tag{5-32}$$

From Eqs. (5-31) and (5-32), it follows that

$$\begin{aligned} \tilde{\boldsymbol{\omega}} \boldsymbol{I} \boldsymbol{\omega} &= \tilde{\boldsymbol{\omega}} \left\{ \boldsymbol{I}' + \sum_i \sum_j \left(\sum_k \boldsymbol{\zeta}_{ik} \tilde{\boldsymbol{\zeta}}_{jk} \right) Q_i Q_j \right\} \boldsymbol{\omega} \\ &= (\boldsymbol{P} - \boldsymbol{p})^{\mathrm{T}} \boldsymbol{\mu} (\boldsymbol{P} - \boldsymbol{p}) + \tilde{\boldsymbol{\omega}} \sum_i \sum_j \left(\sum_k \boldsymbol{\zeta}_{ik} \tilde{\boldsymbol{\zeta}}_{jk} \right) \boldsymbol{\varphi}\, Q_i Q_j. \end{aligned} \tag{5-33}$$

Interchanging the subscripts i and j in Eq. (5-26), and taking the sum of squares over i, we have

$$\begin{aligned}\sum_i \dot{Q}_i^2 &= \sum_i \left(\bar{p}_i^{\,2} - 2\bar{p}_i \sum_j \tilde{\boldsymbol{\zeta}}_{ji} Q_j \boldsymbol{\omega} + \tilde{\boldsymbol{\omega}} \sum_j \sum_k \boldsymbol{\zeta}_{ji} \tilde{\boldsymbol{\zeta}}_{ki} Q_j Q_k \boldsymbol{\omega} \right) \\ &= \sum_i \bar{p}_i^{\,2} - 2\tilde{\boldsymbol{p}}\boldsymbol{\omega} + \tilde{\boldsymbol{\omega}} \left\{ \sum_j \sum_k \left(\sum_i \boldsymbol{\zeta}_{ji} \tilde{\boldsymbol{\zeta}}_{ki} \right) Q_j Q_k \right\} \boldsymbol{\omega},\end{aligned} \tag{5-34}$$

where use has been made of Eq. (5-28). On the other hand, by premultiplying Eq. (5-26) by $2\tilde{\boldsymbol{\omega}}\boldsymbol{\zeta}_{ij} Q_i$, summing over *i* and *j*, and using Eq. (5-28), it follows that

$$\begin{aligned}2\tilde{\boldsymbol{\omega}} \sum_i \sum_j \boldsymbol{\zeta}_{ij} Q_i \dot{Q}_j &= 2\tilde{\boldsymbol{\omega}} \sum_i \sum_j \boldsymbol{\zeta}_{ij} Q_i \left(\bar{p}_j - \sum_k \tilde{\boldsymbol{\zeta}}_{kj} \boldsymbol{\omega} Q_k \right) \\ &= 2\tilde{\boldsymbol{\omega}} \left(\boldsymbol{p} - \sum_i \sum_j \sum_k \boldsymbol{\zeta}_{ij} \tilde{\boldsymbol{\zeta}}_{kj} \boldsymbol{\omega} Q_i Q_k \right) \\ &= 2\tilde{\boldsymbol{p}}\boldsymbol{\omega} - 2\tilde{\boldsymbol{\omega}} \sum_j \sum_k \sum_i \boldsymbol{\zeta}_{ji} \tilde{\boldsymbol{\zeta}}_{ki} \boldsymbol{\omega} Q_j Q_k.\end{aligned} \tag{5-35}$$

By dropping the first term in braces in Eq. (5-22) and rewriting the remainder by using eqs. (5-33–35), the kinetic energy can be expressed only in terms of the momentum as given by

$$T = \tfrac{1}{2} \sum_i \bar{p}_i^{\,2} + \tfrac{1}{2} \left(\tilde{\boldsymbol{P}} - \tilde{\boldsymbol{p}} \right) \boldsymbol{\mu} \left(\boldsymbol{P} - \boldsymbol{p} \right). \tag{5-36}$$

In order to take account of the change in tensor of inertia due to the molecular deformation, the inverse tensor of inertia $\boldsymbol{\mu} = \boldsymbol{I}'^{-1}$ is expanded in powers of the normal coordinates as

$$\boldsymbol{\mu} = \boldsymbol{\mu}^e + \sum_i \boldsymbol{\mu}_i^{(1)} Q_i + \sum_i \sum_j \boldsymbol{\mu}_i^{(2)} Q_i Q_j + \cdots. \tag{5-37}$$

The components of the vector $\boldsymbol{\mu}_i^{(1)}$ and the matrix $\boldsymbol{\mu}_i^{(2)}$ of the coefficients are calculated as follows. Since the product of Eqs. (5-37) and (5-31) is a unit matrix, it follows that

$$\boldsymbol{\mu}\boldsymbol{I}' = \boldsymbol{\mu}^e \boldsymbol{I}^e + \sum_i \left(\boldsymbol{\mu}_i^{(1)} \boldsymbol{I}^e + \boldsymbol{\mu}^e \boldsymbol{a}_i \right) Q_i + \sum_i \sum_j \left(\boldsymbol{\mu}_i^{(2)} \boldsymbol{I}^e + \boldsymbol{\mu}_i^{(1)} \boldsymbol{a}_i + \boldsymbol{\mu}^e \boldsymbol{A}'_{ij} \right) Q_i Q_j + \cdots = \boldsymbol{E}.$$

For the first term of the above equation to be a unit matrix, it should hold that

$$\left(\boldsymbol{\mu}^e \right)_{\alpha\alpha} = 1/I_\alpha^e. \tag{5-38a}$$

Next we equate the coefficients of the first- and the second-order terms in Q_i to zero, and obtain the following:

$$\left(\boldsymbol{\mu}_i^{(1)} \right)_{\alpha\beta} = -\left(\boldsymbol{\mu}^e \boldsymbol{a}_i \boldsymbol{\mu}^e \right)_{\alpha\beta} = -a_i^{\alpha\beta} \Big/ \left(I_\alpha^e I_\beta^e \right) \tag{5-38b}$$

$$\begin{aligned}\left(\boldsymbol{\mu}_i^{(2)} \right)_{\alpha\beta} &= -\left(\boldsymbol{\mu}^e \boldsymbol{a}_i \boldsymbol{\mu}^e \boldsymbol{a}_i \boldsymbol{\mu}^e \right)_{\alpha\beta} + \left(\boldsymbol{\mu}^e \boldsymbol{A}'_{ij} \boldsymbol{\mu}^e \right)_{\alpha\beta} \\ &= \left\{ \sum_\gamma \left(a_i^{\alpha\gamma} a_i^{\gamma\beta} / I_\gamma^e \right) + A_{ij}^{\alpha\beta} \right\} \Big/ \left(I_\alpha^e I_\beta^e \right).\end{aligned} \tag{5-38c}$$

For the purpose of analyzing the vibrational anharmonicity, it is convenient to use the dimensionless normal coordinates[3] introduced in Chapter 4 (Eq. (4-48)). In the dimensionless normal coordinate system, the components of the internal angular momentum $\boldsymbol{p}$ (Eq. (5-28)) are given by

$$\begin{aligned} p_\alpha &= \sum_i \sum_j \zeta_{ij}^\alpha \left(\frac{\hbar}{2\pi c \omega_i} \right)^{1/2} q_i \left(\frac{\partial T}{\partial q_j} \right) \left(\frac{2\pi c \omega_j}{\hbar} \right)^{1/2} \\ &= \sum_i \sum_j \zeta_{ij}^\alpha \left(\frac{\omega_j}{\omega_i} \right)^{1/2} q_i \, p_j \\ &= \sum_i \sum_{<j} \zeta_{ij}^\alpha \left\{ \left(\frac{\omega_j}{\omega_i} \right)^{1/2} q_i \, p_j - \left(\frac{\omega_i}{\omega_j} \right)^{1/2} p_i \, q_j \right\}. \end{aligned} \tag{5-39}$$

In the last formula in Eq. (5-39), the sum over j is limited to the range $i < j$ by noting that the relation $\zeta_{ij}^\alpha = -\zeta_{ji}^\alpha$ follows from the definition of the Coriolis coupling constants, Eq. (5-16). This form will be utilized in later calculations.

On expanding the potential energy in powers of the dimensionless normal coordinates, all the terms can be put in parentheses leaving a common factor hc outside as shown by

$$V = hc \left(\sum_i \omega_i \, q_i^2 + \sum_i \sum_{\le j} \sum_{\le k} k_{ijk} \, q_i q_j q_k + \sum_i \sum_{\le j} \sum_{\le k} \sum_{\le l} k_{ijkl} \, q_i q_j q_k q_l + \cdots \right). \tag{5-40}$$

Since q_1, q_2, ... are dimensionless, the dimension of all the coefficients ω_i, k_{ijk}, k_{ijkl}, ... in parentheses in Eq. (5-40) should be the same as that of V/hc. Since the dimension of hc is given by

$$[hc] = [\text{energy} \times \text{time} \times \text{distance/time}] = [\text{energy} \times \text{distance}]$$

and the dimension of V/hc is accordingly [distance^{-1}], the unit cm^{-1} widely used in spectroscopic experiments can be used for all the potential parameters in Eq. (5-40).

Substituting Eqs. (5-37) and (5-38a–c) into Eq. (5-36), transforming the Q_i's into the corresponding q_i's and collecting the terms of the same powers, we obtain an expression for the kinetic energy in the form

$$\begin{aligned} T = \frac{hc}{2} \sum_i \omega_i \frac{p_i^2}{\hbar^2} &+ \tfrac{1}{2} \sum_\alpha \frac{P_\alpha^{\,2} + 2\, p_\alpha P_\alpha + p_\alpha^{\,2}}{I_\alpha^e} \\ &+ \tfrac{1}{2} \sum_\alpha \sum_\beta \frac{1}{I_\alpha^e I_\beta^e} \left\{ \sum_i a_i^{\alpha\beta} \left(\frac{\hbar}{2\pi c \omega_i} \right)^{1/2} \left(P_\alpha P_\beta q_i + p_\alpha q_i P_\beta + P_\alpha q_i p_\beta \right) \right. \\ &\left. + \sum_i \sum_j \left(\sum_\gamma \frac{a_i^{\alpha\gamma} a_j^{\beta\gamma}}{I_\gamma^e} + A_{ij}^{\alpha\beta} \right) \frac{\hbar}{2\pi c (\omega_i \omega_j)^{1/2}} P_\alpha P_\beta q_i q_j \right\} + \cdots. \end{aligned} \tag{5-41}$$

The next step is to construct the Hamiltonian as the sum of Eqs. (5-40) and (5-41), and rearrange the terms by collecting those with the same powers in each of P_α, p_i and q_i. Note that p_α is quadratic in p_i and q_i, as seen from Eq. (5-39). Since the terms with the least power are quadratic, let the collection of the terms of the nth power be denoted by H_{n-2}. The total Hamiltonian is then expressed as a sum of partial Hamiltonians each consisting of terms in the same order of approximation in the form

$$H = T + V = H_0 + H_1 + H_2 + \cdots \tag{5-42}$$

where

$$H_0 = \tfrac{1}{2}\left\{\sum_\alpha \frac{P_\alpha{}^2}{I_\alpha^e} + hc\sum_i \omega_i\left(\frac{p_i{}^2}{\hbar^2} + q_i{}^2\right)\right\} \tag{5-43a}$$

$$H_1 = -\sum_\alpha \frac{p_\alpha P_\alpha}{I_\alpha^e} - \tfrac{1}{2}\sum_\alpha\sum_\beta \frac{P_\alpha P_\beta}{I_\alpha^e I_\beta^e}\sum_i a_i^{\alpha\beta}\left(\frac{\hbar}{2\pi c\omega_i}\right)^{1/2} q_i + hc\sum_i\sum_{\le j}\sum_{\le k} k_{ijk}\, q_i q_j q_k \tag{5-43b}$$

and

$$H_2 = \tfrac{1}{2}\sum_\alpha \frac{p_\alpha{}^2}{I_\alpha^e} + \tfrac{1}{2}\sum_\alpha\sum_\beta \frac{P_\alpha P_\beta}{I_\alpha^e I_\beta^e}\left\{\sum_i\sum_j\left(\sum_\gamma \frac{a_i^{\alpha\gamma} a_j^{\beta\gamma}}{I_\gamma^e} + A_{ij}^{\alpha\beta}\right)\frac{\hbar}{2\pi c(\omega_i\omega_j)^{1/2}} q_i q_j\right\}$$
$$- \tfrac{1}{2}\sum_i\sum_\alpha\sum_\beta \frac{a_i^{\alpha\beta}}{I_\alpha^e I_\beta^e}\left(\frac{\hbar}{2\pi c\omega_i}\right)^{1/2}(p_\alpha q_i + q_i p_\alpha)P_\beta \tag{5-43c}$$
$$+ hc\sum_i\sum_{\le j}\sum_{\le k}\sum_{\le l} k_{ijkl}\, q_i q_j q_k q_l .$$

5.3 Perturbation Theory of Anharmonic Vibrations

On evaluating improved energies of the nuclear motion by applying the perturbation theory to the vibrational-rotational Hamiltonian, the second-order perturbation energy should be taken into account from the beginning since the first-order energy vanishes. In this case, the process of deriving the formulae for the perturbation energies can be greatly simplified by eliminating the first-order Hamiltonian with a mathematical technique called the contact transformation (Van Vleck transformation).[4, 5]

Following the standard process of the perturbation theory, we write the Hamiltonian as a sum of terms arranged in the order of approximation,

$$H = H_0 + \lambda H_1 + \lambda^2 H_2 + \cdots. \tag{5-44}$$

Taking next an unknown polynomial of the current variables p_i and q_i ($i = 1, 2, \cdots$) and letting it be denoted by S, we introduce an exponential function T defined by

$$T = \exp(i\lambda S) = 1 + i\lambda S - (1/2)\lambda^2 S^2 + \cdots, \tag{5-45}$$

and transform the Hamiltonian H into the form

$$H' = THT^{-1} = e^{i\lambda S} He^{-i\lambda S}. \tag{5-46}$$

Replacing S in Eq. (5-45) with $-S$ to obtain

$$T^{-1} = \exp(-i\lambda S) = 1 - i\lambda S - (1/2)\lambda^2 S^2 + \cdots,$$

substituting the result, together with eqs. (5-44) and (5-45), into Eq. (5-46) and rearranging the terms in ascending order of λ after expanding, we obtain the transformed Hamiltonian,

$$\begin{aligned} H' &= (1 + i\lambda S - \tfrac{1}{2}\lambda^2 S^2)(H_0 + \lambda H_1 + \lambda^2 H_2)(1 - i\lambda S - \tfrac{1}{2}\lambda^2 S^2) \\ &= H_0 + \{H_1 + i(SH_0 - H_0 S)\}\lambda + \{H_2 + i(SH_1 - H_1 S) \\ &\quad - \tfrac{1}{2}(S^2 H_0 - 2SH_0 S + H_0 S^2)\}\lambda^2 + \cdots \\ &= H_0 + H_1' \lambda + H_2' \lambda^2 + \cdots. \end{aligned} \tag{5-47}$$

The operator of the type $(AB - BA)$ which appears repeatedly in Eq. (5-47) is called the commutator of A and B, and is denoted by

$$[A, B] \equiv AB - BA.$$

A pair of operators A and B for which the commutator $[A, B]$ vanishes is said to be commutable. An important commutator is that between a coordinate q_i and its conjugate momentum p_i. By operating $[p_i, q_i]$ on an arbitrary function of p_i and q_i, $f(p_i, q_i)$, and replacing p_i with the operator $-i\hbar(\partial/\partial q_i)$, we have

$$\begin{aligned} (p_i q_i - q_i p_i) f(p_i, q_i) &= -i\hbar\left\{\frac{\partial}{\partial q_i} q_i f(p_i, q_i) - q_i \frac{\partial}{\partial q_i} f(p_i, q_i)\right\} \\ &= -i\hbar f(p_i, q_i). \end{aligned}$$

Hence operating $[p_i, q_i]$ on a function of p_i and q_i is equivalent to multiplying the same function by $-i\hbar$. Thus we may write it as

$$[p_i, q_i] = -i\hbar. \tag{5-48}$$

By using the definition of a commutator, the coefficient of λ in Eq. (5-47), which shall be called the transformed first-order Hamiltonian and be denoted by H_1', is written as

$$H_1' = H_1 + i(SH_0 - H_0 S) = H_1 + i[S, H_0]. \tag{5-49}$$

Accordingly, if we can find a suitable function S for which the relation

$$i[H_0, S] = -i[S, H_0] = H_1 \tag{5-50}$$

holds, H_1' can be eliminated from the transformed Hamiltonian H' according to Eq. (5-49). The coefficient of λ^2 in Eq. (5-47), that is, the transformed second-order Hamiltonian, H_2', can now be greatly simplified as shown in the form

$$\begin{aligned} H_2' &= H_2 + i(SH_1 - H_1 S) - \tfrac{1}{2}\{S(SH_0 - H_0 S) - (SH_0 - H_0 S)S\} \\ &= H_2 + i\{S(H_1 + \tfrac{1}{2}i[S, H_0])\} - i\{(H_1 + \tfrac{1}{2}i[S, H_0])S\} \\ &= H_2 + \tfrac{1}{2}i[S, H_1]. \end{aligned} \tag{5-51}$$

The terms in the first order Hamiltonian H_1 in Eq. (5-43b) are classified into five types according to the number and order of powers of the coordinates and momenta involved. Let the terms involving the factors $p_a q_b$ and $q_a p_b$, which appear on replacing p_α in the first sum in the right side with Eq. (5-39), be denoted by $H_{1;1}(a\ b\ 0)$ and the terms involving q_a in the second sum by $H_{1;2}(a\ 0\ 0)$. Similarly, the terms involving the factors of types q_a^3, $q_a^2 q_b$ and $q_a q_b q_c$ will be denoted by $H_{1;3}(a\ a\ a)$, $H_{1;4}(a\ a\ b)$ and $H_{1;5}(a\ b\ c)$, respectively. Then the first-order Hamiltonian H_1 is written in the form

$$\begin{aligned} H_1 &= \sum_a \sum_{\neq b} H_{1;1}(a b 0) + \sum_a H_{1;2}(a 0 0) + \sum_a H_{1;3}(a a a) \\ &+ \sum_a \sum_{\neq b} H_{1;4}(a b 0) + \sum_a \sum_{<b} \sum_{<c} H_{1;5}(a b c). \end{aligned} \tag{5-52}$$

Since the linearity in operating the commutator,

$$\left[A, \sum_i B_i\right] = \sum_i (A B_i) - \left(\sum_i B_i\right) A = \sum_i [A, B_i],$$

is obvious from the definition, the total S-function which satisfies Eq. (5-50) is obtained by choosing such an appropriate partial S-function $S_\rho(a\ b\ c)$ for each type of $\rho = 1$–5 that the relation

$$i\left[H_0, S_\rho(a b c)\right] = H_{1;\rho}(a b c). \tag{5-53}$$

holds, and collecting them together in the form

$$S = \sum_\rho \sum_{a,b,c} S_\rho(a b c). \tag{5-54}$$

The range of the sum over a, b and c for each ρ in Eq. (5-54) is the same as that for the corresponding type in Eq. (5-52). The five types of terms $H_{1;\rho}(a\ b\ c)$ involved in the first order Hamiltonian and the partial S-functions $S_\rho(a\ b\ c)$ used for eliminating them are listed in Table 5.1.

For deriving these S-functions, according to Herman and Schaffer,[4] we need to evaluate preliminarily certain commutators including the dimensionless normal coordinates, their conjugate momenta and their products as follows. From a repeated use of Eq. (5-48), it follows that

$$\begin{aligned} [p, q^2] &= p q^2 - q^2 p = p q^2 - q p q + q p q - q^2 p \\ &= (p q - q p) q + q (p q - q p) = -2 i \hbar q \end{aligned} \tag{5-55a}$$

$$[p^2, q] = p^2 q - p q p + p q p - q p^2 = -2 i \hbar p. \tag{5-55b}$$

Note that the total power of p and q diminishes by two on commuting the operators.

The construction of S-functions will be illustrated below in the case of $S_1(ab0)$ used for eliminating $H_{1;1}(ab0)$, which consists only of quadratic terms involving the variable factors of the type $q_a p_b$ as shown in Table 5.1. The vibrational part of H_0 in Eq. (5-43a) has quadratic terms involving the variable factors p_a^2 and q_a^2, from which the commutators of

Table 5.1 S-functions used for elimination of the first-order Hamiltonian.

ρ	$S_\rho(abc)$	a_ρ	$H_{1;\rho}(abc)$
1	$-\dfrac{a_1\{(\omega_a{}^2+\omega_b{}^2)q_aq_b+2\omega_a\omega_b p_a p_b/\hbar^2\}}{hc(\omega_a\omega_b)^{1/2}(\omega_a{}^2-\omega_b{}^2)}$	$\sum_\alpha \dfrac{\zeta_{ab}^\alpha}{I_\alpha^e}P_\alpha$	$a_1\left\{\left(\dfrac{\omega_b}{\omega_a}\right)^{1/2}q_a p_b-\left(\dfrac{\omega_a}{\omega_b}\right)^{1/2}p_a q_b\right\}$
2	$-\dfrac{a_2}{\hbar hc\omega_a}p_a$	$\sum_\alpha\sum_\beta \dfrac{a_a^{\alpha\beta}}{I_\alpha^e I_\beta^e}\left(\dfrac{\hbar}{2\pi c\omega_a}\right)^{1/2}P_\alpha P_\beta$	$a_2\, q_a$
3	$\dfrac{2a_3}{\hbar hc\,\omega_a}\left(\dfrac{p_a{}^3}{3\hbar^2}+\dfrac{q_a p_a q_a}{2}\right)$	$hc\,k_{aaa}$	$a_3 q_a{}^3$
4	$\dfrac{a_4}{\hbar hc}\{-a^{aab}\dfrac{p_a{}^2p_b}{\hbar^2}+b_{aa}{}^b q_a{}^2 p_b + b_{ab}{}^a(p_aq_a+q_ap_a)q_b\}$	$hc\,k_{aab}$	$a_4 q_a{}^2 q_b$
5	$\dfrac{a_5}{\hbar hc}\left(-a^{abc}\dfrac{p_ap_bp_c}{\hbar^2}+b_{bc}{}^a p_aq_bq_c + b_{ac}{}^b q_ap_bq_c+b_{ab}{}^c q_aq_bp_c\right)$	$hc\,k_{abc}$	$a_5 q_aq_bq_c$

the desired types

$$[p_a{}^2,\, q_a\, q_b] = -2i\hbar p_a\, q_b \quad\text{and}\quad [q_a{}^2,\, p_a\, p_b] = -2i\hbar q_a\, p_b$$

are formed. Accordingly, the same quadratic terms as those in $H_{1;1}(ab0)$ will be generated in the commutator $i[H_0, S_1(ab0)]$, if appropriate quadratic terms involving the variable factors q_aq_b and p_ap_b are introduced in $S_1(ab0)$. The complete commutators between these factors in $S_1(ab0)$ and H_0 are then given by

$$\frac{i}{\hbar hc}\left[H_0,\frac{-p_a\,p_b}{\hbar^2}\right]=\frac{-i}{2\hbar^2}\{\omega_a[q_a{}^2,p_a]p_b+\omega_b[q_b{}^2,p_b]p_a\}$$
$$=\frac{1}{\hbar}(\omega_a q_a\,p_b+\omega_b p_a\,q_b) \tag{5-56a}$$

and

$$\frac{i}{\hbar hc}[H_0,q_a\,q_b]=\frac{i}{2\hbar^2}\{\omega_a[p_a{}^2,q_a]q_b+\omega_b[p_b{}^2,q_b]q_a\}$$
$$=\frac{1}{\hbar}(\omega_a p_a\,q_b+\omega_b q_a\,p_b). \tag{5-56b}$$

Let the desired S-function be given in the form

$$S_{1;1}(ab0)=\frac{1}{\hbar hc}\left(-\alpha\frac{p_a\,p_b}{\hbar^2}+\beta\, q_a\, q_b\right). \tag{5-57}$$

Substitute Eq. (5-57) into Eq. (5-53), and substitute further Eqs. (5-56a,b) into the left side and $H_{1;1}(ab0)$ in Table 5.1 into the right side. Comparing the coefficients of the terms with the same powers of variables between the right and left sides of the resulting equation, we obtain simultaneous equations with respect to the unknowns α and β in the form

$$\left.\begin{aligned}\omega_a\,\alpha+\omega_b\,\beta&=-a_1(\omega_b/\omega_a)^{1/2}\\ \omega_b\,\alpha+\omega_a\,\beta&=-a_1(\omega_a/\omega_b)^{1/2}\end{aligned}\right\}\tag{5-58}$$

Solving Eq. (5-58), we have

$$\alpha=\frac{a_1}{\omega_a^{\,2}-\omega_b^{\,2}}\left\{\omega_a\left(\frac{\omega_b}{\omega_a}\right)^{1/2}+\omega_b\left(\frac{\omega_a}{\omega_b}\right)^{1/2}\right\}=\frac{2a_1\,\omega_a\,\omega_b}{(\omega_a^{\,2}-\omega_b^{\,2})(\omega_a\,\omega_b)^{1/2}}$$

$$\beta=\frac{-a_1}{\omega_a^{\,2}-\omega_b^{\,2}}\left\{\omega_b\left(\frac{\omega_b}{\omega_a}\right)^{1/2}+\omega_a\left(\frac{\omega_a}{\omega_b}\right)^{1/2}\right\}=\frac{-a_1(\omega_a^{\,2}+\omega_b^{\,2})}{(\omega_a^{\,2}-\omega_b^{\,2})(\omega_a\,\omega_b)^{1/2}}$$

The S-function $S_5(abc)$ to eliminate $H_{1;5}(abc)$, the cubic potential term involving three coordinates $k_{abc}\,q_a\,q_b\,q_c$, is quite complicated. On introducing the factor $p_a\,q_b\,q_c$ in $S_5(abc)$, the commutator with q_a^2 in H_0 is found to include the desired factor $q_a\,q_b\,q_c$, but unnecessary commutators involving factors $p_a\,p_b\,q_c$ and $p_a\,q_b\,p_c$ are generated from p_b^2 and p_c^2 in H_0, respectively. Fortunately, factors $p_a\,p_b\,q_c$ and $p_a\,q_b\,p_c$ are also involved in commutators $[q_c^2, p_a\,p_b\,p_c]$ and $[q_b^2, p_a\,p_b\,p_c]$, respectively, and both q_c^2 and q_b^2 are involved in certain terms in H_0. These facts suggest that the undesired terms generated in $[H_0, p_a\,q_b\,q_c]$ are removable on supplementing $S_5(abc)$ with an appropriate term involving the factor $p_a\,p_b\,p_c$. Considering cyclically, we find immediately that four types of cubic variable factors, $q_a\,q_b\,q_c$, $q_a\,p_b\,p_c$, $p_a\,q_b\,p_c$ and $p_a\,p_b\,q_c$, can be generated in $[H_0, S_5(abc)]$ from four terms in $S_5(abc)$ involving factors $p_a\,p_b\,p_c$, $p_a\,q_b\,q_c$, $q_a\,p_b\,q_c$ and $q_a\,q_b\,p_c$. Hence the coefficients of the four terms in $S_5(abc)$ can be so chosen as to generate $i[H_0, S_5(abc)]$ in which the coefficient of the term involving $q_a\,q_b\,q_c$ is k_{abc} while those of the other three terms vanish altogether. To accomplish this calculation, the following commutators are necessary.

$$\begin{aligned}\frac{i}{\hbar hc}\left[H_0,-\frac{p_ap_bp_c}{\hbar^2}\right]&=\frac{-i}{2\hbar^2}\left\{\omega_a\left[q_a^{\,2},p_a\right]p_bp_c+\omega_b\left[q_b^{\,2},p_b\right]p_ap_c+\omega_c\left[q_c^{\,2},p_c\right]p_ap_b\right\}\\&=(\omega_aq_ap_bp_c+\omega_bp_aq_bp_c+\omega_cp_ap_bq_c)/\hbar^2\end{aligned}\tag{5-59a}$$

$$\begin{aligned}\frac{i}{\hbar hc}[H_0,p_aq_bq_c]&=\frac{i}{2}\left\{\omega_a\left[q_a^{\,2},p_a\right]q_bq_c+\omega_b\left[\frac{p_b^{\,2}}{\hbar^2},q_b\right]p_aq_c+\omega_c\left[\frac{p_c^{\,2}}{\hbar^2},q_c\right]p_aq_b\right\}\\&=(\omega_aq_aq_bq_c+\omega_bp_ap_bq_c+\omega_cp_aq_bp_c)\end{aligned}\tag{5-59b}$$

$$\frac{i}{\hbar hc}[H_0,q_ap_bq_c]=\omega_ap_ap_bq_c+\omega_bq_aq_bq_c+\omega_cq_ap_bp_c\tag{5-59c}$$

$$\frac{i}{\hbar hc}[H_0,q_aq_bp_c]=\omega_ap_aq_bp_c+\omega_bq_ap_bp_c+\omega_cq_aq_bq_c\tag{5-59d}$$

Let $S_5(abc)$ be expressed in the form

$$S_5(abc) = \frac{1}{\hbar hc}\left(-\alpha\frac{p_a p_b p_c}{\hbar^2} + \beta_a p_a q_b q_c + \beta_b q_a p_b q_c + \beta_c q_a q_b p_c\right). \tag{5-60}$$

The coefficients α, β_a, β_b and β_c should be so determined as to satisfy the relation

$$i[H_0, S_5(a\,b\,c)] = a_5\, q_a\, q_b\, q_c. \tag{5-61}$$

Substituting Eq. (5-60) into Eq. (5-61), rewriting the resulting commutators term by term as indicated by Eqs. (5-59a–d) and equating the coefficients of the corresponding terms in the left and the right sides, we obtain the simultaneous equations with respect to four unknowns α, β_a, β_b and β_c in the form

$$\left.\begin{array}{l} \omega_a\beta_a + \omega_b\beta_b + \omega_c\beta_c = -a_5 \\ \omega_a\alpha + \omega_c\beta_b + \omega_b\beta_c = 0 \\ \omega_b\alpha + \omega_c\beta_a + \omega_a\beta_c = 0 \\ \omega_c\alpha + \omega_b\beta_a + \omega_a\beta_b = 0 \end{array}\right\}. \tag{5-62}$$

Equation (5-62) is easily solved by rewriting it in the form

$$\left.\begin{array}{l} \alpha + \beta_a + \beta_b + \beta_c = -a_5(\omega_a + \omega_b + \omega_c)^{-1} \\ \alpha + \beta_a - \beta_b - \beta_c = -a_5(\omega_a - \omega_b - \omega_c)^{-1} \\ \alpha - \beta_a + \beta_b - \beta_c = a_5(\omega_a - \omega_b + \omega_c)^{-1} \\ \alpha - \beta_a - \beta_b + \beta_c = a_5(\omega_a + \omega_b - \omega_c)^{-1} \end{array}\right\}. \tag{5-63}$$

The solution is then substituted into Eq. (5-60) to give the S-function $S_5(abc)$ shown in Table 5.1, where the following abbreviations are used.

$$a^{abc} = \alpha/a_5 = -(1/4)\{(\omega_a + \omega_b + \omega_c)^{-1} + (\omega_a - \omega_b - \omega_c)^{-1} + (-\omega_a + \omega_b - \omega_c)^{-1} + (-\omega_a - \omega_b + \omega_c)^{-1}\}$$

$$b^a_{bc} = \beta_a/a_5 = -(1/4)\{(\omega_a + \omega_b + \omega_c)^{-1} + (\omega_a - \omega_b - \omega_c)^{-1} - (-\omega_a + \omega_b - \omega_c)^{-1} - (-\omega_a - \omega_b + \omega_c)^{-1}\}$$

$$b^b_{ac} = \beta_b/a_5 = -(1/4)\{(\omega_a + \omega_b + \omega_c)^{-1} - (\omega_a - \omega_b - \omega_c)^{-1} + (-\omega_a + \omega_b - \omega_c)^{-1} - (-\omega_a - \omega_b + \omega_c)^{-1}\}$$

$$b^c_{ab} = \beta_c/a_5 = -(1/4)\{(\omega_a + \omega_b + \omega_c)^{-1} - (\omega_a - \omega_b - \omega_c)^{-1} - (-\omega_a + \omega_b - \omega_c)^{-1} + (-\omega_a - \omega_b + \omega_c)^{-1}\}.$$

The commutators between these S-functions and H_1 contribute to the second order transformed Hamiltonian H'_2 according to Eq. (5-51). Since the first order transformed

Hamiltonian H'_1 is eliminated by the contact transformation, the second order perturbation energies are given directly as the diagonal elements of the matrix of H'_2 with respect to the zeroth-order eigenfunctions. In H'_2, the terms which give non-vanishing diagonal elements should be in even powers for each of the variables, P_α ($\alpha = X, Y, Z$), p_i and q_i, because the energy of the system is invariant on the change of sign of any one of these variables. The second-order non-transformed Hamiltonian in which only such terms are retained is given by

$$H_2 = \tfrac{1}{2}\sum_{i}\sum_{\neq j}\sum_{\alpha}\frac{\zeta_{ij}^{\alpha 2}}{2 I_\alpha^e}\frac{\omega_j}{\omega_i}\left(q_i^2 p_j^2 + p_j^2 q_i^2\right)$$
$$+\tfrac{1}{2}\sum_{\alpha}\left(\frac{P_\alpha}{I_\alpha^e}\right)^2\left\{\sum_i\left(\sum_\gamma\frac{a_i^{\alpha\gamma 2}}{I_\gamma^e} + A_{ii}^{\alpha\alpha}\right)\frac{2\pi c\hbar}{\omega_i^2}q_i^2\right\} + hc\sum_i\sum_{\leq j}k_{iijj}\,q_i^2\,q_j^2. \qquad (5\text{-}64)$$

The first term in Eq. (5-64) arises from the first term in Eq. (5-43c) into which Eq. (5-39) is substituted. The commutators between the important S-functions and H_1 in the non-degenerate case are shown in Table 5.2 together with their matrix elements which contribute to the perturbation energy.

5.4 Anharmonicity Constants and Anharmonic Resonances

Among the matrix elements shown in Table 5.2, those quadratic in the factor depending on the vibrational quantum number, (v_i + 1/2), constitute the terms representing the second order perturbation to the vibrational energy. The improved vibrational energy levels are given by adding these terms to the unperturbed energy levels in Eq. (4-69), in the form[6)]

$$G(v_1, v_2, \cdots) = \sum_i \omega_i(v_i + \tfrac{1}{2}) + \sum_i\sum_{\leq j} x_{ij}(v_i + \tfrac{1}{2})(v_j + \tfrac{1}{2}). \qquad (5\text{-}65)$$

The constant x_{ij} is called the anharmonicity constant, and is given in terms of the cubic and the quartic force constants in the dimensionless normal coordinates as

$$x_{ii} = \frac{1}{4}\left\{6\,k_{iiii} - 15\frac{k_{iii}^2}{\omega_i} - \sum_j\frac{k_{iij}^2}{\omega_i}\left(\frac{8\omega_i^2 - 3\omega_j^2}{4\omega_i^2 - \omega_j^2}\right)\right\} \qquad (5\text{-}66a)$$

and

$$x_{ij} = \frac{1}{2}k_{iijj} - 6\frac{k_{iii}\,k_{ijj}}{\omega_i} - 4\frac{k_{iij}^2\,\omega_i}{4\omega_i^2 - \omega_j^2} - \sum_k\frac{k_{iik}\,k_{jjk}}{\omega_k}$$
$$-\frac{1}{8}\sum_k k_{ijk}^2\left(\frac{1}{\omega_i + \omega_j + \omega_k} + \frac{1}{\omega_i + \omega_j - \omega_k} + \frac{1}{\omega_i - \omega_j + \omega_k} - \frac{1}{\omega_i - \omega_j - \omega_k}\right)$$
$$+2\sum_\alpha\frac{(\zeta_{ij}^\alpha)^2 h}{8\pi^2 c I_\alpha^e}\frac{\omega_j}{\omega_i}. \qquad (5\text{-}66b)$$

The first and last terms in Eq. (5-66b) are derived from the matrix elements of the third and first terms in Eq. (5-64), respectively. The remaining terms in Eq. (5-66b) arise from the relevant matrix elements shown in Table 5.2.

Table 5.2 The terms in the second-order Hamiltonian derived from S-functions and the corresponding matrix elements

ρ	τ	$H'_{2;\tau,\rho}=(i/2)[S_\rho,H_{1;\tau}]$	$\langle\psi^{(0)}\|H'_{2;\tau,\rho}\|\psi^{(0)}\rangle$
1	1	$\frac{a_1{}^2\hbar^2}{2hc}\left(J_{ab}\frac{q_a{}^2}{\omega_a}+J_{ba}\frac{q_b{}^2}{\omega_b}\right)$	$\frac{a_1{}^2\hbar^2}{2hc}\left\{\frac{J_{ab}}{\omega_a}(v_a+\frac{1}{2})+\frac{J_{ba}}{\omega_b}(v_b+\frac{1}{2})\right\}$
2	2	$-\frac{a_2{}^2}{2hc\omega_a}$	$-\frac{a_2{}^2}{2hc\omega_a}$
2	3	$\frac{3a_2a_3q_a{}^2}{2hc\omega_a}$	$\frac{3a_2a_3}{2hc\omega_a}(v_a+\frac{1}{2})$
3	2	$\frac{3a_3a_2}{hc\omega_a}\left(\frac{p_a{}^2}{\hbar^2}+\frac{q_a{}^2}{2}\right)$	$\frac{3a_3a_2}{2hc\omega_a}(v_a+\frac{1}{2})$
3	3	$\frac{3a_3{}^2}{2hc\omega_a}\left(\frac{p_a{}^2q_a{}^2+q_a{}^2p_a{}^2}{\hbar^2}+\frac{2}{3}+q_a{}^4\right)$	$-\frac{a_3{}^2}{hc\omega_a}\left\{\frac{15}{4}(v_a+\frac{1}{2})^2+\frac{7}{16}\right\}$
2[b)]	4	$-\frac{a_2'a_4q_a{}^2}{2hc\omega_{a'}}$	$-\frac{a_2'a_4}{2hc\omega_{a'}}(v_a+\frac{1}{2})$
4[b)]	2′	$\frac{a_4a_2'}{2hc\omega_{a'}}\left(-a^{aab}\frac{p_a{}^2}{\hbar^2}+b_{aa}{}^bq_a{}^2\right)$	$\frac{a_4a_2'}{2hc\omega_{a'}}(-a^{aab}+b_{aa}{}^b)(v_a+\frac{1}{2})$
3	4′	$-\frac{a_3a_4'}{hc\omega_{a'}}\left(\frac{p_a{}^2}{\hbar^2}+\frac{q_a{}^2}{2}\right)q_{a'}{}^2$	$-\frac{3a_3a_4'}{2hc\omega_{a'}}(v_a+\frac{1}{2})(v_{a'}+\frac{1}{2})$
4[b)]	4	$-\frac{a_4{}^2}{2hc}\left(a^{aab}\frac{p_a{}^2q_a{}^2+q_a{}^2p_a{}^2}{\hbar^2}+b_{aa}{}^bq_a{}^4+b_{ab}{}^aq_a{}^2q_b{}^2\right)$	$-\frac{a_4{}^2}{4hc}\left\{(a^{aab}+3b_{aa}{}^b)(v_a+\frac{1}{2})^2+b_{ab}{}^a(v_a+\frac{1}{2})(v_b+\frac{1}{2})\right\}$
4[b)]	4″	$-\frac{a_4a_4''}{2hc}\left(a^{aab}\frac{p_a{}^2q_{a''}{}^2}{\hbar^2}+b_{aa}{}^bq_a{}^2q_b{}^2\right)$	$-\frac{a_4a_4''}{2hc}(a^{aab}+b_{aa}{}^b)(v_a+\frac{1}{2})(v_{a''}+\frac{1}{2})$
5	5	$-\frac{a_5{}^2}{2hc}\left(\frac{a^{abc}}{4}+b_{bc}{}^aq_b{}^2q_c{}^2+b_{ac}{}^bq_a{}^2q_c{}^2+b_{ab}{}^cq_a{}^2q_b{}^2\right)$	$\frac{a_5{}^2}{2hc}\left\{\left(\frac{a^{abc}}{4}+b_{bc}{}^a(v_b+\frac{1}{2})(v_c+\frac{1}{2})+b_{ac}{}^b(v_a+\frac{1}{2})(v_c+\frac{1}{2})+b_{ab}{}^c(v_a+\frac{1}{2})(v_b+\frac{1}{2})\right\}$

a) $J_{ab}=(3\omega_a{}^2+\omega_b{}^2)/(\omega_a{}^2-\omega_b{}^2)$

b) a_ρ' and a_ρ'' (ρ = 2, 4) are obtained from a_ρ in Table 5.1 by changing the suffix a to a' and a'', respectively.

On taking account of the effect of anharmonicity up to the second-order, the zero point energy is given, by substituting $v_1=v_2=\cdots=0$ into Eq. (5-65), in the form

$$G(0,0,\cdots)=\tfrac{1}{2}\sum_i\omega_i+\tfrac{1}{4}\sum_i\sum_{\le j}x_{ij}. \tag{5-67}$$

Let the transition energy from the zero point level to the level specified by the vibrational quantum numbers v_1, v_2, $\cdots$ be denoted by $(v_1\nu_1+v_2\nu_2+\cdots)$ in wavenumbers. This energy is given in the general form

$$
\begin{aligned}
(v_1\nu_1 + v_2\nu_2 + \cdots) &= G(v_1, v_2, \cdots) - G(0,0,\cdots) \\
&= \sum_i v_i\omega_i + \sum_i \sum_{\leq j} x_{ij}\left\{v_i v_j + \tfrac{1}{2}(v_i + v_j)\right\} \\
&= \sum_i \left\{\omega_i + x_{ii}(1 + v_i) + \tfrac{1}{2}\sum_{k\neq i} x_{ik}(1 + v_k)\right\} v_i .
\end{aligned}
\tag{5-68}
$$

For the fundamental levels, which are defined by a single quantum number being 1 and all the others vanishing, we have

$$
(\nu_i) = \omega_i + 2x_{ii} + \tfrac{1}{2}\sum_{k\neq i} x_{ik} . \tag{5-69}
$$

The levels called the first overtones are specified by a single quantum number being 2 and the others being 0. These levels have the transition energies

$$
(2\nu_i) = 2\omega_i + 6x_{ii} + \sum_{k\neq i} x_{ik} . \tag{5-70}
$$

It follows from Eqs. (5-69) and (5-70) that the anharmonicity constant x_{ii} can be calculated from the experimental data of fundamental and first overtone frequencies by simple arithmetic in the form

$$
x_{ii} = \{(2\nu_i) - 2\times(\nu_i)\}/2 . \tag{5-71}
$$

For the combination tones specified by two quantum numbers being 1 and the others being 0, we have

$$
(\nu_i + \nu_j) = \omega_i + \omega_j + 2(x_{ii} + x_{ii}) + \tfrac{1}{2}\sum_{k\neq i,j}(x_{ik} + x_{jk}) + x_{ij} . \tag{5-72}
$$

By combining Eq. (5-72) with the expressions for the corresponding two fundamentals, the formula for estimating x_{ij} experimentally is obtained to be

$$
x_{ij} = (\nu_i + \nu_j) - (\nu_i) - (\nu_j).
$$

The anharmonicity constants x_{ii} and x_{ij} amount to about 5–7 % of the corresponding fundamental frequencies ν_i and ν_j when the normal coordinates Q_i and Q_j involve the stretching of a bond attached to a hydrogen atom at an end, while they are mostly less than 2 % of the corresponding ν_i and ν_j in the other cases. In the case of highly excited vibrational levels related to the internal rotation around a single bond, the nuclear displacements are so large that the perturbation theory based on the small displacements of nuclei cannot be applied. Such a large distortion should be treated separately as a special case. The parameters in the torsional potential in Eq. (2-58) may be determined provided that more than one transition frequency is observed.[7] In such cases, the parameters used in the analysis of experimental data are available in the literature usually with the observed frequencies, and can be directly referred to in constructing the model potential function. It should be noted, however, that the effect of these parameters corresponds to the combined effect of the torsional potential terms and the nonbonded interaction terms in the potential function of the whole molecule used in molecular mechanics calculations.

For most vibrational modes other than the internal rotation modes, the anharmonicity constants or simply the observed vibrational frequencies are reported in the literature of experimental spectroscopy. To derive higher order potential parameters from these data, we must rely on the second-order perturbation theory or some other method of analyzing the vibrational anharmonicity.

If a normal frequency ω_j happens to nearly coincide with twice another normal frequency $2\omega_i$ or with the sum of two other frequencies $\omega_i + \omega_k$, the procedure of the perturbation theory must be changed partly.[6] In the case when both the conditions $\omega_j \approx 2\omega_i$ and $k_{iij} \neq 0$ hold, a term with a nearly vanishing denominator appears in each of Eqs. (5-66a) and (5-66b), leading to abnormally large anharmonicity constants x_{ii} and x_{ij}, respectively. In this case, the wave functions for the states $(v_i,v_j) = (0,1)$ and $(v_i,v_j) = (2,0)$ mix each other considerably, giving rise to an unusually large intensity of the overtone band in the infrared absorption and/or Raman spectra due to the transition from v_i to $v_i + 2$. This phenomenon is called the Fermi-Dennison resonance.[6] Similarly, if $\omega_j \approx \omega_i + \omega_k$, and $k_{ijk} \neq 0$, the anharmonicity constant x_{ij} is blown up by the nearly vanishing denominator in a term in Eq. (5-66b), and an abnormal behavior of the combination band $(\nu_i + \nu_k)$ is observed in the vibrational spectra. This phenomenon is called the Fermi resonance. The Fermi-Dennison and the Fermi resonances are jointly called first-order resonances. The nearly vanishing denominators appearing in Eqs. (5-66a,b) in the cases of first-order resonances are called the resonance denominators.

Since the perturbation theory is based on the assumption that the disturbance due to higher order terms is small, it cannot be applied to the case when any anharmonicity constants diverge. The divergence of anharmonicity constants can be avoided simply by dropping the terms containing the resonance denominators from the S-functions listed in Table 5.1. In this case, some of the terms containing the cubic force constants k_{iij} and k_{ijk} in H_1' are left unremoved, whence the anharmonicity corrections expressed as the diagonal matrix elements of H_2' in terms of the anharmonicity constants should be revised by using the related off-diagonal matrix elements of the unremoved terms in H_1'. In the case of the Fermi-Dennison resonance, $\omega_j \approx 2\omega_i$ for example, the eigenvalues of the two-dimensional block on the diagonal

$$\begin{bmatrix} \langle 2\nu_i|H_2'|2\nu_i\rangle & \langle 2\nu_i|H_1'|\nu_j\rangle \\ \langle \nu_j|H_1'|2\nu_i\rangle & \langle \nu_j|H_2'|\nu_j\rangle \end{bmatrix}$$

are calculated numerically, and are adopted as corrections to the energies instead of the diagonal elements $\langle 2\nu_i|H_2'|2\nu_i\rangle$ and $\langle \nu_j|H_2'|\nu_j\rangle$ themselves. In this case, it follows from the assumption that

$$\langle 2\nu_i|H_2'|2\nu_i\rangle \cong \langle \nu_j|H_2'|\nu_j\rangle,$$

whence the revised vibrational levels are separated from each other approximately by $2\langle 2\nu_i|H_1'|\nu_j\rangle$. Note that the mixing between the two levels $(2\nu_i)$ and (ν_j) approaches 1 : 1 as the difference between the two diagonal elements diminishes.

As clarified from the above discussion, if the cubic term $k_{iij}\, q_i^2\, q_i$ in the potential function is absent, the resonance between $(2\nu_i)$ and (ν_j) does not take place even if $\omega_j \approx 2\omega_i$. Since the Hamiltonian H is totally symmetric and all cubic terms appearing in H are also totally symmetric, the cubic constant k_{iij} must vanish if the product $q_i^2q_j$ is not totally symmetric. In order for $q_i^2q_j$ to be totally symmetric, the wave functions $|2v_i\rangle$ and $|v_j\rangle$ must belong to the same irreducible representation as each other. Similarly, if the wave functions $|v_i + v_k\rangle$ and $|v_j\rangle$ belong to the same irreducible representation as each other, the cubic constant k_{ijk} may appear in H.

The symmetry property of the product of two normal coordinates q_i and q_j, q_iq_j, is described by the product of the irreducible representations of q_i and q_j. Consider first the case when both the irreducible representations of q_i and q_j are one-dimensional. In order for the product of two one-dimensional irreducible representations for a given symmetry operation to be unity, the two representations, being either +1 or −1, must be the same as each other in every symmetry operation. In other words, the normal coordinates q_i and q_j must belong to the same irreducible representation as each other in order for their product to be totally symmetric. The product of two degenerate irreducible representations is a reducible representation in general. On reducing it, a totally symmetric irreducible representation appears only when and always when the two degenerate representations are the same as each other. Hence it is generally concluded that the necessary and sufficient condition for the appearance of the totally symmetric product q_iq_j is that q_i and q_j belong to the same irreducible representation as each other. Finally, the form of the wave functions of harmonic oscillators in Eq. (4-53) ascertains that q_i^2 and $|2v_i\rangle$ belong to the same irreducible representation as each other, and similarly that q_iq_k and $|v_i + v_k\rangle$ have the same symmetry.

From the above consideration, it is concluded that anharmonic resonances can occur only between vibrational levels having the same symmetry. The occurrence of anharmonic resonances often complicates the interpretation of spectra and the procedure of theoretical calculation, but sometimes it enables the estimation of the magnitude of a special cubic constant separately from the other potential constants.

In the second-order perturbation theory, only the diagonal matrix elements of H_2' are taken into account in principle. It sometimes happens, however, that certain off-diagonal matrix elements of H_2' give rise to large shifts of vibrational levels and mixing of vibrational wave functions. This phenomenon is called second-order anharmonic resonance. A well-known type of second-order anharmonic resonance is that which occurs between the first overtone levels of the symmetric and the antisymmetric stretching modes of the XH_2 type triatomic molecules or the XH_2 groups of more-than-three atomic molecules. This resonance, observed first for water vapor by Darling and Dennison,[3] is called the Darling-Dennison resonance.

Let the two X–H stretching coordinates of an XH_2 group be denoted briefly by $\Delta R_1 = \Delta r(XH_1)$ and $\Delta R_2 = \Delta r(XH_2)$, and assume that the X–H stretching potential function in terms of these coordinates lacks the higher order cross terms as given by

$$V = K_{rr}\left(\Delta R_1^{\,2} + \Delta R_2^{\,2}\right) + K'\,\Delta R_1\,\Delta R_2 + K_{rrr}\left(\Delta R_1^{\,3} + \Delta R_2^{\,3}\right) + K_{rrrr}\left(\Delta R_1^{\,4} + \Delta R_2^{\,4}\right).$$

Now we define the symmetric and the antisymmetric stretching coordinates of the XH_2 group as

$$\Delta S_s = (\Delta R_1 + \Delta R_2)/\sqrt{2} \quad \text{:(XH}_2\text{ symmetric stretching)}$$

and

$$\Delta S_a = (\Delta R_1 - \Delta R_2)/\sqrt{2} \quad : (\mathrm{XH_2} \text{ antisymmetric stretching})$$

respectively. The stretching potential function in terms of ΔS_s and ΔS_a is given by

$$\begin{aligned} V = {} & (K_{rr} + K')\Delta S_s^2 + (K_{rr} - K')\Delta S_a^2 + K_{rrr}\left(\Delta S_s^3 + 3\,\Delta S_s\,\Delta S_a^2\right)/\sqrt{2} \\ & + K_{rrrr}\left(\Delta S_s^4 + 6\,\Delta S_s^2\,\Delta S_a^2 + \Delta S_a^4\right)/2. \end{aligned} \tag{5-73}$$

The mixing of the XH_2 stretching vibrations with the other vibrations is so small that the coordinates ΔS_s and ΔS_a are approximately regarded as normal coordinates. In this case, it follows from Eq. (5-73) that[8)]

$$\left.\begin{aligned} k_{saa} &\cong 3\,k_{sss} \\ k_{aaaa} &\cong k_{ssss} \\ k_{ssaa} &\cong 6\,k_{ssss} \end{aligned}\right\}. \tag{5-74}$$

If all terms related to the modes other than ΔS_s and ΔS_a are neglected, the anharmonicity constants are calculated from Eqs. (5-66a,b) to be

$$x_{ss} = \frac{3k_{ssss}}{2} - \frac{15k_{sss}^2}{4\omega_s} \tag{5-75a}$$

$$x_{aa} = \frac{3k_{aaaa}}{2} - \frac{k_{saa}}{8}\left(\frac{1}{\omega_s + 2\omega_a} + \frac{1}{\omega_s - 2\omega_a} + \frac{4}{\omega_s}\right) \tag{5-75b}$$

$$x_{sa} = k_{ssaa} - \frac{3\,k_{sss}\,k_{saa}}{\omega_s} - \frac{k_{saa}^2}{2}\left(\frac{1}{\omega_s + 2\omega_a} - \frac{1}{\omega_s - 2\omega_a}\right). \tag{5-75c}$$

Writing ω_a as $\omega_a = \omega_s + \delta$, expanding the quantities in parentheses in Eqs. (5-75b,c) in powers of δ / ω_s and substituting Eq. (5-74), we have

$$x_{aa} = \frac{3k_{ssss}}{2} - \frac{15k_{sss}^2}{4\omega_s} + \frac{2\delta\,k_{sss}^2}{\omega_s^2} \tag{5-76a}$$

and

$$x_{sa} = 4\left(\frac{3k_{ssss}}{2} - \frac{15k_{sss}^2}{4\omega_s}\right) + \frac{10\delta\,k_{sss}^2}{\omega_s^2}. \tag{5-76b}$$

Neglecting the last terms in Eqs. (5-76a,b) leads to an approximate relation, $x_{as} \approx 4x_{aa}$.[9, 10)]

Tanaka and Machida analysed the anharmonicity of the NH_2 stretching potential of aniline by using a four-atomic model consisting of the NH_2 group and the carbon atom attached to the nitrogen atom.[10)] The N–H stretching potential function was expressed by Eq. (5-73). The higher order diagonal terms K_{rrr} and K_{rrrr} were estimated from the quadratic constant K_{rr} by assuming the Morse function with $a_{\mathrm{N-H}} = 20\ \mathrm{nm}^{-1}$. In Fig. 5.2, the observed frequencies of aniline in the first overtone region of the NH_2 stretching modes are compared with the calculated frequencies at various levels of approximation. Obviously, the calculated

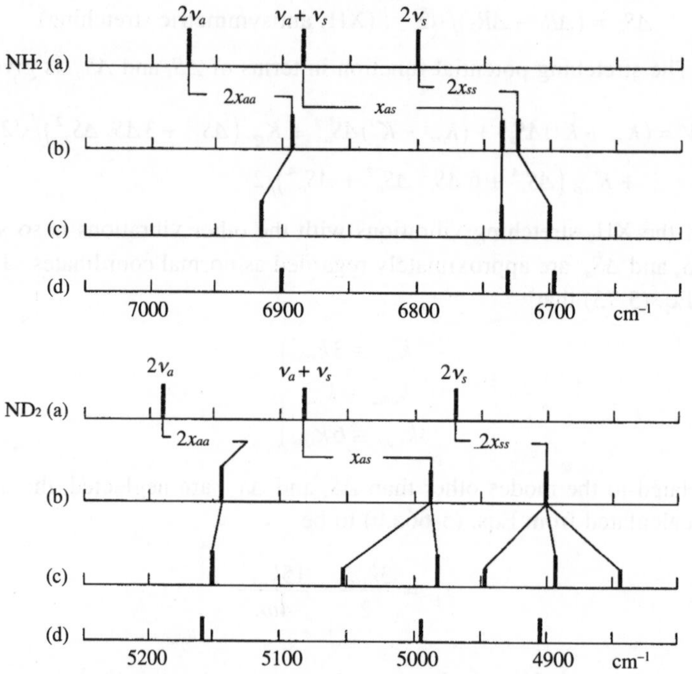

Fig. 5.2 First overtone and combination frequencies of NH_2 stretching vibrations of aniline: (a) sum of the calculated fundamental frequencies, (b) corrected for the first order anharmonicity, (c) corrected for the second-order resonance, (d) observed frequencies.
(Reproduced with permission from Y. Tanaka and K. Machida, *J. Mol. Spectrosc.*, **51**, 508 (1971), p. 518 (Fig. 4))

frequencies are greatly improved by taking account of the effect of the Darling–Dennison resonance.

The effect of substituents on the anharmonicity of the NH_2 stretching potential of substituted anilines was investigated by using the same four-atomic model.[11] From a similar analysis of the spectra in the first overtone region, the parameters $a_{\text{N-H}}$ of 22 *m*- and *p*-substituted anilines[12] were revealed to fall within the range 19.6–20.6 nm^{-1}. In addition, the deviations of K_{rr} and K' from those of aniline, ΔK_{rr} and $\Delta K''$, were found to satisfy an approximate relationship $\Delta K' \approx -0.2\,\Delta K_{rr}$. Based on this result, the constraints $a_{\text{N-H}} = 20$ nm^{-1} and $\Delta K' = -\Delta K_{rr}/5$ were introduced so that ΔK_{rr} was left as a single adjustable parameter to fit the data of (ν_s), (ν_a), $(2\nu_s)$, $(\nu_s + \nu_a)$ and $(2\nu_a)$ for each substituted aniline. The fit was quite successful as shown in Fig. 5.3. The values of ΔK_{rr} thus obtained were found to correlate well with the Hammet constant, σ, by

$$\Delta K_{rr} = 0.064\sigma. \tag{5-77}$$

On letting the rate constants of a reaction for the standard and the substituted compounds be denoted by k_0 and k, respectively, the Hammet rule is stated in the form[13]

$$\ln(k/k_0) = \rho\sigma. \tag{5-78}$$

The relation between the activation energy ΔE and the rate constant is given by

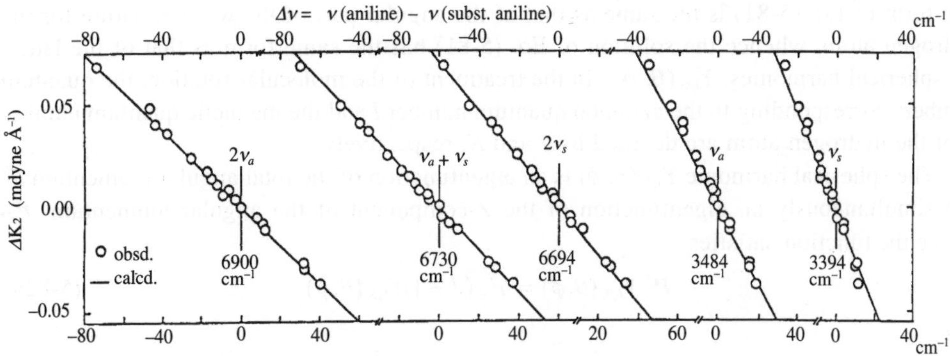

Fig. 5.3 Fundamental frequencies and first overtone and combination frequencies of NH_2 stretching vibrations of *m*- and *p*-substituted anilines.
(Reproduced with permission from Y. Tanaka and K. Machida, *J. Mol. Spectrosc.*, **55**, 435 (1975), p. 439 (Fig. 1))

$$k = A\exp(-\Delta E/RT). \tag{5-79}$$

Applying Eq. (5-79) to the standard and the substituted compounds, taking the difference and replacing ΔE with the parameter D_e in the Morse function, we have

$$-(\Delta E - \Delta E_0)/RT = -\Delta K_{rr}/a^2RT = \rho\sigma. \tag{5-80}$$

Elimination of σ from Eqs. (5-77) and (5-80) gives the reaction constant, –3.9 at 25 °C and –3.7 at 35 °C. According to the comprehensive review on Hammet's rule by Jaffe,[14] the reaction which shows the constant ρ nearest to these values among the reactions of anilines is the following reaction of substituted anilines

$$\text{Ar–NH}_2 + \text{Cl–Ar}'(\text{NO}_2) \rightarrow \text{Ar–NH–Ar}'(\text{NO}_2) : (\rho = -3.9 \sim -4.1)$$

For reference, the whole range of ρ cited by Jaffe is 5.0 ~ – 4.2, of which the substitution reactions at the NH_2 group of anilines cover the range –1.2 ~ – 4.2 (except for the diazotization reactions for which ρ = 2.3 ~ 2.7). The agreement between the spectroscopic and the kinetic data suggests that the substituents which raise the fundamental and the first overtone frequencies of the NH_2 stretching vibrations of anilines raise also the activation energy of dissociation of the N–H bond.

5.5 Rotational Energy Levels and Vibration-Rotation Interaction

If a molecule is a perfectly rigid body, and the relative positions of nuclei are rigorously fixed, the principal moments of inertia I_X, I_Y and I_Z are constant, and the Hamiltonian consists of the first term in braces in Eq. (5-43a). The solution of the wave equation derived from this Hamiltonian takes various forms according to relative magnitude of I_X, I_Y and I_Z.

A molecule with three principal moments of inertia being equal to one another is called a spherical top molecule. The wave equation of a rigid spherical top molecule having the principal moment of inertia I is given by

$$(2I)^{-1}\boldsymbol{P}^2\psi = E\psi. \tag{5-81}$$

The form of Eq. (5-81) is the same as that of the angular part of the wave equation for the hydrogen atom, whence the solution of Eq. (5-81) has the same form as that of the latter, the spherical harmonics, $Y_{l,m}(\theta, \phi)$. In the treatment of the molecular rotation, the quantum numbers corresponding to the azimutal quantum number l and the magnetic quantum number m of the hydrogen atom are denoted by J and K, respectively.

The spherical harmonic $Y_{J,K}(\theta, \phi)$ is an eigenfunction of the total angular momentum $\boldsymbol{P}^2$ and simultaneously an eigenfunction of the Z-component of the angular momentum, P_Z. Hence the function satisfies

$$\boldsymbol{P}^2\, Y_{J,K}(\theta,\phi) = \hbar^2 J(J+1) Y_{J,K}(\theta,\phi) \tag{5-82a}$$

and

$$P_Z\, Y_{J,K}(\theta,\phi) = \hbar\, K\, Y_{J,K}(\theta,\phi) \tag{5-82b}$$

simultaneously.[15] Substitution of $Y_{J,K}(\theta, \phi)$ into the wave function ψ in Eq. (5-81) followed by the use of Eq. (5-82a) leads to the energy eigenvalue in the form

$$E_J = \frac{\hbar^2}{2I} J(J+1) = h\frac{h}{8\pi^2 I} J(J+1). \tag{5-83}$$

Here we introduce the rotational constant defined by

$$B = h/(8\pi^2 I) \tag{5-84}$$

and rewrite Eq. (5-83) as

$$E_J = hBJ(J+1). \tag{5-83'}$$

The rotational constant B has the dimension of the frequency, and is given in the unit of MHz in microwave spectroscopy. On dividng Eq. (5-84) by the speed of light, c in·cm s^{-1}, the rotational constant in cm^{-1} is obtained.

In the case of linear molecules, it is customary to put the molecule on the Z-axis, so it holds that $P_Z = 0$ and $I_X = I_Y$. The rotational Hamiltonian in this case is written as

$$H = \frac{1}{2I}\left(P_X{}^2 + P_Y{}^2\right) = \frac{1}{2I}\boldsymbol{P}^2, \tag{5-85}$$

which is formally the same as that for spherical top molecules, and the eigenvalues are given by Eq. (5-83).

In the case of symmetric top molecules, the single more-than-two fold axis is customarily taken as the Z-axis, giving

$$0 < I_X = I_Y \neq I_Z > 0.$$

It is convenient to write the rotaional Hamiltonian in this case as

$$H = \frac{1}{2I_X}\left(P_X{}^2 + P_Y{}^2\right) + \frac{1}{2I_Z}P_Z{}^2 = \frac{1}{2I_X}\boldsymbol{P}^2 + \left(\frac{1}{2I_Z} - \frac{1}{2I_X}\right)P_Z{}^2 \tag{5-86}$$

By operating Eq. (5-86) on $Y_{J,K}(\theta, \phi)$, and substituting Eqs. (5-82a,b), the energy eigenvalues are obtained as

$$E_{J,K} = hc\left\{\frac{\hbar}{4\pi\, I_X} J(J+1) + \frac{\hbar}{4\pi}\left(\frac{1}{I_Z} - \frac{1}{I_X}\right)K^2\right\}. \tag{5-87}$$

For nonlinear molecules, the Cartesian axes are customarily so chosen that the magnitudes of the three rotational constants A, B and C with respect to the three principal moments of inertia, I_X, I_Y and I_Z, respectively, are in the order $A \geq B \geq C$. The symmetric top molecules are then classified into the prolate tops for which $A > B = C$ and the oblate tops for which $A = B > C$. The rotational energy levels in terms of the rotational constants are given by

$$\left.\begin{aligned} E_{J,K} &= hc\{B\,J(J+1) + (A-B)K^2\} \ : \ \text{prolate top} \\ E_{J,K} &= hc\{B\,J(J+1) + (C-B)K^2\} \ : \ \text{oblate top} \end{aligned}\right\}. \tag{5-88}$$

In the case of asymmetric top molecules for which the three rotational constants are all different from one another, no expression of the rotational levels in closed analytical forms can be derived in a rigorous sense. If B is very close to either A or C, the perturbation theory is applied by taking the eigenfunctions of a hypothetical symmetric top molecule as the zeroth-order wave functions, and the energy levels are calculated by adding the higher order corrections to the zeroth-order levels given by Eqs. (5-88). Under a favorable experimental condition, many rotational lines are observed to be utilized for the fitting so that the rotational constants A, B and C can be determined very accurately.

In the above discussion, the molecules have been regarded as rigid bodies, and only the rotational part of the zeroth-order Hamiltonian has been taken into account. In the next step of approximation, the second-order Hamiltonian after the contact transformation, H_2' in Eq. (5-51), will be taken up, since the first-order Hamiltonian H_1 (Eq. 5-43b)) has been eliminated by the contact transformation, so that $H'_1 = 0$.

Among the terms in H_2', only those in the commutator $(i/2)[S_2, H_{1;2}]$ introduced by the contact transformation include the total angular momenta P_α but do not include any vibrational variables q_i and p_i. This term is derived by substituting a_2 in Table 5.1 into the equation in Table 5.2 in the form[16)]

$$H'_{2;2,2} = \tfrac{i}{2}[S_2, H_{1;2}] = \tfrac{1}{4}\sum_\alpha\sum_\beta\sum_\gamma\sum_\delta \tau_{\alpha\beta\gamma\delta} P_\alpha P_\beta P_\gamma P_\delta \tag{5-89}$$

where

$$\tau_{\alpha\beta\gamma\delta} = \frac{-\hbar}{2\pi\, c^2 I_\alpha^e I_\beta^e I_\gamma^e I_\delta^e} \sum_i \frac{a_i^{\alpha\beta} a_i^{\gamma\delta}}{\omega_i^{\,2}}. \tag{5-90}$$

The coefficient $\tau_{\alpha\beta\gamma\delta}$ is called a centrifugal distortion constant. The constant $a_i^{\alpha\beta}$ defined by Eq. (5-21) is rewritten with the help of Eqs. (3-80), (3-13) and (4-21) in the form

$$\begin{aligned} a_i^{\alpha\beta} &= \sum_k \left(\frac{\partial I_{\alpha\beta}}{\partial x_k}\right)_0 \left(\frac{\partial x_k}{\partial Q_i}\right)_0 = \sum_k\sum_j \left(\frac{\partial I_{\alpha\beta}}{\partial R_j}\right)_0 \left(\frac{\partial R_j}{\partial x_k}\right)_0 L_{ki}^x \\ &= \sum_j \left(\frac{\partial I_{\alpha\beta}}{\partial R_j}\right)_0 \sum_k B_{jk} L_{ki}^x = \sum_j \left(\frac{\partial I_{\alpha\beta}}{\partial R_j}\right)_0 L_{ji}. \end{aligned} \tag{5-91}$$

Substituting Eq. (5-91) into Eq. (5-90) and using Eq. (4-14), we have

$$\begin{aligned}\tau_{\alpha\beta\gamma\delta} &= -\frac{h}{I_\alpha^e I_\beta^e I_\gamma^e I_\delta^e}\sum_i\sum_j\sum_k\left(\frac{\partial I_{\alpha\beta}}{\partial R_j}\right)_0\frac{L_{ji}\,L_{ki}}{4\pi^2c^2\omega_i^2}\left(\frac{\partial I_{\gamma\delta}}{\partial R_k}\right)_0 \\ &= -\frac{h}{I_\alpha^e I_\beta^e I_\gamma^e I_\delta^e}\sum_j\sum_k\left(\frac{\partial I_{\alpha\beta}}{\partial R_j}\right)_0\left(\sum_i L_{ji}\,\Lambda_i^{-1}L_{ki}\right)\left(\frac{\partial I_{\gamma\delta}}{\partial R_k}\right)_0.\end{aligned} \tag{5-92}$$

On the other hand, substituting Eq. (4-21) into Eq. (4-19) and taking the inverse, we have

$$\boldsymbol{\Lambda}^{-1} = \boldsymbol{L}^{-1}(\boldsymbol{F}^{RR})^{-1}\tilde{\boldsymbol{L}}^{-1} \qquad \therefore \quad \boldsymbol{L}\boldsymbol{\Lambda}^{-1}\tilde{\boldsymbol{L}} = (\boldsymbol{F}^{RR})^{-1} \equiv \boldsymbol{C}.$$

Equation (5-92) is then rewritten as

$$\tau_{\alpha\beta\gamma\delta} = -\frac{h}{I_\alpha^e I_\beta^e I_\gamma^e I_\delta^e}\sum_j\sum_k\left(\frac{\partial I_{\alpha\beta}}{\partial R_j}\right)_0 C_{jk}\left(\frac{\partial I_{\gamma\delta}}{\partial R_k}\right)_0, \tag{5-93}$$

which connects the centrifugal distortion constants with the elements of the matrix $\boldsymbol{F}^{-1}$, *i.e.*, the compliance matrix introduced in section 2.4.2. Experimentally, the centrifugal distortion constants are obtained by a careful analysis of rotational spectra in the gas phase.

Linear molecules have no Z-component of the angular momentum, so that only P_X and P_Y appear in the four-fold sum in Eq. (5-85). Since $X_n^0 = Y_n^0 = 0$ for linear molecules, $a_i^{XX} = a_i^{YY}$ and $a_i^{XY} = 0$ from Eq. (5-21), whence Eq. (5-85) becomes

$$H = \tfrac{1}{4}\tau_{XXXX}\sum_\alpha\sum_\beta P_\alpha{}^2 P_\beta{}^2 = \tfrac{1}{4}\tau_{XXXX}\boldsymbol{P}^4. \tag{5-94a}$$

Furthermore, since $\boldsymbol{P}^4 = \boldsymbol{P}^2\boldsymbol{P}^2$, operating $\boldsymbol{P}^2$ on Eq. (5-82a) and substituting the original equation into the right side, we have

$$\begin{aligned}P^4\,Y_{J,K}(\theta,\phi) &= \hbar^2 J(J+1)P^2 Y_{J,K}(\theta,\phi) \\ &= \hbar^4 J^2(J+1)^2\,Y_{J,K}(\theta,\phi)\ .\end{aligned} \tag{5-94b}$$

By combining Eqs. (5-83′) and (5-94a,b) and using an abbreviated notation $D = -(\hbar^4/4h)\tau_{XXXX}$, the rotational energy of a linear molecule in the second order perturbation theory is given by

$$E_J = h\{B\,J(J+1) - D\,J^2(J+1)^2\} = h\,B'\,J(J+1) \tag{5-95a}$$

where

$$B' = B - D\,J(J+1). \tag{5-95b}$$

From Eq. (5-95b) in which $D > 0$, it is seen that the speed up of rotational motion causes more or less expansion of the molecule and thereby an apparent increase in the moment of inertia. Similar effects are also known for symmetric and asymmetric top molecules, though the expressions of energy levels are much more complicated.[17, 18]

To distinguish the rotational levels stacked on the excited vibrational levels from those stacked on the vibrational ground level, the former and the latter are called vibrotational levels and pure rotational levels, respectively. In the zeroth-order approximation in which the rotation and the vibration of a molecule is completely separated, all the vibrotational levels are obtained by superposing a set of pure rotational levels on each vibrational level. In actual molecules, because of the perturbation due to the vibration–rotation interaction terms in Eqs. (5-43b,c), the intervals between the corresponding rotational levels superposed on different vibrational levels are not equal to each other. Accordingly, the rotational constants obtained by fitting the observed vibrotational levels depend on the vibrational quantum numbers.

In the non-degenerate case, the dependence of the rotational constants on the vibrational quantum numbers is expressed by

$$B^\alpha(v_1, v_2, \cdots) = B_e^\alpha - \sum_i \alpha_i^\alpha (v_i + 1/2). \tag{5-96}$$

The coefficient α_i^α is called the vibration–rotation interaction constant,[6] and is expressed in the wavenumber units as

$$\alpha_i^\alpha = \alpha_{i\ \mathrm{harm}}^\alpha + \alpha_{i\ \mathrm{anh}}^\alpha \tag{5-97}$$

where

$$\alpha_{i\ \mathrm{harm}}^\alpha = \frac{\hbar^2}{8\pi^2 c^2 I_\alpha^{e\,2} \omega_i}\left[A_{ii}^{\alpha\alpha} - \sum_\gamma \frac{\left(a_i^{\alpha\gamma}\right)^2}{I_\gamma^e} - \sum_{j\neq i} \left(\zeta_{ij}^\alpha\right)^2 \frac{3\omega_i^2 - \omega_j^2}{\omega_i^2 - \omega_j^2}\right] \tag{5-98a}$$

and

$$\alpha_{i\ \mathrm{anh}}^\alpha = -\frac{h}{8\pi^2 c I_\alpha^{e\,2}}\left(\frac{\hbar}{2\pi c}\right)^{1/2}\left(\frac{3k_{iii}\, a_i^{\alpha\alpha}}{\omega_i^{3/2}} + \sum_j \frac{k_{iij}\, a_j^{\alpha\alpha}}{\omega_j^{3/2}}\right). \tag{5-98b}$$

The third term in the brackets in Eq. (5-98a) and the total right side of Eq. (5-98b) represent the energy terms in Table 5.2 containing the product of the total angular momentum and the vibrational factor, $P_\alpha^2(v_i + 1/2)$, in which the operator P_α^2 is replaced by the constant $\hbar^2/hc$. The remaining terms in Eq. (5-98a) arise from the terms containing the factor $P_\alpha^2 q_i^2$ in H_2 in Eq. (5-43c).

The rotational constant B_0^α in the vibrational ground state is given, from Eq. (5-96), in the form

$$B_0^\alpha \equiv B^\alpha(0, 0, \cdots) = B_e^\alpha - \sum_i \alpha_i^\alpha / 2. \tag{5-99}$$

Since B_0^α does not coincide with B_e^α, the structure parameters R_0 obtained from B_0^α are different from the corresponding values at the equilibrium, R_e. From Eqs. (5-96) and (5-99), the rotational constants in the fundamental level, where $v_i = 1$ and the other vibrational quantum numbers are all zero, are given by

$$B^\alpha(0, \cdots, v_i = 1, 0, \cdots) = B_0^\alpha - \alpha_i^\alpha. \tag{5-100}$$

The rotational constants at the equilibrium B_e^α can be determined if the vibration–rotation interaction constants α_i^a for all i's are obtained from experiments and substituted into Eq. (5-99) with B_0^a. This sort of analysis has been conducted for few molecules.

5.6 Average Internuclear Distances

On comparing the theoretical value of a given physical quantity f of a molecule with the experimental data, it is important to understand how to estimate the thermal average of that quantity, which must generally be a function of the relative positions of the nuclei, at the temperature T, $\langle f \rangle^T$. If the quantity f can be expanded in powers of the normal coordinates, $\langle f \rangle^T$ is given in terms of the thermal averages of the normal coordinates and their powers in the form

$$\langle f \rangle^T = f_e + \sum_i \left(\frac{\partial f}{\partial q_i} \right)_0 \langle q_i \rangle^T + \tfrac{1}{2} \sum_i \sum_j \left(\frac{\partial^2 f}{\partial q_i \partial q_j} \right)_0 \langle q_i q_j \rangle^T + \cdots, \tag{5-101}$$

Here, the thermal average is defined as

$$\langle f \rangle^T = \sum_n \langle f \rangle_n \exp(-E_n/kT) \Big/ \left\{ \sum_m \exp(-E_m/kT) \right\} \tag{5-102}$$

where E_n is the energy eigenvalue of the quantum state n, and $\langle f \rangle_n$ is the expectation value of f in the state n.

Within the framework of the harmonic approximation, the variable q_i takes positive and negative values homogeneously so that the average $\langle q_i \rangle^T$ vanishes, but the average of the square, which cannot be negative, $\langle q_i^2 \rangle^T$, does not vanish. When the potential function contains any odd-order terms with respect to q_i, the distribution of q_i becomes not symmetric with respect to the origin so that $\langle q_i \rangle^T$ does not necessarily vanish.

According to the perturbation theory, the first-order wave function $\psi_n^{(1)}$ can be expanded in terms of the unperturbed wave functions as

$$\psi_n^{(1)} = \sum_k a_{nk} \psi_k^{(0)} = \psi_n^{(0)} - \sum_{k \neq n} \{ \langle \psi_k^{(0)} | H_1 | \psi_n^{(0)} \rangle / (E_k - E_n) \} \psi_k^{(0)}. \tag{5-103}$$

If the vibrational quantum numbers of $\psi_n^{(0)}$ are v_1, v_2, $\cdots$, v_i, $\cdots$, the zeroth-order wave functions $\psi_k^{(0)}$ for which the matrix element $\langle \psi_k^{(0)} | q_i | \psi_n^{(0)} \rangle$ does not vanish are only those with the quantum numbers v_1, v_2, $\cdots$, $v_i \pm 1$, $\cdots$. On writing these wave functions as $|v_i + 1\rangle$ and $|v_i - 1\rangle$, the non-vanishing coefficients in the series of Eq. (5-103) are calculated to be

$$\langle v_i + 1 | H_1 | v_i \rangle / hc\omega_i = \frac{(v_i + 1)^{1/2}}{2^{1/2} \omega_i} \left\{ 3k_{iii}(v_i + 1) + \sum_{j \neq i} k_{ijj}(v_j + 1/2) \right\} \tag{5-104a}$$

and

$$\langle v_i - 1 | H_1 | v_i \rangle / hc\omega_i = \frac{v_i^{1/2}}{2^{1/2} \omega_i} \left\{ 3k_{iii} v_i + \sum_{j \neq i} k_{ijj}(v_j + 1/2) \right\}, \tag{5-104b}$$

where the effect of the molecular rotation is neglected by dropping the terms involving the angular momenta.

By using the first equality in Eq. (5-103), the expectation value of q_i with respect to the first-order wave function can be expanded in terms of those with respect to the unperturbed wave functions as

$$\begin{aligned}\langle\psi_n^{(1)}|q_i|\psi_n^{(1)}\rangle &= \langle\psi_n^{(0)}|q_i|\psi_n^{(0)}\rangle + \sum_{k\neq n}\left\{a_{kn}\langle\psi_n^{(0)}|q_i|\psi_k^{(0)}\rangle + a_{kn}^*\langle\psi_k^{(0)}|q_i|\psi_n^{(0)}\rangle\right\} \\ &+ \sum_{k,\ m\neq n}\sum a_{kn}a_{kn}^*\langle\psi_k^{(0)}|q_i|\psi_m^{(0)}\rangle.\end{aligned} \tag{5-105}$$

In the right side of Eq. (5-105), the first term vanishes while the third term is cubic in infinitesimal quantities (quadratic in a_{kn} and linear in q_i). Calculation of the remaining second term with the help of Eqs. (5-104a,b) and Eqs. (A2-4a,b) in Appendix 2 gives

$$\begin{aligned}\langle q_i\rangle &= \langle\psi_n^{(1)}|q_i|\psi_n^{(1)}\rangle \\ &= \left\{\langle v_i+1|H_1|v_i\rangle\langle v_i+1|q_i|v_i\rangle + \langle v_i-1|H_1|v_i\rangle\langle v_i-1|q_i|v_i\rangle\right\}/hc\omega_i \\ &= -\frac{1}{\omega_i}\left\{3k_{iii}(v_i+\tfrac{1}{2}) + \sum_{j\neq i}k_{ijj}(v_j+\tfrac{1}{2})\right\}.\end{aligned} \tag{5-106}$$

By substituting Eq. (5-106) into the transformation relation between the normal and the Cartesian coordinates

$$X_p = X_p^0 + \sum_i L_{pi}^X Q_i = X_p^0 + \frac{\hbar}{(hc)^{1/2}}\sum_i \frac{L_{pi}^X}{\omega_i^{1/2}}q_i,$$

the expectation value of the Cartesian coordinate X_p of the nucleus p is obtained in the form

$$\langle X_p\rangle_n = X_p^0 - \frac{\hbar}{(hc)^{1/2}}\sum_i \frac{L_{pi}^X}{\omega_i^{3/2}}\left\{3k_{iii}(v_i+\tfrac{1}{2}) + \sum_{j\neq i}k_{ijj}(v_j+\tfrac{1}{2})\right\} \tag{5-107}$$

The expectation value of the Z-component of the principal moment of inertia is given by

$$\langle I_Z\rangle_n = \sum_p m_p\left(\langle X_p\rangle_n{}^2 + \langle Y_p\rangle_n{}^2\right). \tag{5-108a}$$

Substituting Eq. (5-107) and a similar expression for the Y-component into Eq. (5-108a), and dropping the higher order terms, we obtain

$$\begin{aligned}\langle I_Z\rangle_n &= \sum_p m_p\left(X_p^{02} + Y_p^{02}\right) - \frac{\hbar}{(hc)^{1/2}}\sum_i \frac{1}{\omega_i^{3/2}}\sum_p m_p\left(X_p^0 L_{pi}^X + Y_p^0 L_{pi}^Y\right) \\ &\times\left\{3k_{iii}(v_i+\tfrac{1}{2}) + \sum_{j\neq i}k_{ijj}(v_j+\tfrac{1}{2})\right\}.\end{aligned} \tag{5-108b}$$

Substitution of Eq. (5-21) into Eq. (5-108b) and exchanging the subscripts i and j followed by a little rearrangement leads finally to

$$\langle I_Z\rangle_n = I_Z^e - \frac{\hbar}{(hc)^{1/2}}\sum_i \frac{a_i^{ZZ}}{\omega_i^{3/2}}\left\{3k_{iii}(v_i+\tfrac{1}{2}) + \sum_{j\neq i} k_{ijj}(v_j+\tfrac{1}{2})\right\}$$

$$= I_Z^e - \frac{\hbar}{(hc)^{1/2}}\left\{\sum_i \frac{3a_i^{ZZ}k_{iii}}{\omega_i^{3/2}}(v_i+\tfrac{1}{2}) + \sum_j\sum_{i\neq j}\frac{a_j^{ZZ}k_{iij}}{\omega_j^{3/2}}(v_i+\tfrac{1}{2})\right\} \quad (5\text{-}109)$$

$$= I_Z^e - \frac{\hbar}{(hc)^{1/2}}\sum_i\left(\frac{3a_i^{ZZ}k_{iii}}{\omega_i^{3/2}} + \sum_{j\neq i}\frac{a_j^{ZZ}k_{iij}}{\omega_j^{3/2}}\right)(v_i+\tfrac{1}{2}).$$

The expectation values of the X and Y components can be obtained in similar ways.

Following the definition of the rotational constants as Eq. (5-84) divided by c, the expectation value of the rotational constant corresponding to $\langle I_\alpha\rangle_n$ in the first order approximation is given by

$$\langle B^\alpha\rangle_n = \frac{\hbar}{4\pi c\, I_\alpha^e}\left\{1 + \frac{\hbar}{(hc)^{1/2} I_\alpha^e}\sum_i\left(\frac{3a_i^{\alpha\alpha}k_{iii}}{\omega_i^{3/2}} + \sum_{j\neq i}\frac{a_j^{\alpha\alpha}k_{iij}}{\omega_j^{3/2}}\right)(v_i+\tfrac{1}{2})\right\}.$$

By referring to Eqs. (5-96, 97, 98b), $\langle B^\alpha\rangle_n$ can also be written in the form

$$\langle B^\alpha\rangle_n = B^\alpha(v_1, v_2, \cdots) + \sum_i \alpha_{i\ \text{harm}}^\alpha (v_i+\tfrac{1}{2}). \quad (5\text{-}110)$$

The expectation value of the components of the principal moment of inertia calculated by using the wave function for the vibrational ground state is denoted by $\langle I_z\rangle$, and the molecular structure derived from $\langle I_z\rangle$ is called the r_z structure.[19,20] As seen from Eq. (5-108), the r_z structure has a clear physical meaning as "the structure represented by the average coordinates in the vibrational ground state". Furthermore, the r_z structure has the advantage that it can be derived from the experimental data only by using knowledge of the harmonic part of the potential function. On the other hand, it must be noted that a correction for the anharmonicity of the potential is necessary for estimating the r_z structure from the calculated equilibrium structure according to the second term in Eq. (5-109).

The interatomic distance obtained from experimental data of electron diffraction in the gas phase, r_g, means "the center of gravity of the probability distribution function of the internuclear distance r, $P(r)$," *i.e.*, the thermal average of the internuclear distance, $\langle r\rangle^T$.[21]

$$r_g = \langle r\rangle^T = \int_0^\infty r\,P(r)\,dr \quad (5\text{-}111)$$

The function $P(r)$ is defined in such a way that $P(r)dr$ is the probability for the internuclear distance to be in the range between r and $r + dr$.

For the calculation of $\langle r\rangle^T$, the average of the vibrational levels, $\langle v_i + 1/2\rangle^T$, must be expressed in an analytical form. Before deriving such an expression, we introduce the abbreviations $u = hc\omega_i/kT$ and $v = v_i$, and differentiate both sides of the identity for a geometrical series

$$\sum_{v=0}^\infty e^{-vu} = (1-e^{-u})^{-1} \quad (5\text{-}112)$$

with respect to u, obtaining

$$\sum_{v=0}^{\infty} v e^{-vu} = e^{-u}(1-e^{-u})^{-2}. \tag{5-113}$$

Substituting eqs. (5-112) and (5-113) into Eq. (5-102) leads to

$$\begin{aligned}
\langle v_i + 1/2\rangle^T &= \frac{\sum_{v_i=0}^{\infty}(v_i+1/2)\exp\{-hc\omega_i(v_i+1/2)/kT\}}{\sum_{v_i=0}^{\infty}\exp\{-hc\omega_i(v_i+1/2)/kT\}} \\
&= \frac{1}{2} + \frac{\sum_{v_i=0}^{\infty} v_i\exp\{-hc\omega_i(v_i+1/2)/kT\}}{\sum_{v_i=0}^{\infty}\exp\{-hc\omega_i(v_i+1/2)/kT\}} \\
&= \frac{1}{2}\frac{1+\exp(-hc\omega_i/kT)}{1-\exp(-hc\omega_i/kT)} = \frac{1}{2}\coth\left(\frac{hc\omega_i}{2kT}\right).
\end{aligned} \tag{5-114}$$

By using Eq. (5-114) and the same expression for v_j, the thermal average of Eq. (5-106) is obtained to be

$$\langle q_i\rangle^T = -\frac{1}{2\omega_i}\left\{3k_{iii}\coth\left(\frac{hc\omega_i}{2kT}\right) + \sum_j k_{ijj}\coth\left(\frac{hc\omega_j}{2kT}\right)\right\}. \tag{5-115}$$

The expectation value of a second-order term in Eq. (5-101) is derived, from the integral with respect to the unperturbed wave functions, in the form

$$\langle \psi_n^{(0)}|q_i q_j|\psi_n^{(0)}\rangle = \delta_{ij}(v_i + 1/2).$$

The thermal average of a second-order term is then obtained from Eq. (5-114) as

$$\langle q_i q_j\rangle^T = \frac{\delta_{ij}}{2}\coth\left(\frac{hc\omega_i}{2kT}\right). \tag{5-116}$$

By using the Cartesian displacement coordinates introduced in Chapter 3, and letting the sum over X, Y and Z be denoted by Σ_α, the averaged distance between the nuclei p and q is given by

$$\begin{aligned}
\langle r_{pq}\rangle^T &= \left\langle\left\{\sum_\alpha(\alpha_q-\alpha_p)^2\right\}^{1/2}\right\rangle^T = \left\langle\left\{\sum_\alpha(\alpha_{pq}^e + \Delta\alpha_q - \Delta\alpha_p)^2\right\}^{1/2}\right\rangle^T \\
&= \left\langle\left\{(r_{pq}^e)^2 + 2\sum_\alpha \alpha_{pq}^e(\Delta\alpha_q - \Delta\alpha_p) + \sum_\alpha(\Delta\alpha_q-\Delta\alpha_p)^2\right\}^{1/2}\right\rangle^T,
\end{aligned} \tag{5-117}$$

where r_{pq}^e is the equilibrium distance between the nuclei p and q. Since the displacements $\Delta\alpha_p$ and $\Delta\alpha_q$ are small compared to r_{pq}^e, Eq. (5-117) can be expanded in powers of the displacements of each nucleus as given by

$$\langle r_{pq}\rangle^T \approx r_{pq}^e + \frac{1}{r_{pq}^e}\sum_\alpha \alpha_{pq}^e\left(\langle\Delta\alpha_q\rangle^T - \langle\Delta\alpha_p\rangle^T\right) + \frac{1}{2\,r_{pq}^e}\sum_\alpha\left(\langle\Delta\alpha_q{}^2\rangle^T - 2\langle\Delta\alpha_p\,\Delta\alpha_q\rangle^T + \langle\Delta\alpha_p{}^2\rangle^T\right). \tag{5-118}$$

The Cartesian displacement coordinates are related to the normal coordinates by Eq. (3-80), which must also hold for the averages of these coordinates as

$$\langle\Delta x_k\rangle^T = \sum_i L_{ki}^x\langle Q_i\rangle^T = \frac{\hbar}{(hc)^{1/2}}\sum_i\frac{L_{ki}^x}{\omega_i^{1/2}}\langle q_i\rangle^T. \tag{5-119}$$

The thermal average of a second-order term is obtained by using Eq. (5-116) with the condition that $i = j$. The result is

$$\langle\Delta x_k\,\Delta x_l\rangle^T = \sum_i\sum_j L_{ki}^x\,L_{lj}^x\langle Q_i\,Q_j\rangle^T = \frac{\hbar^2}{hc}\sum_i\frac{L_{ki}^x\,L_{li}^x}{\omega_i}\langle q_i{}^2\rangle^T. \tag{5-120}$$

Substituting eqs. (5-115) and (5-116) into eqs. (5-119) and (5-120), respectively, to calculate the second and the third terms of Eq. (5-118), and subtracting the result from $\langle r_{pq}\rangle^T$, we obtain the equilibrium internuclear distance, r_{pq}^e. Since knowledge of the cubic terms of the potential function is necessary for calculating Eq. (5-115), the correction to the first-order term in Eq. (5-119) cannot easily be obtained for molecules containing more than three nuclei.

On the other hand, the correction to the second-order terms in Eq. (5-120), which is described within the framework of the harmonic approximation, can be calculated rather easily. By subtracting the third term contributed by the second-order terms from the right side of Eq. (5-118), and extrapolating the result into the temperature 0 K, the internuclear distance r_g, which corresponds to r_z derived from the rotational constants, is obtained. In this case, the difference between r_g and r_z represents the difference between the average of the distance (r_g) and the distance between the averaged positions (r_z). As seen from the examples in Table 5.3, the differences between r_e and r_g or between r_e and r_z are generally greater than those between r_g and r_z.

The concept of thermal averages can be extended straightforwardly to any structure parameters other than bond lengths by using the mean square amplitude matrix developed by Cyvin.[22] For a set of parameters S_1, S_2, $\cdots$ for describing any small changes of molecular structure, the mean square amplitude matrix at the temperature T, $\boldsymbol{\Sigma}^S(T)$, is defined as consisting of the elements

$$\Sigma_{ij}^S(T) = \langle S_i S_j\rangle^T. \tag{5-121}$$

It is seen from Eqs. (4-48) and (5-116) that the normal coordinates Q_1, Q_2, $\cdots$ have a diagonal mean square amplitude matrix $\boldsymbol{\Sigma}^Q(T)$ with the elements

$$\Sigma_{ii}^Q(T) = \langle Q_i^2\rangle^T = h\coth(hc\,\omega_i/2kT)/(8\pi^2 c\,\omega_i). \tag{5-122}$$

From the definition, Eq. (5-120) is just the (i, j) element of the mean square amplitude matrix for the nuclear Cartesian displacements. The corresponding matrix representation of Eq. (5-120) is obtained by combining Eqs. (3-80) and (5-122) in the form

Table 5.3 Average bond lengths and equilibrium bond lengths (pm).

Molecule	Ref.	Bond	r_0	r_g	r_z	r_e
CH_4	a)	C–H	109.40	110.68	110.0	108.5
CD_4	a)	C–D	109.18	110.27	109.7	108.5
H_2O	b)	H–O	95.6	97.4	97.14	95.72
D_2O	b)	D–O	95.7	97.0	96.79	95.7
SO_2	c)	S=O	143.36		143.49	143.08

a) K. Kuchitsu and L. S. Bartell, *J. Chem. Phys.*, **36**, 2470 (1962).
b) K. Kuchitsu and L. S. Bartell, *J. Chem. Phys.*, **36**, 2460 (1962).
c) Y. Morino, Y. Kikuchi, S. Saito and E. Hirota, *J. Mol. Spectrosc.*, **13**, 95 (1964).

$$\boldsymbol{\Sigma}^{x}(T) = \langle \Delta\boldsymbol{x}\,\Delta\tilde{\boldsymbol{x}} \rangle^{T} = \boldsymbol{L}^{x}\langle \boldsymbol{Q}\tilde{\boldsymbol{Q}} \rangle^{T} \tilde{\boldsymbol{L}}^{x} = \boldsymbol{L}^{x}\,\boldsymbol{\Sigma}^{Q}(T)\,\tilde{\boldsymbol{L}}^{x}. \tag{5-123}$$

Similarly, the mean square amplitude matrix for the internal coordinates is derived from $\boldsymbol{\Sigma}^{Q}(T)$ through the transformation in Eq. (4-23) as

$$\boldsymbol{\Sigma}^{R}(T) = \langle \boldsymbol{R}_{\text{in}}\,\tilde{\boldsymbol{R}}_{\text{in}} \rangle^{T} = \boldsymbol{L}\langle \boldsymbol{Q}\tilde{\boldsymbol{Q}} \rangle^{T} \tilde{\boldsymbol{L}} = \boldsymbol{L}\,\boldsymbol{\Sigma}^{Q}(T)\,\tilde{\boldsymbol{L}}. \tag{5-124}$$

The thermal average of the ith internal coordinate at T K is given by the square root of the ith diagonal element of $\boldsymbol{\Sigma}^{R}(T)$ in Eq. (5-124). The forms of the mean square amplitude matrix at the limits $T \to 0$ K and $T \to \infty$ are quite simple. Since coth $(hc\omega_i/2kT)$ is 1 at 0 K, it follows from Eq. (5-122) that

$$\boldsymbol{\Sigma}^{Q}(0) = (h/4\pi)\boldsymbol{\Lambda}^{-1/2},$$

where $\boldsymbol{\Lambda}^{-1/2}$ is the diagonal matrix with the ith element $\lambda_i^{-1/2} = (2\pi c\omega_i)^{-1}$. Hence we have

$$\boldsymbol{\Sigma}^{R}(0) = \boldsymbol{L}\,\boldsymbol{\Sigma}^{Q}(0)\,\tilde{\boldsymbol{L}} = (h/4\pi)\boldsymbol{L}\boldsymbol{\Lambda}^{-1/2}\tilde{\boldsymbol{L}}. \tag{5-125}$$

On the other hand, since

$$\lim_{T\to\infty} \coth\{hc\omega_i/(2kT)\} = 2kT/(hc\omega_i), \tag{5-126}$$

we have, at the high temperature limit, the relation

$$\lim_{T\to\infty}\{\boldsymbol{\Sigma}^{R}(T)/kT\} = \lim_{T\to\infty}\{\boldsymbol{L}\,\boldsymbol{\Sigma}^{Q}(T)\,\tilde{\boldsymbol{L}}/kT\} = \boldsymbol{L}\boldsymbol{\Lambda}^{-1}\tilde{\boldsymbol{L}} = \boldsymbol{C}, \tag{5-127}$$

where $\boldsymbol{C}$ is the compliance matrix defined in section 2.4.2. A detailed treatise of the properties and the application of the mean square amplitude matrices has been given by Cyvin.[23)]

5.7 Nonlinear Transformation of Anharmonic Potential

For discussing the anharmonicity of the molecular potential function, higher order terms must be retained in the expansion of the internal coordinates in powers of the Cartesian displacements of nuclei according to the order of the perturbation calculation.[24~27)] In the normal coordinate analysis in which the potential function is taken to be a quadratic form of the normal coordinates, each internal coordinate is expressed as a linear combination of the Cartesian displacement coordinates by Eq. (3-13). The internal coordinates expanded in powers of the nuclear Cartesian displacements up to any order higher than the first are called the curvilinear internal coordinates. The curvilinear internal coordinates must be distinguished from those defined by Eq. (3-13) which are called the linearized internal coordinates. To avoid any confusion between the two types of internal coordinates, the symbol $\mathfrak{R}$ is proposed for the curvilinear internal coordinates.[27,28)] Hereafter the symbol ΔR_a shall be used exclusively to represent the linearized internal coordinates, and the ath curvilinear internal coordinate is written as

$$\mathfrak{R}_a = \sum_i B_{ai}\Delta x_i + \tfrac{1}{2}\sum_i\sum_j B_{aij}\Delta x_i \Delta x_j + \tfrac{1}{6}\sum_i\sum_j\sum_k B_{aijk}\Delta x_i \Delta x_j \Delta x_k + \cdots. \tag{5-128}$$

How high the order of powers in the expansion must be taken is determined by the order of approximation. According to the Born–Oppenheimer approximation, the molecular potential function must be kept invariant under any isotopic substitution, whereas the terms higher than quadratic in the potential function expressed as a power series of ΔR_a starting from the quadratic terms are mass-dependent as shown below.

Let the transformation from ΔR_a to $\mathfrak{R}_a$ be expressed as

$$\mathfrak{R}_a = \sum_p A_{ap}\Delta R_p + \tfrac{1}{2}\sum_p\sum_q A_{apq}\Delta R_p \Delta R_q + \cdots \tag{5-129}$$

and let us try to determine the coefficients A_{ap}, A_{apq}, $\cdots$. Since the transformation must be done in the principal system of inertia, in which the external variables are fixed at zero, the portion related to the internal coordinates in the inverse transformation of Eq. (3-13),

$$\Delta x_i = \sum_p (B^{-1})_{ip}\Delta R_p, \tag{5-130}$$

is substituted into Eq. (5-128), and the result is compared with Eq. (5-129) to give

$$A_{ap} = \sum_i B_{ai}(B^{-1})_{ip} = \delta_{ap} \tag{5-131a}$$

and

$$A_{apq} = \sum_i\sum_j B_{aij}(B^{-1})_{ip}(B^{-1})_{jq}. \tag{5-131b}$$

The curvilinear internal coordinates represent the changes in the internal variables due to any nuclear displacement irrespective of nuclear mass. Hence the potential function in terms of the curvilinear internal coordinates in the form

$$V = \tfrac{1}{2}\sum_i \sum_j F_{ij}\ \mathcal{S}_i \mathcal{S}_j + \tfrac{1}{6}\sum_i \sum_j \sum_k F_{ijk}\ \mathcal{S}_i \mathcal{S}_j \mathcal{S}_k$$

$$+ \tfrac{1}{24}\sum_i \sum_j \sum_k \sum_l F_{ijkl}\ \mathcal{S}_i \mathcal{S}_j \mathcal{S}_k \mathcal{S}_l + \cdots \qquad (5\text{-}132)$$

is isotopically invariant. Substituting Eqs. (5-129) and (5-131a) into Eq. (5-132), and collecting the coefficients of the cubic terms as given by

$$f_{ijk} = F_{ijk} + \sum_m \left(F_{im}\, A_{mjk} + F_{jm}\, A_{mik} + F_{km}\, A_{mij} \right), \qquad (5\text{-}133)$$

we obtain

$$V = \tfrac{1}{2}\sum_i \sum_j F_{ij}\ \Delta R_i\ \Delta R_j + \tfrac{1}{6}\sum_i \sum_j \sum_k f_{ijk}\ \Delta R_i\ \Delta R_j\ \Delta R_k + \cdots. \qquad (5\text{-}134)$$

The coefficient f_{ijk} depends on the elements of $\boldsymbol{B}^{-1}$ through eqs. (5-133) and (5-131), and the elements of $\boldsymbol{B}^{-1}$ depend generally on the nuclear mass according to Eq. (3-43). Hence f_{ijk} is not ascertained to be isotopically invariant, and great care must be taken on utilizing the data for isotopic molecules when estimating the potential parameters.

For simplicity, we shall consider here a one-dimensional problem. Let the expansion of $\mathcal{S}$ in powers of the Cartesian displacements Δx terminated at the nth order term be denoted by $\mathcal{S}_n$. In this case, the difference between the true value of the curvilinear internal coordinate and $\mathcal{S}_n$ is given as a power series of Δx starting from the $(n + 1)$th order term, so that we can write it as

$$\mathcal{S} = \mathcal{S}_n + A_n\ \mathcal{S}_n^{\,n+1} + \cdots = \mathcal{S}_n(1 + A_n\, \mathcal{S}_n^{\,n}) + \cdots .$$

If the expansion of the potential function in powers of $\mathcal{S}_n$ starts from the mth order term, the term of the lowest order is written, besides the potential constant, in the form

$$\mathcal{S}^m = \mathcal{S}_n^{\,m}(1 + A_n\, \mathcal{S}_n^{\,n})^m + \cdots = \mathcal{S}_n^{\,m} + mA_n\, \mathcal{S}_n^{\,n+m} + \cdots,$$

which shows that the isotopic dependence starts from the $(n + m)$th order term. On writing the highest order of the terms in the potential function necessary for the calculation as the kth, the highest order n of the terms to be included in the expansion in Eq. (5-128) is obtained to be

$$n = k - m + 1$$

from the relation $k + 1 = n + m$. In the second-order perturbation theory, $k = 4$. Accordingly, if a potential function expanded at the equilibrium is used without redundant coordinates, $m = 2$, whence $n = 3$. In molecular mechanics calculation, the potential function is not always expanded at the potential minima of individual molecules. Accordingly, if the normal coordinate analysis is to be included in the calculation, $k = 2$ and $m = 1$, whence $n = 2$.

In calculating the expansion formula of the internal coordinates in powers of nuclear Cartesian displacements, exchanging the subroutines to perform the arithmetics of power series and calculate the power series of basic functions of power series enables one to control the order of expansion in the numerical calculation. It is also possible to control the order of expansion by using an integer argument of the subroutines, but an increase in conditional branches in the main stream of the program is not desirable for speed-up of the calculation.

Once the normal coordinates and the normal frequencies are deduced from the quadratic part of the potential function expanded in powers of nuclear Cartesian displacements, $\boldsymbol{\Delta x}$, the potential function can be expanded in powers of dimensionless normal coordinates, $\boldsymbol{q}$, as shown in Eq. (5-40), by using the transformation relations from $\boldsymbol{\Delta x}$ to $\boldsymbol{Q}$ in Eq. (3-80) and from $\boldsymbol{Q}$ to $\boldsymbol{q}$ in Eq. (4-48).

By defining the diagonal matrix $\boldsymbol{\alpha}$ involving α_i in Eq. (4-47) as the ith diagonal element, the inverse transformation of Eq. (4-48) is written in the matrix form

$$\boldsymbol{Q} = \boldsymbol{\alpha}^{-1}\, \boldsymbol{q}. \tag{5-135}$$

Combining Eqs. (3-80) and (5-135) leads to the transformation relation from $\boldsymbol{q}$ to $\boldsymbol{\Delta x}$ in the form

$$\boldsymbol{\Delta x} = \boldsymbol{m}^{-1/2}\, \boldsymbol{L}^{mx}\, \boldsymbol{\alpha}^{-1}\, \boldsymbol{q} = \boldsymbol{L}^{q}\, \boldsymbol{q}. \tag{5-136}$$

On substituting each element of Eq. (5-136) into Eq. (5-128), we obtain

$$\mathcal{R}_a = \sum_r L^q_{ar}\, q_r + \sum_r \sum_s L^q_{ars}\, q_r q_s + \sum_r \sum_s \sum_t L^q_{arst}\, q_r q_s q_t, \tag{5-137}$$

where

$$L^q_{ar} = \sum_i B_{ai}\, L^x_{ir}$$

$$L^q_{ars} = \sum_i \sum_j B_{aij}\, L^x_{ir}\, L^x_{js}$$

and

$$L^q_{arst} = \sum_i \sum_j \sum_k B_{aijk}\, L^x_{ir}\, L^x_{js}\, L^x_{kt}.$$

Finally, substitution of Eq. (5-137) into Eq. (5-132) gives the expansion of the potential function in powers of dimensionless normal coordinates in the form

$$V = \tfrac{1}{2} \sum_r \omega_r\, q_r^{\,2} + \sum_r \sum_{\le s} \sum_{\le t} k_{rst}\, q_r q_s q_t + \sum_r \sum_{\le s} \sum_{\le t} \sum_{\le u} k_{rstu}\, q_r q_s q_t q_u + \cdots, \tag{5-138}$$

where the coefficients are given by

$$\omega_r = \sum_i \sum_j F_{ij}\, L^q_{ir} L^q_{jr}$$

$$k_{rst} = \sum_i \sum_j F_{ij} \left(L^q_{irs} L^q_{jt} + L^q_{irt} L^q_{js} + L^q_{ist} L^q_{jr} \right) + \sum_i \sum_j \sum_k F_{ijk}\, L^q_{ir}\, L^q_{js}\, L^q_{kt}$$

and

$$\begin{aligned} k_{rstu} = &\sum_i \sum_j F_{ij} \left(L^q_{irst} L^q_{ju} + L^q_{irsu} L^q_{jt} + L^q_{irtu} L^q_{js} + L^q_{istu} L^q_{jr} + L^q_{irs} L^q_{jtu} + L^q_{irt} L^q_{jsu} + L^q_{iru} L^q_{jst} \right) \\ &+ \sum_i \sum_j \sum_k F_{ijk} \left(L^q_{irs}\, L^q_{jt}\, L^q_{ku} + L^q_{irt}\, L^q_{js}\, L^q_{ku} + L^q_{iru}\, L^q_{js}\, L^q_{kt} \right) \end{aligned} \tag{5-139}$$

Table 5.4 Fundamental frequencies and vibration–rotation interaction constants of formaldehyde (cm^{-1}).

	ν_i		α_A		α_B		α_C	
	calculated	observed	calculated	observed	calculated	observed	calculated	observed
H_2CO								
ν_1	2779.4	2782.4[a]	0.1834	0.1609[a]	0.0005	–0.0007[a]	0.0025	0.0016[a]
ν_2	1744.4	1746.1[b]	0.0057	0.0057[c]	0.0071	0.0076[c]	0.0094	0.0088[c]
ν_3	1499.4	1500.1[b]	–0.0639	–0.0598[b]	0.0000	–0.0046[d]	0.0038	0.0048[d]
ν_4	1165.0	1167.2[d]	0.0276	0.0278[d]	0.0101	0.0133[d]	–0.0007	–0.0017[d]
ν_5	2847.2	2843.2[a]	0.1500	0.1896[a]	0.0010	0.0033[a]	0.0021	0.0037[a]
ν_6	1250.0	1249.1[d]	0.0175	0.0172[d]	–0.0026	–0.0030[d]	0.0067	0.0067[d]
D_2CO								
ν_1	2057.5	2057.1[e]	0.0778		0.0020		0.0047	
ν_2	1693.4	1701.6[f]	–0.0016	–0.0014[c]	0.0072	0.0072[c]	0.0062	0.0062[c]
ν_3	1096.7	1101.3[e]	–0.0263	–0.023[g]	–0.0142	–0.0150[g]	–0.0011	–0.0009[g]
ν_4	938.9	937.8[e]	–0.0012	0.0684[h]	0.0179	0.0138[h]	–0.0007	–0.0019[h]
ν_5	2161.9	2160.3[e]	0.0573		0.0027		0.0020	
ν_6	992.8	987.6[e]	0.0056	–0.0571[h]	–0.0014	–0.0033[h]	0.0069	0.0063[h]

[a] K. Yamada, T. Nakagawa, K. Kuchitsu and Y. Morino, *J. Mol. Spectrosc.*, **38**, 70 (1971).
[b] T. Nakagawa, H. Kashiwagi, H. Kurihara and Y. Morino, *J. Mol. Spectrosc.*, **31**, 436 (1969).
[c] J.W.C. Johns and A.R.W. McKeller, *J. Mol. Spectrosc.*, **48**, 354 (1973).
[d] T. Nakagawa and Y. Morino, *J. Mol. Spectrosc.*, **38**, 84 (1971).
[e] S. Hemple, Dissertation, University of Pennsylvania (1970); cited in H. Khoshkhoo and E.R. Nixon, *Spectrochim. Acta*, **A29**, 603 (1973).
[f] S. Tatematsu, T. Nakagawa, K. Kuchitsu and J. Overend, *Spectrochim. Acta*, **A30**, 1585 (1974).
[g] T. Takahashi, T. Nakagawa and K. Kuchitsu, Symposium on Molecular Structure, Chemical Society of Japan, Osaka, November, 1975.
[h] D. Coffey, Jr., D. Yamada and E. Hirota, *J. Mol. Spectrosc.*, **64**, 98 (1975).

Table 5.5 Centrifugal distortion constants[a] and Coriolis coupling constants of formaldehyde (MHz)

	τ_{aaaa}		τ_{bbbb}		τ_{cccc}		ζ_{46}^{a}	
	calculated	observed	calculated	observed	calculated	observed	calculated	observed
H_2CO	83.9	82.9[b]	0.392	0.390[b]	0.225	0.223[b]	0.537	0.54[c]
D_2CO	21.0	21.3[d]	0.305	0.306[d]	0.121	0.121[d]	0.508	0.498[e]
HDCO	48.0	57.2[f]	0.352	0.361[f]	0.165	0.172[f]	0.431	0.44[g]
$H_2{}^{13}CO$	83.9	86.2[h]	0.370	0.363[h]	0.216	0.209[h]	0.540	

[a] Notation is taken from J.K.G. Watson, *J. Chem. Phys.*, **46**, 1935 (1967).
[b] F.Y. Chu, S.M. Freund, J.W.C. Johns and T. Oka, *J. Mol. Spectrosc.*, **48**, 328 (1973).
[c] H.H. Blau, Jr. and H.H. Nielsen, *J. Mol. Spectrosc.*, **1**, 124 (1957).
[d] Footnote [g] Table 5.4.
[e] J.C. Brand, *J. Chem. Soc.*, **1956**, 858.
[f] T. Oka, H. Hirakawa and K. Shimoda, *J. Phys. Soc. Jpn.*, **15**, 2265 (1960).
[g] V. Sethuraman, V.A. Job and K.K. Innes, *J. Mol. Spectrosc.*, **33**, 189 (1970).
[h] D.R. Johnson, F.J.Lovasand W.H. Kirchhoff, *J. Phys. Chem. Ref. Data*, **1**, 1011 (1972).

Table 5.6 Force constants of formaldehyde.

	K_{ij}[a]			K_{ijk}[a]			K_{ijkl}[a]	
	empirical[b]	theoretical[c]		empirical[b]	theoretical[c]		empirical[b]	theoretical[c]
K_{RR}[d]	6.325	6.651	K_{RRR}	−14.627	−14.725	K_{RRRR}	28.789	22.217
K_{rr}[d]	2.473	2.608	K_{rrr}	−5.247	−5.131	K_{rrrr}	6.495	6.338
$K_{rr'}$[d]	0.044	0.066	$K_{RR\phi}$	−0.134	−0.469	$K_{\phi\phi\phi\phi}$	0.152	0.028
K_{Rr}	0.631	0.503	$K_{R\phi\phi}$	−0.963	−0.655	$K_{\phi\phi\phi'\phi'}$	−0.419	0.205
$K_{\phi\phi}$[d]	0.614	0.651	$K_{R\phi\phi'}$	0.239	−0.006	$K_{RR\phi\phi}$	2.0	0.395
$K_{\phi\phi'}$[d]	0.396	0.456	$K_{\phi\phi\phi}$	0.153	−0.038	$K_{rr\phi\phi}$	3.75	−0.308
$K_{R\phi}$	0.415	0.461	$K_{\phi\phi\phi'}$	0.172	0.309	$K_{rr\phi'\phi'}$	−1.05	−0.276
$K_{r\phi}$	0.019	−0.057	$K_{R\gamma\gamma}$	−0.103	−0.117	$K_{rr\phi\phi'}$	1.34	−0.476
$K_{r\phi'}$	−0.152	−0.164	$K_{\phi\gamma\gamma}$	0.234	0.132	$K_{rr'\phi\phi'}$	1.56	0.511
$K_{\gamma\gamma}$[d]	0.200	0.205	$K_{r\phi\phi}$	−0.50	−0.151	$K_{rr'\phi\phi}$	0.78	0.366
						$K_{\gamma\gamma\gamma\gamma}$	0.124	0.039
						$K_{\phi\phi\gamma\gamma}$	0.10	−0.108
						$K_{RR\gamma\gamma}$	0.6	−0.141

a) Coefficients in the expansion of the potential function in the form

$$V = \sum_{i}\sum_{\le j} K_{ij}\, \text{Я}_i \text{Я}_j + \sum_{i}\sum_{\le j}\sum_{\le k} K_{ijk}\, \text{Я}_i \text{Я}_j \text{Я}_k + \sum_{i}\sum_{\le j}\sum_{\le k}\sum_{\le l} K_{ijkl}\, \text{Я}_i \text{Я}_j \text{Я}_k \text{Я}_l + \cdots$$

The numerals are given in the units derived as combinations of the following basic units. force : mdyne(= 10^{-8} N), length : Å (= 10^{-10} m), angle : radian.

b) The assignment of the overtone $2\nu_1$ used in the empirical estimation of force constants was revised later from 5430 to 5507 cm^{-1}.[e]

c) Among the *ab initio* force constants in GVFF quartic force field, only those constants corresponding to non-zero empirical constants are shown.

d) Coordinate symbols : R : C=O stretch, r : C–H_1 stretch, r': C–H_2 stretch, ϕ : O=C–H_1 bend, ϕ': O=C–H_2 bend, γ: out-of-plane bend.

e) D.E. Reisner, R.W. Field, J.L. Kinsey and H.L. Dai, *J. Chem. Phys.*, **80**, 5968 (1984).

$$+ L^q_{itu} L^q_{jr} L^q_{ks} + L^q_{isu} L^q_{jr} L^q_{kt} + L^q_{ist} L^q_{jr} L^q_{ku} \Big)$$

$$+ \sum_i \sum_j \sum_k \sum_l F_{ijkl} L^q_{ir} L^q_{js} L^q_{kt} L^q_{lu}.$$

By substituting the coefficients in Eq. (5-138) into the formulae of the perturbation theory, the higher order molecular constants x_{ij} in Eqs.(5-66a,b) and α_i^A in Eqs.(5-98a,b) are obtained. The form of Eq. (5-139) is rather complicated, but the calculation can be accomplished very quickly because the number of higher-order force constants can be made so small in terms of the curvilinear internal coordinates that in most cases only those with the same suffixes must be retained.[29]

As an example of the second-order perturbation calculation by an empirical potential function, the calculated values of data for vibration–rotation spectra of formaldehyde and its deuterium isotopomers are listed together with the observed values in Tables 5.4 and 5.5.[30] For the hydrogen stretching frequencies, the deviation from the observed values is reduced to about one tenth on introducing the anharmonic correction to the harmonic approximation. In Table 5.6, the force constants used in the empirical calculation are compared with the values calculated from the *ab initio* MO method.[31] The deviation between the two methods is large for higher-order parameters, indicating that the estimation becomes difficult as the order of parameters rises.

References

1) H. Margenau and G.M. Murphy, *The Mathematics of Physics and Chemistry*, van Nostrand, Princeton (1943), Chapt. 9.
2) M. Goldsmith, G. Amat and H.H. Nielsen, *J. Chem. Phys.*, **24**, 1178 (1956).
3) B.T. Darling and D.M. Dennison, *Phys. Rev.*, **57**, 128 (1950).
4) R. C. Herman and W. H. Schaffer, *J. Chem. Phys.*, **16**, 453 (1948).
5) G. Amat, M. Goldsmith and H.H. Nielsen, *J. Chem. Phys.*, **27**, 838 (1957).
6) H.H. Nielsen, *Handbuch der Physik*, vol. 38, p.173, Springer, Berlin (1957).
7) W.G. Fateley and F.A. Miller, *Spectrochim. Acta*, **17**, 85 (1961).
8) S. Reichman and J. Overend, *J. Chem. Phys.*, **48**, 3095 (1968).
9) A. Burneau and J. Corset, *J. Chim. Phys. Physicochim. Biol.*, **69**, 142, 153, 171 (1972).
10) Y. Tanaka and K. Machida, *J. Mol. Spectrosc.*, **51**, 508 (1974).
11) Y. Tanaka and K. Machida, *J. Mol. Spectrosc.*, **55**, 435 (1975).
12) J.H. Lady and K.B. Whetsel, *Spectrochim. Acta*, **21**, 1669 (1965).
13) L.P. Hammet, *J. Am. Chem. Soc.*, **59**, 96 (1937); *Trans. Faraday Soc.*, **34**, 156 (1938).
14) H.H. Jaffe, *Chem. Rev.*, **53**, 191 (1953).
15) H. Eyring, J. Walter and G.E. Kimball, *Quantum Chemistry*, John Wiley & Sons, New York (1944), Chapts. 3 and 4.
16) E.B. Wilson, Jr. and J.B. Howard, *J. Chem. Phys.*, **4**, 260 (1936).
17) D. Kivelson and E.B. Wilson, Jr., *J. Chem. Phys.*, **20**, 1575 (1952); *ibid.*, **21**, 1229 (1953).
18) J.K.G. Watson, *J. Chem. Phys.*, **45**, 1360 (1966); *ibid.*, **46**, 1935 (1967).
19) T. Oka, *J. Phys. Soc. Jpn.*, **15**, 2274 (1960).
20) M. Toyama, T. Oka and Y. Morino, *J. Mol. Spectrosc.*, **13**, 193 (1964).
21) K. Kuchitsu and L.S. Bartell, *J. Chem. Phys.*, **35**, 1945 (1961).
22) S.J. Cyvin, *Spectrochim. Acta*, **15**, 828, 835, 958 (1959); *Acta Chem. Scand.*, **13**, 1397, 1400, 1809 (1959).
23) S.J. Cyvin, *Molecular Vibrations and Mean Square Amplitudes*, Elsevier, Amsterdam (1968).
24) J. Pliva, *Collect. Czech. Chem. Commun.*, **23**, 777, 1839,1846, 1852 (1958).
25) K. Kuchitsu and Y. Morino, *Bull. Chem. Soc. Jpn.*, **38**, 805, 814 (1965).
26) M. Pariseau, I. Suzuki and J. Overend, *J. Chem. Phys.*, **42**, 2335 (1965); I. Suzuki, M. Pariseau and J. Overend, *J. Chem. Phys.*, **44**, 3561 (1966).
27) A.R. Hoy, I.M. Mills and G. Strey, *Mol. Phys.*, **24**, 1265 (1972).
28) I. Suzuki, *Appl. Spectrosc. Rev.*, **9**, 249 (1975).
29) K. Machida, *J. Chem. Phys.*, **44,** 4186 (1966).
30) Y. Tanaka and K. Machida, *J. Mol. Spectrosc.*, **64**, 429 (1977).
31) L.B. Harding and W.C. Ermler, *J. Comput. Chem.*, **6**, 13 (1985).

Chapter 6

Thermodynamic Functions

6.1 Thermodynamic Functions and Molecular Force Field

To describe the temperature dependence of chemical equilibria, we must calculate the free energy from the enthalpies and entropies of each constituent molecule of the initial and final systems within the specified range of temperature. According to the formulae of statistical mechanics, the thermodynamic functions of an isolated molecule can be calculated from its rotational and vibrational levels as certain functions of temperature and pressure.[1–3] In turn, the rotational and the vibrational levels of an isolated molecule are determined solely from its force field. Thus, molecular mechanics provides a general method of calculating thermodynamic functions of polyatomic molecules in the gas phase.

Even when the formulae of statistical mechanics cannot be used because of lacking data on vibrational levels, we can calculate thermodynamic functions by using certain additivity rules based on empirical contributions from the local structure units of the molecule. Many of earlier force fields which did not aim at calculating vibrational frequencies adopted the empirical approach for the calculation of heats of formation. The time and memory required for the normal coordinate analysis of an N-atomic molecule increase in proportion to N^3–N^4, while the time and memory required for simple addition increase at most in proportion to N. Thus, the advantage of empirical method in dealing with large systems is expected to become more prominent with progress in the capacity and speed of the computer in the future.

This chapter describes the principles underlying the statistical mechanical calculation of thermodynamic functions of isolated molecules and their equilibrium mixtures. The data required for this calculation are the energies of dissociation into the atomic states, the principal moments of inertia and the normal frequencies. The empirical methods of calculating thermodynamic functions based on various additivity rules are then outlined. Finally, how to deduce the anharmonic corrections to the thermodynamic functions under the harmonic approximation is described.

6.2 Partition Function

Calculation of any thermodynamic functions of a molecule can be accomplished if we have complete knowledge of the partition function of that molecule. According to the principle of quantum mechanics, the energy of a given molecule takes discrete values, E_1, E_2, $\cdots$. Each of these energy values is associated with either a single or more than one wave function. A wave function represents an "eigenstate", or simply a state of that molecule. A molecule in any molecular assembly transfers from one state to another by exchanging certain amounts of energy with the surrounding molecules.

Let g_i be the number of states to which the ith energy value E_i is associated. Then the probability that a molecule has the energy E_i at a moment is proportional to $g_i \exp(-E_i/kT)$ according to Boltzmann's distribution law. Let P_i be this probability, and let us write it in the form

$$P_i = g_i \exp(-E_i/kT)/Q. \tag{6-1}$$

From the condition that the sum of all the probabilities is unity, it follows that

$$Q = \sum_i g_i \exp(-E_i/kT). \tag{6-2}$$

Note that the ratio P_i at a non-zero T remains invariant on replacing E_i in Eqs. (6-1) and (6-2) with $E_i - E_c$, where E_c is any constant, whence the energy may be measured from an arbitrarily chosen origin. At $T = 0$ K, both the denominator and the numerator in Eq. (6-1) vanish, so that the ratio is indeterminate. How to avoid this difficulty will be shown later. The quantity Q is called the partition function of the molecule. The energy of this molecular assembly per molecule, U, is the average of energy eigenvalues E_i weighted by its population P_i. Thus we have

$$U = \sum_i P_i E_i = \sum_i E_i \, g_i \exp(-E_i/kT) \Big/ Q. \tag{6-3}$$

Differentiating Eq. (6-2) with respect to T gives

$$\frac{\partial Q}{\partial T} = \sum_i \frac{E_i g_i}{kT^2} \exp\left(-\frac{E_i}{kT}\right). \tag{6-4}$$

Dividing Eq. (6-4) by Q, and substituting Eq. (6-3) into the right side, we have

$$\frac{1}{Q}\frac{\partial Q}{\partial T} = \frac{\partial \ln Q}{\partial T} = \frac{1}{KT^2}\frac{\sum_i E_i g_i \exp(-E_i/kT)}{Q} = \frac{U}{KT^2},$$

whence

$$U = kT^2(\partial \ln Q/\partial T). \tag{6-5}$$

The internal energy U is expressed in terms of the Helmholtz free energy A and the entropy S in the form

$$U = A + TS. \tag{6-6}$$

According to the second law of thermodynamics, the change in internal energy dU on an adiabatic change, in which the only work done on the system is of a mechanical type associated with the change of volume dV, is given by

$$dU = -PdV = dA + SdT.$$

Hence we have

$$dA = -PdV - SdT. \tag{6-7}$$

Under the isobaric and adiabatic conditions, we can take V and T as independent variables. Then the change in A is given by

$$dA = (\partial A/\partial V)_T dV + (\partial A/\partial T)_V dT. \tag{6-8}$$

From Eqs.(6-7), (6-8) and (6-6), we obtain

$$(\partial A/\partial T)_V = -S = (A-U)/T. \tag{6-9}$$

The quantity A/T is then differentiated with respect to T under the constant volume, giving

$$\left\{\frac{\partial(A/T)}{\partial T}\right\}_V = \frac{1}{T}\left(\frac{\partial A}{\partial T}\right)_V - \frac{A}{T^2} = \frac{1}{T}\left\{\left(\frac{\partial A}{\partial T}\right)_V - \frac{A}{T}\right\}. \tag{6-10}$$

Substituting Eq. (6-9) into Eq. (6-10) and solving for U, we have

$$U = -T^2\{\partial(A/T)/\partial T\}_V, \tag{6-11}$$

which is a form of the Gibbs–Helmholtz equation.[2] Eliminating U from Eqs. (6-5) and (6-11) leads to

$$\{\partial(A/T)/\partial T\}_V = -k(\partial \ln Q/\partial T)_V,$$

or

$$\left\{\frac{\partial}{\partial T}\left(\frac{A}{T} + k\ln Q\right)\right\}_V = 0.$$

Introducing the integration constant C,[3] the Helmholtz free energy A is expressed as

$$A = -kT\ln Q + CT. \tag{6-12}$$

By differentiating Eq.(6-12) with respect to T under a constant V, and substituting the result into Eq. (6-9), the relation between the entropy and the partition function is derived as

$$S = k\ln Q + kT(\partial \ln Q/\partial T)_V - C. \tag{6-13}$$

Before evaluating the integration constant C, let the partition function be redefined in such a way that the probability P_i has a definite value at 0 K. By expressing the energy of the lowest state as E_1, and the energy of the ith state measured from E_1 as E_i', we have $E_i = E_i' + E_1$. The partition function defined by taking E_1 as the origin of the energy is then given by

$$Q' = g_1 + \sum_{2\le i} g_i \exp(-E_i'/kT). \tag{6-14}$$

In terms of Q', Q in Eq. (6-2) and S in Eq. (6-13) are given, respectively, as

$$Q = Q'\exp(-E_1/kT) \tag{6-15}$$

and

$$S = k\ln Q' - \frac{E_1}{T} + kT\left\{\left(\frac{\partial \ln Q'}{\partial T}\right)_V + \frac{E_1}{kT^2}\right\} - C$$
$$= k\ln Q' + kT\left(\frac{\partial \ln Q'}{\partial T}\right)_V - C. \tag{6-16}$$

On the other hand, the limits of Eq. (6-14) and its temperature derivative on $T \to 0$ K are calculated to be

$$\lim_{T\to 0} Q' = g_1, \tag{6-17}$$

and

$$\lim_{T\to 0} \frac{\partial Q'}{\partial T} = \lim_{T\to 0} \frac{1}{kT^2} \sum_i \frac{g_i E_i'}{\exp(E_i'/kT)} = 0, \tag{6-18}$$

respectively. The last equality in Eq. (6-18) results from the relation

$$\lim_{T\to 0} kT^2 \exp(E_i'/kT) = \infty.$$

In terms of Q', the probability that the molecule has the energy E_i' is given by $P_i' = g_i \exp(-E_i'/kT)/Q'$, which leads to definite limits in the forms

$$\lim_{T\to 0} P_1' = 1 \quad \text{and} \quad \lim_{T\to 0} P_i' = 0 \;\; (i > 0).$$

Combining Eqs. (6-17) and (6-18), we have

$$\lim_{T\to 0} \frac{\partial \ln Q'}{\partial T} = \lim_{T\to 0} \frac{1}{Q'}\frac{\partial Q'}{\partial T} = \frac{0}{g_1} = 0. \tag{6-19}$$

According to the third law of thermodynamics, the entropy of a perfectly ordered molecular assembly with a single state of the lowest energy is taken to be zero. Thus, substituting Eqs. (6-17) and (6-19) into the limit of Eq. (6-16) at 0 K and replacing g_1 with 1, we see that[3)]

$$C = k\ln g_1 = k\ln 1 = 0.$$

For such a system, Eqs. (6-12) and (6-13) turn out to be

$$A = -kT\ln Q \tag{6-20}$$

and

$$S = k\ln Q + kT(\partial \ln Q/\partial T)_V, \tag{6-21}$$

respectively. Substitution of Eq. (6-5) into Eq. (6-21) leads to

$$S = k\ln Q + U/T. \tag{6-22}$$

In the definition of a partition function given by Eq. (6-2), each energy state of a molecule is simply specified by a single suffix i. Under the first approximation described in Chapter 4, the nuclear state i of the molecule consists of mutually independent substates, $i_1, i_2, \cdots$, brought about by the separation of variables. Correspondingly, the energy of the state i is expressed as the sum of the energies associated with individual substates in the form

$$E_i \equiv E(i_1, i_2, \cdots) = E(i_1) + E(i_2) + \cdots = \sum_p E(i_p).$$

Since the degeneracy of the state i is the product of the degeneracies of the substates, $g(i_1)$, $g(i_2)$, ..., the partition function of the whole molecule is factored into the product of the partition functions of individual substates as given by

$$Q = \sum_{i_1}\sum_{i_2}\cdots\sum_{i_{3N}}\left\{\prod_{p=1}^{3N} g(i_p)\right\}\exp\left\{-\sum_{p=1}^{3N} E(i_p)/kT\right\}$$
$$= \left[\prod_{p=1}^{3N}\sum_{i_p} g(i_p)\exp\{-E(i_p)/kT\}\right]. \tag{6-23}$$

6.3 Translation, Rotation and Vibration

The form of the partition function for a given substate is determined by the distribution pattern of the energy levels in that substate. The energy levels are obtained by solving the Schrödinger equation for the variables which enter the wave function of that substate as the arguments. There are three types of substates of nuclear motion, *i.e.*, translations, rotations and vibrations. For the first two types, the formulae of energy levels will be used here without showing the details of derivation.

Suppose that a molecule with the mass M is freely moving within a box in the shape of a rectangular parallelepiped with three edges of lengths a, b and c. The energy levels of this molecule is given by

$$E_T = \frac{h^2}{8M}\left(\frac{i^2}{a^2} + \frac{j^2}{b^2} + \frac{k^2}{c^2}\right),$$

where i, j and k are positive integers.[4] By using Eq. (6-2), the partition function for the translation of this molecule is given as the product of three factors in the form

$$Q_T = \left\{\sum_i \exp\left(-\frac{h^2 i^2}{8M a^2 kT}\right)\right\}\left\{\sum_j \exp\left(-\frac{h^2 j^2}{8M b^2 kT}\right)\right\}\left\{\sum_k \exp\left(-\frac{h^2 k^2}{8M c^2 kT}\right)\right\}. \tag{6-24}$$

In order to briefly estimate the spacing of the translational energy levels of typical organic molecules in the gaseous state, the dimensionless coefficient of i^2 in Eq. (6-24) is calculated by taking the mass of a benzene molecule, $(156.1/6.02214) \times 10^{-23}$ g, as M and $a = 1$ cm. The result is

$$\frac{h^2}{8Ma^2kT} = \frac{(6.6261\times10^{-34})^2 \times 6.02214\times10^{23}}{8\times156.1\times10^{-3-4}\times1.3807\times10^{-23}\times273.15}\ \frac{(\mathrm{J\ s})^2}{\mathrm{kg\ m^2 J}} = 5.614\times10^{-19}.$$

Because of the smallness of the spacing, the sum in the first factor in the right side of Eq. (6-24) may be replaced with an integral in the form

$$\sum_i \exp(-\alpha^2 i^2) \cong \int_0^\infty \exp(-\alpha^2 i^2)di = \sqrt{\pi}/\alpha,$$

where $\alpha = h/(8Ma^2\,kT)^{1/2}$. Calculating the factors involving b and c in similar ways, multiplying the three factors by one another and replacing the product abc with the volume V, we obtain the translational partition function in the form

$$Q_T = \frac{(2\pi\, M k\, T)^{3/2}}{h^3} a\, b\, c = \frac{(2\pi\, M k\, T)^{3/2}}{h^3} V \quad . \tag{6-25}$$

The partition function of rotation is given different expressions between linear and nonlinear molecules.[3] Since the rotational energy of a linear molecule with the quantum number K is $(2J + 1)$-fold degenerate over the range $-J \le K \le J$, the partition function is given, according to Eq.(6-2), by

$$Q_R = \sum_{J=1}^{\infty} (2J+1)\exp\{-h\, B\, J(J+1)/kT\}. \tag{6-26}$$

For molecules with very small rotational constant B, the quantum number J may be regarded as a continuous variable. In this case, it is convenient to introduce a new variable y by

$$y = J\,(J + 1). \tag{6-27a}$$

Differentiating Eq. (6-27a) yields

$$dy = (2J + 1)\, dJ. \tag{6-27b}$$

The summation in Eq. (6-26) is then replaced by an integral, which can be evaluated by using Eqs. (6-27a,b) as given by

$$\int_0^{\infty} (2\, J+1)\exp\{-h\, B\, J(J+1)/k\, T\} dJ = \int_0^{\infty} \exp(-h\, B\, y/k\, T)\, dy = \frac{k\, T}{h\, B}. \tag{6-28}$$

By combining Eq. (6-28) with the definition of the rotational constant (Eq. (5-84)), the rotational partition function of a linear molecule is expressed in the form

$$Q_R = 8\pi^2 I\, k\, T/\sigma\, h^2 = 4\pi\,(2\,\pi\, I\, k\, T)/\sigma\, h^2\,, \tag{6-29}$$

where σ is called the symmetry number, which counts the number of possible orientations of the molecule indistinguishable unless different nuclei of the same kind can be discriminated from each other. Thus, σ is 2 for centrosymmetrical linear molecules and 1 for non-centrosymmetrical linear molecules.

A linear molecule is an isotropic two-dimensional rotator. The Hamiltonian of a one-dimensional rotator is the term in the first sum in the braces in the right side of Eq. (5-43a), $(P_\alpha^2/2I_\alpha)$. Formally, this Hamiltonian has the same form as that of the one-dimensional free particle, $(P_x^2/2m)$, in which the mass m is replaced by the moment of inertia. Thus the partition function of a one-dimensional rotator is written in the same form as that of a one-dimensional free particle on which the periodic boundary condition $\psi\,(2\pi) = \psi\,(0)$ is imposed, that is,

$$Q_R^{(1)} = 2\,\pi\,(2\,\pi\, I\, k\, T)^{1/2}/h. \tag{6-30}$$

This expression corresponds to the one-dimensional factor of the translational partition function (Eq. (6-25)), $(2\,\pi M k T)^{1/2} a/h$, in which M and a are replaced by I and 2π, respectively.

Similarly, the partition function of a two-dimensional rotator (Eq. (6-29)) is obtained from the two-dimensional factor $(2\pi MkT)ab/h^2$ in Eq. (6-25) by the substitutions $M \rightarrow I$ and $ab \rightarrow 4\pi$. Note that the spaces allowed for the motion of the one- and the two-dimensional free particles, a and ab, respectively, are replaced by the space allowed for the motion of a particle fixed at the unit distance from the origin, that is, the circumference of a circle and the surface area of a sphere with the unit radii. Derivation of the rotational partition function of a nonlinear polyatomic molecule is quite complicated. Here we show only the result as

$$Q_R = 8\pi^2(I_A I_B I_C)^{1/2}(2\pi kT)^{3/2}/(\sigma h^3) \tag{6-31}$$

where I_A, I_B and I_C are the three principal moments of inertia. Formally, Eq. (6-31) is obtained by taking the product of Eqs. (6-29) and (6-30), after replacing I in the former by $(I_A I_B)^{1/2}$ and that in the latter by I_C. Through linear and nonlinear molecules, the symmetry number σ indicates the number of ways in which the molecule can be superposed on itself in the course of possible rotations around its center of gravity.

The vibrational partition function within the framework of the harmonic approximation takes the form

$$Q_i^{HV} = \sum_n \exp\left\{-\frac{hc\omega_i}{kT}\left(n+\tfrac{1}{2}\right)\right\} = \frac{\exp(-hc\omega_i/2kT)}{1-\exp(-hc\omega_i/kT)}. \tag{6-32}$$

Equation (6-32) is obtained by substituting the energy of the harmonic oscillator with the frequency ω_i, Eq. (4-69), into the definition of the partition function, Eq. (6-2), and using Eq. (5-112) for the sum of a geometrical series. If the energy is measured from the vibrational ground state, the vibrational partition function turns out to be

$$Q_i^{HV} = \sum_n \exp\left(-\frac{hc\omega_i}{kT}n\right) = \frac{1}{1-\exp(-hc\omega_i/kT)}. \tag{6-33}$$

In the calculation of thermodynamic functions by the statistical mechanical method, the natural logarithms of the partition functions and their temperature derivatives are used repeatedly. They are listed in Table 6.1 for convenience.

The translational, rotational and vibrational energies per molecule are given, by

Table 6.1 $\ln Q$ and $\partial \ln Q/\partial T$.

Type of motion	$\ln Q$	$\partial \ln Q/\partial T$
translation	$\frac{3}{2}\ln T + \ln\frac{(2\pi Mk)^{3/2}}{h^3}V$	$\frac{3}{2T}$
rotation (linear molecules)	$\ln T + \ln\frac{8\pi^2 Ik}{\sigma h^2}$	$\frac{1}{T}$
rotation (non-linear molecules)	$\frac{3}{2}\ln T + \ln\frac{8\pi^2(I_A I_B I_C)(2\pi k)^{3/2}}{\sigma h^2}$	$\frac{3}{2T}$
vibration	$-\frac{hc\omega_i}{2kT} - \ln\left\{1-\exp\left(\frac{hc\omega_i}{kT}\right)\right\}$	$\frac{hc\omega_i}{kT^2}\left\{\frac{1}{2} - \frac{\exp(hc\omega_i/2kT)}{1-\exp(hc\omega_i/kT)}\right\}$

σ : symmetry number

substituting the entries of the second column of Table 6.1 into Eq. (6-5), in the forms

$$U_T = (3/2)kT \tag{6-34a}$$

$$\left.\begin{aligned} U_R &= kT, && \text{(linear molecules)} \\ U_R &= (3/2)kT && \text{(nonlinear molecules)} \end{aligned}\right\} \tag{6-34b}$$

$$U_V = \sum_i hc\omega_i \left\{ \frac{1}{2} + \frac{\exp(-hc\omega_i/2kT)}{1 - \exp(-hc\omega_i/kT)} \right\}. \tag{6-34c}$$

Similarly, the entropies are obtained from the first column in Table 6.1 and Eq. (6-21). For the translation, we have

$$S_T = k\left[\frac{3}{2}\ln M + \frac{5}{2}\ln T - \ln P\right] - k\left[\frac{5}{2} + \ln\left\{\frac{1}{N_A}\left(\frac{2\pi k}{h^2 N_A}\right)^{3/2}\right\} + \ln R'\right], \tag{6-35a}$$

where P is the pressure and R' is the gas constant. The latter enters Eq. (6-35a) on eliminating V by the law of ideal gases, $PV = R'T$. For the rotation of linear molecules, we have

$$S_R = k[\ln T + \ln I - \ln\sigma] + k\left[1 + \ln\left(\frac{8\pi^2 k}{h^2 N_A}\right)\right] \tag{6-35b}$$

and for that of nonlinear molecules, we have

$$S_R = k\left[\frac{3}{2}\ln T + \frac{1}{2}\ln I_A I_B I_C - \ln\sigma\right] + k\left[\frac{3}{2} + \ln\left\{8\pi^2\left(\frac{2\pi k}{h^2 N_A}\right)^{3/2}\right\}\right]. \tag{6-35c}$$

In each of Eqs. (6-35a–c), the term depending on the molecular structure is given separately from the constant term. The entropy of molecular vibrations under the harmonic aproximation is given by

$$S_V = k\sum_i\left[\frac{hc\omega_i}{kT}\frac{\exp(-hc\omega_i/kT)}{1 - \exp(-hc\omega_i/kT)} - \ln\left\{1 - \exp\left(-\frac{hc\omega_i}{kT}\right)\right\}\right]. \tag{6-35d}$$

Note that the entropies given by Eqs. (6-35a–d) are the values per molecule.

The heat capacity is defined as the energy required for raising the temperature of the sample by 1 K. There are two types of heat capacities according to whether the volume or the pressure is kept constant during the temperature raising. The heat capacity at constant volume, C_v, is the temperature derivative of the internal energy in the form

$$C_v = (\partial U/\partial T)_V, \tag{6-36}$$

while the heat capacity at constant pressure, C_p, is the temperature derivative of the enthalpy, $H = U + PV$, in the form

$$C_p = (\partial H/\partial T)_P. \tag{6-37}$$

By differentiating Eqs. (6-34a–c), the contributions from the translation, the rotation and the vibration to C_v are obtained as follows:

$$C_v^T = (3/2)k \tag{6-38a}$$

$$\left.\begin{array}{l} C_v^R(\text{linear molecules}) = k \\ C_v^R(\text{nonlinear molecules}) = (3/2)k \end{array}\right\} \tag{6-38b}$$

$$C_v^{HV} = k\sum_i \left(\frac{hc\omega_i}{kT}\right)^2 \frac{\exp(hc\omega_i/kT)}{\{1-\exp(hc\omega_i/kT)\}^2} \tag{6-38c}$$

6.4 Internal Rotation

The vibrational partition function, Eq. (6-32), which has been derived under the harmonic approximation with the assumption that the nuclear displacements are small, diverges when the normal frequency approaches zero. Such a case is encountered when the coefficient V_n^0 in the torsional potential in Eq. (2-114) decreases, leading to an increase in the amplitude and a decrease in the frequency of the torsional vibration. Special care is required for evaluating the partition function in this case.

If the nonbonded atom–atom interaction between the parts separated by a single bond in a molecule is negligible and if $V_n = 0$ in addition, the internal rotation around this bond becomes completely free. In this case, the partition function can be estimated by using the one-dimensional free rotor model as follows. Let 1 and 2 be the parts separated by the single bond, and I_1 and I_2 the moments of inertia of the parts 1 and 2, respectively, evaluated with respect to the bond axis. A quantity called the reduced moment of inertia, I_m is then defined in terms of I_1 and I_2 in the form

$$I_m = I_1 I_2/(I_1 + I_2), \tag{6-39}$$

from which the partition function is calculated by

$$Q_m^{FR} = \frac{2\pi}{\sigma h}(2\pi I_m kT)^{1/2}. \tag{6-40}$$

On the other hand, Eq. (6-32) is used if the barrier height is high enough for the torsional motion to be regarded as a harmonic vibration.

The nuclear motion in the case of an intermediate barrier height is called a hindered rotation. The standard method of calculating thermodynamic functions from a precise treatment of partition function for systems including hindered rotors has been developed by Pitzer *et al.*[5,6] In this method, the solution is obtained from certain numerical tables constructed by an involved mathematical procedure. To the author's knowledge, this method has not been widely incorporated in the current programs for automated molecular mechanics calculations.

Truhlar proposed a convenient method of calculating thermodynamic functions of hindered rotors by interpolation of those of a free rotor and a harmonic oscillator.[7] In this method, an empirical factor,

$$f_m = \tanh(Q_m^{FR}\, u_m), \tag{6-41}$$

where

$$u_m = hc\omega_m/kT, \tag{6-42}$$

is introduced, and the partition function of a hindered rotor is calculated as the product of the partition function of a harmonic oscillator and the factor f_m in the form

$$Q_m^{HR} = Q_m^{HV} f_m = Q_m^{HV} \tanh(Q_m^{FR}\, u_m). \tag{6-43}$$

The factor f_m approaches $Q_m^{FR}\, u_m$ as u_m diminishes. On the other hand, it is seen from Eqs. (6-33) and (6-42) that Q_m^{HV} approaches u_m^{-1} on the limit that $u_m \to 0$. The limit of the product of Q_m^{HV} and f_m becomes thus

$$\lim_{u_m \to 0} Q_m^{HR} = Q_m^{FR}\, u_m \lim_{u_m \to 0} Q_m^{HV} = Q_m^{FR}.$$

On the other hand, we have

$$\lim_{u_m \to \infty} Q_m^{HR} = Q_m^{HV},$$

since f_m approaches unity on increase of u_m. Thus the factor f_m interpolates between Q_m^{HV} and Q_m^{FR} in accordance with the barrier height. Now we introduce the quantity

$$a_m = \left(\frac{8\pi^3 I_m}{k}\right)^{1/2} \frac{c\,\omega_m}{\sigma_m}, \tag{6-44}$$

and rewrite the product of Eqs. (6-40) and (6-42) in the form

$$Q_m^{FR}\, u_m = \frac{2\pi}{\sigma_m}\left(\frac{2\pi I_m k T}{h^2}\right)^{1/2} \frac{h c \omega_m}{k T} = a_m\, T^{-1/2}. \tag{6-45}$$

Substituting Eq. (6-45) into Eq. (6-41), taking the logarithm and differentiating with respect to temperature, we have

$$\begin{aligned} \frac{\partial \ln f_m}{\partial T} &= \frac{1}{f_m}\frac{\partial}{\partial T}\tanh(a_m T^{-1/2}) = -\frac{a_m T^{-3/2}}{2}\frac{\mathrm{sech}^2(a_m T^{-1/2})}{\tanh(a_m T^{-1/2})} \\ &= \frac{-a_m T^{-3/2}}{2\sinh(a_m T^{-1/2})\cosh(a_m T^{-1/2})} = -\frac{a_m T^{-3/2}}{\sinh(2 a_m T^{-1/2})}. \end{aligned} \tag{6-46}$$

By substituting Eqs. (6-43) and (6-46) into Eq. (6-5), the contribution of the hindered rotation to the internal energy is obtained in the form

$$U_m^{HR} = k T^2 \frac{\partial \ln Q_m^{HV} f_m}{\partial T} = U_m^{HV} - k T \frac{a_m T^{-1/2}}{\sinh(2 a_m T^{-1/2})}. \tag{6-47}$$

Similarly, the contribution to the entropy is obtained from Eq. (6-22) as

$$\begin{aligned} S_m^{HR} &= \frac{U_m^{HR}}{T} - k \ln Q_m^{HR} \\ &= S_m^{HV} - k \frac{a_m T^{-1/2}}{\sinh(2 a_m T^{-1/2})} - k \ln \tanh(a_m T^{-1/2}), \end{aligned} \tag{6-48}$$

and the contribution to the heat capacity at the constant volume is obtained from Eq. (6-36) as

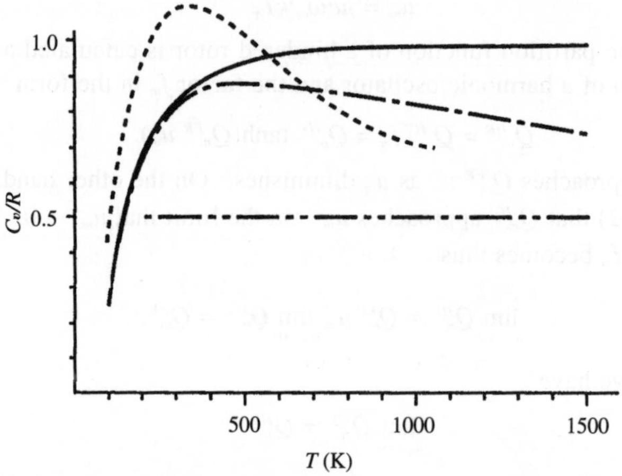

Fig. 6.1 Temperature dependence of heat capacity at constant volume of ethane: (solid line) harmonic oscillator model, (dotted line) Pitzer's method, (dot and dash) Truhlar's method, (double dot and dash) free rotor model.

$$
\begin{aligned}
(C_v^{HR})_m &= (C_v^{HV})_m - \frac{\partial}{\partial T}\left\{ kT \frac{a_m T^{-1/2}}{\sinh(2a_m T^{-1/2})} \right\} \\
&= (C_v^{HV})_m - k\left\{ \frac{a_m T^{-1/2}}{2\sinh(2a_m T^{-1/2})} - \frac{a_m{}^2 T^{-1}\cosh(2a_m T^{-1/2})}{\sinh^2(2a_m T^{-1/2})} \right\} \\
&= (C_v^{HV})_m - k\frac{a_m T^{-1/2}}{\sinh(2a_m T^{-1/2})}\left\{ \frac{1}{2} + \frac{a_m T^{-1/2}}{\tanh(2a_m T^{-1/2})} \right\}.
\end{aligned}
\tag{6-49}
$$

The heat capacities of ethane calculated by various models are compared with each other in Fig. 6.1.

6.5 Thermodynamic Functions of Equilibrium Mixtures

The effect of mixing should be taken into account in calculating thermodynamic functions of an assemby of isomeric molecules in thermal equilibrium. A simple additive law holds for the internal energy and enthalpy of an equilibrium mixture of n-isomers.

Let X_i be the molar fraction of the ith isomer, and U_i and H_i be the energy and the enthalpy, respectively, carried by this isomer. The energy, U_M, and the enthalpy, H_M, of the mixture are given by

$$U_M = \sum_{i=1}^{n} X_i U_i \tag{6-50a}$$

and

$$H_M = \sum_{i=1}^{n} X_i H_i \ , \tag{6-50b}$$

respectively. The heat capacity of an equilibrium mixture is obtained by differentiating

Eq. (6-50b) with the temperature T. In this case, we must pay attention to the fact that the molar fraction X_i depends on T. The heat capacity at constant pressure is written as

$$(C_p)_M = \sum_{i=1}^{n} X_i \frac{\partial H_i}{\partial T} + \sum_{i=1}^{n} \frac{\partial X_i}{\partial T} H_i. \tag{6-51}$$

Given the Gibbs free energy of the ith isomer measured from that of the most stable isomer by ΔG_i, it then follows from Boltzmann's distribution law that

$$X_i = \exp(-\Delta G_i/kT) \Big/ \left\{ \sum_{j=1}^{n} \exp(-\Delta G_j/kT) \right\}. \tag{6-52}$$

Differentiating Eq. (6-52) with respect to T yields

$$\frac{\partial X_i}{\partial T} = -X_i \frac{\partial}{\partial T}\left(\frac{\Delta G_i}{kT}\right) + X_i \sum_{j=1}^{n} X_j \frac{\partial}{\partial T}\left(\frac{\Delta G_j}{kT}\right). \tag{6-53}$$

From the definition of the Gibbs free energy and the enthalpy, we have

$$G = U + PV - TS = H - TS. \tag{6-54}$$

Let us take the differential of Eq. (6-54) under the isochoric and adiabatic conditions. Noting that dU, dV and dS vanish in this case, we have

$$dG = (\partial G/\partial P)_T dP + (\partial G/\partial T)_P dT = V dP - S dT. \tag{6-55}$$

Comparing the second and third formulae in Eq. (6-55) leads to

$$(\partial G/\partial T)_P = -S. \tag{6-56}$$

Differentiating the quotient G/T at constant pressure, and substituting Eq. (6-56) into the result, we obtain

$$\left\{\frac{\partial}{\partial T}\left(\frac{G}{T}\right)\right\}_P = \frac{1}{T}\left(\frac{\partial G}{\partial T}\right)_P - \frac{G}{T^2} = -\frac{1}{T}\left(S + \frac{G}{T}\right). \tag{6-57}$$

In terms of the enthalpy $H = G + TS$, Eq. (6-57) is rewritten as

$$\left\{\frac{\partial}{\partial T}\left(\frac{G}{T}\right)\right\}_P = -\frac{H}{T^2}. \tag{6-58}$$

Equation (6-58) is a form of the Gibbs–Helmholtz equation in terms of the G–H pair.[2] Write down Eq. (6-58) for the most stable isomer and the ith isomer, take the difference and divide by k, and finally let $G_i - G_0$ and $H_i - H_0$ be denoted by ΔG_i and ΔH_i, respectively. Then the result is

$$\frac{\partial}{\partial T}\left(\frac{\Delta G_i}{kT}\right) = -\frac{\Delta H_i}{kT^2}. \tag{6-59}$$

It follows from substitution of Eq. (6-59) into Eq. (6-53) that

$$\frac{\partial X_i}{\partial T} = \frac{X_i \Delta H_i}{kT^2} - X_i \sum_{j=1}^{n} \frac{X_j \Delta H_j}{kT^2}. \tag{6-60}$$

Multiplying Eq. (6-60) by H_i and summing over i, and exchanging the subscripts i and j in the second term of the right side, we obtain

$$\sum_{i=1}^{n} H_i \frac{\partial X_i}{\partial T} = \frac{1}{kT^2}\left(\sum_{i=1}^{n} H_i X_i \Delta H_i - \sum_{j=1}^{n} H_j X_j \sum_{i=1}^{n} X_i \Delta H_i\right)$$

$$= \frac{1}{kT^2}\sum_{i=1}^{n} X_i \Delta H_i\left(H_i - \sum_{j=1}^{n} H_j X_j\right) = \frac{1}{kT^2}\sum_{i=1}^{n} X_i \Delta H_i\left(\Delta H_i - \sum_{j=1}^{n} X_j \Delta H_j\right). \tag{6-61a}$$

The last equality in Eq. (6-61a) is derived as follows by utilizing the fact that the sum of molar fractions is unity:

$$H_i - \sum_{j=1}^{n} H_j X_j = H_i - H_0 + \left(\sum_{j=1}^{n} X_j\right) H_0 - \sum_{j=1}^{n} H_j X_j$$

$$= \Delta H_i - \sum_{j=1}^{n} X_j (H_j - H_0).$$

By making use of the fact that

$$H_i - H_j = \Delta H_i - \Delta H_j,$$

Eq. (6-61a) can be rewritten as

$$\sum_{i=1}^{n} H_i \frac{\partial X_i}{\partial T} = \frac{1}{kT^2}\sum_{i=1}^{n} X_i \Delta H_i\left(H_i \sum_{j=1}^{n} X_j - \sum_{j=1}^{n} H_j X_j\right)$$

$$= \frac{1}{kT^2}\sum_{i=1}^{n} X_i \Delta H_i \sum_{j=1}^{n} (H_i - H_j) X_j$$

$$= \frac{1}{2kT^2}\left\{\sum_{i=1}^{n} X_i \Delta H_i \sum_{j=1}^{n} (H_i - H_j) X_j + \sum_{j=1}^{n} X_j \Delta H_j \sum_{i=1}^{n} (H_j - H_i) X_i\right\}$$

$$= \frac{1}{kT^2}\sum_{i=1}^{n}\sum_{j=i}^{n} X_i X_j (H_i - H_j)^2. \tag{6-61b}$$

Finally, substitution of Eq. (6-61b) into Eq. (6-51) yields the desired expression for the heat capacity at constant pressure of an equilibrium mixture in the form

$$(C_p)_M = \sum_{i=1}^{n} (C_p)_i X_i + \frac{1}{kT^2}\sum_{i=1}^{n}\sum_{j=i}^{n} X_i X_j (H_i - H_j)^2. \tag{6-62}$$

The second term in the right side of Eq. (6-62) represents the effect of the temperature change of the composition. This term has the same form as the expression given by Aston *et al.* for the case of the two-component system.[8] Note that this correction term is always positive as seen from the form of Eq. (6-61b), and may lead to a systematic error at low temperatures if omitted. In the actual calculation, Eq. (6-61a) is more easily coded than Eq. (6-61b), since the double sum in Eq. (6-61a) can be carried out by repeating the single sum twice while that in Eq. (6-61b) cannot. The heat capacity of n-butane calculated from Eq. (6-62) is compared with the experimental data in Fig. 6.2.

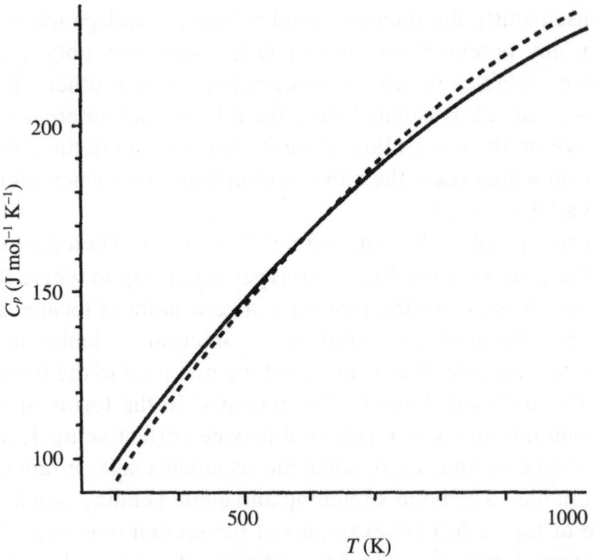

Fig. 6.2 Temperature dependence of heat capacity at constant pressure of equilibrium mixture of rotational isomers of *n*-butane : (solid line) observed, (dotted line) calculated.

The entropy of an equilibrium mixture of isomers in the gaseous state can be calculated as follows. Let the partial pressure and the entropy of the component i in the mixture be denoted by P_i and S_i, respectively. Writing down S_i as the sum of contributions from the translation, rotation and vibration in the forms of Eqs. (6-35a–d), and collecting the terms which do not depend on the pressure into a single notation W_i, we obtain

$$S_i = W_i - k \ln P_i. \tag{6-63}$$

Since the partial pressure is related to the total pressure by

$$P_i = X_i P,$$

Eq. (6-63) may be rewritten as

$$S_i = W_i - k (\ln X_i + \ln P) = S_i^t - k \ln X_i, \tag{6-64}$$

where S_i^t is the entropy of the component i per molecule when $X_i = 1$, that is, when the component i forms a pure gas at the pressure P. The entropy of the mixture is then given by

$$S_M = \sum_{i=1}^{n} X_i S_i = \sum_{i=1}^{n} X_i S_i^t - k \sum_{i=1}^{n} X_i \ln X_i. \tag{6-65}$$

The second term in Eq. (6-65) is called the entropy of mixing.

6.6 Internal Energies Measured from Dissociated States

In general, the empirical potential functions assumed in molecular mechanics calculations include a larger number of internal variables than the number of degrees of freedom of the nuclear motion, and consequently the internal variables are not independent of each other. Each individual term in any potential function includes, however, only a single internal variable or a few of those variables which are independent of each other. If a single term in a potential function is regarded as isolated from the others, each internal variable in that term can take any value within the range allowed for it. Such a term in the potential function is usually written in a form which takes the value zero at least once when all the arguments are changed over their variable ranges.

Now let us call x_0 a zero point of the function $f(x)$ if $f(x_0) = 0$. Then the model potential functions given in Chapter 2 are classified into two types according to where the zero points of the potential energy are located. In the first type, a zero point is located at the limit of infinite separation between the nuclei consituting the internal variables involved in the potential function. This type includes the bond stretching potential in the form of the Morse function in Eq. (2-78), the nonbonded atom–atom potential in the forms of the exp–6 and the n–6 types and the Coulomb interaction potential between effective nuclear charges. An energy term of the second type becomes zero when the arguments take certain values specific to the type of internal variable. The bond stretching and angle bending potentials written as polynomials in the form of Eq. (2-52) are examples of the second type with the zero points corresponding to the potential minima. The Morse potential expressed as Eq. (2-78) plus dissociation energy D_{ij}^e also belongs to this type. Physically, the zero point of the first type is located at the dissociation limit of the molecule into individual atoms, while that of the second type is located at an equilibrium value of internal variables in the strainless state.

Since the number of internal variables used in molecular mechanics calculations is usually larger than the number of internal degrees of freedom of the nuclear motion, all the energy terms in the potential function cannot vanish simultaneously at a given configuration point in the nuclear coordinate space. For the classical mechanical model of an isolated molecule at the limit of the vanishing kinetic energy of nuclei, the value of each internal variable is so determined as to minimize the sum of all the energy terms. The nuclear coordinates at this point give an equilibrium structure as described in detail in Chapter 3.

It follows from the above discussion that, when the bond stretching energy terms are written in the first type, the total potential function calculated for an equilibrium structure gives the internal energy U_0 of the classical model molecule measured from the state perfectly dissociated into the constituent atoms. If the bond stretching energy terms are written in the second type, the internal energy U_0 is given by the potential function at the equilibrium minus the sum of the dissociation energies of all the bonds in the molecule. The increment for the bond i–j in the latter case corresponds to D_{ij}^e in Eqs. (2-78) and (2-81). Adding this classical internal energy to the translational, rotational and vibrational energies calculated as described in the preceding section, we obtain the internal energy per molecule of an ideal gas obeying the law of quantum statistics at the given temperature.

6.7 Calculation of Thermodynamic Functions by Empirical Rules of Additivity

The statistical mechanical method of calculating thermodynamic functions is theoretically correct within the framework of the adopted approximation, but practical use of this method can be hardly extended beyond relatively small molecules because of lack of data on normal frequencies. To avoid such a difficulty of the statistical mechanical method, various methods of calculating thermodynamic functions from the empirical parameters attributed to chemical bonds and special local structures have been proposed.

The atomic heats of formation of unconjugated compounds without any strain can be crudely estimated as the sum of all the bond energies. In a strict sense, the increment of the heat of formation per bond is more or less affected by the environment of each bond. This effect can be taken into account, at least partly, by using certain assemblies of several bonds or "groups" instead of individual bonds. Even in cases where a bond or a group is involved in a strained structure, fairly accurate heats of formation are obtained by applying appropriate corrections for such structures determined empirically.[9,10] The method of calculating thermodynamic properties of compounds from this standpoint is called the group contribution method.

Suppose that each term in the potential function of Eq. (2-51) is expressed as the energy gained on the changes in the internal variables involved in that term from their intrinsic equilibrium values, *i.e.*, the equilibrium values in a strainless molecule. In this case, the potential function evaluated for an equilibrium structure of a given molecule, V_0, in Eq. (4-1), represents the energy correction for the strains in individual structure parameters and the interaction between nonbonded atoms. This V_0 corresponds to the energy correction for the strain in the group contribution method. Accordingly, the potential function can be used in combination with appropriate bond increments for evaluating the heats of formation without any knowledge of the normal frequencies. In this method, the standard heat of formation of a conformer is calculated by

$$\Delta H_f^0 = \sum_i n_i \, BE_i + E_s + 4RT, \qquad (6\text{-}66)$$

where n_i is the number of the ith bond or structure-type, BE_i is the bond increment for the ith bond, and E_s is the steric energy given by the value of the potential function for the equilibrium structure of the conformer. The term $4RT$ in Eq. (6-66) arises from the high-temperature limit of the partition functions of the translation and the rotation ($6 \times RT/2$)) plus the PV term of the ideal gas. In the program MM2, the parameters BE_i have been determined by the method of least squares with reference to the experimental standard heats of formation minus ($E_s + 4RT$) of 52 hydrocarbons including normal, branched and three- to six-membered cyclic compounds.

We must note here that the standard heat of formation is defined as the heat absorbed on the formation of a compound from its constituent elements not in the atomic state but in a specifically defined standard state. In the case of hydrocarbons, the standard states are crystalline graphite for carbon and hydrogen gas for hydrogen. The equation of this formation reaction of an alkane containing n carbon atoms is written as

$$n\,\mathrm{C(graphite)} + (n+1)\,\mathrm{H_2} = \mathrm{C}_n\mathrm{H}_{2n+2}\,.$$

Since there are $n-1$ C–C bonds and $2n+2$ C–H bonds in a molecule of C_nH_{2n+2}, the heat balance of the reaction is given by

$$\begin{aligned}(n-1)B_{\text{C-C}} + (2n+2)B_{\text{C-H}} - n\,B_{\text{C}_{\text{graph}}} - (n+1)B_{\text{H-H}} \\ = (n-1)(B_{\text{C-C}} - \tfrac{1}{2}B_{\text{C}_{\text{graph}}}) + (n+1)(2B_{\text{C-H}} - B_{\text{H-H}} - \tfrac{1}{2}B_{\text{C}_{\text{graph}}})\end{aligned} \tag{6-67}$$

where $B_{\text{X-Y}}$ is the heat absorbed on the formation of the bond X–Y from the atoms X and Y. Comparing Eqs. (6-66) and (6-67) for C_nH_{2n+2}, we have

$$BE_{\text{CC}} = B_{\text{C-C}} - \tfrac{1}{2}B_{\text{C}_{\text{graph}}}. \tag{6-68a}$$

and

$$BE_{\text{CH}} = B_{\text{C-H}} - \tfrac{1}{2}B_{\text{H-H}} - \tfrac{1}{4}B_{\text{C}_{\text{graph}}}. \tag{6-68b}$$

The method of group contribution may be converted to the method of statistical thermodynamics if the vibrational contribution is subtracted from BE_i and calculated separately from Eq. (6-34c).

The bond increments used in the program MM*n* series by Allinger *et al.*[11–15] are listed in Table 6.2 together with a few supplementary increments which reflect secondary differences in the environment of the bonds. The bond increments in MM1,[11] MM2[12] and the first paper on MM3[13] are comparable with the corresponding quantities in Eqs. (6-68a,68b) calculated from the bond energies quoted by Pauling.[16] In the second paper on MM3[14] and in MM4,[15] the heats of formation are calculated by the standard method of statistical thermodynamics. In this case, the bond increments are equivalent to the dissociation energies estimated from the spectroscopically determined potential curves. The use of the dissociation energy itself as BE_i is favorable in making its physical meaning clearer. In this case, however, we must note that the standard heat of formation of the product is obtained by subtracting from the result the atomic heat of formation of the standard reactant system. In order to minimize

Table 6.2 Heat of formation parameters in MM1 through MM4 (kJ mol^{-1}).

Type	MM1[a]	MM2[b]	MM3(1)[c]	MM3(2)[d]	MM4[e]
Bond parameters					
C–C[f]	12.43	−0.017	10.238	−376.983	−373.217
C–H[f]	−18.70	−13.410	−19.205	−441.630	−442.358
Structure increments					
C–CH$_3$	3.10	−6.318	4.372	−0.590	−0.4033
C–CHC$_2$(iso)	−3.10	0.326	−10.991	−5.427	−5.2614
C–CC$_3$(neo)	−7.322	−4.686	−27.786	−13.862	−11.4131
4-ring	0.	0.	−7.448	0.	0.
5-ring	0.	0.	−23.045	6.276	6.293
ring	0.	0.	0.	18.700	20.6882
torsion	0.	0.	1.757	3.138	2.3912

[a] Ref. 11. [b] Ref. 12. [c] Ref. 13. [d] Ref. 14. [e] Ref. 15.

[f] In the Morse function used in RISE[g], the following dissociation energies are used.

$$D^e_{\text{CH}_2\text{-CH}_2} = 359.62 \text{ kJ mol}^{-1}, \quad D^e_{\text{C-H}} = 440.09 \text{ kJ mol}^{-1}.$$

[g] Y. Miwa and K. Machida, *J. Am. Chem. Soc.*, **110**, 5183 (1988).

the loss of significant digits due to taking a small difference between two large quantities, the actual program is recommended to be coded in the sense of Eqs. (6-68a,68b) irrespective of the definition of BE_i. This sort of consideration is particularly important in dealing with large molecules.

In the program MMP2, which is an extended version of MM2 to the conjugated hydrocarbon system,[17] the standard heat of formation is calculated by

$$\Delta H_f^0 = \sum_i n_i \, BE_i + E_s + 4RT - E_{\text{conj}}.$$

The bond increments BE_i related to the π bonds have been estimated from the data of ethylene, benzene and many other conjugated hydrocarbons on the asumption that they can be used together with the parameters for the non-conjugated compounds determined in MM2. The parameter list has been supplemented with three new structure increments for three types of branched hydrocarbon skeleton including the sp^2 hybridization. The parameter E_{conj} is the steric energy arising from the shifts of sp^2–sp^2 bonds, and is estimated by using the Morse function. This method has proved to predict successfully the structures and the heats of formation of aromatic compounds, cyclophanes and annulenes which deviate from the planar structure because of the steric hindrance due to bulky substituents.[17]

6.8 Correction for Vibrational Anharmonicity

Theromodynamic functions calculated under the harmonic approximation are the product of an idealization of the actual molecular system. The reliability of these quantities may be estimated by comparison with the result of calculations under any higher order approximations. To improve accuracy of the calculations, we need only to reformulate the calculation of partition functions, while the formulae connecting the partition functions to thermodynamic functions need not be altered from those in the harmonic approximation at all.

On starting the higher order calculation, it is convenient to rewrite the vibrational transition energy in terms of the fundamental frequency ν_i instead of the normal frequency ω_i by substituting Eq. (5-69) into Eq. (5-68).[18] In the absence of degeneracy, the result is given in the form

$$G(v_1, v_2, \cdots) - G(0, 0, \cdots) = \sum_i v_i \left\{ \nu_i + (v_i - 1)x_{ii} + \sum_{j>i} x_{ij} v_j \right\}, \tag{6-69}$$

By taking the exponential functions of the products of Eq. (6-69) and the factor $-hc/kT$ and summing them up over all the combinations of vibrational quantum numbers, the vibrational partition function within the framework of the second-order perturbation theory is obtained in the form

$$\begin{aligned} Q^{AV} &= \sum_{v_1, v_2, \cdots} \exp\left[-\frac{hc}{kT} \sum_i v_i \left\{ \nu_i + (v_i - 1)x_{ii} + \sum_{j>i} v_j x_{ij} \right\} \right] \\ &= \sum_{v_1, v_2, \cdots} \exp\left(-\frac{hc}{kT} \sum_i v_i \nu_i \right) \exp\left[-\frac{hc}{kT} \sum_i v_i \left\{ (v_i - 1)x_{ii} + \sum_{j>i} v_j x_{ij} \right\} \right]. \end{aligned} \tag{6-70}$$

In contrast to the case of harmonic approximation, there appear factors containing two vibrational quanta in Eq. (6-70), so we cannot factor the vibrational partition function into those for individual normal modes. Since the anharmonicity constants x_{ii} and x_{ij} are small quantities, the second exponential factor involving the brackets, exp $[\cdots]$, in Eq. (6-70) may be expanded according to the relation

$$\exp(x) \approx 1 + x\,,$$

giving rise to many factors of the forms

$$\exp\left\{\frac{hc}{kT}(v_i - 1)v_i\, x_{ii}\right\} \approx 1 + \frac{hc}{kT}(v_i - 1)v_i\, x_{ii}$$

and

$$\exp\left(\frac{hc}{kT}v_i\, v_j\, x_{ij}\right) \approx 1 + \frac{hc}{kT}v_i\, v_j\, x_{ij}\,.$$

Taking the product of these factors in the form

$$\exp[\cdots] \cong 1 + \sum_i \frac{hc}{kT}(v_i - 1)v_i\, x_{ii} + \sum_i \sum_{<j} \frac{hc}{kT} v_i\, v_j\, x_{ij}\,. \tag{6-71}$$

and substituting it back into Eq. (6-70), we obtain

$$\begin{aligned} Q^{AV} &= \sum_{v_1, v_2, \cdots} \left\{ \prod_i \exp\left(-\frac{hc}{kT} v_i \nu_i\right) \right\} \left\{ 1 + \sum_i \frac{hc}{kT}(v_i - 1)v_i\, x_{ii} + \sum_i \sum_{<j} \frac{hc}{kT} v_i\, v_j\, x_{ij} \right\} \\ &= Q^{HV} + \sum_i \frac{Q^{HV}}{Q_i^{HV}} \frac{hc}{kT} x_{ii} \sum_{v_i} (v_i - 1)v_i \exp\left(-\frac{hc}{kT} v_i \nu_i\right) \\ &\quad + \sum_i \sum_{<j} \frac{Q^{HV}}{Q_i^{HV} Q_j^{HV}} \frac{hc}{kT} x_{ij} \sum_{v_i, v_j} v_i\, v_j \exp\left\{-\frac{hc}{kT}(v_i \nu_i + v_j \nu_j)\right\}. \end{aligned} \tag{6-72}$$

The second and third terms in Eq. (6-72) represent the anharmonic correction to the partition function in the harmonic approximation. The factor Q^{HV}/Q_i^{HV} in the second term is obtained by dropping the factor involving the quantum number v_i from the harmonic partition function Q^{HV}. Similarly, dropping the factors involving v_i and v_j from Q^{HV} gives the factor $Q^{HV}/Q_i^{HV} Q_j^{HV}$ in the third term. Equation (6-72) can be rewritten in a simpler form as follows.

Differentiating both the sides of Eq. (5-113) with respect to u, and subtracting Eq. (5-113) from the result, we have

$$\begin{aligned} \frac{d}{du} \sum_v v \exp(-vu) - \sum_v v \exp(-vu) &= \sum_v (v^2 - v) \exp(-vu) \\ &= \frac{2\exp(-2u)}{\{1 - \exp(-u)\}^3} = \frac{2}{\{1 - \exp(-u)\}(\exp u - 1)^2}\,. \end{aligned} \tag{6-73}$$

On the other hand, the product of two forms of Eq. (5-113) with v labeled by the suffixes i and j is written in the form

$$\sum_{v_i}\sum_{v_j} v_i\, v_j \exp\{-(u_i\, v_i + u_j\, v_j)\} = \frac{\exp\{-(u_i + u_j)\}}{\{1-\exp(-u_i)\}^2\{1-\exp(-u_j)\}^2}$$
$$= \left[(\exp u_i - 1)(\exp u_j - 1)\{1-\exp(-u_i)\}\{1-\exp(-u_j)\}\right]^{-1}. \tag{6-74}$$

The sums over the quantum numbers in the second and the third terms in the last formula of Eq. (6-72) can be rewritten in closed forms by using Eqs. (6-73) and (6-74), respectively. To express these terms compactly, it is useful to introduce an abbreviated notation

$$f_{ij} = \frac{(1+\delta_{ij})(hc\, x_{ij}/kT)}{\{\exp(hc\,\nu_i/kT)-1\}\{\exp(hc\,\nu_j/kT)-1\}}. \tag{6-75}$$

By using Eq. (6-75), Eq. (6-72) is greatly simplified in the form

$$Q^{AV} = \left\{\prod_i \frac{1}{1-\exp(-hc\,\nu_i/kT)}\right\}\left(1+\sum_{j\geq i} f_{ij}\right). \tag{6-76}$$

The vibrational partition function in the framework of the second-order perturbation theory should involve the correction terms arising from the interaction between the rotations and the vibrations of the molecule. Let the moments of inertia of an asymmetric top molecule involved in the rotational partition function in Eq. (6-31) be rewritten in terms of the rotational constants given by Eq. (5-84), and introduce the dependence on the vibrational quantum numbers according to Eq. (5-96). The rotational constant B^A is given a dimensionless expression in the form

$$\sigma_A = \frac{hc}{kT}B^A = \frac{hc}{kT}\left(B_0^A - \sum_i \alpha_i^A v_i\right) = \sigma_A^0\left(1-\sum_i \delta_i^A v_i\right),$$

where B_0^A is the rotational constant for the vibrational ground state (Eq. (5-99)). Defining σ_B and σ_C in a similar way for the B- and the C- axes, respectively, we obtain the rotational partition function in the form

$$Q^R = \frac{1}{8^{1/2}\pi\sigma}\left(\frac{2\pi\, kT}{hc\, B^A}\right)^{1/2}\left(\frac{2\pi\, kT}{hc\, B^B}\right)^{1/2}\left(\frac{2\pi\, kT}{hc\, B^C}\right)^{1/2}$$
$$= \frac{1}{\sigma}\left(\frac{\pi}{\sigma_A^0\sigma_B^0\sigma_C^0}\right)^{1/2}\left(1+\tfrac{1}{2}\sum_i \delta_i v_i\right), \tag{6-77}$$

where

$$\delta_i = \delta_i^A + \delta_i^B + \delta_i^C = \frac{hc}{kT}\left(\frac{\alpha_i^A}{\sigma_A^0}+\frac{\alpha_i^B}{\sigma_B^0}+\frac{\alpha_i^C}{\sigma_C^0}\right).$$

Let Eq. (6-77) be multiplied by the vibrational partition function in Eq. (6-72), and take out the terms involving the vibrational quantum numbers v_i. These terms are rewritten, with the help of Eq. (5-113), in the form

$$\tfrac{1}{2}\sum_i \delta_i \sum_{v_i} v_i \exp(-u_i\, v_i) = \tfrac{1}{2}\sum_i \frac{\delta_i}{\{1-\exp(-u_i)\}(\exp u_i - 1)}\,. \tag{6-78}$$

Adding Eq. (6-78) to Eq. (6-76), we obtain the vibrational partition function, corrected for the anharmonicity in the potential function and the vibration–rotation interaction up to the second order, in the form

$$Q^{AV} = \left\{\prod_i \frac{1}{1-\exp(-hc\nu_i/kT)}\right\}\left\{1+\sum_i\sum_{\le j} f_{ij} + \tfrac{1}{2}\sum_i \frac{\delta_i}{\exp(hc\nu_i/kT)-1}\right\}. \tag{6-79}$$

In the case of water vapor, the difference between the heat capacity at constant pressure calculated from the partition function in Eq. (6-79), C_p, and that calculated from the partition function for the rigid rotor–harmonic oscillator model, C_p^0, amounts to 0.2–0.3 J K^{-1} mol^{-1} at 1000 °C and about 0.9 J K^{-1} mol^{-1} at 2000 °C.[18] The difference is much smaller than these values at lower temperatures.

Another method of calculating thermodynamic functions as accurately as possible is to sum up the contributions of all the vibrotational levels to the partition function as given by the definition in Eq. (6-2). The levels may be obtained by either the theoretical calculation or the spectroscopic measurement. This method has been applied to water vapor.[19]

References

1) G.N. Lewis and M. Randall, Revised by K.S. Pitzer and L. Brewer, *Thermodynamics*, 2nd ed., Chapter 27, McGraw-Hill, New York (1961).
2) W.J. Moore, *Physical Chemistry*, 4th ed., Chapter 3, Prentice Hall, Englewood Cliffs (1972).
3) J. H. Knox, *Molecular Thermodynamics, An Introduction to Statistical Mechanics for Chemists*, John Wiley & Sons, New York (1971).
4) L. Pauling and E.B. Wilson, Jr., *Introduction to Quantum Mechanics with Applications to Chemistry*, Chapter 2, McGraw-Hill New York (1935).
5) K.S. Pitzer and W.D. Gwinn, *J. Chem. Phys.*, **10**, 428 (1942).
6) K.S. Pitzer, *J. Chem. Phys.*, **14**, 239 (1946).
7) D.G. Truhlar, *J. Comput. Chem.*, **12**, 266 (1991).
8) J.G. Aston, G. Szasz, H.W. Wooley and F.G. Brickwedde, *J. Chem. Phys.*, **14**, 67 (1946).
9) R.C. Reid, J.M. Prausnitz and T.K. Sherwood, *The Properties of Gases and Liquids,* 3rd ed., Chapter 7, McGraw-Hill, New York (1977).
10) S.W. Benson, F.R. Cruickshank, D.M. Golden, G.R. Haugen, H.E. O'Neal, A.S. Rodgers, R. Shaw and R. Walsh, *Chem. Rev.*, **69**, 279 (1969).
11) N.L. Allinger, M.T. Tribble, M.A. Miller and D.H. Wertz, *J. Am. Chem. Soc.*, **93**, 1637 (1971).
12) N.L. Allinger, *J. Am. Chem. Soc.*, **99**, 8127 (1977).
13) N.L. Allinger, Y.H. Yuh and J.H. Lii, *J. Am. Chem. Soc.*, **111**, 8551 (1989).
14) J.H. Lii and N.L. Allinger, *J. Am. Chem. Soc.*, **111**, 8566 (1989).
15) N.L. Allinger, K. Chen and J.H. Lii, *J. Comput. Chem.*, **17**, 642 (1996).
16) L. Pauling, *The Nature of Chemical Bonds*, 3rd ed., p. 85, Cornell University Press, New York (1960).
17) J.T. Sprague, J.C. Tai, Y. Yuh and N.L. Allinger, *J. Comput. Chem.*, **8**, 581 (1987).
18) A.S. Friedman and L. Haar, *J. Chem. Phys.*, **22**, 2051 (1954).
19) J.M.L. Martin, J.P. Francois and R. Gijbels, *J. Chem. Phys.*, **96**, 7633 (1992).

Chapter 7

Electric Properties of Molecules

7.1 Point Charge Model of Molecules

Since a molecule is an assembly of positively charged nuclei and negatively charged electrons, a certain electric field is formed around a molecule. If a static or oscillating electric field is applied on a molecule, it shows its own response as the result of interaction with the applied field. Because of this response, the interaction between molecules and the electric field is useful as an indispensable source of information in all fields of chemistry. Obviously, the behavior of a molecule in an applied electric field is closely related to the distribution of charged particles constituting the molecule. On the other hand, as long as a molecule is an assembly of charged particles, the force field to be used in the molecular mechanics calculation should contain the terms representing the Coulombic forces acting between the particles in any form.

Thus the intramolecular charge distribution is related to the electric properties and the force field of the molecule simultaneously. This chapter deals with the methods of describing the intramolecular charge distribution as a function of nuclear coordinates and calculating various molecular properties derivable therefrom.

7.2 Multipole-Multipole Interaction

On evaluating the net electric force acting between any nuclei in a molecule, the effect of electron clouds around each nucleus should be taken into account. The electron density is a real continuous quantity given by the product of the electronic wave function and its complex conjugate. To simplify the discussion, we partition the space around a nucleus into such small cells that the electron density in each cell may be taken as constant, regarding the electron cloud around the nucleus to be an assembly of point charges each of which is given by the product of the electron density and the cell volume.[1)]

In Fig. 7.1, let A be an assembly of point charges distributed within a finite portion of space around the origin O, and B be another assembly of charged particles around a point $P(X, Y, Z)$. The distance between O and P is assumed to be large enough compared to the extent of either the assembly A or B. In the following, the ith point charge of the assembly A is denoted by q_{Ai}, and the locations of the point P and the point charge q_{Ai} are specified by the position vectors

$$\boldsymbol{R} = [X \;\; Y \;\; Z]^{\mathrm{T}}$$

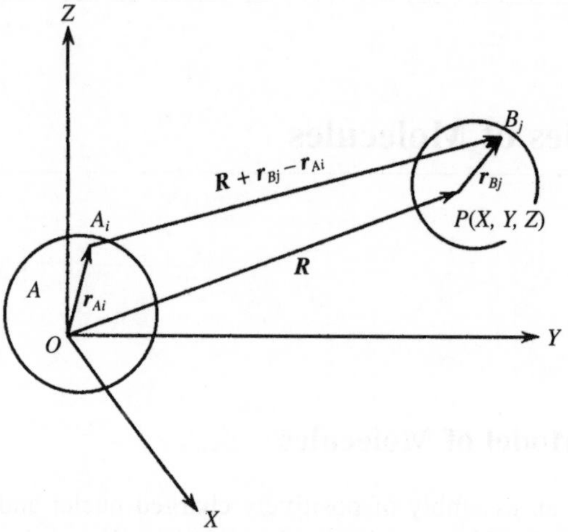

Fig. 7.1 Distance vector connecting the point charges A_i and B_j belonging to charge assemblies A and B, respectively.

and

$$\boldsymbol{r}_{Ai} = [X_{Ai} \;\; Y_{Ai} \;\; Z_{Ai}]^{\mathrm{T}},$$

respectively. Then the position vector of the jth point charge of the assembly B, q_{Bj}, is denoted by

$$\boldsymbol{R} + \boldsymbol{r}_{Bj} = [X + X_{Bj} \;\; Y + Y_{Bj} \;\; Z + Z_{Bj}]^{\mathrm{T}},$$

where

$$\boldsymbol{r}_{Bj} = [X_{Bj} \;\; Y_{Bj} \;\; Z_{Bj}]^{\mathrm{T}}$$

is the vector directed from the point P to the point charge q_{Bj}. Let $R_{Ai,Bj}$ be the distance between the point charges q_{Ai} and q_{Bj}. Then the electric potential at the point $\boldsymbol{r}_{Ai}$ generated by the charge assembly B is given by

$$\phi(\boldsymbol{R},\boldsymbol{r}_{Ai}) = \sum_j q_{Bj}/R_{Ai,Bj} = \sum_j q_{Bj}\left|\boldsymbol{R} + \boldsymbol{r}_{Bj} - \boldsymbol{r}_{Ai}\right|^{-1}. \tag{7-1}$$

Then the total electrostatic energy between the charge assemblies A and B is given by

$$V = \sum_i \sum_j q_{Ai}\, q_{Bj}/R_{Ai,Bj} = \sum_i q_{Ai}\, \phi(\boldsymbol{R},\boldsymbol{r}_{Ai}). \tag{7-2}$$

Since it has been assumed that the extents of the charge assemblies A and B are much smaller than the distance between the two assemblies, it should hold that $|\boldsymbol{r}_{Ai}| \ll |\boldsymbol{R}|$ and $|\boldsymbol{r}_{Bj}| \ll |\boldsymbol{R}|$, and we can expand Eqs. (7-1) and (7-2) in powers of the components of $\boldsymbol{r}_{Ai}$ and $\boldsymbol{r}_{Bj}$. This calculation involves repeated differentiations of $R^{-n} = |\boldsymbol{R}|^{-n}$ with respect to X, Y and Z, whereby the differential operator introduced in Chapter 3,

$$\boldsymbol{\nabla} \equiv [\partial/\partial X \quad \partial/\partial Y \quad \partial/\partial Z]^{\mathrm{T}},$$

gives convenient expressions in the forms

$$\boldsymbol{\nabla}\left(\frac{1}{R^n}\right) = -\frac{n}{R^{n+2}}\boldsymbol{R} \tag{7-3}$$

and

$$\boldsymbol{\nabla}\tilde{\boldsymbol{\nabla}}\left(\frac{1}{R^n}\right) = \frac{n}{R^{n+2}}\left\{-\boldsymbol{E} + \frac{(n+2)\boldsymbol{R}\tilde{\boldsymbol{R}}}{R^2}\right\}. \tag{7-4}$$

Setting $\boldsymbol{r}_{Ai}$ in Eq. (7-1) to be $\boldsymbol{0}$, and expanding $|\boldsymbol{R} + \boldsymbol{r}_{Bj}|$ in powers of X_{Bj}, Y_{Bj} and Z_{Bj} at the point P up to the quadratic terms, we have

$$\begin{aligned} \left|\boldsymbol{R} + \boldsymbol{r}_{Bj}\right|^{-n} &= |\boldsymbol{R}|^{-n} + \tilde{\boldsymbol{\nabla}}|\boldsymbol{R}|^{-n}\boldsymbol{r}_{Bj} + \tfrac{1}{2}\tilde{\boldsymbol{r}}_{Bj}\left(\boldsymbol{\nabla}\tilde{\boldsymbol{\nabla}}|\boldsymbol{R}|^{-n}\right)\boldsymbol{r}_{Bj} + \cdots \\ &= \frac{1}{R^n} - \frac{n}{R^{n+2}}\left(\tilde{\boldsymbol{R}}\boldsymbol{r}_{Bj}\right) + \tfrac{1}{2}\frac{n}{R^{n+2}}\left\{-\tilde{\boldsymbol{r}}_{Bj}\boldsymbol{r}_{Bj} + \frac{n+2}{R^2}\left(\tilde{\boldsymbol{r}}_{Bj}\boldsymbol{R}\tilde{\boldsymbol{R}}\boldsymbol{r}_{Bj}\right)\right\}. \end{aligned} \tag{7-5}$$

Since the matrix product obeys the combination law, $\boldsymbol{A}(\boldsymbol{BC}) = (\boldsymbol{AB})\boldsymbol{C}$, and the order of multiplication may be inverted if the result is a (1×1) matrix (a scalar), the factor $\tilde{\boldsymbol{r}}_{Bj}\boldsymbol{R}\tilde{\boldsymbol{R}}\boldsymbol{r}_{Bj}$ in the last term in Eq. (7-5) may be rewritten as

$$\tilde{\boldsymbol{r}}_{Bj}\boldsymbol{R}\tilde{\boldsymbol{R}}\boldsymbol{r}_{Bj} = \left(\tilde{\boldsymbol{r}}_{Bj}\boldsymbol{R}\right)\left(\tilde{\boldsymbol{R}}\boldsymbol{r}_{Bj}\right) = \left(\tilde{\boldsymbol{R}}\boldsymbol{r}_{Bj}\right)\left(\tilde{\boldsymbol{r}}_{Bj}\boldsymbol{R}\right) = \tilde{\boldsymbol{R}}\left(\boldsymbol{r}_{Bj}\tilde{\boldsymbol{r}}_{Bj}\right)\boldsymbol{R}\ .$$

Substituting $n = 1$ into Eq. (7-5) after this revision, multiplying the result by q_{Bj} and summing over j, we obtain the electric potential at the origin generated by the charge assembly B in the form

$$\phi(\boldsymbol{R},0) = \frac{1}{R}\sum_j q_{Bj} - \frac{\tilde{\boldsymbol{R}}}{R^3}\sum_j q_{Bj}\boldsymbol{r}_{Bj} + \frac{1}{2R^3}\left\{-\sum_j q_{Bj}\left(\tilde{\boldsymbol{r}}_{Bj}\boldsymbol{r}_{Bj}\right) + \frac{3}{R^2}\tilde{\boldsymbol{R}}\sum_j q_{Bj}\left(\boldsymbol{r}_{Bj}\tilde{\boldsymbol{r}}_{Bj}\right)\boldsymbol{R}\right\}. \tag{7-6}$$

Let q_i be the ith member of an assembly of point charges, and let the position vector of q_i be given by

$$\boldsymbol{r}_i = [X_i \ \ Y_i \ \ Z_i]^{\mathrm{T}}$$

Then the tensor of rank n consisting of 3^n components of the form

$$M_{X\cdots Y\cdots Z\cdots} = \sum_i q_i X_i^{\,k} Y_i^{\,l} Z_i^{\,m}, \tag{7-7}$$

where k, l and m are zero or positive integers satisfying the relation $k + l + m = n$, is called the 2^n-pole moment of the charge assembly. In the actual expressions of tensor components in the left side of Eq. (7-7), the suffixes X, Y and Z appear k, l and m times respectively. Of these tensors, that of rank zero is a scalar representing the net charge of the assembly, that of rank one is a vector known as the dipole moment,

$$\boldsymbol{\mu} \equiv [\mu_X \quad \mu_Y \quad \mu_Z]^{\mathrm{T}} = \left[\sum_i q_i X_i \quad \sum_i q_i Y_i \quad \sum_i q_i Z_i\right]^{\mathrm{T}}, \tag{7-8}$$

and that of rank two is the quadrupole moment,

$$\boldsymbol{\Theta}' \equiv \begin{bmatrix} \Theta'_{XX} & \Theta'_{XY} & \Theta'_{XZ} \\ \Theta'_{YX} & \Theta'_{YY} & \Theta'_{YZ} \\ \Theta'_{ZX} & \Theta'_{ZY} & \Theta'_{ZZ} \end{bmatrix} = \begin{bmatrix} \sum_i q_i X_i^2 & \sum_i q_i X_i Y_i & \sum_i q_i X_i Z_i \\ \sum_i q_i Y_i X_i & \sum_i q_i Y_i^2 & \sum_i q_i Y_i Z_i \\ \sum_i q_i Z_i X_i & \sum_i q_i Z_i Y_i & \sum_i q_i Z_i^2 \end{bmatrix}. \tag{7-9}$$

The vector representation of Eq. (7-8) and the matrix representaion of Eq. (7-9) are

$$\boldsymbol{\mu} = \sum_i q_i \boldsymbol{r}_i$$

and

$$\boldsymbol{\Theta}' = \sum_i q_i (\boldsymbol{r}_i \tilde{\boldsymbol{r}}_i),$$

respectively.[1] Let the net charge, the dipole moment and the quadrupole moment of the charge assembly B be denoted by Q_B, $\boldsymbol{\mu}_B$ and $\boldsymbol{\Theta}_B$, respectively. Then we can rewrite Eq. (7-6) as

$$\phi(\boldsymbol{R}, \boldsymbol{0}) = \frac{Q_B}{R} - \frac{1}{R^3} \tilde{\boldsymbol{R}} \boldsymbol{\mu}_B - \frac{\mathrm{Tr}(\boldsymbol{\Theta}'_B)}{2R^3} + \frac{3}{2R^5} \tilde{\boldsymbol{R}} \boldsymbol{\Theta}'_B \boldsymbol{R}. \tag{7-10}$$

Note that Tr ($\boldsymbol{\Theta}'_B$) means the trace, that is, the sum of the diagonal elements of $\boldsymbol{\Theta}'_B$, and is brought into Eq. (7-10) from the following relation.

$$\sum_j q_{Bj} (\tilde{\boldsymbol{r}}_{Bj} \boldsymbol{r}_{Bj}) = \sum_j q_{Bj} (X_{Bj}^2 + Y_{Bj}^2 + Z_{Bj}^2) = (\Theta'_B)_{XX} + (\Theta'_B)_{YY} + (\Theta'_B)_{ZZ} .$$

Since the quadrupole moment is a symmetric tensor of rank two, one may suppose it to have six independent components. However, on redefining the quadrupole moment by

$$\boldsymbol{\Theta}_B = \tfrac{1}{2} \{ 3\boldsymbol{\Theta}'_B - \mathrm{Tr}(\boldsymbol{\Theta}'_B) \boldsymbol{E} \}, \tag{7-11}$$

the diagonal elements of Eq. (7-11) turn out to be

$$(\Theta_B)_{XX} = \tfrac{1}{2} \sum_j q_{Bj} (3X_{Bj}^2 - r_{Bj}^2) = \tfrac{1}{2} \sum_j q_{Bj} (2X_{Bj}^2 - Y_{Bj}^2 - Z_{Bj}^2)$$

$$(\Theta_B)_{YY} = \tfrac{1}{2} \sum_j q_{Bj} (3Y_{Bj}^2 - r_{Bj}^2) = \tfrac{1}{2} \sum_j q_{Bj} (2Y_{Bj}^2 - Z_{Bj}^2 - X_{Bj}^2)$$

$$(\Theta_B)_{ZZ} = \tfrac{1}{2} \sum_j q_{Bj} (3Z_{Bj}^2 - r_{Bj}^2) = \tfrac{1}{2} \sum_j q_{Bj} (2Z_{Bj}^2 - X_{Bj}^2 - Y_{Bj}^2)$$

for which it holds that Tr ($\boldsymbol{\Theta}_B$) = 0. In this case, the diagonal elements of $\boldsymbol{\Theta}_B$ are not independent of each other. By using the relation $R^2 = \tilde{\boldsymbol{R}} \boldsymbol{R} = \tilde{\boldsymbol{R}} \boldsymbol{E} \boldsymbol{R}$, Eq. (7-10) can be simplified in terms of $\boldsymbol{\Theta}_B$ as

$$\phi(\boldsymbol{R}, \boldsymbol{0}) = \frac{Q_B}{R} - \frac{1}{R^3} \tilde{\boldsymbol{R}} \boldsymbol{\mu}_B + \frac{1}{R^5} \tilde{\boldsymbol{R}} \boldsymbol{\Theta}_B \boldsymbol{R}. \tag{7-12}$$

According to Eq. (7-12), the electric field produced by the assembly of point charges B is represented approximately by the net charge Q_B, the dipole moment $\boldsymbol{\mu}_B$ and the quadrupole moment $\boldsymbol{\Theta}_B$, all located at the point P. The energy of interaction between the charge assemblies A and B is now estimated by expanding the potential $\phi(\boldsymbol{R}, \boldsymbol{r}_{Ai})$ at the origin in powers of the components of $\boldsymbol{r}_{Ai}$. From the definition of $\phi(\boldsymbol{R}, \boldsymbol{r}_{Ai})$ in Eq. (7-1), it follows that

$$\phi(\boldsymbol{R}, \boldsymbol{r}_{Ai}) = \phi(\boldsymbol{R} - \boldsymbol{r}_{Ai}, \boldsymbol{0}).$$

Then, replacing $\boldsymbol{R}$ in Eq. (7-12) with $\boldsymbol{R} - \boldsymbol{r}_{Ai}$ leads to

$$\phi(\boldsymbol{R},\boldsymbol{r}_{Ai}) = Q_B|\boldsymbol{R}-\boldsymbol{r}_{Ai}|^{-1} - |\boldsymbol{R}-\boldsymbol{r}_{Ai}|^{-3}(\tilde{\boldsymbol{R}}-\tilde{\boldsymbol{r}}_{Ai})\boldsymbol{\mu}_B + |\boldsymbol{R}-\boldsymbol{r}_{Ai}|^{-5}(\tilde{\boldsymbol{R}}-\tilde{\boldsymbol{r}}_{Ai})\boldsymbol{\Theta}_B(\boldsymbol{R}-\boldsymbol{r}_{Ai}). \quad (7\text{-}13)$$

Substituting Eq.(7-5) with $n = 1$, 3, and 5 into Eq. (7-13), we obtain

$$\begin{aligned}
\phi(\boldsymbol{R},\boldsymbol{r}_{Ai}) &= \phi(\boldsymbol{R},\boldsymbol{0}) + \frac{Q_B}{R^3}\left\{\tilde{\boldsymbol{R}}\boldsymbol{r}_{Ai} + \tfrac{1}{2}\tilde{\boldsymbol{r}}_{Ai}\left(-\boldsymbol{E} + \frac{3\boldsymbol{R}\tilde{\boldsymbol{R}}}{R^2}\right)\boldsymbol{r}_{Ai}\right\} \\
&\quad - \frac{3\tilde{\boldsymbol{R}}\boldsymbol{\mu}_B}{R^5}\left\{\tilde{\boldsymbol{R}}\boldsymbol{r}_{Ai} + \tfrac{1}{2}\tilde{\boldsymbol{r}}_{Ai}\left(-\boldsymbol{E} + \frac{5\boldsymbol{R}\tilde{\boldsymbol{R}}}{R^2}\right)\boldsymbol{r}_{Ai}\right\} + \left(\frac{1}{R^3} + \frac{3}{R^5}\tilde{\boldsymbol{r}}_{Ai}\boldsymbol{R}\right)\tilde{\boldsymbol{\mu}}_B\,\boldsymbol{r}_{Ai} \\
&\quad + \frac{5\tilde{\boldsymbol{R}}\boldsymbol{\Theta}_B\boldsymbol{R}}{R^7}\left\{\tilde{\boldsymbol{R}}\boldsymbol{r}_{Ai} + \frac{\tilde{\boldsymbol{r}}_{Ai}}{2}\left(-\boldsymbol{E} + \frac{7\boldsymbol{R}\tilde{\boldsymbol{R}}}{R^2}\right)\boldsymbol{r}_{Ai}\right\} \\
&\quad - 2\left(\frac{1}{R^5} + \frac{5}{R^7}\tilde{\boldsymbol{R}}\boldsymbol{r}_{Ai}\right)\tilde{\boldsymbol{R}}\,\boldsymbol{\Theta}_B\,\boldsymbol{r}_{Ai} + \frac{1}{R^5}\tilde{\boldsymbol{r}}_{Ai}\,\boldsymbol{\Theta}_B\,\boldsymbol{r}_{Ai}. \qquad (7\text{-}14)
\end{aligned}$$

Multiplying Eq.(7-14) by the charge q_{Ai} and summing over i, we obtain the energy of Coulomb interaction, V_C between the charge assemblies A and B. By defining the total charge Q_A, the dipole moment $\boldsymbol{\mu}_A$ and the quadrupole moment $\boldsymbol{\Theta}_A$ for the charge assembly A, this sum is written as

$$\begin{aligned}
V_C &= Q_A\,\phi(\boldsymbol{R},\boldsymbol{0}) + \frac{Q_B}{R^3}\left(\tilde{\boldsymbol{R}}\boldsymbol{\mu}_A + \frac{\tilde{\boldsymbol{R}}\,\boldsymbol{\Theta}_A\boldsymbol{R}}{R^2}\right) - \frac{\tilde{\boldsymbol{R}}\boldsymbol{\mu}_B}{R^5}\left(3\tilde{\boldsymbol{R}}\boldsymbol{\mu}_A + \frac{5\tilde{\boldsymbol{R}}\,\boldsymbol{\Theta}_A\boldsymbol{R}}{R^2}\right) + \frac{1}{R^3}\tilde{\boldsymbol{\mu}}_B\,\boldsymbol{\mu}_A \\
&\quad + \frac{2}{R^5}\tilde{\boldsymbol{R}}\boldsymbol{\Theta}_A\,\boldsymbol{\mu}_B + \frac{5}{R^7}\tilde{\boldsymbol{R}}\boldsymbol{\Theta}_B\boldsymbol{R}\left(\tilde{\boldsymbol{R}}\boldsymbol{\mu}_A + \frac{7\tilde{\boldsymbol{R}}\,\boldsymbol{\Theta}_A\boldsymbol{R}}{3R^2}\right) \qquad (7\text{-}15) \\
&\quad - \frac{2}{R^5}\tilde{\boldsymbol{R}}\,\boldsymbol{\Theta}_B\left(\boldsymbol{\mu}_A + \frac{10\boldsymbol{\Theta}_A\boldsymbol{R}}{3R^2}\right) + \frac{2}{3R^5}\sum_u\sum_v(\Theta_A)_{uv}(\Theta_B)_{uv}.
\end{aligned}$$

On deriving Eq. (7-15), use has been made of the relations

$$\begin{aligned}
\sum_i q_i\,\tilde{\boldsymbol{r}}_{Ai}\left(-\boldsymbol{E} + \frac{n\boldsymbol{R}\tilde{\boldsymbol{R}}}{R^2}\right)\boldsymbol{r}_{Ai} &= -\sum_i q_{Ai}\,\tilde{\boldsymbol{r}}_{Ai}\,\boldsymbol{r}_{Ai} + \frac{n}{R^2}\tilde{\boldsymbol{R}}\boldsymbol{\Theta}\boldsymbol{R} \\
&= \frac{1}{R^2}\tilde{\boldsymbol{R}}\{-(\mathrm{Tr}\,\boldsymbol{\Theta}'_A)\boldsymbol{E} + n\,\boldsymbol{\Theta}'_A\}\boldsymbol{R} = \frac{n}{3R^2}\tilde{\boldsymbol{R}}\left\{3\,\boldsymbol{\Theta}'_A - \frac{3}{n}(\mathrm{Tr}\,\boldsymbol{\Theta}'_A)\boldsymbol{E}\right\}\boldsymbol{R} \\
&= \frac{n}{3R^2}\tilde{\boldsymbol{R}}\left\{2\boldsymbol{\Theta}_A - \frac{3-n}{n}\left(\sum_i q_i\,\tilde{\boldsymbol{r}}_{Ai}\,\boldsymbol{r}_{Ai}\right)\boldsymbol{E}\right\}\boldsymbol{R},
\end{aligned}$$

$$-\tilde{\boldsymbol{R}}\boldsymbol{\mu}_B\left\{\frac{3}{2R^5}\frac{3}{5}\tilde{\boldsymbol{R}}\frac{2}{5}\left(\sum_i q_i\,\tilde{\boldsymbol{r}}_{Ai}\,\boldsymbol{r}_{Ai}\right)\boldsymbol{E}\right\}\boldsymbol{R}+\frac{3}{R^5}\sum_i q_i\tilde{\boldsymbol{r}}_{Ai}\,\boldsymbol{R}\,\tilde{\boldsymbol{\mu}}_B\,\boldsymbol{r}_{Ai}$$

$$=\frac{1}{R^5}\tilde{\boldsymbol{R}}\{3\boldsymbol{\Theta}_A'-(\mathrm{Tr}\,\boldsymbol{\Theta}_A')\boldsymbol{E}\}\boldsymbol{\mu}_B=\frac{2}{R^5}\tilde{\boldsymbol{R}}\,\boldsymbol{\Theta}_A\,\boldsymbol{\mu}_B,$$

$$\frac{5}{2R^7}\tilde{\boldsymbol{R}}\,\boldsymbol{\Theta}_B\,\boldsymbol{R}\frac{7}{3}\tilde{\boldsymbol{R}}\frac{4}{7}\left(\sum_i q_i\,\tilde{\boldsymbol{r}}_{Ai}\,\boldsymbol{r}_{Ai}\right)-\frac{10}{R^7}\tilde{\boldsymbol{R}}\left(\sum_i q_i\,\boldsymbol{r}_{Ai}\,\tilde{\boldsymbol{r}}_{Ai}\right)\boldsymbol{\Theta}_B\,\boldsymbol{R}$$

$$=\frac{10}{3R^7}\tilde{\boldsymbol{R}}\{-3\boldsymbol{\Theta}_A'+(\mathrm{Tr}\,\boldsymbol{\Theta}_A')\boldsymbol{E}\}\boldsymbol{\Theta}_B\,\boldsymbol{R}=-\frac{20}{3R^7}\tilde{\boldsymbol{R}}\,\boldsymbol{\Theta}_A\,\boldsymbol{\Theta}_B\,\boldsymbol{R}.$$

The last term in Eq. (7-15) is derived by rewriting the last term in Eq. (7-14) as

$$\tilde{\boldsymbol{r}}_{Ai}\,\boldsymbol{\Theta}_B\,\boldsymbol{r}_{Ai}=\sum_{u=X,}^{Y,Z}\sum_{v=X,}^{Y,Z}u_{Ai}(\Theta_B)_{uv}\,v_{Ai},$$

multiplying it by q_{Ai} and summing over i. The result is written as

$$\sum_i q_{Ai}\,\tilde{\boldsymbol{r}}_{Ai}\,\boldsymbol{\Theta}_B\,\boldsymbol{r}_{Ai}=\sum_u\sum_v\sum_i q_{Ai}\,u_{Ai}\,v_{Ai}\,(\Theta_B)_{uv}$$

$$=\sum_u\sum_v(\Theta_A')_{uv}(\Theta_B)_{uv}=2/3\sum_u\sum_v(\Theta_A)_{uv}(\Theta_B)_{uv}.$$

Let $\boldsymbol{A}$ and $\boldsymbol{B}$ be two arbitrary tensors of the same order. Then the sum of the products of the corresponding components of $\boldsymbol{A}$ and $\boldsymbol{B}$ over all the combinations of suffixes is called the inner product of $\boldsymbol{A}$ and $\boldsymbol{B}$. The simplest example of inner products of two tensors is the scalar product of two vectors. Generally, the inner product of two tensors is kept invariant under any rotation of the axes of the Cartesian coordinate system. This statement is easily proved by transforming the inner product with an orthogonal matrix $\boldsymbol{L}$ for which it holds that

$$\sum_u L_{ui}\,L_{uk}=\delta_{ik}\quad\text{and}\quad\sum_v L_{vj}\,L_{vl}=\delta_{jl}.$$

The result for the case of rank two is

$$\sum_u\sum_v(\Theta_A)_{uv}(\Theta_B)_{uv}=\sum_u\sum_v\left\{\sum_i\sum_j L_{ui}\,L_{vj}\,(\Theta_A^0)_{ij}\right\}\left\{\sum_k\sum_l L_{uk}\,L_{vl}\,(\Theta_B^0)_{kl}\right\}$$

$$=\sum_i\sum_j\sum_k\sum_l\left(\sum_u L_{ui}\,L_{uk}\right)\left(\sum_v L_{vj}\,L_{vl}\right)(\Theta_A^0)_{ij}(\Theta_B^0)_{kl}$$

$$=\sum_i\sum_j(\Theta_A^0)_{ij}(\Theta_B^0)_{ij}.$$

The distance vector $\boldsymbol{R}$ in Eq. (7-15) is written in terms of its absolute value, $R=|\boldsymbol{R}|$, and the unit vector along $\boldsymbol{R}$, $\boldsymbol{e}=\boldsymbol{R}/R$, as

$$\boldsymbol{R}=R\,\boldsymbol{e}. \tag{7-16}$$

On substituting Eq. (7-16) into Eq. (7-15) and collecting the terms involving the same type of pairs of multipole moments, the Coulomb interaction energy V_C is expressed as a sum of various types of the interaction between multipoles in the form

$$V_C = V_{CC} + V_{CD} + V_{CQ} + V_{DD} + V_{DQ} + V_{QQ} + \cdots. \tag{7-17}$$

The terms in Eq. (7-17) are shown below together with the order of the pairing multipole moments involved, m and n, and the name of the interaction.

The charge–charge interaction ($n = m = 0$)

$$V_{CC} = Q_A Q_B / R \tag{7-18a}$$

The charge–dipole interaction ($n = 0$, $m = 1$)

$$V_{CD} = -\frac{1}{R^2} Q_A (\tilde{e}\, \boldsymbol{\mu}_B) \tag{7-18b}$$

The charge–quadrupole interaction ($n = 0$, $m = 2$)

$$V_{CQ} = \frac{1}{R^3} Q_A\, \tilde{\boldsymbol{e}} \boldsymbol{\Theta}_B\, \boldsymbol{e} \tag{7-18c}$$

The dipole–dipole interaction ($n = m = 1$)

$$V_{DD} = \frac{1}{R^3}\left\{ \tilde{\boldsymbol{\mu}}_A\, \boldsymbol{\mu}_B - \frac{3}{R^2}(\tilde{\boldsymbol{R}}\, \boldsymbol{\mu}_A)(\tilde{\boldsymbol{R}}\, \boldsymbol{\mu}_B) \right\} \tag{7-18d}$$

The dipole–quadrupole interaction ($n = 1$, $m = 2$)

$$V_{DQ} = \frac{1}{R^4}\left\{ 5\, \tilde{\boldsymbol{\mu}}_A\, \boldsymbol{e}(\tilde{\boldsymbol{e}}\, \boldsymbol{\Theta}_B\, \boldsymbol{e}) - 2\, \tilde{\boldsymbol{\mu}}_A\, \boldsymbol{\Theta}_B\, \boldsymbol{e} \right\} \tag{7-18e}$$

The quadrupole–quadrupole interaction ($n = m = 2$)

$$V_{QQ} = \frac{1}{R^5}\left\{ \frac{2}{3}\sum_u \sum_v (\Theta_A)_{uv} (\Theta_B)_{uv} - \frac{20}{3}\tilde{\boldsymbol{e}}\, \boldsymbol{\Theta}_A\, \boldsymbol{\Theta}_B\, \boldsymbol{e} + \frac{35}{3}(\tilde{\boldsymbol{e}}\, \boldsymbol{\Theta}_A\, \boldsymbol{e})(\tilde{\boldsymbol{e}}\, \boldsymbol{\Theta}_B\, \boldsymbol{e}) \right\} \tag{7-18f}$$

According to Eq. (7-18a–f), the energy of the Coulomb interaction between the nth moment of A and the mth moment of B is inversely proportional to the $(n + m + 1)$th power of the distance between A and B. The interaction between higher order moments becomes less important as the charge assemblies are more separated from each other. Usually, only the interaction between the moments of the least orders is taken into account in the first-order approximation. For a given assembly of point charges, the moment of the least order is determined as a proper quantity not dependent of the choice of the origin of the coordinate system.

The uniqueness of the moment of the least order can be confirmed as follows. On shifting the origin to the point (X_0, Y_0, Z_0), the coordinates X_i, Y_i and Z_i, are replaced by $X_i - X_0$, $Y_i - Y_0$, and $Z_i - Z_0$, respectively. Equation (7-7) is then rewritten as

$$\begin{aligned} M_{X\cdots Y\cdots Z} &= \sum_i q_i (X_i - X_0)^k (Y_i - Y_0)^l (Z_i - Z_0)^m \\ &= \sum_i q_i X_i^k Y_i^l Z_i^m + k X_0 \sum_i q_i X_i^{k-1} Y_i^l Z_i^m + l\, Y_0 \sum_i q_i X_i^k Y_i^{l-1} Z_i^m \\ &\quad + m\, Z_0 \sum_i q_i X_i^k Y_i^l Z_i^{m-1} + \cdots. \end{aligned} \tag{7-19}$$

If the $(k + l + m)$th moment of the system is that of the least order, all the terms with the powers of less than $k + l + m$ in Eq. (7-19) should vanish, leaving only the first term, which is equal to Eq. (7-7).

The presumption in deducing the multipole model implies that the model works well when applied to the intermolecular interaction in dilute gases or in dilute solutions in non-polar solvents by regarding each molecule as an assembly of charged particles. On the other hand, the approximation used in this model is rather crude when applied to the intramolecular Coulomb interaction between the charge assemblies each of which is regarded to consist of a nucleus and the electrons surrounding it. In this case, the extent of an assembly of charged particles is often comparable to the distances between the neighboring nuclei. It is a great advantage of the multipole model, however, that any complicated assemblies of charges can be represented by a small number of parameters and the calculation is simplified considerably by reducing the number of force centers. The multipole model has been the object of growing attention as a useful method of estimating the Coulomb interaction energy between charges separated far apart in such macromolecules as proteins.[2)]

7.3 Estimation of Effective Atomic Charges

Generally, the effective atomic charge of the nucleus i is defined as the algebraic sum of the positive charge of the nucleus i and the total negative charge of electrons distributed in the neighborhood of i. Since there are no clearcut boundaries between neighboring nuclei, there remains some ambiguity in the definition of the effective atomic charge. Nevertheless, the effective atomic charge model has been frequently used in describing various molecular properties because of the convenient nature of the point charge approximation.

The effective atomic charge may be estimated either theoretically by using the molecular orbital (MO) method or empirically by adjusting appropriate parameters to fit any experimental data. One of the well-known theoretical methods is that based on the population analysis by Mulliken.[3)] Earlier MO calculations were carried out with the empirical or semiempirical scheme, but more elaborate *ab initio* MO calculations are now being carried out since the cost of computation is quickly decreasing. The effect of the choice of basis set functions has been investigated by many authors. As a result, it is generally recognized that modification of basis functions does not necessarily improve the calculated dipole moment.

It has long been pointed out that the dipole moment calculated from the atomic charges based on the population analysis does not agree with the expected value of the dipole moment calculated by using the electronic wave functions explicitly. The most orthodox way of representing the electrostatic influence of a molecule on its surroundings is to form a grid around the molecule and calculate the electrostatic potential at each grid point. A set of atomic charges estimated so as to give the best reproduction of this electrostatic potential may be regarded as a charge distribution which exerts the same electric effect on its surroundings as the molecule itself. The charge distribution determined in this way has been known to reproduce the molecular dipole moment fairly well. The atomic charges calculated by this method combined with the constraint of the charge conservation have been used frequently in recent works.[4)]

Various methods of evaluating the atomic charges empirically without recourse to any standard theoretical analyses of electronic states have been proposed. Generally, an empirical method uses a set of parameters selected under a given model for deriving the effective atomic charges. The parameters are optimized beforehand to fit the experimental data or the charge distribution calculated theoretically. An advantage of the empirical method is its applicability without serious limitation from the molecular size as long as a parameter file and a program for computation is available. The empirical method can be roughly classified into two methods, one using the bond polarizability and the other using the electronegativity. The latter will be taken up first while the former will be discussed in detail in the succeeding section.

The prototype of the latter methods, being cited quite often by later investigators, is the one proposed in 1958 by Del Re.[5] In this method, the attractivity of electrons by the atom i in a molecule is expressed in terms of a parameter specific to the atomic species and a scale of the sensitivity to the effect of an atom k bonded to i; that is

$$\delta_i = \delta_i^0 + \sum_k \gamma_{i(k)}\, \delta_k . \tag{7-20}$$

The parameter δ_i^0 is given as the relative value of the electronegativity of the atom i to that of hydrogen atom,

$$\delta_i^0 = k(\chi_i - \chi_H)/\chi_H , \tag{7-21}$$

and $\gamma_{i(k)}$ is determined so as to fit the experimental value of the dipole moment. The number of unknowns δ_i and also the number of conditions given by Eq. (7-20) are the same as the number of non-equivalent nuclei in the molecule, forming a set of simultaneous linear equations. From δ_i's obtained by solving the simultaneous equations, the quantities called the bond charge, q_{ij}, are calculated by

$$q_{ij} = (\delta_j - \delta_i)/2\varepsilon_{ij} , \tag{7-22}$$

where ε_{ij} is the effective dielectric constant. Physically, q_{ij} means half the charge transferred from the atom i to the atom j on the bond formation between i and j. The bond moment associated with the bond i-j is defined as the product of q_{ij} and the bond vector $\boldsymbol{r}_{ij}$ in the form

$$\boldsymbol{\mu}_{ij} = q_{ij}\, \boldsymbol{r}_{ij} = q_{ij}\left(\boldsymbol{r}_j - \boldsymbol{r}_i\right). \tag{7-23}$$

Since the positive direction of $\boldsymbol{\mu}_{ij}$ is taken to be from i to j, the bond charge q_{ij} is positive when $\boldsymbol{\mu}_{ij}$ and $\boldsymbol{r}_{ij}$ have the same direction as each other, and it should hold that

$$q_{ij} = -q_{ji} . \tag{7-24}$$

Note that Eqs. (7-22) and (7-24) are consistent with each other. It follows from this convention that the effective nuclear charge q_j is obtained simply by adding up all the bond charges for the bonds formed by j in the form

$$q_j = \sum_i q_{ij} . \tag{7-25}$$

By using this method, Del Re successfully reproduced the observed dipole moments of a number of alkyl halides, alcohols, amines and ethers from a small number of optimized parameters.[5]

A simplified modification of Del Re's method has been adopted in a molecular mechanics calculation including the simulation of vibrational spectra.[6] In this method, the bond charge is regarded as a linear function of the difference between Pauling's electronegativities[7] of the end atoms i and j in the form

$$q_{ij} = q_{ij}^0 + \beta_{ij}(\chi_j - \chi_i). \tag{7-26}$$

Equation (7-26) can also be obtained by neglecting $\gamma_{i(k)}$ in Eq. (7-20) to adopt the approximation that $\delta_i = \delta_i^0$, substituting Eq. (7-21) into Eq. (7-22) and adding the constant term q_{ij}^0 to the result. In practice, q_{ij}^0 is introduced only if any charge transfer through the bond is expected for the case where $\chi_j = \chi_i$. The coefficient β_{ij} of the linear term is an adjustable parameter corresponding to $k/2\varepsilon_{ij}\,\chi_H$, and is called the bond charge parameter. No simultaneous equations are to be solved in this simplification, and the individual effective atomic charges can be estimated independently from a small number of parameters. The charge distribution of the *cis* and the *trans* forms of formic acid monomer calculated by this method[8] reproduced well both the directions and the magnitudes of the dipole moments determined from microwave spectra,[9] as shown in Fig. 7.2.

A comment is necessary as to the transferability of β_{ij}'s for alkanes. If we use a common β_{CH} for the C–H bonds of methyl, methylene and methine groups, a large difference in the effective atomic charges is brought about between the ends of a C–C bond connecting two carbon atoms in different classes. Such an imbalance in the charge distribution results in an abnormally large value of the calculated intensity of any infrared absorption band due to a normal vibration involving the stretching of the C–C bond in question. On the other hand, too large polarization of any C–C bonds is avoided and reasonable intensities of the relevant infrared absorption bands are obtained when β_{CH} is set equal to 0.100, 0.125 and 0.150 for the methyl, the methylene and the methine C–H bonds in alkyl groups, respectively. Furthermore, attributing different values of β_{CH} to the three classes of C–H bonds leads to non-vanishing dipole moments 0.78Debye and 0.125Debye of propane and isobutane, respectively. These values agree very well with the experimental values 0.83Debye and 0.132Debye obtained by microwave spectroscopy.[10]

A method called Partial Equalization of Orbital Electronegativity (PEOE) was proposed by Gasteiger and Marsili based on the concept of averaging electronegativities.[11] In this method, the iteration is performed on the basis of the following two principles until the effective charge of every atom converges to a constant value.

(1) The electronegativity of a given atom in the molecule changes in a manner dependent on the charge carried by that atom. Since the charges in the molecule migrate through the bonds in such a way as to minimize the difference in the electronegativities between the two end atoms of every bond, the electronegativities of the constituent atoms of the molecule tend to be equalized.

(2) The migration of charges is resisted by a force which increases in proportion to the polarization brought about by the charge migration, so that the equalization of electronegativities proceeds only partially and stops when the two opposite trends balance each other.

The electronegativity of a given atomic species A is expressed as a quadratic function of the charge Q in the form

$$\chi_A(Q) = a_A + b_A Q + c_A Q^2. \tag{7-27}$$

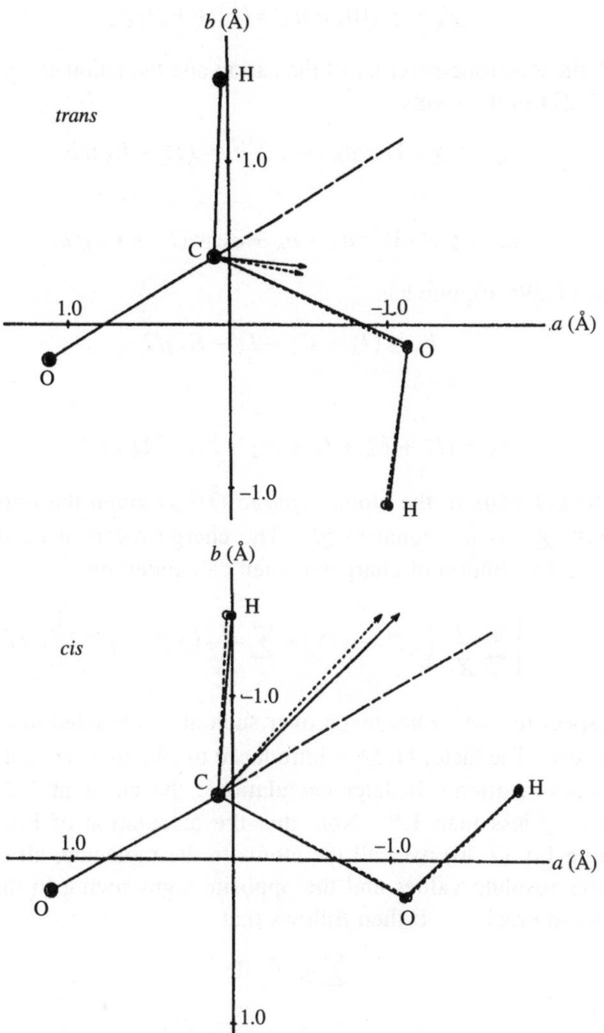

Fig. 7. 2 Atomic coordinates and dipole moment vectors of formic acid monomer in *trans* and *cis* forms ; (solid line) calculated, (dotted line) observed.
(Reproduced with permission from I. Yokoyama, Y. Miwa and K. Machida, *J. Am. Chem. Soc.*, **113**, 6458 (1991), p. 6459 (Fig. 1))

The coefficients a_A, b_A and c_A are determined from the data for the neutral atom A and the monovalent cation A^+ and anion A^- of the electronegativities defined by Mulliken in the form

$$\chi_A = (I_A + E_A)/2, \tag{7-28}$$

where I_A is the ionization potential and E_A is the electron affinity of A.[12] Since $Q = 0$ for an isolated neutral atom, it follows from Eqs. (7-27) and (7-28) that

$$\chi_A^0 = \chi_A(0) = a_A = (I_A^0 + E_A^0)/2. \tag{7-29a}$$

On the other hand, the electronegativities of the cation and the anion are given by substituting $Q = \pm 1$ into Eq. (7-27) in the forms

$$\chi_A^+ = \chi_A(1) = a_A + b_A + c_v = (I_A^+ + E_A^+)/2 \tag{7-29b}$$

and

$$\chi_A^- = \chi_A(-1) = a_A - b_A + c_A = (I_A^- + E_A^-)/2. \tag{7-29c}$$

Thus, solving Eqs. (7-29a–c), one has

$$b_A = (I_A^+ + E_A^+ - I_A^- - E_A^-)/2$$

and

$$c_A = (I_A^+ + E_A^+ + I_A^- + E_A^- - 2I_A^0 - 2E_A^0)/2.$$

On starting the calculation, the atomic charge $Q_i^{\langle 0\rangle}$ is given the initial value zero, and the electronegativity $\chi_i^{\langle 0\rangle}$ is put equal to χ_i^0. The charge received by the atom i from its neighbors in the nth distribution of charges is then calculated by

$$q_i^{\langle n\rangle} = \left[\sum_j \frac{1}{\chi_i^+}\left(\chi_j^{\langle n\rangle} - \chi_i^{\langle n\rangle}\right) + \sum_k \frac{1}{\chi_k^+}\left(\chi_k^{\langle n\rangle} - \chi_i^{\langle n\rangle}\right)\right](1/2)^n. \tag{7-30}$$

The sums with respect to j and k are taken over such atoms bonded to i that $\chi_j^{\langle n\rangle} > \chi_i^{\langle n\rangle}$ and $\chi_k^{\langle n\rangle} < \chi_i^{\langle n\rangle}$, respectively. The factor $(1/2)^n$ is introduced to take into account the damping effect against too much polarization. In later calculations, the constant 1/2 is replaced by an adjustable parameter f less than 1.[13] Note that the calculation of Eq. (7-30) starts with $n = 0$. On summing Eq. (7-30) over all the atoms in the molecule, all the terms appear in pairs with the same absolute values and the opposite signs owing to the condition on the ranges of two sums in brackets. It then follows that

$$\sum_i q_i^{\langle n\rangle} = 0,$$

and the conservation of the total charges in the molecule is ascertained. The atomic charge of the nucleus i in the nth iteration is changed according to

$$Q_i^{\langle n\rangle} = Q_i^{\langle n-1\rangle} + q_i^{\langle n-1\rangle},$$

and the electronegativity to be used in the next cycle of iteration is calculated by

$$\chi_i^{\langle n\rangle} = \chi_i^0 + b_i Q_i^{\langle n\rangle} + c_i Q_i^{\langle n\rangle 2}. \tag{7-31}$$

The calculation of Eqs. (7-30) through (7-31) is repeated until Q_i converges. Five to six iterations are sufficient for convergence in most cases. The calculated atomic charges on carbon and hydrogen nuclei were found to be correlated well with the 1s electron binding energies derived from ESCA experiments and the acidity constants pK_a, respectively.[11] The quadratic coefficient c_A in Eq. (7-27) was dropped in later calculations.[13] Scheraga and his coworkers modified the PEOE method by using different damping factors for different types of bonds,[14] and determined the parameters for a wide variety of compounds including polyatomic ions.[15]

7.4 Induced Dipole Moments and Polarizabilities

When a molecule is placed in an electric field, a force in the same direction as the electric field vector is imposed on each of the negatively charged particles in the molecule, while each of the positively charged particles receive a force in the opposite direction. Because of these forces, any polar molecule placed in an electric field tends to be so oriented against the averaging due to the thermal turbulence that its permanent dipole moment vector forms as large an angle with the applied field vector as possible. In addition, the charge distribution in each molecule tends to be more biased in the presence of an applied electric field than in free space. Accordingly, the dipole moment of a molecule increases, irrespective of whether the molecule is polar or not, when placed in an electric field. The increment of the dipole moment brought about by the applied field is called the induced dipole moment. As long as the applied field strength is small enough to be tractable within the first-order approximation, the components of the induced dipole moment, $\boldsymbol{\mu}$, are regarded to be expressible as linear combinations of the components of the applied field, $\boldsymbol{E}$ in the form

$$\left.\begin{aligned} \mu_X &= \alpha_{XX} E_X + \alpha_{XY} E_Y + \alpha_{XZ} E_Z \\ \mu_Y &= \alpha_{YX} E_X + \alpha_{YY} E_Y + \alpha_{YZ} E_Z \\ \mu_Z &= \alpha_{ZX} E_X + \alpha_{ZY} E_Y + \alpha_{ZZ} E_Z \end{aligned}\right\} \tag{7-32}$$

or, in the matrix representation, in the form

$$\boldsymbol{\mu} = \boldsymbol{\alpha} \boldsymbol{E}, \tag{7-33}$$

where $\boldsymbol{\alpha}$ is called the polarizability tensor, or simply the polarizability, of the molecule. Generally, the direction of the induced dipole moment does not necessarily coincide with that of the applied field. Hence the polarizability is not expressible as a scalar α multiplied by a unit matrix except for the special case where the molecule is isotropic. The concept of induced dipole moments is first introduced for describing the behavior of any molecular assemblies placed in an external electric field, and incorporated later in the theories for predicting electric properties of various molecules.

Smith, Ree, Magee and Eyring introduced the polarizability theory for elucidating the inductive effect through chemical bonds, and tried to predict the atomic charges and dipole moments of polar molecules based on a simple atomic model.[16] In this model, each atom in a molecule is assumed to be a rigid sphere centered at the nucleus with the covalent radius, and the charges of the nucleus and the surrounding electrons are concentrated at the center to form a point net charge called the effective atomic charge.

Suppose that two atoms a and b with covalent radii R_a and R_b, respectively, contact each other to form a bond a–b. Let the effective charges of a and b be denoted by q_a^e and q_b^e, respectively. Then the electric field formed by these charges at the point of contact is given by

$$E = \frac{q_a^e}{R_a^2} - \frac{q_b^e}{R_b^2}.$$

The bond a–b is polarized by this electric field, giving rise to induced charges Q_a^b and Q_a^b ($= -\ Q_a^b$) on the nuclei a and b, respectively, and also to the bond dipole moment μ_{ab} of the bond a–b. If the component of the bond polarizability along the bond is denoted by b_{ab}, it follows from Eq. (7-33) that

$$\mu_{ab} = Q_a^b\, r_{ab} = b_{ab} E = b_{ab}\left(\frac{q_a^e}{R_a^{\ 2}} - \frac{q_b^e}{R_b^{\ 2}}\right), \tag{7-34}$$

where r_{ab} is the bond distance.

Let S_a be Slater's shielding constant of the atom a,[17] and q_a be the total net charge of a. The effective charge of a is then given by

$$q_a^e = q_a^0 + S_a\, q_a. \tag{7-35}$$

Substituting Eq. (7-35) into Eq. (7-34) and introducing the notations

$$\alpha_{ab} = -\frac{b_{ab}}{R_{ab}}\left(\frac{q_a^e}{R_a^{\ 2}} - \frac{q_b^e}{R_b^{\ 2}}\right), \quad \beta_a^b = \frac{S_a\, b_{ab}}{R_{ab}\, R_a^{\ 2}}, \quad \beta_b^a = \frac{S_b\, b_{ab}}{R_{ab}\, R_b^{\ 2}}, \tag{7-36}$$

we have

$$Q_a^b = \alpha_{ab} + \beta_a^b\, q_b - \beta_b^a\, q_a. \tag{7-37}$$

The total net charge of the nucleus a is an unknown, which is defined to be the sum of Eq. (7-37) over all the nuclei b bonded to a, given in the form

$$q_a = \sum_b Q_a^b = \sum_b \alpha_{ab} + \sum_b \beta_a^b\, q_b - \left(\sum_b \beta_b^a\right) q_a. \tag{7-38}$$

Writing down Eq. (7-38) for all the nuclei a, and introducing the abbreviated notations for the constant terms and the coefficients of the unknowns in the linear terms given by

$$b_{aa} = 1 + \sum_b \beta_b^a, \quad b_{ba} = -\beta_b^a, \quad a_a = \sum_b \alpha_{ab}, \tag{7-39}$$

one obtains a set of simultaneous equations in the form

$$\left.\begin{aligned} b_{11}\, a_1 + b_{12}\, a_2 + \cdots + b_{1N}\, a_N &= a_1 \\ b_{21}\, a_1 + b_{22}\, a_2 + \cdots + b_{2N}\, a_N &= a_2 \\ \cdots \quad \cdots \quad \cdots \quad \cdots & \quad \cdots \\ b_{N1}\, a_1 + b_{N2}\, a_2 + \cdots + b_{NN}\, a_N &= a_N \end{aligned}\right\}. \tag{7-40}$$

The total net charge of each atom q_a is obtained by solving Eq. (7-40). According to Eqs. (7-39), (7-36) and (7-37), the coefficients in the left sides and the constants in the right sides of Eq. (7-40) satisfy the conditions

$$\sum_{i=1}^{N} b_{ia} = 1 \tag{7-41a}$$

and

$$\alpha_{ab} = -\alpha_{ba},$$

whence we have

$$\sum_i a_i = 0. \tag{7-41b}$$

By adding together N equations in Eq. (7-40) and using Eqs. (7-41a,b), it is verified that the total charge in the molecule is conserved; thus

$$q_1 + q_2 + \cdots + q_N = 0 \tag{7-42}$$

The dipole moments of halomethanes and ethyl halides calculated by this method, known as the SE method, agree well with the observed values.[13]

In the SE method, the charge transfer through the bond *a–b* is taken to be induced only by the charges of the atoms directly bonded to *a* and *b*, and the effects of remote atoms are neglected. Allinger and Wuesthoff took account of the polarization of the bond *a–b* induced by the atoms bonded to neither *a* nor *b*, formulating a new method of estimating atomic charges called the modified SE (MSE) method.[18] In this method, it is assumed that the induced bond moment μ_{ab} is affected by the charge q_i of an atom i not bonded to either a or b, and changes by the amount

$$\Delta\mu_{ab} = -b_{ab}\, q_i \cos\theta_i / \left(D R_i^2\right),$$

and a correction term

$$\Delta q_a = -\Delta\mu_{ab}/R_{ab} = -b_{ab} \sum_i q_i \cos\theta_i / \left(D R_{ab} R_i^2\right)$$

is added to the induced charge on a in Eq. (7-38). In this case, the coefficients in the simultaneous equations in Eq. (7-40) do not necessarily satisfy the condition given by Eq. (7-41a), hence the conservation of the total charge (Eq. (7-42)) is not ascertained. To avoid this difficulty, an equation in the simultaneous equations Eq.(7-40) for a specified atom A_0 is replaced by Eq. (7-42) so as to give the solution in which the charge is conserved. Some uncertainty in the result may arise from the arbitrariness in the choice of A_0 in setting up the simultaneous equations. It has been reported, however, that this uncertainty does not affect the calculated dipole moments within an accuracy of 10^{-3} Debye.

Applequist *et al.* regarded a molecule to be an assembly of polarizable point dipoles placed at the nuclear positions, and applied the polarizability theory to this system.[19] Dosen-Micovic *et al.* used the bond diopoles instead of the atom dipoles in the method of Applequist *et al.* to represent the molecular dipole moment as the sum of the bond moments.[20] The method of Dosen-Micovic, which is called the IDME (Induced Dipole Moment and Energy) method, is formulated as follows.

The bond i is considered to have a bond dipole moment $\boldsymbol{\mu}_i$, which is represented as the sum of the permanent bond moment $\boldsymbol{\mu}_i^0$ and the induced bond moment $\boldsymbol{\mu}_i^{\mathrm{I}}$ in the form

$$\boldsymbol{\mu}_i = \boldsymbol{\mu}_i^0 + \boldsymbol{\mu}_i^{\mathrm{I}}. \tag{7-43}$$

The induced bond moment $\boldsymbol{\mu}_i^{\mathrm{I}}$ arises from the polarization of the bond i in the internal electric field $\boldsymbol{E}'$ formed by the bond moments of the bonds other than i itself. Let $\boldsymbol{\alpha}_i$ be the bond polarizability tensor of the bond i in the molecule-fixed Cartesian coordinate system. The induced bond moment of the bond i is then given by

$$\boldsymbol{\mu}_i^{\mathrm{I}} = -\boldsymbol{\alpha}_i \boldsymbol{E}'. \tag{7-44}$$

The bond polarizability tensor of the bond i is diagonal in the principal axes system defined for the bond, and is given in the form

$$\boldsymbol{\alpha}_i^0 = \begin{bmatrix} \alpha_{Li}^0 & 0 & 0 \\ 0 & \alpha_{Ti}^0 & 0 \\ 0 & 0 & \alpha_{Vi}^0 \end{bmatrix}. \tag{7-45}$$

The bond polarizability tensor for the molecule-fixed system $\boldsymbol{\alpha}_i$ is then related to $\boldsymbol{\alpha}_i^0$ by the congruence transformation

$$\boldsymbol{\alpha}_i = \boldsymbol{D}_i \alpha_i^0 \tilde{\boldsymbol{D}}_i,$$

where $\boldsymbol{D}_i$ is the matrix of the transformation from the principal axes system of $\boldsymbol{\alpha}_i^0$ to the molecule-fixed system. The internal field $\boldsymbol{E}'$ is calculated as follows. Let the distance vector between the bond moments $\boldsymbol{\mu}_i$ and $\boldsymbol{\mu}_j$ be denoted by

$$\boldsymbol{R}_{ij} = \begin{bmatrix} X_{ij} & Y_{ij} & Z_{ij} \end{bmatrix}^{\mathrm{T}}.$$

Since Eq. (7-18d) can also be applied to the interaction between induced dipoles, the energy of interaction between $\boldsymbol{\mu}_i$ and $\boldsymbol{\mu}_j$ is given by

$$V_{DD} = \tilde{\boldsymbol{\mu}}_i \left\{ \frac{1}{R_{ij}^{\;3}} \left(\boldsymbol{E} - \frac{3}{R_{ij}^{\;2}} \boldsymbol{R}_{ij} \tilde{\boldsymbol{R}}_{ij} \right) \right\} \boldsymbol{\mu}_j = -\tilde{\boldsymbol{\mu}}_i \boldsymbol{T}_{ij} \boldsymbol{\mu}_j, \tag{7-46}$$

where $\boldsymbol{E}$ is a (3×3) unit matrix. The matrix $\boldsymbol{T}_{ij}$ is called the dipole field tensor, and is written in the form

$$\boldsymbol{T}_{ij} = \frac{1}{R_{ij}^{\;5}} \begin{bmatrix} 3X_{ij}^{\;2} - R_{ij}^{\;2} & 3X_{ij}Y_{ij} & 3X_{ij}Z_{ij} \\ 3X_{ij}Y_{ij} & 3Y_{ij}^{\;2} - R_{ij}^{\;2} & 3Y_{ij}Z_{ij} \\ 3X_{ij}Z_{ij} & 3Y_{ij}Z_{ij} & 3Z_{ij}^{\;2} - R_{ij}^{\;2} \end{bmatrix}. \tag{7-47}$$

The sum of Eq. (7-46) over j ($\neq i$) corresponds to the energy of the interaction between the bond moment $\boldsymbol{\mu}_i$ and the internal field $\boldsymbol{E}'$, $\tilde{\boldsymbol{\mu}}_i \boldsymbol{E}'$. The internal field is therefore expressed as

$$\boldsymbol{E}' = -\sum_{j \neq i} \boldsymbol{T}_{ij} \boldsymbol{\mu}_j. \tag{7-48}$$

Combining Eqs. (7-43), (7-44) and (7-48), we obtain

$$\boldsymbol{\mu}_i = \boldsymbol{\mu}_i^0 - \boldsymbol{\alpha}_i \sum_{j \neq i} \boldsymbol{T}_{ij} \boldsymbol{\mu}_j. \tag{7-49}$$

Let N be the number of the bonds, and let us introduce a $(3N \times 3N)$ matrix

$$\boldsymbol{A} = \begin{bmatrix} \boldsymbol{E} & \boldsymbol{\alpha}_1\boldsymbol{T}_{12} & \cdots & \boldsymbol{\alpha}_1\boldsymbol{T}_{1N} \\ \boldsymbol{\alpha}_2\boldsymbol{T}_{21} & \boldsymbol{E} & \cdots & \boldsymbol{\alpha}_2\boldsymbol{T}_{2N} \\ \vdots & \vdots & \ddots & \vdots \\ \boldsymbol{\alpha}_N\boldsymbol{T}_{N1} & \boldsymbol{\alpha}_N\boldsymbol{T}_{N2} & \cdots & \boldsymbol{E} \end{bmatrix} \tag{7-50}$$

and a 3N-dimensional vector

$$\boldsymbol{\mu}_A = [\tilde{\boldsymbol{\mu}}_1 \quad \tilde{\boldsymbol{\mu}}_2 \quad \cdots \quad \tilde{\boldsymbol{\mu}}_N]^{\mathrm{T}}. \tag{7-51}$$

Equation (7-49) is then rewritten simply as

$$\boldsymbol{A}\boldsymbol{\mu}_A = \boldsymbol{\mu}_A{}^0. \tag{7-52}$$

On solving Eq. (7-52) as $\boldsymbol{\mu}_A = \boldsymbol{A}^{-1}\boldsymbol{\mu}_A{}^0$ to obtain the elements of $\boldsymbol{\mu}_i$, the molecular dipole moment is calculated by

$$\boldsymbol{\mu} = \sum_i \boldsymbol{\mu}_i.$$

The permanent component of the bond moment $\boldsymbol{\mu}_i$, $\boldsymbol{\mu}_i{}^0$, is calculated by Del Re's method. As the initial guess of polarizabilities, the experimental data obtained by Le Febre *et al.* from the Kerr effect[21] were used with a cylindrical model in which the component along the bond was different from the other two components as given by

$$\alpha_{Li}^0 \neq \alpha_{Ti}^0 = \alpha_{Vi}^0$$

in Eq. (7-45). The IDME method is carried out by fixing the nuclear positions at the coordinates optimized by the MM2 method based on the bond moments. The molecular dipole moment is then calculated from the bond moment recalculated by the IDME method. The energy is calculated by the MM2 method except that the bond dipole moments used for estimating the dipole–dipole interaction energies are replaced with those recalculated by the IDME method. The conformational energies of a number of dihalolakanes and dihaloalkanones have been successfully reproduced by the IDME method.[20]

Abraham *et al.* proposed a simple scheme of estimating the effective atomic charge at the ith nucleus by utilizing both the electronegativity and the polarizability.[22] In this method, the effects of the nuclei connected to i through two and three bonds are taken into account in addition to the directly bonded nuclei, leading to the atomic charge in the form

$$q_i = \sum_j a(E_j - E_i) + \sum_j bP_i(E_j - E_H) + \sum_j cP_i(E_j - E_H), \tag{7-53}$$

where the first, second and third sums are taken over the nuclei at the α, the β and the γ positions from the ith nucleus, respectively, a, b and c are constants, E_j is the electronegativity of the jth atom given by Hinze and Jaffe,[23] and P_i is the polarizability of the ith atom. In contrast to the PEOE method, the electronegativities are kept constant in the course of evaluating the atomic charges, whereas the charge dependence of the polarizability is introduced in the form

$$P_i = P_i^0\{1 + \alpha(q_i^0 - q_i)\}, \tag{7-54}$$

where $P_i{}^0$ and $q_i{}^0$ are the polarizability and the charge of the ith atom in the standard state, respectively, and α is a constant. The standard atomic polarizabilities are estimated from the atomic hybrid component τ_i, which has been introduced by Miller and Savchik[24] as a parameter in an additive scheme of atomic refractivities, according to

$$P_i^0 = 4\tau_i^2/N_A .$$

To ascertain the neutrality of the molecular charge, the calculation of q_i is skipped for every carbon atom, and the charge of a carbon atom is given as minus the sum of the charges of the atoms bonded to it. This method succeeded in predicting the dipole moments of many haloalkanes, and was later extended to unsaturated systems[25a,b] and peptides.[26]

7.5 Atomic Charge Model and Bond Moment Model

Historically, there are two ways of modeling the charge distribution within a molecule in molecular mechanics, one using the atomic point charges and the other using the bond dipole moments. In the multipole expansion of a given charge distribution, the point charges and dipoles generally constitute the zeroth- and first-order terms, respectively. This does not mean, however, that the approximation in the bond dipole model is one order higher than that in the atomic charge model. In conventional multipole expansion, the first-order dipole terms are introduced as the correction to the zeroth order point charge terms. On the other hand, the effects of point charges in the atomic charge model are not supplemented with, but are replaced in principle by those of bond dipoles in the bond moment model. Because of the equivalence in the order of approximation, it is possible under certain conditions to transform the parameters describing the charge distribution between the two models. Any prescription for such a transformation may be useful on collecting potential parameters in molecular mechanics from the literature.

The transformation of the charge parameters from the bond moment model to the atomic charge model can always be carried out by using Eq. (7-25), while the inverse transformation is not always possible. First of all, for a pure bond moment model to be applicable, the system must have a vanishing net charge. The bond moment is given as the product of the bond charge and the bond length.

In the case of neutral molecules having no ring structure, the bond charges can be uniquely determined from the atomic charges as follows.

(1) The bond charge q_{ji} of the end bond i–j formed by an end nucleus i is set equal to the nuclear charge of the end nucleus, q_i. Assign the bond charges of all end bonds in this way, and remove all the end nuclei from the molecule. Some skeletal nuclei formally become end nuclei by this process.

(2) Add the nuclear charge of each removed end nucleus to the nuclear charge of the skeletal nucleus to which the removed end nucleus has been bonded. On writing the nuclear charge of a skeletal nucleus j after this process as q'_j, we have

$$q'_j = q_j + \sum_i q_i \ , \tag{7-55}$$

where the sum is taken over all the end nuclei bonded to j.

(3) Repeat procedures (1) and (2) for the formal end bonds successively formed this way. If the molecule is neutral and has no ring structure, there should remain two nuclei s and t such that $q_s = -q_t$. Then the process is finished by setting the bond charge q_{ts} equal to q_s.

If the molecule has a ring which lacks any symmetry, the above process cannot be continued further when all the exocyclic nuclei have been removed. Suppose, for example, the case where nuclei 1 through 5 form a five-membered ring. The generation of the nuclear charges from a given set of five bond charges is expressed in the matrix form

$$\begin{bmatrix} q_1 \\ q_2 \\ q_3 \\ q_4 \\ q_5 \end{bmatrix} = \begin{bmatrix} -1 & 0 & 0 & 0 & 1 \\ 1 & -1 & 0 & 0 & 0 \\ 0 & 1 & -1 & 0 & 0 \\ 0 & 0 & 1 & -1 & 0 \\ 0 & 0 & 0 & 1 & -1 \end{bmatrix} \begin{bmatrix} q_{12} \\ q_{23} \\ q_{34} \\ q_{45} \\ q_{51} \end{bmatrix} \tag{7-56}$$

and can be carried out straightforwardly. The inverse process cannot be performed, however, since the (5×5) transformation matrix in Eq. (7-56) has a vanishing determinant and cannot be inverted. If an arbitrary value is given to one of the bond charges, say q_{51}, the other bond charges can be determined successively as $q_{12} = q_{51} - q_1$, $q_{23} = q_{12} - q_2$, $\cdots$. Thus we have an infinite number of sets of bond charges for a given set of nuclear charges.

If the ring has any symmetry so that two members of the ring nuclei A and C bonded to a common nucleus B are equivalent to each other, the problem is reduced to the acyclic case by setting first both q_{AB} and q_{CB} equal to $q_B/2$, changing q_A and q_C to

$$q'_A = q'_C = q_A + q_B/2$$

and removing the nucleus B. If the nucleus B at the bridgehead of a condensed ring lies on an n-fold axis, one may proceed in a similar way by attributing first the bond charge q_B/n to the n equivalent bonds connecting B to the adjacent ring nuclei. If two ring nuclei A and B bonded by a homogeneous bond are equivalent, the bond charge q_{AB} is set equal to zero, and the bond A–B is removed to lead to the acyclic case without changing the nuclear charges of A and B.

7.6 Nuclear Charges and Charge Fluxes

In the atomic charge model, the contribution of the Coulomb interaction to the nonbonded atom–atom interaction energy is calculated by

$$V_C(r_{ij}) = \sum_i \sum_{>j} \frac{q_i q_j}{\varepsilon_a r_{ij}}, \tag{7-57}$$

where q_i is the effective atomic charge on the nucleus i, ε_a is the apparent permittivity of the medium surrounding the nuclei, and r_{ij} is the internuclear distance. Since the distribution of electrons changes more or less with any distortion of the molecule, the effective charge q_i must be a certain function of the internal variables R_1, R_2, $\cdots$. As long as the molecular distortion is small, we may take the power series of this function up to the linear terms, and express the dependence of the effective charge q_i on the molecular structure in the form

$$q_i = q_i^0 + \sum_k (\partial q_i / \partial R_k)_0 \Delta R_k, \tag{7-58}$$

where q_i^0 is the standard value of the effective charge q_i to be realized when all the structure parameters around the nucleus i take the standard values. The first derivative $(\partial q_i/\partial R_k)_0$ is called the charge flux, and was first introduced by Decius in order to elucidate the infrared absorption intensities of polyatomic molecules.[27)]

Substituting Eq. (7-58) into Eq. (7-57) and differentiating with respect to the nuclear Cartesian displacements, we have

$$\frac{\partial V_C}{\partial x_m} = \frac{1}{\varepsilon_a}\sum_i \sum_{\neq j}\left\{\left(\sum_p \frac{\partial q_i}{\partial R_p} B_{pm}\right)\frac{q_j^0}{r_{ij}} + \tfrac{1}{2} q_i q_j\left(\frac{\partial}{\partial x_m}\frac{1}{r_{ij}}\right)\right\} \tag{7-59}$$

and

$$\begin{aligned}\frac{\partial^2 V_C}{\partial x_m \partial x_n} = \frac{1}{\varepsilon_a}\sum_i \sum_{\neq j}\Bigg\{&\tfrac{1}{2}\left(\sum_p \frac{\partial q_i}{\partial R_p} B_{pm}\right)\left(\sum_q \frac{\partial q_j}{\partial R_q} B_{qn}\right)\frac{1}{r_{ij}} \\ &+ q_i\left(\sum_p \frac{\partial q_j}{\partial R_p} B_{pm}\right)\left(\frac{\partial}{\partial x_n}\frac{1}{r_{ij}}\right) + \tfrac{1}{2} q_i q_j\left(\frac{\partial^2}{\partial x_m \partial x_n}\frac{1}{r_{ij}}\right)\Bigg\}\end{aligned} \tag{7-60}$$

Equations (7-59) and (7-60) enable calculation of the contributions of the Coulomb interaction between the dynamical effective charges of the nuclei to the gradients and the Hessian matrix. On any distortion of an isolated molecule, the effective charges of individual nuclei may change but the total net charge of the molecule must remain invariant. This requirement can be fulfilled by assuming that the change in effective atomic charges arises only from charge flows through the bonds and calculating the atomic charge flux of the nucleus i associated with the change in the internal coordinate R_p according to

$$\partial q_i/\partial R_p = \sum_j \partial q_{ji}/\partial R_p, \tag{7-61}$$

where the sum is taken over the nuclei j bonded to i. The derivative $\partial q_{ij}/\partial R_p$ is called the bond charge flux and defined as the charge flowing from j to i through the bond j–i on the unit increase of R_p. From the definition, it is obvious that

$$\partial q_{ji}/\partial R_p = -\partial q_{ij}/\partial R_p\,.$$

If the molecule has a symmetry element and the internal coordinate ΔR_p defined as the increment of the internal variable R_p changes the sign on a symmetry operation, the bond charge flux $\partial q_{ji}/\partial R_p$ must vanish for any bonds j–i, since q_{ji} is a scalar quantity and must remain invariant on any symmetry operation. The out-of-plane bending coordinates, the torsional coordinates and the linear bending coordinates usually lie on certain symmetry elements for simple basic molecules, so that charge fluxes are rarely introduced for these coordinates.

If there is a set of internal coordinates ΔR_1, ΔR_2, $\cdots$, ΔR_{Nr} satisfying a redundancy condition

$$\sum_{p=1}^{Nr} a_p\, \Delta R_p = 0,$$

the normal coordinate analysis may be carried out by constructing a non-redundant set of linear combinations of internal coordinates. There is some arbitrariness in the choice of the non-redundant set. If the coordinates ΔR_1, ΔR_2, $\cdots$, ΔR_{Nr} are not equivalent to each other at all, a non-redundant set may be obtained by simply dropping one of these coordiantes. If some of them, say, ΔR_1, ΔR_2, $\cdots$, ΔR_{Ns}, form an equivalent set, $N_s - 1$ linear combinations orthogonal to the simple sum $\sum_{p=1}^{Ns} \Delta R_p$ may be adopted as members of the non-redundant set, so the choice of the non-equivalent set is not unique if more than one equivalent set of internal coordiantes are involved in a redundancy. For the calculated intensities to be independent of the choice of the non-redundant set, the charge fluxes must be chosen so as to satisfy the condition

$$\sum_{p=1}^{Nr} a_p(\partial q_{ji}/\partial R_p) = 0$$

A pairwise Coulomb potential function described in terms of the charges and charge fluxes has been reported to reproduce the Coulomb potential for water dimers determined from the *ab initio* calculation, and has been shown to be important for studying the intermolecular interaction.[28)]

References

1) J.O. Hirschfelder, C.F. Curtiss and R.B. Bird., *Molecular Theory of Gases and Liquids*, John Wiley & Sons, New York (1964).
2) M. Saito, *Mol. Simul.*, **8**, 321 (1992).
3) R.S. Mulliken, *J. Chem. Phys.*, **23**, 1833 (1955).
4) S.R. Cox and D.E. Williams, *J. Comput. Chem.*, **2**, 304 (1981).
5) G. Del Re, *J. Chem. Soc.*, **1958**, 4031.
6) Y. Miwa and K. Machida, *J. Am. Chem. Soc.*, **110**, 5183 (1988).
7) L. Pauling, *J. Am. Chem. Soc.*, **54**, 3570 (1932).
8) I. Yokoyama, Y. Miwa and K. Machida, *J. Am. Chem. Soc.*, **113**, 6458 (1991).
9) a) H. Kim, R. Keller and W. D. Gwinn, *J. Chem. Phys.*, **37**, 2748 (1962); b) W. H. Hocking, *Z. Naturforsch.*, **31A**, 1113 (1976).
10) D.R. Lide, *J. Chem. Phys.*, **29**, 914 (1958); *ibid.*, **33**, 1514 (1960).
11) J. Gasteiger and M. Marsili, *Tetrahedron*, **36**, 3219 (1980).
12) R.S. Mulliken *J. Chem. Phys.*, **2**,782 (1934).
13) L.-G. Hammarstrom, T. Liljefors and J. Gasteiger, *J. Comput. Chem.*, **9**, 424 (1988).
14) K.T. No, J.A. Grant and H.A. Scheraga, *J. Phys. Chem.*, **94**, 4732. (1990).
15) K.T. No, J.A. Grant, M.S. Jhon and H.A. Scheraga, *J. Phys. Chem.*, **94**, 4740 (1990).
16) L.P. Smith, T.Ree, J.L. Magee and H. Eyring, *J. Am. Chem. Soc.*, **73**, 2263 (1951).
17) J.C. Slater, *Phys. Rev.*, **36**, 57 (1930).
18) N.L. Allinger and M.T. Wuesthoff, *Tetrahedron*, **33**, 3 (1977).
19) J. Applequist, J.R. Carl and K.-K. Fung, *J. Am. Chem. Soc.*, **94**, 2952 (1972).
20) L. Dosen-Micovic, D. Jeremic and N.L. Allinger, *J. Am. Chem. Soc.*, **105**, 1716, 1723 (1983).
21) R.J.W. Le Fevre, B.J. Orr and G.L.D. Ritachie, *J. Chem. Soc.* (B), **273**, 220 (1966).
22) R.J. Abraham, L. Griffiths and P. Loftus, *J. Comput. Chem.*, **3**,407 (1982).
23) J. Hinze and H.H. Jaffe, *J. Am. Chem. Soc.*, **84**, 540 (1962); *idem.*, *J. Phys. Chem.*, **67**, 1501 (1963).
24) K.A. Miller and J. A. Savchik, *J. Am. Chem. Soc.*, **101**, 7206 (1979).
25) a) R. J. Abraham and B. Hudson, *J. Comput. Chem.*, **5**, 562 (1984); b) R.J. Abraham and P.E. Smith, *J. Comput. Chem.*, **9**, 288 (1988).
26) R.J. Abraham and B. Hudson, *J. Comput. Chem.*, **6**, 173 (1985).
27) J.C. Decius, *J. Mol. Spectrosc.*, **57**, 348 (1975).
28) U. Dinur and A.T. Hagler, *J. Chem. Phys.*, **97**, 9161 (1992).

Chapter 8

Simulation of Vibrational Spectra

8.1 Interaction between Molecules and Electromagnetic Waves

The term "vibrational spectra" usually indicates infrared absorption and Raman spectra in which transitions between vibrational states are observed. Suppose that vibrational transition from the state n to the state m is caused by the absorption of light polarized along the Cartesian axis α. According to the quantum mechanical theory of the interaction between molecules and electromagnetic waves, the integrated band intensity due to this transition is given by

$$A_{nm} = \frac{2N_A(f_n - f_m)\,\omega_{nm}\,\pi^2}{3\,\varepsilon_0 hc}\langle m|\mu_\alpha|n\rangle^2, \tag{8-1}$$

where N_A is the Avogadro number, f_n is the fraction of the molecules in the state n, ω_{nm} is the energy difference in wavenumber between the states n and m given by

$$\omega_{nm} = (E_m - E_n)/hc,$$

ε_0 is the permittivity of vacuum, h is the Planck constant, c is the speed of light, μ_α is the α component of the dipole moment, and $\langle m|\mu_\alpha|n\rangle$ is the α component of the transition moment between states n and m.[1] The last quantity is defined as an integral,

$$\langle m|\mu_\alpha|n\rangle \equiv \int \psi_m^* \mu_\alpha \psi_n d\tau, \tag{8-2}$$

where ψ_m and ψ_n are the wave functions of states m and n, respectively.

Next we consider the Raman scattering process of a gaseous sample associated with the transition from the state n to the state m under irradiation of light of intensity I.[2] If the electric vectors of the incident and the scattered lights are taken along the α and the β axes, respectively, the energy of the scattered light collected within a solid angle Ω around the scattering direction during a unit time is given by

$$A_{nm}^r = \frac{N_n \pi^2 (\omega_0 - \omega_{nm})^4}{\varepsilon_0{}^2}\langle m|\alpha_{\alpha\beta}|n\rangle^2\,\Omega\,I, \tag{8-3}$$

where N_n is the number of molecules in the state n, ω_0 is the wavenumber of the incident light, and $\langle m|\alpha_{\alpha\beta}|n\rangle$ is the $\alpha\beta$ component of the transition polarizability defined as an integral

$$\langle m|\alpha_{\alpha\beta}|n\rangle \equiv \int \psi_m^* \alpha_{\alpha\beta} \psi_n d\tau. \tag{8-4}$$

In Eqs. (8-3) and (8-4), the axes α and β are taken to be fixed in the space. Equation (8-3) holds regardless of whether $\alpha = \beta$ or not.

Since the dipole moment of any molecule is a continuous function of relative nuclear positions, its α component can be expanded as a power series of the normal coordinates as

$$\mu_\alpha = \mu_\alpha^0 + \sum_i (\partial \mu_\alpha / \partial Q_i)_0 Q_i + \cdots. \tag{8-5}$$

Substitution of Eq. (8-5) into the definition of the transition moment in Eq. (8-2), leads to

$$\langle m|\mu_\alpha|n\rangle = \mu_\alpha \int \psi_m^* \psi_n d\tau + \sum_i (\partial \mu_\alpha / \partial Q_i) \int \psi_m^* Q_i \psi_n d\tau + \cdots. \tag{8-6}$$

The approximation in which terms higher than the first in the expansion in Eq. (8-5) are neglected and the wave functions in Eq. (8-6) are taken to be those for harmonic oscillators is called the double harmonic approximation. On employing the double harmonic approximation, the first term in the right side of Eq. (8-6) vanishes when $m \neq n$ by virtue of the orthogonality of the wave functions. In addition, Appendix 2 shows that the integral including Q_i in the sum in Eq. (8-6) has a non-vanishing value only when the quantum number v_i in $|m\rangle$ differs by one from that in $|n\rangle$ and all the other quantum numbers in $|m\rangle$ are the same as those in $|n\rangle$. When v_i in $|m\rangle$ is greater than that in $|n\rangle$ by one, the non-vanishing integral is given by

$$\langle v_i + 1|\mu_\alpha|v_i\rangle = \left\{\frac{(v_i + 1)h}{8\pi^2 c\omega_i}\right\}^{1/2} \left(\frac{\partial \mu_\alpha}{\partial Q_i}\right)_0. \tag{8-7}$$

The double harmonic approximation may also be used in calculating the components of transition polarizabilities. Neglecting the terms of higher order than the first in the Taylor series expansion of polarizability components,

$$\alpha_{\alpha\beta} = \alpha_{\alpha\beta}^0 + \sum_i \left(\partial \alpha_{\alpha\beta} / \partial Q_i\right)_0 Q_i + \cdots, \tag{8-8}$$

and using the wave functions for harmonic oscillators, we obtain the expression for a non-vanishing component of the transition polarizability in the form

$$\langle v_i + 1|\alpha_{\alpha\beta}|v_i\rangle = \left\{\frac{(v_i + 1)h}{8\pi^2 c\omega_i}\right\}^{1/2} \left(\frac{\partial \alpha_{\alpha\beta}}{\partial Q_i}\right)_0. \tag{8-9}$$

Although the normal frequencies constitute an important group of objects of molecular mechanics calculations, the difficulty in interpreting vibrational spectra for the correct assignment of normal frequencies increases rapidly with molecular size. The opportunity for comparing the result of calculation with reliable experimental data is thus limited to relatively simple molecules. To overcome this difficulty, it is recommended to calculate both the normal frequencies and the transition probabilities by using the potential parameters and the spectral intensity parameters simultaneously. Then the simulated vibrational spectra may be compared directly with the observed spectra regardless of either the availability of vibrational assignments or the complexity of the molecule.

In this case, it is worth paying attention to the fact that the infrared absorption intensities are determined by the changes in the molecular dipole moment on the structure changes along individual normal coordinates. The intensity parameters for infrared absorptions should be consistent with the effective charge distribution from which part of the nonbonded atom–atom interaction potential arises.

8.2 Infrared Absorption Intensity Parameters

The total absorption intensity due to the vibrational transitions allowed under the double harmonic approximation is calculated by substituting Eq. (8-7) into Eq. (8-1) where the states $|n\rangle$ and $\langle m|$ stand for $|v_i\rangle$ and $\langle v_i+1|$, respectively, and summing up over non-negative v_i. By using the population f_i, which is derived from Eq. (5-112) based on the Boltzmann distribution in the form

$$f_n = \frac{\exp(-nhc\omega_i/kT)}{\sum_{v=0}^{\infty}\exp(-vhc\omega_i/kT)} = \frac{1-\exp(-hc\omega_i/kT)}{\exp(nhc\omega_i/kT)},$$

the integrated intensity of the infrared absorption band due to the ith normal mode is calculated to be

$$A_i = \frac{1}{nl}\int \ln\frac{I_o}{I}d\nu = \frac{N_A\,g_i}{12\,\varepsilon_0\,c^2}\left\{\left(\frac{\partial\mu_X}{\partial Q_i}\right)^2+\left(\frac{\partial\mu_Y}{\partial Q_i}\right)^2+\left(\frac{\partial\mu_Z}{\partial Q_i}\right)^2\right\}, \tag{8-10}$$

where g_i is the degeneracy of the ith normal mode. The dipole moment derivatives in Eq. (8-10) are calculated by employing the effective nuclear point charge approximation which is frequently used in constructing the model potential function. In the point charge approximation, the components of the dipole moment are defined to be

$$\mu_X = \sum_k q_k X_k. \tag{8-11}$$

On assuming that the effective nuclear charge q_k depends on the molecular structure, differentiation of Eq. (8-11) with the normal coordinate Q_i leads to[3)]

$$\frac{\partial\mu_X}{\partial Q_i} = \sum_k\left(q_k^e\frac{\partial X_k}{\partial Q_i}+\sum_j\frac{\partial q_k}{\partial R_j}\frac{\partial R_j}{\partial Q_i}\right). \tag{8-12}$$

The derivatives $\partial X_k/\partial Q_i$ and $\partial R_j/\partial Q_i$ are obtained as suitable elements of the transformation matrices $\boldsymbol{L}^x$ and $\boldsymbol{L}$, respectively, in the course of normal coordinate analysis. The charge flux $\partial q_k/\partial R_j$ has also been introduced in the dynamical point charge model for describing the Coulomb energy between the nonbonded nuclei.[4)] By using q_k^e estimated from Eq. (7-55) at the equilibrium position of the molecule, the infrared absorption intensities can be calculated from the same parameters representing the charge distribution as used in the potential energy calculations.

The infrared absorption intensities can also be calculated by using the bond dipole moments and their derivatives with respect to the internal coordinates. These quantities as well as the bond polarizabilities and their derivatives used in the Raman intensity calculation were introduced by Eliashevich and Wolkenstein.[5)] The bond dipole moments, the bond polarizabilities and their derivatives with respect to internal coordinates have long been known as the electrooptical parameters and have been widely used in the calculation of infrared absorption and Raman intensities.[6,7)]

As parameters for describing the Coulomb part of the potential energy, however, the bond moments and their derivatives with respect to internal coordinates are not as convenient as the effective atomic point charges and charge fluxes. There are two reasons for this difference. First, the expression of the Coulomb energy in terms of the bond moments, Eq. (7-18d), is much more complicated than that in terms of the atomic point charges, Eq. (7-18a). Second, the centers of interaction in the bond moment scheme are usually taken at the midpoints of bonding nuclei, making the calculation of the distances of interaction not as easy as in the case of the point charge model.

On assuming that the band shape is represented by a Lorentzian function, the absorbance at ν cm^{-1}, $A(\nu)$, is given by

$$A(\nu) \equiv \left(\ln \frac{I_0}{I} \right)_\nu = n\,l \sum_i \left[\frac{2\omega_i}{\pi} \frac{A_i}{4(\nu - \nu_i)^2 + \omega_i^2} \right], \tag{8-13}$$

where n is the concentration of the sample, l is the cell length, and ω_i is the half bandwidth of the infrared absorption band due to the normal mode i. The sum is taken over all the normal modes. As a working hypothesis, it is reasonable to assume that the bandwidth ω_i is a weighted average of certain bandwidths inherent to individual internal coordinates,

$$\omega_i = \left(\sum_k \eta_k L_{ki}^2 \Big/ \sum_k L_{ki}^2 \right) \omega_0, \tag{8-14}$$

where η_k is an empirical parameter assigned to the internal coordinate k, L_{ki} is the the (k,i) element of the $\boldsymbol{L}$ matrix, and ω_0 is the standard bandwidth depending on the sample state and the instrumental condition.[8)] In principle, the bandwidth is affected in a complicated way by various factors such as the lifetime of the excited states, the vibrational anharmonicity and the intermolecular interaction. At present, it is quite difficult to analyze the effect of each of these factors quantitatively. It seems reasonable, however, to assume that the extent of the contribution of each factor to the bandwidth depends on what parts of the molecule vibrate significantly on the vibrational excitation. Accordingly, one can expect the bandwidth to reflect the nature of the local groups vibrating with a large amplitude on the excitation of the normal mode. The parameter η_k in Eq. (8-14) represents the relative bandwidth of a hypothetical normal mode to which only the vibration of the kth internal coordinate, R_k, contributes, and is called the half bandwidth parameter. For actual normal modes, the squared relative amplitude of R_k in the normal mode Q_i, L_{ki}^2, is used as the weight with which η_k contributes to the bandwidth A_i of Q_i. The spectrum is drawn by calculating the absorbance or the transmittance (I/I_0) from Eq. (8-12) at appropriate intervals of wavenumbers, 1 cm^{-1}, for example, through the assigned spectral range.

8.3 Raman Scattering Intensity Parameters

In order to describe the Raman band intensities in terms of suitable parameters reflecting the local structures of molecules, it is useful to represent the molecular polarizability $\boldsymbol{\alpha}$ as the sum of the bond polarizabilities in the form

$$\boldsymbol{\alpha} = \sum_i \sum_j \boldsymbol{\alpha}_{ij}, \tag{8-15}$$

where $\boldsymbol{\alpha}_{ij}$ is the bond polarizability of the bond i–j and, as in the case of the infrared absorption intensities, to expand the components of each bond polarizability as the power series of internal coordinates. Bond polarizabilities and their derivatives with respect to internal coordinates have long been known as empirical Raman intensity parameters. As in the case of infrared absorption intensities, they were introduced as the electrooptical parameters by Eliashevich and Wolkenstein.[5)]

Generally, the bond polarizability tensor $\boldsymbol{\alpha}_{ij}$ is an ellipsoid with three principal components of different magnitudes. For simplicity, let us take up first the case where the tensor has the rotational symmetry, with the axis of rotation along the bond i–j. Let α_{ij}^E be the principal component of $\boldsymbol{\alpha}_{ij}$ in any direction perpendicular to the bond i–j, and $\alpha_{ij}^E + \gamma_{ij}^L$ be the principal component of $\boldsymbol{\alpha}_{ij}$ along the bond i–j. The bond polarizability tensor $\boldsymbol{\alpha}_{ij}$ is then expressed as

$$\boldsymbol{\alpha}_{ij} = \gamma_{ij}^L \boldsymbol{e}_{ij} \tilde{\boldsymbol{e}}_{ij} + \alpha_{ij}^E \boldsymbol{E}, \tag{8-16}$$

where $\boldsymbol{e}_{ij}$ is the unit vector along the bond i–j directed from i to j, and $\boldsymbol{E}$ is a (3×3) unit matrix. The axial symmetry of $\boldsymbol{\alpha}_{ij}$ is retained unless certain nonlinear bond angles are spanned by

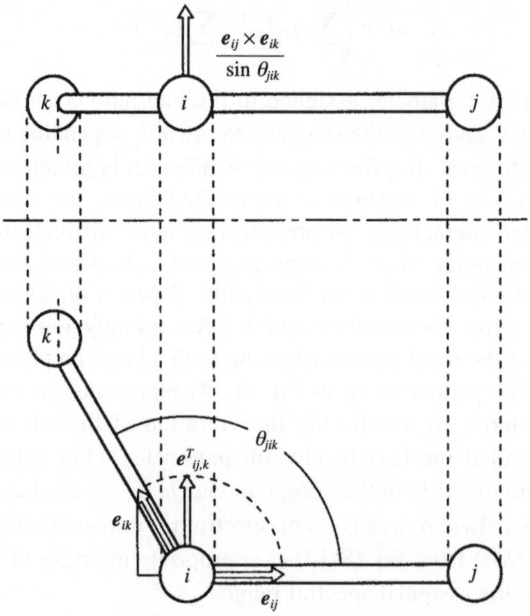

Fig. 8.1 Unit vectors along principal axes of anisotropic partial bond polarizability $\boldsymbol{\alpha}_{ij}$, $\boldsymbol{e}_{ij}$, $\boldsymbol{e}^T_{ij,k}$ and $\boldsymbol{e}_{ij} \times \boldsymbol{e}_{ik} / \sin\theta_{ijk}$.

either the nucleus i or j at the equilibrium.

The axial symmetry of $\boldsymbol{\alpha}_{ij}$ may be lost if the nucleus i is bonded to the nuclei j and k, and the bond angle θ_{jik} at the equilibrium is not 180°. It is convenient to define the partial bond polarizability, $\boldsymbol{\alpha}_{ij,k}$, for each of such nonlinear bond angles.[9] In Fig. 8.1, let $\boldsymbol{e}^T_{ij,k}$ be the unit vector perpendicular to the bond i–j and coplanar with the angle θ_{jik}. Then the partial bond polarizability $\boldsymbol{\alpha}_{ij,k}$ is defined by

$$\boldsymbol{\alpha}_{ij,k} = \gamma^L_{ij,k}\, \boldsymbol{e}_{ij}\, \tilde{\boldsymbol{e}}_{ij} + \gamma^T_{ij,k}\, \boldsymbol{e}^T_{ij,k}\, \tilde{\boldsymbol{e}}^T_{ij,k} + \alpha^E_{ij,k} \boldsymbol{E}. \tag{8-17}$$

The unit vector $\boldsymbol{e}^T_{ij,k}$ is expressed, as seen from Fig. 8.1, in terms of the unit vectors along the bonds i–j and i–k in the form

$$\boldsymbol{e}^T_{ij,k} = \left(-\boldsymbol{e}_{ij} \cos\theta_{jik} + \boldsymbol{e}_{ik}\right)/\sin\theta_{jik}. \tag{8-18}$$

The vector $\boldsymbol{e}^T_{ij,k}$ is minus the $\boldsymbol{s}$ vector of the angle bending coordinate $\Delta\theta_{jik}$ for the nucleus j multiplied by the bond length r_{ij}. The sign convention is rather trivial for later discussion.

In order to express the derivatives of all the independent components of the bond polarizability with respect to the nuclear Cartesian displacements collectively, it is convenient to introduce a (1×6) matrix by arranging the six independent components of the bond polarizability in a row as

$$\boldsymbol{V}^\alpha_{ij} = \left[(\alpha_{ij})_{XX} \quad (\alpha_{ij})_{YY} \quad (\alpha_{ij})_{ZZ} \quad (\alpha_{ij})_{YZ} \quad (\alpha_{ij})_{ZX} \quad (\alpha_{ij})_{XY}\right]. \tag{8-19}$$

Similarly, we rearrange the elements of the (3×3) matrices $\boldsymbol{e}_{ij}\, \tilde{\boldsymbol{e}}_{ij}$ and $\boldsymbol{E}$ to define the (1×6) matrices,

$$\boldsymbol{V}_{ij} = r_{ij}^{-2}\left[x_{ij}^2 \quad y_{ij}^2 \quad z_{ij}^2 \quad y_{ij}\, z_{ij} \quad z_{ij}\, x_{ij} \quad x_{ij}\, y_{ij}\right] \tag{8-20a}$$

and

$$\boldsymbol{V}_E = [1 \quad 1 \quad 1 \quad 0 \quad 0 \quad 0], \tag{8-20b}$$

respectively. The matrix $\boldsymbol{V}_{ij}$ can be written as a product of the (1×3) matrix (row vector) $\tilde{\boldsymbol{e}}_{ij}$ and a (3×6) matrix

$$\boldsymbol{A}_{ij} = \frac{1}{2\, r_{ij}} \begin{bmatrix} 2x_{ij} & 0 & 0 & 0 & z_{ij} & y_{ij} \\ 0 & 2y_{ij} & 0 & z_{ij} & 0 & x_{ij} \\ 0 & 0 & 2z_{ij} & y_{ij} & x_{ij} & 0 \end{bmatrix} \tag{8-21}$$

in the form

$$\boldsymbol{V}_{ij} = \tilde{\boldsymbol{e}}_{ij}\, \boldsymbol{A}_{ij}.$$

In analogy with the case of forming $\boldsymbol{V}_{ij}$ from $\boldsymbol{e}_{ij}\, \tilde{\boldsymbol{e}}_{ij}$, we form a (1×6) matrix $\boldsymbol{V}^T_{ij,k}$ by arranging the six elements of the (3×3) symmetric matrix $\boldsymbol{e}^T_{ij,k}\, \tilde{\boldsymbol{e}}^T_{ij,k}$ in a row. It follows from Eq. (8-18) that

$$\boldsymbol{e}^T_{ij,k}\, \tilde{\boldsymbol{e}}^T_{ij,k} = \left\{\boldsymbol{e}_{ij}\, \tilde{\boldsymbol{e}}_{ij} \cos^2\theta_{jik} - \left(\boldsymbol{e}_{ij}\, \tilde{\boldsymbol{e}}_{ik} + \boldsymbol{e}_{ik}\, \tilde{\boldsymbol{e}}_{ij}\right)\cos\theta_{jik} + \boldsymbol{e}_{ik}\, \tilde{\boldsymbol{e}}_{ik}\right\}/\sin^2\theta_{jik}$$

or, in terms of $\boldsymbol{V}^T_{ij,k}$, that

$$\boldsymbol{V}^T_{ij,k} = \frac{\cos^2\theta_{jik}}{\sin^2\theta_{jik}}\boldsymbol{V}_{ij} - \frac{2\cos\theta_{jik}}{\sin^2\theta_{jik}}\boldsymbol{V}_{ij,ik} + \frac{1}{\sin^2\theta_{jik}}\boldsymbol{V}_{ik} \tag{8-22}$$

where

$$\boldsymbol{V}_{ij,ik} = \left(\tilde{\boldsymbol{e}}_{ij}\,\boldsymbol{A}_{ik} + \tilde{\boldsymbol{e}}_{ik}\,\boldsymbol{A}_{ij}\right)/2.$$

By using the (1×6) matrices $\boldsymbol{V}_{ij}$, $\boldsymbol{V}^T_{ij,k}$, and $\boldsymbol{V}_E$, the total bond polarizability of the bond i–j is expressed as the sum of partial bond polarizabilities in the form

$$\boldsymbol{V}^\alpha_{ij} = \sum_k \boldsymbol{V}_{ij,k} + \sum_m \boldsymbol{V}_{ji,m} = \gamma^L_{ij}\,\boldsymbol{V}_{ij} + \sum_k \gamma^T_{ij,k}\,\boldsymbol{V}^T_{ij,k} + \sum_m \gamma^T_{ji,m}\,\boldsymbol{V}^T_{ji,m} + \alpha^E_{ij}\,\boldsymbol{V}_E. \tag{8-23}$$

The sums with respect to k and m in the last formula of Eq. (8-21) are taken over the nuclei bonded to i and over the nuclei bonded to j, respectively. The axial component γ^L_{ij} and the isotropic component α^E_{ij} of the bond polarizability $\boldsymbol{\alpha}_{ij}$ are given by summing up the corresponding components of the partial bond polarizabilities in the forms

$$\gamma^L_{ij} = \sum_k \gamma^L_{ij,k} + \sum_m \gamma^L_{ji,m}$$

and

$$\alpha^E_{ij} = \sum_k \alpha^E_{ij,k} + \sum_m \alpha^E_{ji,m},$$

respectively. The principal axes of the non-cylindrical bond polarizability $\boldsymbol{\alpha}_{ij}$ calculated in this way follow the instantaneous symmetry of the environment of the bond i–j automatically on any distortion of the molecule. For instance, the C–Cl bond of the methyl chloride molecule in Fig. 8.2 has three partial bond polarizabilities with reference to the three hydrogen nuclei, H_1, H_2 and H_3, $\gamma_{\text{C–Cl,H}i}$ ($i = 1$–3). If each of these partial bond polarizabilities has a larger in-plane component than the out-of-plane component, the section of each polarizability ellipsoid perpendicular to the C–Cl bond is an ellipse with the long axis on the reference plane, Cl–C–H$_i$. However, since the long axes intersect one another at an angle

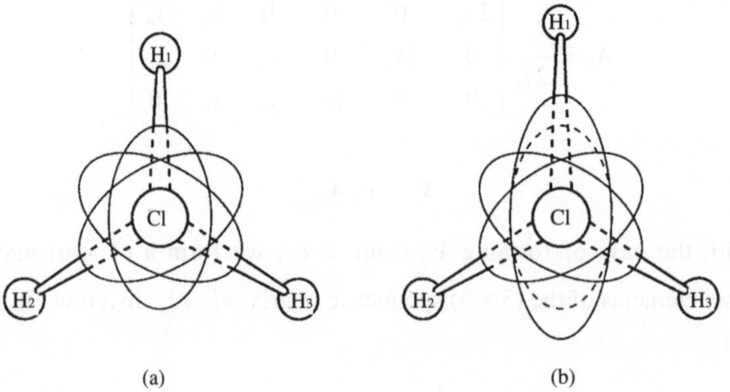

Fig. 8.2 Partial bond polarizabilities of C–Cl bond of CH_3Cl. (a) at equilibrium, (b) deformed on stretching of a C–H bond.

of 60°, the sum of the partial bond polarizabilities recovers the cylindricity at the equilibrium. On the other hand, if the in-plane component $\gamma^{L}_{\text{C-Cl,H1}}$ increases on any stretching of the C–H$_1$ bond, the symmetry of the bond polarizability $\boldsymbol{\alpha}_{\text{C-Cl}}$ changes in the same way as the molecular symmetry as a whole, and the plane of symmetry through H$_1$ is retained while the C_3 axis is lost as shown in Fig. 8.2(b).

By differentiating the elements of the (1 × 6) matrix $\boldsymbol{V}^{\alpha}_{ij}$ in Eq. (8-19) with respect to the Cartesian coordinates of the nucleus n, X_n, Y_n and Z_n (or the Cartesian displacement coordinates), and arranging the resulting derivatives columnwise, we obtain a (3 × 6) matrix in the form

$$\boldsymbol{\nabla}_n \boldsymbol{V}^{\alpha}_{ij} = \sum_p \boldsymbol{s}_{pn} \left\{ \left(\frac{\partial \gamma^{L}_{ij}}{\partial R_p} \right)_0 \boldsymbol{V}_{ij} + \sum_k \left(\frac{\partial \gamma^{T}_{ij,k}}{\partial R_p} \right)_0 \boldsymbol{V}^{T}_{ij,k} + \sum_m \left(\frac{\partial \gamma^{T}_{ji,m}}{\partial R_p} \right)_0 \boldsymbol{V}^{T}_{ji,m} + \left(\frac{\partial \alpha^{E}_{ij}}{\partial R_p} \right)_0 \boldsymbol{V}_E \right\}$$
$$+ \gamma^{L}_{ij} \boldsymbol{\nabla}_n \boldsymbol{V}_{ij} + \sum_k \gamma^{T}_{ij,k} \boldsymbol{\nabla}_n \boldsymbol{V}^{T}_{ij,k} + \sum_m \gamma^{T}_{ji,m} \boldsymbol{\nabla}_n \boldsymbol{V}^{T}_{ji,m}, \tag{8-24}$$

where $\boldsymbol{s}_{pn}$ is the $\boldsymbol{s}$ vector defined by Eq. (3-47). In Eq. (8-24), the terms including the derivatives of polarizability components in the braces represent the contribution from the deformation of polarizability tensor on any changes in internal variables, while the remaining three terms give the contribution from rotations of the bond polarizability tensors associated with the change in bond directions. The former reflects the influence of the electron redistribution on the molecular distortion, while the latter is determined purely from the geometrical arrangement of nuclei in the molecule. Calculations of all the terms in Eq. (8-24) can be coded in a computer program as multiplications of input parameters by certain functions of bond vector components.

Suppose that any physical quantities derivable from the electronic structure of a molecule can be written in the form of a tensor $\boldsymbol{A}$. Within the framework of the Born–Oppenheimer approximation, the direction of the principal axes of $\boldsymbol{A}$ and the components for the principal axes, $A^{\alpha\beta}$, are invariant under any isotopic substitution. On the other hand, a derivative of $A^{\alpha\beta}$ with respect to an internal coordinate R_i, which is orthogonal to the external coordinates, is given as the sum of the products of the isotopic invariants $(\partial A^{\alpha\beta}/\partial x_k)_0$ and the mass-dependent coefficients of transformation $(\partial x_k/\partial R_i)_0$ in the form

$$\left(\partial A^{\alpha\beta}/\partial R_i\right)_0 = \sum_k \left(\partial A^{\alpha\beta}/\partial x_k\right)_0 \left(\partial x_k/\partial R_i\right)_0. \tag{8-25}$$

Thus, the isotopic invariance of $(\partial A^{\alpha\beta}/\partial R_i)_0$ is not assured except in certain cases where, for example, the principal axes do not rotate on an isotopic substitution because of the symmetry.[10] Here, x_k implies a Cartesian coordinate in a local coordinate system defined with reference to the direction of a bond or the plane of a bond angle. This coordinate system, being uniquely defined regardless of the type of the isotopic substitution, does not necessarily coincide with the principal axes of inertia. If we write the matrix of transformation from the system of the principal axes to this coordinate system as $\boldsymbol{C}$, the mass-dependent factor $(\partial x_k/\partial R_i)_0$ is expressed as

$$\left(\partial x_k/\partial R_i\right)_0 = \left\{ \boldsymbol{C}(\boldsymbol{B}_{\text{in}})^{-1} \right\}_{ki}.$$

According to the above discussion, the derivatives of the principal Cartesian components of partial bond polarizabilities $\boldsymbol{\alpha}_{ij,k}$ with respect to internal coordinates do not satisfy the condition for the isotopic invariance. On the other hand, the parameters $\partial\gamma_{ij}{}^{L}/\partial R_p$, $\partial\gamma_{ij,k}{}^{T}/\partial R_p$ and $\partial\alpha_{ij}{}^{E}/\partial R_p$ introduced in Eq. (8-24) give $(\partial A^{\alpha\beta}/\partial x_k)_0$ if multiplied by the components of $\boldsymbol{s}$ vectors written in the mass-independent local coordinate systems, and thus satisfy the invariance condition.

A more generalized theory of non-cylindrical bond polarizabilities was proposed by Gussoni *et al.*[11] In this method, the directions of the principal axes of the bond polarizability perpendicular to the bond are not explicitly defined. In this sense, the theory of Gussoni *et al.* is more flexible than the formulation based on the partial bond polarizabilities for which the principal axes are fixed. Instead, a new quantity named the bond orientation parameter, which is not invariant under the isotopic substitution, is introduced for describing the orientation of the principal axes of each bond polarizability.

In the method adopted in RISE, the derivatives of partial bond polarizabilities with respect to the Cartesian displacements need to be calculated only for the nuclei entering Eq. (8-25) as the suffixes of $\boldsymbol{V}_{ij}$, $\boldsymbol{V}_{ij,k}^{T}$ and $\boldsymbol{V}_{ji,m}^{T}$. For this purpose it is convenient to use the differential operator $\boldsymbol{\nabla}_{ij}$ introduced in Chapter 3 (Eq. (3-64)). Operating $\boldsymbol{\nabla}_{ij}$ on each element of $\boldsymbol{V}_{ij}$ and using the basic relation in Eq. (3-65a), we have

$$\boldsymbol{\nabla}_{ij}\,\boldsymbol{V}_{ij} = 2\left(\boldsymbol{A}_{ij} - \boldsymbol{e}_{ij}\,\boldsymbol{V}_{ij}\right)/r_{ij}\,. \tag{8-26}$$

Similarly, by operating $\boldsymbol{\nabla}_{ij}$ and $\boldsymbol{\nabla}_{ik}$ on $\boldsymbol{V}_{ij,k}^{T}$ in Eq. (8-22) and carrying out the differentiation of the right sides by using Eq. (4-33) together with the relations

$$\boldsymbol{\nabla}_{ij}\,\frac{\cos^2\theta_{jik}}{\sin^2\theta_{jik}} = \boldsymbol{\nabla}_{ij}\left(\frac{1}{\sin^2\theta_{jik}} - 1\right) = -\frac{2\cos\theta_{jik}}{\sin^3\theta_{jik}}\,\boldsymbol{\nabla}_{ij}\,\theta_{jik}$$

and

$$\boldsymbol{\nabla}_{ij}\,\frac{\cos\theta_{jik}}{\sin^2\theta_{jik}} = -\frac{\left(1+\cos^2\theta_{jik}\right)}{\sin^3\theta_{jik}}\,\boldsymbol{\nabla}_{ij}\,\theta_{jik},$$

we obtain

$$\boldsymbol{\nabla}_{ij}\,\boldsymbol{V}_{ij,k}^{T} = 2\left(r_{ij}\sin\theta_{jik}\right)^{-1}\left\{\cos\theta_{jik}\left(\boldsymbol{e}_{ij,k}^{T}\boldsymbol{V}_{ij,k}^{T} - \boldsymbol{A}_{ij,k}\right) + \left(\boldsymbol{e}_{ij,k}^{T} - \boldsymbol{e}_{ij}\cot\theta_{jik}\right)\left(\boldsymbol{V}_{ij}\cos\theta_{jik} - \boldsymbol{V}_{ij,ik}\right)\right\} \tag{8-27}$$

and

$$\boldsymbol{\nabla}_{ik}\,\boldsymbol{V}_{ij,k}^{T} = 2\left(r_{ik}\sin\theta_{jik}\right)^{-1}\left\{\boldsymbol{e}_{ij,k}^{T}\boldsymbol{V}_{ij,k}^{T} + \boldsymbol{A}_{ij,k} + \boldsymbol{e}_{ij}\left(\boldsymbol{V}_{ij}\cos\theta_{jik} - \boldsymbol{V}_{ij,ik}\right)/\sin\theta_{jik}\right\}. \tag{8-28}$$

In Eqs. (8-27) and (8-28), the (3×6) matrix $\boldsymbol{A}_{ij,k}$ is defined by

$$\boldsymbol{A}_{ij,k} = \left(\boldsymbol{A}_{ik} - \boldsymbol{A}_{ij}\cos\theta_{jik}\right)/\sin\theta_{jik}\,.$$

The derivatives with respect to the Cartesian displacements of the nucleus i are obtained by calculating Eqs. (8-27) and (8-28) for all the bond vectors starting or terminating at i, $\boldsymbol{r}_{ij}$ and $\boldsymbol{r}_{ki}$, respectively, and making use of the simple relation

$$\boldsymbol{\nabla}_{i} = \sum_j \boldsymbol{\nabla}_{ji} - \sum_k \boldsymbol{\nabla}_{ik}\,.$$

Once the derivatives of polarizability tensor components with respect to the Cartesian displacements of all the nuclei are obtained, the variables for differentiation can be transformed into the dimensionless normal coordinates, thus leading to the components of the transition polarizability tensor in the form

$$(\partial \alpha_{uv}/\partial q_a) = (h/4\pi^2 c\omega_a)^{1/2} \sum_n (\partial \alpha_{uv}/\partial x_n) L^x_{na}. \tag{8-29}$$

The transition polarizability tensor is characterized by two quantities in the forms

$$\alpha'_a = \sum_u (\partial \alpha_{uu}/\partial q_a) \tag{8-30}$$

and

$$\gamma_a = \left\{ 3\sum_u \sum_v (\partial \alpha_{uv}/\partial q_a)^2 - 9\alpha'^2_a \right\} \Big/ 2. \tag{8-31}$$

These quantities, which are invariant under any rotation of the Cartesian coordinate axes, play an important role in the theory of Raman scattering intensities.[12] In the case of 90°-scattering, the intensity of the parallel component of the scattered light, *i.e.*, the light polarized in the same direction as the polarization of the incident light, is given by

$$S^{\mathrm{I}}_a = A_a\left(45\alpha'^2_a + 4\gamma_a{}^2\right), \tag{8-32a}$$

while the intensity of the perpendicular component, *i.e.*, the light polarized perpendicularly to the parallel component, is given by

$$S^{\mathrm{A}}_a = A_a(3\gamma_a^2). \tag{8-32b}$$

The depolarization ratio of the ath normal mode, ρ_a, is defined as the ratio of the perpendicular component to the parallel component in the form

$$\rho_a = S^{\mathrm{A}}_a \,/\, S^{\mathrm{I}}_a.$$

The constant A_a is obtained by substituting Eq. (8-9) into Eq. (8-3), taking the sum over the quantum number v_a and using Eq. (5-113). The result is

$$A_a = \frac{h(\omega_0 - \omega_a)^4}{360\,\varepsilon_0{}^2\, c\,\omega_a\{1 - \exp(-hc\omega_a/kT)\}}. \tag{8-33}$$

If the sample is transparent, the light scattered once by a molecule in the sample can be considered to leave the sample without further reflection. In the case of powdered samples, on the other hand, the light experiencing multiple reflections at the boundaries of crystallites before and after the Raman scattering may enter the spectrophotometer. The intensity of the scattered light measured without the analyzer is proportional to $S^{\mathrm{I}}_a + S^{\mathrm{A}}_a$ in the case of transparent samples, while the anisotropic component S^{A}_a contributes twice to the intensity of the light scattered by powdered samples. Including both cases, the intensity of the unpolarized scattered light is given by

$$S_a = A_a\left(45\alpha'^2_a + p\gamma_a{}^2\right), \tag{8-34}$$

where p is 7 for transparent samples and 10 for turbid samples.

When the Raman band shape is assumed to be Lorentzian, the scattering intensity at every 1 cm^{-1} within a specified wavenumber range is calculated in such a way that the integrated intensity is proportional to Eq. (8-33). By using the generated table of wavenumber *vs.* intensity, the Raman spectrum in a relative intensity scale can be drawn with an appropriate graphic routine.

8.4 Spectral Intensity Distributions

It has been stated in Chapter 4 that the potential energy distribution $(PED)_{ji}$ is used for representing the ratio with which the distortion of the *j*th internal variable contributes to the energy of excitation of the *i*th normal mode. Similarly, it is useful for understanding the vibrational spectrum of a compound to represent the extent to which an internal variable contributes to the infrared absorption or the Raman scattering intensity of a normal mode. According to the types of normal modes, the internal variables with large potential energy distribution do not necessarily contribute much to the intensities. Hence it is desirable to establish a measure with which the spectral intensities can be decomposed into the contributions of local structures for investigating the correlation between the molecular structure and the vibrational spectra. Such a device is particularly important when we try to adjust the intensity parameters by comparing the experimental and the calculated spectra.

On truncating the expansion of an internal symmetry coordinate in powers of normal coordinates at the linear terms in the form

$$S_i = \sum_a (\partial S_i/\partial Q_a)_0 Q_a = \sum_a L^S_{ia} Q_a,$$

the derivatives of each component of the molecular dipole moment with respect to normal coordinates are given by

$$(\partial \mu_\alpha/\partial Q_a)_0 = \sum_i (\partial \mu_\alpha/\partial S_i)_0 (\partial S_i/\partial Q_a)_0 = \sum_i L^S_{ia} (\partial \mu_\alpha/\partial S_i)_0. \tag{8-35}$$

Abbreviating the constant factor in the expression for the infrared absorption intensity in Eq. (8-10) as $C_a = N\pi g_a/3c^2$, and substituting Eq. (8-35) into a factor of $(\partial\mu_\alpha/\partial Q_i)_0{}^2$, one has

$$A_a = C_a \sum_\alpha (\partial \mu_\alpha/\partial Q_a)_0{}^2 = C_a \sum_\alpha \left\{ \sum_i L^S_{ia} (\partial \mu_\alpha/\partial S_i)_0 \right\} (\partial \mu_\alpha/\partial Q_a)_0, \tag{8-36}$$

where the summation index α runs over *X*, *Y* and *Z*. The infrared absorption intensity distribution, $(IID)_{ia}$, is now defined by[13)]

$$(IID)_{ia} = C_i L^S_{ia} \sum_\alpha (\partial \mu_\alpha/\partial S_i)_0 (\partial \mu_\alpha/\partial Q_a)_0 \Big/ A_a. \tag{8-37}$$

According to Eq. (8-37), $(IID)_{ia}$ is the projection of the weighted vector of dipole moment derivative $C_a L^S_{ia}(\partial\boldsymbol{\mu}/\partial S_i)_0$ onto the transition moment vector $(\partial\boldsymbol{\mu}/\partial Q_a)_0$ made dimensionless by scaling with A_a. If these two vectors form an obtuse angle, $(IID)_{ia}$ becomes negative, indicating that the infrared absorption intensity of the *a*th normal mode is partially compensated by the change in the *i*th internal coordinate (see Fig. 8.3). The numerator in the right side of Eq. (8-37) was independently defined by Qian and Krimm as the dipole

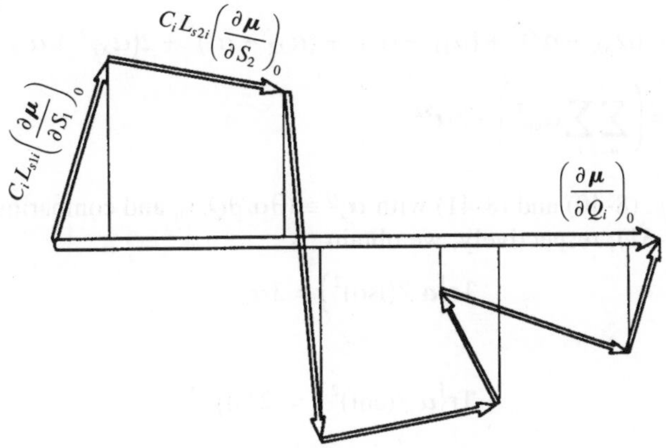

Fig. 8.3 Transition dipole moment vector given as the sum of contributions from individual internal coordinates.

moment derivative distribution, $(\mathrm{DDD})_{ia}$,

$$(\mathrm{DDD})_{ia} = C_a L_{ia}^S \sum_{\alpha} (\partial \mu_\alpha / \partial S_i)_0 (\partial \mu_\alpha / \partial Q_a)_0. \tag{8-38}$$

and was used for the same purpose as $(\mathrm{IID})_{ji}$.[14)]

In the case of Raman spectra, the situation is complicated since the theoretical expression of the intensity varies depending on the method of the intensity measurement.[12)] For convenience of expression, the polarizability tensor $\boldsymbol{\alpha}$ is divided into the isotropic component $\boldsymbol{\alpha}$(iso) in the form

$$\boldsymbol{\alpha}\,(\mathrm{iso}) = \begin{bmatrix} \alpha' & 0 & 0 \\ 0 & \alpha' & 0 \\ 0 & 0 & \alpha' \end{bmatrix} \tag{8-39a}$$

and the anisotropic component $\boldsymbol{\alpha}$(ani) in the form

$$\boldsymbol{\alpha}\,(\mathrm{ani}) = \begin{bmatrix} \alpha_{XX} - \alpha' & \alpha_{XY} & \alpha_{XZ} \\ \alpha_{XY} & \alpha_{YY} - \alpha' & \alpha_{YZ} \\ \alpha_{XZ} & \alpha_{YZ} & \alpha_{ZZ} - \alpha' \end{bmatrix} \tag{8-39b}$$

where

$$\alpha' = (\alpha_{XX} + \alpha_{YY} + \alpha_{ZZ})/3. \tag{8-40}$$

The total polarizability tensor is written as

$$\boldsymbol{\alpha} = \boldsymbol{\alpha}(\mathrm{iso}) + \boldsymbol{\alpha}(\mathrm{ani}).$$

From Eqs. (8-39b) and (8-40), the trace of the square of the anisotropic part is calculated to be

$$\mathrm{Tr}\{\boldsymbol{\alpha}\,(\mathrm{ani})^2\} = (\alpha_{XX} - \alpha')^2 + (\alpha_{YY} - \alpha')^2 + (\alpha_{ZZ} - \alpha')^2 + 2(\alpha_{XY}{}^2 + \alpha_{YZ}{}^2 + \alpha_{ZX}{}^2)$$
$$= \left(\sum_u \sum_v \alpha_{uv}{}^2\right) - 3\alpha'^2. \tag{8-41}$$

Replacing $\boldsymbol{\alpha}$ in Eqs. (8-40) and (8-41) with $\boldsymbol{\alpha}_a{}^Q \equiv (\partial\boldsymbol{\alpha}/\partial Q_a)_0$, and comparing the results with Eqs. (8-30) and (8-31), respectively, we obtain

$$\mathrm{Tr}\{\boldsymbol{\alpha}_a^Q(\mathrm{iso})^2\} = 3\alpha_a'^2 \tag{8-42}$$

and

$$\mathrm{Tr}\{\boldsymbol{\alpha}_a^Q(\mathrm{ani})^2\} = (2/3)\gamma_a{}^2. \tag{8-43}$$

Similarly, Eqs. (8-39a) and (8.40) lead to

$$\mathrm{Tr}\{\boldsymbol{\alpha}_a^Q(\mathrm{iso})\boldsymbol{\alpha}_a^Q(\mathrm{ani})\} = \alpha_a'\{(\alpha_a^Q)_{XX} + (\alpha_a^Q)_{YY} + (\alpha_a^Q)_{ZZ}\} - 3\alpha_a'^2 = 0. \tag{8-44}$$

To derive a simple expression for the contribution of the change in each internal symmetry coordinate to the unpolarized Raman band intensities measured for the 90°-scattering, a linear combination of the isotropic and the anisotropic components of the polarizability tensor are defined in the form

$$\boldsymbol{\Phi}_a = \boldsymbol{\alpha}_a^Q(\mathrm{iso}) + (p/10)^{1/2}\boldsymbol{\alpha}_a^Q(\mathrm{ani}).$$

The trace of $\boldsymbol{\Phi}_a{}^2$ is calculated, with the help of Eqs. (8-40), (8-42) and (8-43), to be

$$\mathrm{Tr}(\boldsymbol{\Phi}_a{}^2) = \mathrm{Tr}\{\boldsymbol{\alpha}_a^Q(\mathrm{iso})^2\} + \frac{p}{10}\mathrm{Tr}\{\boldsymbol{\alpha}_a^Q(\mathrm{ani})^2\} = \frac{45\alpha_a' + p\gamma_a}{15}. \tag{8-45}$$

Using the transformation relation between the polarizability derivatives with respect to the normal coordinates and those with respect to the internal symmetry coordinates,

$$\boldsymbol{\alpha}_a^Q(\mathrm{iso}) = \sum_i L_{ia}\boldsymbol{\alpha}_i^S(\mathrm{iso}), \quad \boldsymbol{\alpha}_a^Q(\mathrm{ani}) = \sum_i L_{ia}\boldsymbol{\alpha}_i^S(\mathrm{ani})$$

and introducing the abbreviated notation

$$K_{ia} = \mathrm{Tr}\{\boldsymbol{\alpha}_i^S(\mathrm{iso})\boldsymbol{\alpha}_a^Q(\mathrm{iso})\} + \frac{p}{10}\mathrm{Tr}\{\boldsymbol{\alpha}_i^S(\mathrm{ani})\boldsymbol{\alpha}_a^Q(\mathrm{ani})\}, \tag{8-46}$$

we can rewrite Eq. (8-45) as

$$\mathrm{Tr}(\boldsymbol{\Phi}_a{}^2) = \mathrm{Tr}\left\{\sum_i \boldsymbol{\alpha}_i^S(\mathrm{iso})\,L_{ia}\,\boldsymbol{\alpha}_a^Q(\mathrm{iso})\right\} + \frac{p}{10}\mathrm{Tr}\left\{\sum_i \boldsymbol{\alpha}_i^S(\mathrm{ani})\,L_{ia}\,\boldsymbol{\alpha}_a^Q(\mathrm{ani})\right\}$$
$$= \sum_i K_{ia}\,L_{ia}. \tag{8-47}$$

Thus the Raman scattering intensity distributions (RID) in the relative scale are given by

$$(\mathrm{RID})_{ia} = K_{ia}\, L_{ia} / \mathrm{Tr}(\boldsymbol{\Phi}_a{}^2).$$

Note that the definition of $(\mathrm{RID})_{ia}$ given here holds for the cases where the molecules in the sample are randomly oriented, and should be modified in the cases involving oriented samples.

8.5 Selection Rules

In order that the vibrational transition from the state n to the state m is caused by absorption of infrared radiation, at least one of the components of the transition moment

$$(\mu_u)_{mn} = \int \phi_m^* \, \mu_u \, \phi_n d\tau, \qquad (u = X, Y, Z) \tag{8-48}$$

must be different from zero.[15] If the molecule has any symmetry element, the vibrational wave functions ϕ_1, ϕ_2, $\cdots$ and the components of the dipole moment μ_X, μ_Y and μ_Z are classified into certain irreducible representations according to their symmetry properties. Then a product of these quantities, $\phi_m^* \mu_u \phi_n$, belongs also to an irreducible representation which may be found immediately by using the multiplication table of the point group of the molecule. If this irreducible representation is not totally symmetric, the definite integral in Eq. (8-48) must vanish rigorously. In order for $\phi_m^* \mu_u \phi_n$ to be totally symmetric, the product of the wavefunctions $\phi_m^* \phi_n$ and the component of the dipole moment μ_u must belong to the same irreducible representation. Hence there should be some component of the dipole moment belonging to the same irreducible representation as $\phi_m^* \phi_n$ in order for the vibrational transition $n \to m$ to be observable in the infrared absorption spectrum.

As already discussed, possible transitions from the ground state under the harmonic approximation are restricted to those of the type $0 \to 1$. Since the wave function of the vibrational ground state ϕ_0 is totally symmetric, the symmetry of the product $\phi_1^* \phi_0$ is the same as that of ϕ_1. As seen from Eq. (4-65), the wave function of the first excited state of the normal mode i is proportional to the product of a totally symmetric exponential function and the normal coordinate Q_i. Hence the symmetry of the wave function $\phi_1 \equiv |v_i\rangle$ coincides with that of Q_i. Accordingly, if the symmetry of Q_i coincides with that of any component of the dipole moment, the transition from the vibrational ground state to the first excited state of the normal mode i is observable in the infrared absorption spectra.

The components of the dipole moment μ_X, μ_Y and μ_Z behave in the same way as the external coordinates describing the translational motion of the whole molecule, T_X, T_Y and T_Z in Eqs. (3-29a–c), on the transformation of the Cartesian coordinates associated with each symmetry operation. In other words, the components μ_X, μ_Y and μ_Z belong to the same irreducible representation as the coordinates T_X, T_Y and T_Z, respectively. In conclusion, a normal mode belonging to an irreducible representation to which any translational coordinate belongs is infrared active.

In order for the transition $n \to m$ to be observable in the Raman spectrum (or to be Raman active), the integral

$$(\alpha_{uv})_{mn} = \int \phi_m^* \, \alpha_{uv} \, \phi_n d\tau, \qquad (u, v = X, Y, Z) \tag{8-49}$$

must have a non-vanishing value. From a completely analogous consideration with the case of infrared absorption spectra, it is concluded that a normal vibration belonging to the same irreducible representation as a component of the polarizability tensor is Raman active.

Classification of the components of polarizability tensor into the irreducible representations of a given point group is somewhat more complicated than the case of the components of the dipole moment, since the polarizability components themselves form the bases of a reducible representation. Let $\boldsymbol{\mu}_i$, $\boldsymbol{\alpha}$ and $\boldsymbol{E}$ be the induced dipole moment, the polarizability tensor and the external electric field, respectively, in the system (X, Y, Z), and let $\boldsymbol{\mu}_i'$, $\boldsymbol{\alpha}'$ and $\boldsymbol{E}'$ be the corresponding quantities in the system (X', Y', Z'). In the systems (X, Y, Z) and (X', Y', Z'), we have

$$\boldsymbol{\mu}_i = \boldsymbol{\alpha}\,\boldsymbol{E} \tag{8-50a}$$

and

$$\boldsymbol{\mu}_i' = \boldsymbol{\alpha}'\boldsymbol{E}', \tag{8-50b}$$

respectively. Let the matrix of transformation from the system (X, Y, Z) to the system (X', Y', Z') be denoted by $\boldsymbol{C}$. Then the vectors $\boldsymbol{\mu}_i'$ and $\boldsymbol{E}'$ are expressed as

$$\boldsymbol{\mu}_i' = \boldsymbol{C}\boldsymbol{\mu}_i \tag{8-51a}$$

and

$$\boldsymbol{E}' = \boldsymbol{C}\,\boldsymbol{E}, \tag{8-51b}$$

respectively. Substituting Eqs. (8-51a,b) into Eq. (8-50b) and using the orthogonality relation, $\tilde{\boldsymbol{C}}\boldsymbol{C} = \boldsymbol{E}$, one has

$$\boldsymbol{\alpha}' = \boldsymbol{C}\,\boldsymbol{\alpha}\,\tilde{\boldsymbol{C}}. \tag{8-52}$$

For a molecule having an n-fold axis with $n \geq 2$, the Z-axis is customarily chosen as the n-fold axis. On rotating the molecule around the Z-axis by the angle θ, the transformation relation among the components of $\boldsymbol{\alpha}$ is compactly expressed as follows in terms of the abbreviations $C = \cos\theta$ and $S = \sin\theta$.

By noting that the symmetric tensor $\boldsymbol{\alpha}$ has six independent components, Eq. (8-52) is expanded as

$$\boldsymbol{C}\,\boldsymbol{\alpha}\,\tilde{\boldsymbol{C}} = \begin{bmatrix} C & S & 0 \\ -S & C & 0 \\ 0 & 0 & 1 \end{bmatrix} \begin{bmatrix} \alpha_{XX} & \alpha_{XY} & \alpha_{XZ} \\ \alpha_{XY} & \alpha_{YY} & \alpha_{YZ} \\ \alpha_{XX} & \alpha_{YZ} & \alpha_{ZZ} \end{bmatrix} \begin{bmatrix} C & -S & 0 \\ S & C & 0 \\ 0 & 0 & 1 \end{bmatrix}$$

$$= \begin{bmatrix} C^2\alpha_{XX} + S^2\alpha_{YY} & -CS\alpha_{XX} + CS\alpha_{YY} & 0 \\ -CS\alpha_{XX} + CS\alpha_{YY} & S^2\alpha_{XX} + C^2\alpha_{YY} & 0 \\ 0 & 0 & \alpha_{ZZ} \end{bmatrix} \tag{8-53}$$

$$+ \begin{bmatrix} 2CS\alpha_{XY} & (C^2 - S^2)\alpha_{XY} & C\alpha_{XZ} + S\alpha_{YZ} \\ (C^2 - S^2)\alpha_{XY} & -2CS\alpha_{XY} & -S\alpha_{XZ} + C\alpha_{YZ} \\ C\alpha_{XZ} + S\alpha_{YZ} & -S\alpha_{XZ} + C\alpha_{YZ} & 0 \end{bmatrix}.$$

The transformation from the components of $\boldsymbol{\alpha}$ into those of $\boldsymbol{\alpha}'$ is then given by the matrix representation in the form

$$\begin{bmatrix} \alpha'_{ZZ} \\ \alpha'_{XX} \\ \alpha'_{YY} \\ \alpha'_{XY} \\ \alpha'_{ZX} \\ \alpha'_{YZ} \end{bmatrix} = \begin{bmatrix} 1 & 0 & 0 & 0 & 0 & 0 \\ 0 & C^2 & S^2 & 2CS & 0 & 0 \\ 0 & S^2 & C^2 & -2CS & 0 & 0 \\ 0 & -CS & CS & C^2 - S^2 & 0 & 0 \\ 0 & 0 & 0 & 0 & C & S \\ 0 & 0 & 0 & 0 & -S & C \end{bmatrix} \begin{bmatrix} \alpha_{ZZ} \\ \alpha_{XX} \\ \alpha_{YY} \\ \alpha_{XY} \\ \alpha_{ZX} \\ \alpha_{YZ} \end{bmatrix}. \tag{8-54}$$

The (3 × 3) diagonal block related to the (*XX*), (*YY*) and (*XY*) components in Eq. (8-54) can be decomposed further into a (1 × 1) and a (2 × 2) blocks by the similarity transformation in the form

$$\begin{bmatrix} 1/\sqrt{2} & 1/\sqrt{2} & 0 \\ 1/\sqrt{2} & -1/\sqrt{2} & 0 \\ 0 & 0 & \sqrt{2} \end{bmatrix} \begin{bmatrix} C^2 & S^2 & 2CS \\ S^2 & C^2 & -2CS \\ -CS & CS & C^2 - S^2 \end{bmatrix} \begin{bmatrix} 1/\sqrt{2} & 1/\sqrt{2} & 0 \\ 1/\sqrt{2} & -1/\sqrt{2} & 0 \\ 0 & 0 & 1/\sqrt{2} \end{bmatrix} = \begin{bmatrix} 1 & 0 & 0 \\ 0 & \cos 2\theta & \sin 2\theta \\ 0 & -\sin 2\theta & \cos 2\theta \end{bmatrix}. \tag{8-55}$$

Combining Eqs. (8-54) and (8-55) leads to

$$\begin{bmatrix} \alpha'_{ZZ} \\ (\alpha'_{XX} + \alpha'_{YY})/\sqrt{2} \\ (\alpha'_{XX} - \alpha'_{YY})/\sqrt{2} \\ \sqrt{2}\alpha'_{XY} \\ \alpha'_{ZX} \\ \alpha'_{YZ} \end{bmatrix} = \begin{bmatrix} 1 & 0 & 0 & 0 & 0 & 0 \\ 0 & 1 & 0 & 0 & 0 & 0 \\ 0 & 0 & \cos 2\theta & \sin 2\theta & 0 & 0 \\ 0 & 0 & -\sin 2\theta & \cos 2\theta & 0 & 0 \\ 0 & 0 & 0 & 0 & \cos\theta & \sin\theta \\ 0 & 0 & 0 & 0 & -\sin\theta & \cos\theta \end{bmatrix} \times \begin{bmatrix} \alpha_{ZZ} \\ (\alpha_{XX} + \alpha_{YY})/\sqrt{2} \\ (\alpha_{XX} - \alpha_{YY})/\sqrt{2} \\ \sqrt{2}\alpha_{XY} \\ \alpha_{ZX} \\ \alpha_{YZ} \end{bmatrix}. \tag{8-56}$$

The (6 × 6) matrix of the transformation in Eq. (8-56) is a direct sum of two one-dimensional and two two-dimensional irreducible representations of a molecule having a more-than-three-fold axis with respect to a set of bases formed from the six polarizability components. For molecules having no more-than-three-fold axes, all the off-diagonal elements of the transformation matrix in Eq. (8-54) vanish since sin 180° = 0, so that this matrix is already a direct sum of six one-dimensional irreducible representations.

The behaviors of the polarizability components on the inversion at the center of symmetry or the reflection with respect to a plane of symmetry may be found by using the matrices

$$\boldsymbol{i} = \begin{bmatrix} -1 & 0 & 0 \\ 0 & -1 & 0 \\ 0 & 0 & -1 \end{bmatrix}, \qquad \boldsymbol{\sigma}_{XY} = \begin{bmatrix} 1 & 0 & 0 \\ 0 & 1 & 0 \\ 0 & 0 & -1 \end{bmatrix}$$

$$\boldsymbol{\sigma}_{YZ} = \begin{bmatrix} -1 & 0 & 0 \\ 0 & 1 & 0 \\ 0 & 0 & 1 \end{bmatrix}, \qquad \boldsymbol{\sigma}_{ZX} = \begin{bmatrix} 1 & 0 & 0 \\ 0 & -1 & 0 \\ 0 & 0 & 1 \end{bmatrix}$$

as the transformation matrix for the Cartesian coordinates in Eq. (8-52). If $\boldsymbol{C} = \boldsymbol{i}$, it follows that $\boldsymbol{\alpha}' = \boldsymbol{\alpha}$ at once, showing that the inversion keeps all the polarizability components invariant. If $\boldsymbol{C} = \boldsymbol{\sigma}_{XY}$, the transformed polarizability is given by

$$\boldsymbol{\alpha}' = \boldsymbol{\sigma}_{XY}\,\boldsymbol{\alpha}\,\boldsymbol{\sigma}_{XY} = \begin{bmatrix} \alpha_{XX} & \alpha_{XY} & -\alpha_{XZ} \\ \alpha_{XY} & \alpha_{YY} & -\alpha_{YZ} \\ -\alpha_{XZ} & -\alpha_{YZ} & \alpha_{ZZ} \end{bmatrix},$$

showing that α_{XZ} and α_{YZ} change the sign, while the other four components are kept invariant.

In the character tables in the literature, the irreducible representations to which the components of the dipole moments and the polarizability are usually specified.[16] In that case, the linear combinations of the polarizability components, $\alpha_{XX} \pm \alpha_{YY}$, are sometimes shown instead of individual components. In such a case, the linear combinations form the bases of irreducible representations.

8.6 Illustrative Simulation of Vibrational Spectra

As an example of the simulation, the observed and calculated infrared absorption spectra of acetic acid in the monomer–dimer equilibrium in solution in carbon tetrachloride are shown in Fig. 8.4.[8] The half bandwidth parameters for the monomer were all taken to be 1.0, while those for the dimer were taken from the observed values shown in Table 8.1. The standard bandwidth ω_0 was taken to be 10 cm^{-1}. The entropy and enthalpy of dimerization calculated

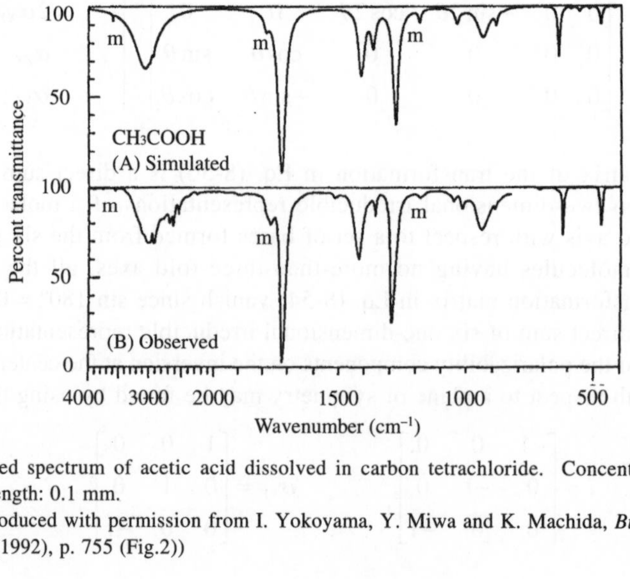

Fig. 8.4 Infrared spectrum of acetic acid dissolved in carbon tetrachloride. Concentration: 0.161 mol dm^{-3}, pathlength: 0.1 mm.
(Reproduced with permission from I. Yokoyama, Y. Miwa and K. Machida, *Bull. Chem. Soc. Jpn.*, **65**, 746, (1992), p. 755 (Fig.2))

Table 8.1 Half bandwidth parameters of acetic acid dimer.

Internal Coordinates	η_i
OH stretching	70.0
COH bending	6.0
OH···O linear bending	8.0
$HC_\alpha H$ bending	2.0
$HC_\alpha C$ bending	2.0

η_i is 1.0 for the internal coordinates not given in the table.

Table 8.2 Enthalpy and entropy changes on dimerization of acetic acid.

Thermodynamic function	calculated	observed
ΔH/kJ mol^{-1}	−63.1	−61.9 ± 4.2 [a]
ΔS/J mol^{-1} K^{-1}	−160.0	−145.6 ± 8.4 [a] −175 ± 11 [b]

[a] Ref. 16

[b] N. Lumbroso-Bader, C. Coupry, D. Baron, D. H. Clague and G. Govil, *J. Magn. Reson.*, **17**, 386 (1973).

from the RISE potential function are compared with the experimental values in Table 8.2. The monomer/dimer ratio used in the calculation of the transmittance was estimated to be 1/23 from the experimental value of ΔG.[17] This ratio estimated from the calculated values of ΔH and ΔS becomes 1/12, which gives too large intensities of the monomer bands. Presumably, very accurate estimation of thermodynamic quantities is required for simulating vibrational spectra of equilibrium mixtures involving the process of association or conformational change.

The charge fluxes of the acetic acid monomer and dimer used in the calculation of the potential functions and the infrared absorption intensities are shown in Table 8.3. The sign of the O–H bond flux $\partial q_{OH}/\partial r_{OH}$ is negative for the monomer and positive for the dimer. This means that the positive charge on the hydrogen atom of the O–H bond diminishes for the monomer while the charge increases for the dimer on stretching of the O–H bond. The increase in the O–H bond distance leads finally to the ionic and the radical dissociations in the cases of the dimer and the monomer, respectively. The sign and magnitude of the charge flux of the monomer agree well with those of methanol, −0.11 eÅ^{-1}, estimated experimentally.[18]

It is well known that the infrared absorption band due to the O–H stretching vibration of carboxylic acid monomers shifts toward the low-frequency side by about 450 to 500 cm^{-1} and is enhanced by about 15 times on dimerization. The large positive charge flux of the O–H bond of carboxylic acid dimers increases the positive charge on the hydrogen atom as the O–H bond stretches, and accelerates the contribution of the Coulomb interaction to the stabilization of the dimer. The charge flow also increases the change in the O–H bond moment, thereby enhancing the infrared absorption intensity of the O–H stretching vibration. The lowering of the O–H stretching frequency arises from the decrease in magnitude of the second derivative of the stretching potential caused by the negative contribution from the cubic term in the expansion of the Morse function, Eq. (2-78).

Table 8.3 Bond charges and bond charge fluxes of acetic acid.

	monomer	dimer		monomer/dimer
Bond charge (e) [a]			Charge flux(bending)[b] ($e\cdot rad^{-1}$)	
$\beta(H–C_\alpha)$	0.1500	0.1500	$\partial q(C_\alpha H)/\partial\theta(HC_\alpha H)$	–0.005
$\beta(C–O)$	0.0193	0.0193	$\partial q(C_\alpha H)/\partial\theta'(HC_\alpha H)$	–0.015
$\beta(C{=}O)$	0.3187	0.3187	$\partial q(C_\alpha H)/\partial\theta(HC_\alpha C)$	–0.015
$\beta(O–H)$	0.2390	0.1300	$\partial q(C_\alpha C)/\partial\theta'(HC_\alpha C)$	0.020
$\beta(C–C_\alpha)$	0.1380	0.1380	$\partial q(C_\alpha C)/\partial\theta(HC_\alpha C)$	–0.004
			$\partial q(C_\alpha C)/\partial\theta'(HC_\alpha H)$	0.004
Charge flux(stretching)[b] ($e\cdot nm^{-1}$)			$\partial q(C{=}O)/\partial\theta(O–C{=}O)$	0.130
$\partial q(C_\alpha H)/\partial r(HC_\alpha)$	-1.00	-1.00	$\partial q(C{=}O)/\partial\theta(C_\alpha–C{=}O)$	–0.100
$\partial q(OH)/\partial r(HO)$	-1.10	7.80	$\partial q(C{=}O)/\partial\theta'(C_\alpha–C–O)$	–0.030
$\partial q(CC_\alpha)/\partial r(CC_\alpha)$	-0.20	-0.20	$\partial q(C–O)/\partial\theta(O–C{=}O)$	0.130
$\partial q(O-C)/\partial r(C-O)$	7.60	11.38	$\partial q(C–O)/\partial\theta'(C_\alpha–C{=}O)$	–0.065
$\partial q(O=C)/\partial r(C=O)$	6.80	5.50	$\partial q(C–O)/\partial\theta(C_\alpha–C–O)$	–0.065
$\partial q(C_\alpha H)/\partial r'(HC_\alpha)$	-0.35	-0.35	$\partial q(OH)/\partial\theta(C–O–H)$	0.050
$\partial q(O\cdots H)/\partial r'(C=O)$	—	0.24	$\partial q(O–C)/\partial\theta(C–C–H)$	0.050

[a] electronic unit

[b] The numerator bond shares an atom and two atoms, respectively, with the primed and unprimed internal coordinates in the denominator.

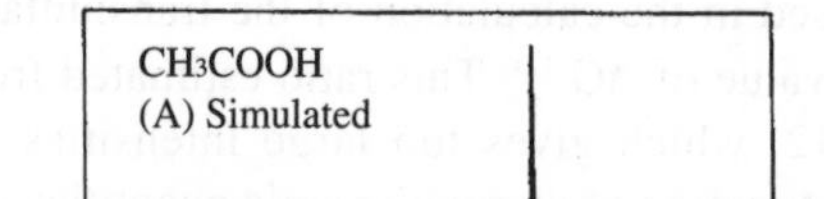

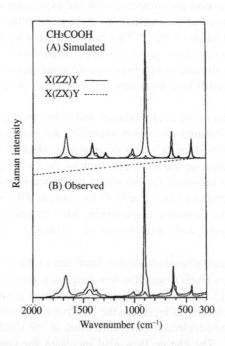

Fig. 8.5 Polarized Raman spectra of liquid acetic acid. A, B, C and D in the symbol A(BC)D indicate the directions of propagation and polarization of the incidence, and the directions of polarization and propagation of the scattered light, respectively.[19] (Reproduced with permission from I. Yokoyama, Y. Miwa and K. Machida, *Bull. Chem. Soc. Jpn.*, **65**, 746 (1992), p. 759 (Fig. 4))

The infrared absorption band due to the C=O stretching vibration of acetic acid also shows an intensity enhancement and a frequency decrease on dimerization. In general, there are two normal modes of carboxylic acid dimers arising mainly from the symmetric (Raman active) and the antisymmetric (infrared active) stretching vibrations of the two C=O bonds. In the case of acetic acid, the Raman active frequency was observed to be 84 cm^{-1} higher than the infrared active frequency. The splitting between the two C=O stretching frequencies, the intensity and the frequency changes on the dimerization and the increase of the C=O bond distance are all reproduced by introducing two charge fluxes, $\partial q_{O=C}/\partial r_{CO}$ and $\partial q_{O\cdots H}/\partial r_{CO}$. The former and the latter fluxes mean that the stretching of the C=O bond favors the polarization of the C=O bond and the charge transfer through the hydrogen bond O···H, respectively.

The observed and simulated Raman spectra of acetic acid in the liquid state are compared with each other in Fig. 8.5. The dotted line in the observed spectra shows the sensitivity curve of the detector according to which the calculated intensity has been scaled. The percentage of the monomer is known to be negligible so that the simulation has been carried out by using only the calculated spectrum of the dimer. On calculating the spectrum, the intermolecular interaction other than that between the dimerized pair is not explicitly taken into account. The same bandwidth parameters as those for the infrared spectra are used for the Raman spectrum with the standard bandwidth of 12 cm^{-1}. The agreement in position, relative intensity and depolarization ratio of each band between the calculated and the observed spectra is satisfactory.

References

1) L. Pauling and E.B. Wilson, Jr., *Introduction to Qunatum Mechanics with Application to Chemistry*, Chapter 11, McGraw-Hill, New York (1935).
2) H. Eyring, J. Walter and G.E. Kimball, *Quantum Chemistry*, Section 8g, John Wiley & Sons, New York (1944).
3) J.C. Decius, *J. Mol. Spectrosc.*, **57**, 348 (1975).
4) Y. Miwa and K. Machida, *J. Am. Chem. Soc.*, **110**, 5183 (1988).
5) M. Eliashevich and M. Wolkenstein, *Zh. Eksp. Teor. Fiz.*, **9**, 101 (1945).
6) M. Gussoni and S. Abbate, *J. Chem. Phys.*, **65**, 3439 (1976).
7) L.M. Sverdlov, M.A. Kovner and E.P. Krainov, *Vibrational Spectroscopy of Polyatomic Molecules*, John Wiley & Sons, New York (1974).
8) I. Yokoyama, Y. Miwa and K. Machida, *Bull. Chem. Soc. Jpn.*, **65**, 746 (1992).
9) K. Machida, H. Noma and Y. Miwa, *Indian J. Pure Appl. Phys.*, **26**, 197 (1988).
10) B.L. Crawford, Jr., *J. Chem. Phys.*, **20**, 977 (1952).
11) M. Gussoni, S. Abbate and G. Zerbi, *J. Raman Spectrosc.*, **6**, 289 (1977).
12) D.A. Long, *Raman Spectroscopy*, McGraw-Hill, New York (1977).
13) H. Noma and K. Machida, *J. Mol. Struct.*, **242**, 207 (1991).
14) W. Qian and S. Krimm, *J. Phys. Chem.*, **98**, 9992 (1994).
15) E.B. Wilson, Jr., J.C. Decius and P.C. Cross, *Molecular Vibrations*, p. 146, McGraw-Hill, New York (1954)
16) Ref. 15, pp. 323-329.
17) A.D.H. Clague and H. J. Bernstein, *Spectrochim. Acta*, **A25**, 593 (1969).
18) M. Gussoni, C. Castiglioni and G. Zerbi, *J. Phys. Chem.*, **88**, 600 (1984).
19) S.P.S Porto, P.A.Fleury and T.C. Damen, *Phys. Rev.*, **154**, 522 (1967).

Chapter 9

Mechanics of Molecular Crystals

9.1 Molecular Force Field and Crystal Force Field

If we want to compare the results of molecular mechanics calculation for an isolated molecule with experimental data, we should refer to data for the gas phase in principle, since the quantities calculated from molecular mechanics such as molecular structures, thermodynamic functions and vibrational spectra are not negligibly affected by the intermolecular interaction in most liquid and solid phases. It is not easy, however, to extract sufficient structural information for complicated molecules from the gas phase only. Hence the development of molecular mechanics methods taking account of intermolecular interaction becomes important.

Correct estimation of the force and energy due to intermolecular interaction is important also for improvement of the potential functions of isolated molecules. It is generally accepted that the force acting between two nonbonded atoms separated by more than 3 or 4 bonds in a molecule is determined purely by the internuclear distance, not depending on the number of bonds between them. In such a case, the force due to the interaction between the two atoms does not depend on whether the atoms belong to the same molecule or not. Thus it is reasonable to use the same set of potential parameters for intra- and intermolecular pairs of nonbonded atoms according to the type of atoms constituting the pair.

Since the forces acting between nonbonded nuclei are much weaker than those required to generate any changes in internal coordinates in general, it is difficult to estimate the magnitude of nonbonded forces exclusively from intramolecular data. In contrast, there are no terms including internal coordinates in intermolecular potential functions except for the cases where the potential for hydrogen bonds or coordination bonds is described in terms of stretching and bending coordinates. The lattice energies, crystal structures and frequencies of lattice vibrations are thus determined principally by the nonbonded atom–atom interactions between different molecules. We may use the mechanical behavior of molecular crystals as a reliable source of information for quantitative estimation of the potential parameters which describe nonbonded atom–atom interaction. This chapter deals with the principles of the methods of calculating equilibrium structures, elastic constants, quantities related to nuclear vibrations, lattice energies and surface energies of molecular crystals.

9.2 Equilibrium Structures and Elastic Constants of Crystals

Although a crystal is an assembly of a large indefinite number of molecules or ions, one can reduce the number of variables to a finite and tractably small number by utilizing the translational symmetry characteristic of the crystal.[1)] Imagine a perfect crystal of infinite size partitioned into identical parallelepiped blocks each of which contains N_a , N_b and N_c unit cells along the *a*-, *b*- and *c*-axes, respectively. The partitioning is accomplished by three sets of parallel lattice planes which intersect the *a*-, *b*- and *c*-axes at integer multiples of $N_a a$, $N_b b$ and $N_c c$, respectively. The total number of unit cells in a parallelepiped block is $N = N_a N_b N_c$. Here it is assumed that N_a, N_b and N_c are all far larger than unity. The parallelepiped blocks are given consecutive indices, $M = -\infty, \cdots, -1, 0, 1, \cdots, +\infty$. The unit cells in the parallelepiped M are numbered in the form $I' = MN + I$, where $1 \le I \le N$. The parallelepiped given the index $M = 0$ shall be called the basic parallelepiped. A unit cell is taken to contain n nuclei, and the nuclei in the Ith unit cell are indexed as $I1, I2, \cdots, In$. Let $X_{Ip}{}^0$, $Y_{Ip}{}^0$ and $Z_{Ip}{}^0$ be the Cartesian coordinates of the nucleus Ip at the equilibrium position, and ΔX_{Ip}, ΔY_{Ip} and ΔZ_{Ip} be the Cartesian displacement coordinates of the same nucleus at a given moment. The instantaneous Cartesian coordinates of the nucleus Ip are then given by

$$X_{Ip} = X_{Ip}^0 + \Delta X_{Ip}, \quad Y_{Ip} = Y_{Ip}^0 + \Delta Y_{Ip}, \quad Z_{Ip} = Z_{Ip}^0 + \Delta Z_{Ip}. \tag{9-1}$$

As in the case of isolated molecules, the Cartesian displacement vector of the nucleus Ip is introduced with the definition

$$\Delta \boldsymbol{r}_{Ip} = \begin{bmatrix} \Delta X_{Ip} & \Delta Y_{Ip} & \Delta Z_{Ip} \end{bmatrix}^{\mathrm{T}}, \tag{9-2}$$

and the Cartesian displacements of all the nuclei in the unit cell I are expressed collectively by a vector with $3n$ elements as

$$\Delta \boldsymbol{x}_I = \begin{bmatrix} \Delta \tilde{\boldsymbol{r}}_{I1} & \Delta \tilde{\boldsymbol{r}}_{I2} & \cdots & \Delta \tilde{\boldsymbol{r}}_{In} \end{bmatrix}^{\mathrm{T}}. \tag{9-3}$$

Because of the translational symmetry of crystals, the Cartesian coordinates of the nucleus Ip at the equilibrium position satisfy the relation

$$\boldsymbol{r}_{Ip}^0 = \boldsymbol{r}_{1p}^0 + \boldsymbol{r}_I,$$

where $\boldsymbol{r}_I$ is the lattice vector of the Ith cell defined as the position vector of the cell origin. Let $\boldsymbol{e}_a$, $\boldsymbol{e}_b$ and $\boldsymbol{e}_c$ be the unit vectors taken along the positive direction of the *a*-, *b*- and *c*-axes of the crystal, and I_a, I_b and I_c be certain integers satisfying the condition

$$0 \le I_a \le N_a - 1, \; 0 \le I_b \le N_b - 1 \quad \text{and} \quad 0 \le I_c \le N_c - 1,$$

respectively. The unit cell vectors $\boldsymbol{a}$, $\boldsymbol{b}$ and $\boldsymbol{c}$ are defined as the vectors along the three unit cell axes as given by $\boldsymbol{a} = a\boldsymbol{e}_a$, $\boldsymbol{b} = b\boldsymbol{e}_b$ and $\boldsymbol{c} = c\boldsymbol{e}_c$, respectively. Then the vector $\boldsymbol{r}_I$ in the basic parallelepiped is written in terms of the unit cell vectors in the form

$$\boldsymbol{r}_I = I_a a\, \boldsymbol{e}_a + I_b b\, \boldsymbol{e}_b + I_c c\, \boldsymbol{e}_c = I_a \boldsymbol{a} + I_b \boldsymbol{b} + I_c \boldsymbol{c}. \tag{9-4}$$

The integer I represents a set of I_a, I_b and I_c ordered in an arbitrary way except that $I = 1$ and N correspond to the sets $(I_a, I_b, I_c) = (0, 0, 0)$ and (N_a-1, N_b-1, N_c-1), respectively. A simple example of such a numbering is

$$I = 1 + I_a + N_a I_b + N_a N_b I_c. \tag{9-5}$$

The potential energy of the crystal per unit cell is expanded in powers of the Cartesian displacement coordinates in the form

$$V = N^{-1}\left(\sum_{I=1}^{N} \Delta\tilde{x}_I\, F_I^x + \tfrac{1}{2}\sum_{I=1}^{N} \sum_{J=-\infty}^{\infty} \Delta\tilde{x}_I\, F_{IJ}^{xx}\, \Delta x_J + \cdots \right), \tag{9-6}$$

where the vector F_I^x and the matrix F_{IJ}^{xx} consist of those elements which are given, for example, by

$$(F_I^x)_{3p-2} = (\partial V/\partial X_{Ip})_0 \quad \text{and} \quad (F_{IJ}^{xx})_{3p-1,3q} = (\partial^2 V/\partial Y_{Ip}\partial Z_{Jq})_0,$$

respectively. The Hessian matrix F_{IJ}^{xx} approaches the zero matrix as the distance between the unit cells I and J increases, and may be neglected when this distance exceeds a certain distance limit $r_{\max}$. Thus the sum over J in Eq. (9-6) in the actual calculation is taken within the range defined by

$$|r_J - r_I| \le r_{\max}. \tag{9-7}$$

The matrix F_{IJ}^{xx} is determined if the indices I and J are given. Because of the translational symmetry of crystals, however, F_{IJ}^{xx} depends only on the direction and magnitude of the vector $r_J - r_I$, and it is of no matter where the vector starts. Accordingly, if we choose such a unit cell index L as to be related to I and J by $r_J - r_I = r_L - r_1$ or, since $r_1 = \mathbf{0}$, by

$$r_J = r_I + r_L, \tag{9-8}$$

it follows that

$$F_{IJ}^{xx} = F_{1L}^{xx}. \tag{9-9}$$

Equation (9-9) applies to the cases where the indices I and J are out of the range from 1 to N, too. Obviously, an indefinitely large number of variables are involved in Eq. (9-6). If we restrict the deformation of a crystal to be homogeneous, the crystal dynamics can be described in terms of a definite number of variables corresponding to the degrees of freedom of motions of the nuclei in a unit cell plus the cell constants, a, b, c, α, β and γ. Here, a homogeneous deformation means a crystal deformation in which the translational symmetry is retained.

When the cell constants change according to a homogeneous strain caused by an applied stress, a visual change in the crystal shape such as an expansion, a contraction or a change in a face angle results. The cell constants may thus be taken as the variables which describe the macroscopic structure of the crystal. On the other hand, changes in the Cartesian displacement coordinates of nuclei or in the internal coordinates give rise to no apparent changes in crystal size or crystal shape. These coordinates are thus the variables which describe the microscopic structure of the crystal.

In the X-ray analysis of single crystals, crystallographers often use the relative coordinate system in which the unit cell axes are taken as the coordinate axes and the length along each unit cell axis is represented as the ratio to the relevant cell constant. Let the coordinates of the nucleus i in this system be given by

$$r_{Ip}^c = \begin{bmatrix} X_{Ip}^c & Y_{Ip}^c & Z_{Ip}^c \end{bmatrix}^{\mathrm{T}},$$

where the X^c-, Y^c- and Z^c-axes are taken to be along the a-, b- and c-axes of the unit cell, respectively. On choosing the X-axis of the Cartesian coordinate system along the a-axis of the unit cell, and placing the Y-axis on the ab-plane, the transformation from the r^c-system

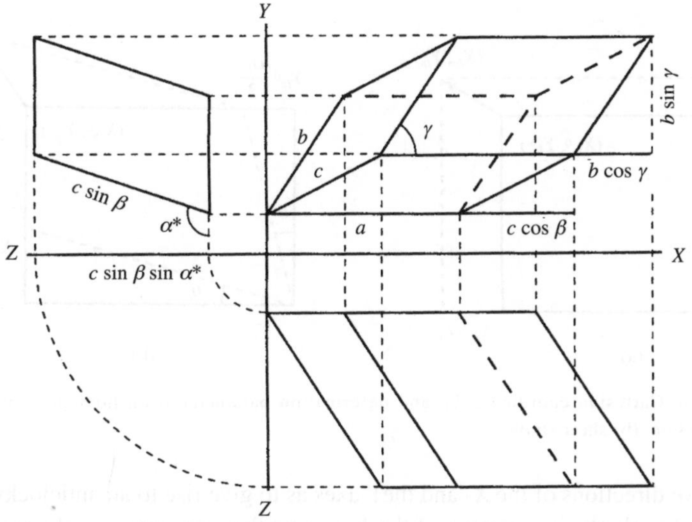

Fig. 9.1 Projections of a unit cell onto the planes of the Cartesian coordinate system.

to the **r**-system is written in the form

$$\boldsymbol{r}_{Ip} = \begin{bmatrix} X_{Ip} \\ Y_{Ip} \\ Z_{Ip} \end{bmatrix} = \boldsymbol{C}_r\, \boldsymbol{r}_{Ip}^c = \begin{bmatrix} a & b\cos\gamma & c\cos\beta \\ 0 & b\sin\gamma & -c\sin\beta\cos\alpha^* \\ 0 & 0 & c\sin\beta\sin\alpha^* \end{bmatrix} \begin{bmatrix} X_{Ip}^c \\ Y_{Ip}^c \\ Z_{Ip}^c \end{bmatrix}, \tag{9-10}$$

where α^* is the dihedral angle between the planes *ab* (containing the negative direction of the *b*-axis) and *ac* (containing the positive direction of the *c*-axis); (see Fig. 9.1 and Appendix 3). According to the cosine theorem in spherical trigonometry (Appendix 1), the angles α^*, α, β and γ satisfy the relation

$$\cos\alpha^* = (\cos\beta\cos\gamma - \cos\alpha)/(\sin\beta\sin\gamma). \tag{9-11}$$

If the nuclei in a crystal are simply buried in an elastic and homogeneous medium as isolated particles, and exert no force on each other, the relative coordinates of nuclei do not change on any of those macroscopic deformations of the crystal which take place homogeneously. Obviously, the translational symmetry along the crystal axes is not lost here. In such a homogeneous macroscopic deformation, what may change in the relative coordinate system are only the scales along the axes, *a*, *b* and *c*, and the angles between them, α, β and γ. The nuclear displacements arising from this type of crystal deformation are expressed as the changes in Cartesian coordinates in the following way.[2)]

For simplicity, let us consider first only the displacements constrained on the *XY*-plane. By fixing the origin, the whole crystal is $(1 + \sigma_{XX})$ and $(1 + \sigma_{YY})$ times expanded in the directions of the *X*- and the *Y*-axes, respectively. From Fig. 9.2(a), the displacement of the nucleus *Ip* on this deformation is given by

$$\Delta X_{Ip} = X_{Ip}^0\, \sigma_{XX}, \qquad \Delta Y_{Ip} = Y_{Ip}^0\, \sigma_{YY}.$$

On the other hand, the nucleus *Ip* is displaced as shown in Fig. 9.2(b) by such sheer strains

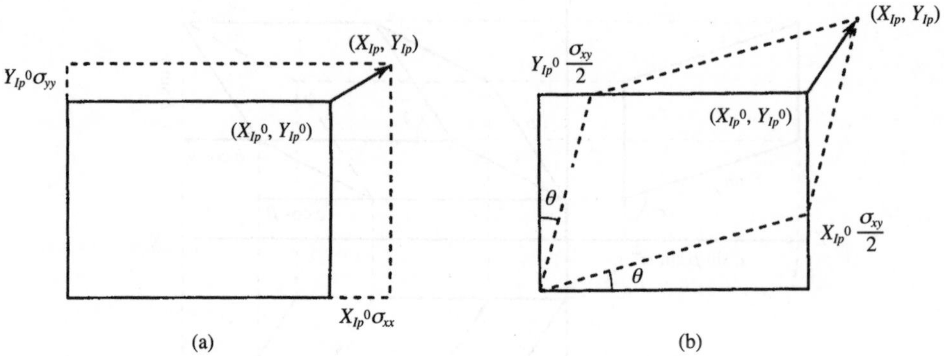

Fig. 9.2 Change in Cartesian coordinate X_{Ip} and deformation parameter σ on homogeneous deformation. (a) expansion, (b) sheer strain.

along the positive directions of the X- and the Y-axes as to give rise to an anticlockwise rotation of the X-axis and a clockwise rotation of the Y-axis by the same amount of angle, $\theta = \arctan \sigma_{XY}/2$. The nuclear displacement in this case is given by

$$\Delta X_{Ip} = Y_{Ip}^0 \, \sigma_{XY}/2, \qquad \Delta Y_{Ip} = X_{Ip}^0 \, \sigma_{XY}/2..$$

The nuclear displacement of each nucleus caused by an arbitrary macroscopic deformation within the XY-plane is the sum of the above two types of displacements, given in the form

$$\Delta \boldsymbol{r}_{Ip} = \begin{bmatrix} \Delta X_{Ip} \\ \Delta Y_{Ip} \end{bmatrix} = \begin{bmatrix} X_{Ip}^0 & 0 & Y_{Ip}^0/2 \\ 0 & Y_{Ip}^0 & X_{Ip}^0/2 \end{bmatrix} \begin{bmatrix} \sigma_{XX} \\ \sigma_{YY} \\ \sigma_{XY} \end{bmatrix}. \tag{9-12}$$

The nuclear displacements in the YZ- and the ZX-planes can be expressed in similar ways by introducing the parameters σ_{ZZ}, σ_{YZ}, σ_{ZX}. Any macroscopic deformation of crystals in the three-dimensional space is then given in terms of the six parameters collected into a vector

$$\boldsymbol{\sigma} = \begin{bmatrix} \sigma_{XX} & \sigma_{YY} & \sigma_{ZZ} & \sigma_{YZ} & \sigma_{ZX} & \sigma_{XY} \end{bmatrix}^{\mathrm{T}}. \tag{9-13}$$

The (2×3) transformation matrix in Eq. (9-12) is now extended into a (3×6) matrix of the form

$$\boldsymbol{D}_{Ip} = \begin{bmatrix} X_{Ip}^0 & 0 & 0 & 0 & Z_{Ip}^0/2 & Y_{Ip}^0/2 \\ 0 & Y_{Ip}^0 & 0 & Z_{Ip}^0/2 & 0 & X_{Ip}^0/2 \\ 0 & 0 & Z_{Ip}^0 & Y_{Ip}^0/2 & X_{Ip}^0/2 & 0 \end{bmatrix}. \tag{9-14}$$

The displacement of the nucleus Ip on the homogeneous macroscopic deformation $\boldsymbol{\sigma}$ is then given by

$$\Delta \boldsymbol{r}_{Ip} = \boldsymbol{D}_{Ip} \, \boldsymbol{\sigma}. \tag{9-15a}$$

Next let us take account of various forces acting between nuclei, and deal with the case where relative coordinates of the nuclei are changed on any macroscopic deformation of the crystal in such a way as to minimize the potential energy of the system. Let the displacement of the nucleus *Ii* on this reorientation of nuclei be denoted in terms of the Cartesian displacement coordinates by

$$\Delta \boldsymbol{r}_{Ip}^{\rho} = \left[\Delta X_{Ip}^{\rho} \quad \Delta Y_{Ip}^{\rho} \quad \Delta Z_{Ip}^{\rho}\right]^{\mathrm{T}}. \tag{9-15b}$$

Adding Eqs. (9-15a) and (9-15b) together, we obtain the total displacement of the nucleus *Ip* associated with the macroscopic deformation $\boldsymbol{\sigma}$ in the form

$$\Delta \boldsymbol{r}_{Ip} = \boldsymbol{D}_{Ip}\, \boldsymbol{\sigma} + \Delta \boldsymbol{r}_{Ip}{}^{\rho}. \tag{9-16}$$

The first term in the right side of Eq. (9-16) represents the nuclear displacement following the macroscopic deformation of the crystal, while the second term arises from the microscopic relaxation of the nuclei by which the internal strain energy due to the macroscopic deformation is minimized.

Now we introduce a matrix $\boldsymbol{D}_{Ip,Jq}$ obtained by subtracting Eq. (9-14) from its counterpart for the nucleus *Jq* in the form

$$\boldsymbol{D}_{Ip,Jq} = \boldsymbol{D}_{Jq} - \boldsymbol{D}_{Ip}. \tag{9-17}$$

The change in the distance vector from *Ip* to *Jq* caused by the macroscopic deformation $\boldsymbol{\sigma}$ is then given by

$$\Delta \boldsymbol{r}_{Ip,Jq} = \Delta \boldsymbol{r}_{Jq} - \Delta \boldsymbol{r}_{Ip} = \boldsymbol{D}_{Ip,Jq}\, \boldsymbol{\sigma} + \Delta \boldsymbol{r}_{Jq}{}^{\rho} - \Delta \boldsymbol{r}_{Ip}{}^{\rho}. \tag{9-18}$$

From the translational symmetry of the crystal, the matrix $\boldsymbol{D}_{Ip,Jq}$ is imposed a condition in the form

$$\boldsymbol{D}_{Ip,Jq} = \boldsymbol{D}_{Lq} - \boldsymbol{D}_{1p} = \boldsymbol{D}_{1p,Lq} \tag{9-19}$$

for the unit cells 1 and *L* so chosen as to satisfy Eq. (9-8). Now we collect the microscopic displacements of all the nuclei in the unit cell *I* in a vector in the form

$$\Delta \boldsymbol{x}_{I}^{\rho} = \left[\Delta \tilde{\boldsymbol{r}}_{I1}^{\rho} \quad \Delta \tilde{\boldsymbol{r}}_{I2}^{\rho} \quad \cdots \quad \Delta \tilde{\boldsymbol{r}}_{In}^{\rho}\right]^{\mathrm{T}}. \tag{9-20}$$

The microscopic part of Eq. (9-18) is then given by

$$\Delta \boldsymbol{r}_{Jq}^{\rho} - \Delta \boldsymbol{r}_{Ip}^{\rho} = \tilde{\boldsymbol{E}}_{q}\, \Delta \boldsymbol{x}_{J}^{\rho} - \tilde{\boldsymbol{E}}_{p}\, \Delta \boldsymbol{x}_{I}^{\rho}. \tag{9-21}$$

where $\boldsymbol{E}_p$ is a $(3n \times 3)$ matrix which consists of a columnwise array of n (3×3) blocks, and is written in terms of the (3×3) unit matrix $\boldsymbol{E}$ and Kronecker's δ in the form

$$\boldsymbol{E}_{p} = \left[\delta_{1p}\, \boldsymbol{E} \quad \delta_{2p}\, \boldsymbol{E} \quad \cdots \quad \delta_{np}\, \boldsymbol{E}\right]^{\mathrm{T}}.$$

Since the translational symmetry of the crystal is maintained under any homogeneous deformation, it follows that

$$\Delta \boldsymbol{x}^{\rho} \equiv \Delta \boldsymbol{x}_{1}^{\rho} = \Delta \boldsymbol{x}_{2}^{\rho} = \cdots = \Delta \boldsymbol{x}_{N}^{\rho}. \tag{9-22}$$

Substituting Eq. (9-22) into Eq. (9-21), and introducing an abbreviated notation, $\boldsymbol{E}_{pq} = \boldsymbol{E}_q - \boldsymbol{E}_p$, we obtain

$$\tilde{\boldsymbol{E}}_{q}\, \Delta \boldsymbol{x}_{J}^{\rho} - \tilde{\boldsymbol{E}}_{p}\, \Delta \boldsymbol{x}_{I}^{\rho} = \left(\tilde{\boldsymbol{E}}_{q} - \tilde{\boldsymbol{E}}_{p}\right) \Delta \boldsymbol{x}^{\rho} = \tilde{\boldsymbol{E}}_{pq}\, \Delta \boldsymbol{x}^{\rho}. \tag{9-23}$$

By using Eq. (9-23), Eq. (9-18) is simplified as

$$\Delta \boldsymbol{r}_{Ip,Jq} = \boldsymbol{D}_{Ip,Jq}\boldsymbol{\sigma} + \tilde{\boldsymbol{E}}_{pq}\,\Delta \boldsymbol{x}^{\rho}. \tag{9-24}$$

Let the nonbonded atom–atom interaction potential between the nuclei Ip and Jq be denoted by $V_N(r_{Ip,Jq})$, and define the differential operator with respect to the components of the distance vector in the form

$$\boldsymbol{\nabla}_{Ip,Jq} = \left[\partial/\partial X_{Ip,Jq} \quad \partial/\partial Y_{Ip,Jq} \quad \partial/\partial Z_{Ip,Jq}\right]^{\mathrm{T}} = \boldsymbol{\nabla}_{Jq} - \boldsymbol{\nabla}_{Ip}\,. \tag{9-25}$$

Then the gradient and the (3×3) Hessian matrix of $V_N(r_{Ip,Jq})$ with respect to the components of $\Delta \boldsymbol{r}_{Ip,Jq}$ are given by

$$\boldsymbol{F}_N^{x}\left(r_{Ip,Jq}\right) = \left\{\partial V_N\left(r_{Ip,Jq}\right)/\partial r_{Ip,Jq}\right\}_0 (\boldsymbol{\nabla}_{Ip,Jq}\, r_{Ip,Jq})_0 \tag{9-26}$$

and

$$\boldsymbol{F}_N^{xx}\left(r_{Ip,Jq}\right) = \left\{\partial V_N\left(r_{Ip,Jq}\right)/\partial r_{Ip,Jq}\right\}_0 (\boldsymbol{\nabla}_{Ip,Jq}\, \tilde{\boldsymbol{\nabla}}_{Ip,Jq}\, r_{Ip,Jq})_0$$

$$+ \left\{\partial^2 V_N\left(r_{Ip,Jq}\right)/\partial r_{Ip,Jq}{}^2\right\}_0 (\boldsymbol{\nabla}_{Ip,Jq}\, r_{Ip,Jq})_0 (\tilde{\boldsymbol{\nabla}}_{Ip,Jq}\, r_{Ip,Jq})_0. \tag{9-27}$$

By using these coefficients, the potential energy $V_N(r_{Ip,Jq})$ is expanded as a power series of the components of $\Delta \boldsymbol{r}_{Ip,Jq}$ in the form

$$V_N\left(r_{Ip,Jq}\right) = \tilde{\boldsymbol{F}}_N^{x}\left(r_{Ip,Jq}\right)\Delta \boldsymbol{r}_{Ip,Jq} + \tfrac{1}{2}\,\Delta \tilde{\boldsymbol{r}}_{Ip,Jq}\,\boldsymbol{F}_N^{xx}\left(r_{Ip,Jq}\right)\Delta \boldsymbol{r}_{Ip,Jq}. \tag{9-28}$$

Under the condition of the homogeneous deformation, the energy of the nonbonded atom–atom interaction per unit cell is given by

$$\begin{aligned} V_N = \tfrac{1}{2}\sum_L \sum_p \sum_q \Big\{ & V_N^0\left(r_{1p,Lq}^0\right) + \tilde{\boldsymbol{F}}_N^{x}\left(r_{1p,Lq}\right)\left(\boldsymbol{D}_{1p,Lq}\boldsymbol{\sigma} + \tilde{\boldsymbol{E}}_{pq}\,\Delta \boldsymbol{x}^{\rho}\right) \\ & + \tfrac{1}{2}\left(\tilde{\boldsymbol{\sigma}}\,\tilde{\boldsymbol{D}}_{1p,Lq} + \Delta \tilde{\boldsymbol{x}}^{\rho}\,\boldsymbol{E}_{pq}\right)\boldsymbol{F}_N^{xx}\left(r_{Ip,Jq}\right)\left(\boldsymbol{D}_{1p,Lq}\boldsymbol{\sigma} + \tilde{\boldsymbol{E}}_{pq}\,\Delta \boldsymbol{x}^{\rho}\right)\Big\}. \end{aligned} \tag{9-29}$$

The sum with respect to L is taken over such unit cells that $|\boldsymbol{r}_L| \le \boldsymbol{r}_{\max}$, and with respect to both p and q from 1 to n. When $L = 1$, that is, when p and q belong to the same unit cell, the nuclei j belonging to the same molecule as p are omitted from the sum. From the linear terms in Eq. (9-29), we define the vectors

$$\boldsymbol{F}_{\mathrm{ex}}^{\sigma} = \sum_L \sum_{p=1}^{n-\delta_{1L}} \sum_{q=p+\delta_{1L}}^{n} (1+\delta_{pq})^{-1}\,\tilde{\boldsymbol{D}}_{1p,Lq}\,\boldsymbol{F}_N^{x}\left(r_{1p,Lq}\right) \tag{9-30a}$$

and

$$\boldsymbol{F}_{\mathrm{ex}}^{x} = \sum_L \sum_{p=1}^{n-\delta_{1L}} \sum_{q=p+\delta_{1L}}^{n} (1+\delta_{pq})^{-1}\,\boldsymbol{E}_{pq}\,\boldsymbol{F}_N^{x}\left(r_{1p,Lq}\right). \tag{9-30b}$$

Similarly, the quadratic terms in Eq. (9-29) are collected into the matrices in the forms

$$\boldsymbol{F}_{\mathrm{ex}}^{\sigma\sigma} = \sum_L \sum_{p=1}^{n-\delta_{1L}} \sum_{q=p+\delta_{1L}}^{n} (1+\delta_{pq})^{-1}\,\tilde{\boldsymbol{D}}_{1p,Lq}\,\boldsymbol{F}_N^{xx}\left(r_{1p,Lq}\right)\boldsymbol{D}_{1p,Lq}, \tag{9-31a}$$

$$\boldsymbol{F}_{\mathrm{ex}}^{x\sigma} = \sum_L \sum_{p=1}^{n-\delta_{1L}} \sum_{q=p+\delta_{1L}}^{n} (1+\delta_{pq})^{-1}\,\boldsymbol{E}_{pq}\,\boldsymbol{F}_N^{xx}\left(r_{1p,Lq}\right)\boldsymbol{D}_{1p,Lq} \tag{9-31b}$$

and

$$\boldsymbol{F}_{\text{ex}}^{xx} = \sum_{L} \sum_{p=1}^{n-\delta_{1L}} \sum_{q=p+\delta_{1L}}^{n} (1+\delta_{pq})^{-1} \boldsymbol{E}_{pq} \, \boldsymbol{F}_{N}^{xx}\left(r_{1p,Lq}\right) \tilde{\boldsymbol{E}}_{pq}. \tag{9-31c}$$

Substituting Eqs. (9-30a,b) and (9-31a–c) into Eq. (9-29), we obtain

$$V_{\text{ex}} = V_{\text{ex}}^{0} + \tilde{\boldsymbol{\sigma}} \, \boldsymbol{F}_{\text{ex}}^{\sigma} + \Delta\tilde{\boldsymbol{x}}^{\rho} \, \boldsymbol{F}_{\text{ex}}^{x} + \tfrac{1}{2} \tilde{\boldsymbol{\sigma}} \, \boldsymbol{F}_{\text{ex}}^{\sigma\sigma} \boldsymbol{\sigma} + \Delta\tilde{\boldsymbol{x}}^{\rho} \, \boldsymbol{F}_{\text{ex}}^{x\sigma} \boldsymbol{\sigma} + \tfrac{1}{2} \Delta\tilde{\boldsymbol{x}}^{\rho} \, \boldsymbol{F}_{\text{ex}}^{xx} \, \Delta\boldsymbol{x}^{\rho} \tag{9-32}$$

where

$$V_{\text{ex}}^{0} = \sum_{L} \sum_{p=1}^{n-\delta_{1L}} \sum_{q=p+\delta_{1L}}^{n} (1+\delta_{pq})^{-1} V_{N}\left(r_{1p,Lq}^{0}\right).$$

To express the intramolecular part of the potential energy as a function of $\boldsymbol{\sigma}$ and $\Delta\boldsymbol{x}^{\rho}$, we construct a matrix $\boldsymbol{D}_{I0}$ by arranging the coordinates of the origin of the unit cell I as given by Eq. (9-14), and define a (3×6) matrix $\boldsymbol{D}_{I0,Ip} = \boldsymbol{D}_{Ip} - \boldsymbol{D}_{I0}$ in analogy with the case of Eq. (9-17). The matrices $\boldsymbol{D}_{I0,Ip}$ $(p = 1, 2, \cdots, n)$ are then collected columnwise into a $(3n \times 6)$ matrix $\boldsymbol{D}$ in the form

$$\boldsymbol{D} = \left[\tilde{\boldsymbol{D}}_{I0,I1} \quad \tilde{\boldsymbol{D}}_{I0,I2} \quad \cdots \quad \tilde{\boldsymbol{D}}_{I0,In}\right]^{\mathrm{T}}. \tag{9-33}$$

The suffix I has been dropped from the left side of Eq. (9-33), since $\boldsymbol{D}_{I0,Ip}$ does not depend on the position of the unit cell I by virtue of the condition of homogeneous deformation. Now the Cartesian displacements of all the nuclei in a unit cell due to both the macroscopic and microscopic deformations are given by

$$\Delta\boldsymbol{x}_{I} = \boldsymbol{D}\boldsymbol{\sigma} + \Delta\boldsymbol{x}^{\rho}. \tag{9-34}$$

Substituting Eq. (9-34) into the intramolecular part of the potential energy, V_{in}, expanded in a similar way as Eq. (9-6), and introducing new notations in the forms

$$\boldsymbol{F}_{\text{in}}^{\sigma} = \tilde{\boldsymbol{D}} \boldsymbol{F}_{\text{in}}^{x}, \quad \boldsymbol{F}_{\text{in}}^{x\sigma} = \boldsymbol{F}_{\text{in}}^{xx} \boldsymbol{D} \quad \text{and} \quad \boldsymbol{F}_{\text{in}}^{\sigma\sigma} = \tilde{\boldsymbol{D}} \boldsymbol{F}_{\text{in}}^{xx} \boldsymbol{D},$$

we obtain

$$V_{\text{in}} = V_{\text{in}}^{0} + \tilde{\boldsymbol{\sigma}} \, \boldsymbol{F}_{\text{in}}^{\sigma} + \Delta\tilde{\boldsymbol{x}}^{\rho} \, \boldsymbol{F}_{\text{in}}^{x} + \tfrac{1}{2} \tilde{\boldsymbol{\sigma}} \, \boldsymbol{F}_{\text{in}}^{\sigma\sigma} \boldsymbol{\sigma} + \Delta\tilde{\boldsymbol{x}}^{\rho} \, \boldsymbol{F}_{\text{in}}^{x\sigma} \boldsymbol{\sigma} + \tfrac{1}{2} \Delta\tilde{\boldsymbol{x}}^{\rho} \, \boldsymbol{F}_{\text{in}}^{xx} \, \Delta\boldsymbol{x}^{\rho}. \tag{9-35}$$

By combining Eqs. (9-32) and (9-35) and writing as

$$V^{0} = V_{\text{in}}^{0} + V_{\text{ex}}^{0}, \quad \boldsymbol{F}^{\sigma} = \boldsymbol{F}_{\text{in}}^{\sigma} + \boldsymbol{F}_{\text{ex}}^{\sigma}, \quad \boldsymbol{F}^{x\sigma} = \boldsymbol{F}_{\text{in}}^{x\sigma} + \boldsymbol{F}_{\text{ex}}^{x\sigma},$$

$$\boldsymbol{F}^{\sigma\sigma} = \boldsymbol{F}_{\text{in}}^{\sigma\sigma} + \boldsymbol{F}_{\text{ex}}^{\sigma\sigma}, \quad \text{and} \quad \boldsymbol{F}^{xx} = \boldsymbol{F}_{\text{in}}^{xx} + \boldsymbol{F}_{\text{ex}}^{xx},$$

the potential energy per unit cell of the crystal under a homogeneous deformation is given in the form

$$V = V^{0} + \tilde{\boldsymbol{\sigma}} \, \boldsymbol{F}^{\sigma} + \Delta\tilde{\boldsymbol{x}}^{\rho} \, \boldsymbol{F}^{x} + \tfrac{1}{2} \tilde{\boldsymbol{\sigma}} \, \boldsymbol{F}^{\sigma\sigma} \boldsymbol{\sigma} + \Delta\tilde{\boldsymbol{x}}^{\rho} \, \boldsymbol{F}^{x\sigma} \boldsymbol{\sigma} + \tfrac{1}{2} \Delta\tilde{\boldsymbol{x}}^{\rho} \, \boldsymbol{F}^{xx} \, \Delta\boldsymbol{x}^{\rho}. \tag{9-36}$$

In analogy with the case of an isolated molecule, we can define the normal coordinates for the homogeneous deformation of a crystal as the variables in terms of which the potential energy matrix can be expressed in a diagonal form $\boldsymbol{\Lambda}\cdot$ If an adequate intermolecular potential is used, the matrix $\boldsymbol{\Lambda}\cdot$ should have three zero elements on the diagonal corresponding to the translational degrees of freedom of the whole crystal. Once the matrix $\boldsymbol{\Lambda}\cdot$ is obtained by diagonalizing $\boldsymbol{F}^{xx}$ in the form

$$\boldsymbol{\Lambda} = \tilde{\boldsymbol{L}}^{x} \boldsymbol{F}^{xx} \boldsymbol{L}^{x}, \tag{9-37}$$

the microscopic displacements of the nuclei on a homogeneous deformation can be expressed in terms of a vector of the normal coordinates, $\boldsymbol{Q}^{\rho}$, as

$$\Delta \boldsymbol{x}^{\rho} = \boldsymbol{L}^{x} \boldsymbol{Q}^{\rho}. \tag{9-38}$$

On substituting Eq. (9-38) into Eq. (9-36) and differentiating with respect to each element of $\boldsymbol{Q}^{\rho}$, the resulting derivative $\partial V/\partial Q_i^{\rho}$ must vanish since the microscopic nuclear displacements in this case arises in such a way as to minimize the stress due to the homogeneous macroscopic strain $\boldsymbol{\sigma}$. From the simultaneous equations thus obtained, it follows that

$$\boldsymbol{Q}^{\rho} = -\boldsymbol{\Lambda}^{-1} \tilde{\boldsymbol{L}}^{x} (\boldsymbol{F}^{x} + \boldsymbol{F}^{x\sigma} \boldsymbol{\sigma}). \tag{9-39}$$

In Eq. (9-39), the matrix $\boldsymbol{\Lambda}^{-1}$ must be constructed from the subblock of $\boldsymbol{\Lambda}$ containing non-zero diagonal elements, and the rows of $\boldsymbol{L}^{x}$ corresponding to the vanishing elements of $\boldsymbol{\Lambda}$ must be excluded from the calculation of $\boldsymbol{\Lambda}^{-1}\tilde{\boldsymbol{L}}^{x}$. The situation is analogous to the case of Eq. (3-87) for isolated molecules. Substituting Eqs. (9-38) and (9-39) into Eq. (9-36) to eliminate $\Delta \boldsymbol{x}^{\rho}$ leads to

$$V = V^0 + \tilde{\boldsymbol{\sigma}} \left(\boldsymbol{F}^{\sigma} - \tilde{\boldsymbol{F}}^{x\sigma} \boldsymbol{L}^{x} \boldsymbol{\Lambda}^{-1} \tilde{\boldsymbol{L}}^{x} \right) + \tfrac{1}{2} \tilde{\boldsymbol{\sigma}} \left(\boldsymbol{F}^{\sigma\sigma} - \tilde{\boldsymbol{F}}^{x\sigma} \boldsymbol{L}^{x} \boldsymbol{\Lambda}^{-1} \tilde{\boldsymbol{L}}^{x} \boldsymbol{F}^{x\sigma} \right) \boldsymbol{\sigma} \tag{9-40}$$

Since Eq. (9-40) is the crystal energy expanded as a power series of the strain components in a homogeneous deformation, it should contain no linear terms in $\boldsymbol{\sigma}$ at the configuration point corresponding to a macroscopic equilibrium structure of the crystal. In order to eliminate these terms by the Newton–Raphson method, the derivatives of Eq. (9-40) with respect to the components of $\boldsymbol{\sigma}$ are equated to zero, and the resulting simultaneous equations are solved to give

$$\boldsymbol{\sigma}' = \left(\boldsymbol{F}^{\sigma\sigma} - \tilde{\boldsymbol{F}}^{x\sigma} \boldsymbol{L}^{x} \boldsymbol{\Lambda}^{-1} \tilde{\boldsymbol{L}}^{x} \boldsymbol{F}^{x\sigma} \right)^{-1} \left(\boldsymbol{F}^{\sigma} - \tilde{\boldsymbol{F}}^{x\sigma} \boldsymbol{L}^{x} \boldsymbol{\Lambda}^{-1} \tilde{\boldsymbol{L}}^{x} \right). \tag{9-41}$$

Equation (9-41) respresents the correction required to achieve the macroscopic equilibrium structure of the crystal given in terms of the strain components. The microscopic correction to the equilibrium may be calculated in terms of the nuclear Cartesian displacements by substituting $\boldsymbol{\Lambda}$, $\boldsymbol{L}^{x}$, $\boldsymbol{F}^{x}$, $\boldsymbol{F}^{x\sigma}$ and $\boldsymbol{\sigma}'$ into Eq. (9-39) and the resulting $\boldsymbol{Q}^{\rho}$ into Eq. (9-38). Alternatively, the corrections in terms of the vector $\Delta \boldsymbol{R}$ defined by Eq. (3-36) are calculated by

$$\Delta \boldsymbol{R}^{\rho} = \boldsymbol{B} \Delta \boldsymbol{x}^{\rho} = -\boldsymbol{B} \boldsymbol{L}^{x} \boldsymbol{\Lambda}^{-1} \tilde{\boldsymbol{L}}^{x} (\boldsymbol{F}^{x} + \boldsymbol{F}^{x\sigma} \boldsymbol{\sigma}),$$

where the matrix $\boldsymbol{L}^{x}\boldsymbol{\Lambda}^{-1}\tilde{\boldsymbol{L}}^{x}$ is taken to correspond to the positive subblock of $\boldsymbol{\Lambda}$. The microscopic structure of the revised crystal is then determined from the components of $\Delta \boldsymbol{R}^{\rho}$, by shifting and rotating the origin and the principal axes, respectively, of molecules in the asymmetric unit as indicated by the external variables, and generating the nuclear Cartesian coordinates successively by using the $\boldsymbol{Z}$ matrix and the internal variables as described in Chapter 3. The latter method is more precise than the former since the curvilinearity of the transformation from external and internal variables into nuclear Cartesian displacements is implicitly taken into account.

On searching the equilibrium structure of a crystal for a given model potential function, the correction in the macroscopic structure at each step is more intuitively expressed in terms of increments to the cell constants than in terms of components of $\boldsymbol{\sigma}'$. We therefore introduce the correction vector for the cell constants in the form

$$\Delta \boldsymbol{a}' = \begin{bmatrix} \Delta a & \Delta b & \Delta c & \Delta \alpha & \Delta \beta & \Delta \gamma \end{bmatrix}^{\mathrm{T}}, \tag{9-42}$$

and derive the transformation relation between $\boldsymbol{\sigma}'$ and $\Delta \boldsymbol{a}'$ as follows.[3] It is shown in Appendix 3 that the inverse of the matrix $\boldsymbol{C}_r$ in Eq. (9-10) in the strain free crystal is given by

$$\boldsymbol{C}_{r0}^{-1} = \begin{bmatrix} \dfrac{1}{a_0} & \dfrac{-\cos\gamma_0}{a_0 \sin\gamma_0} & \dfrac{\cos\beta_0^*}{a_0 \sin\gamma_0 \sin\beta_0^*} \\ 0 & \dfrac{1}{b_0 \sin\gamma_0} & \dfrac{\cos\alpha_0^*}{b_0 \sin\gamma_0 \sin\alpha_0^*} \\ 0 & 0 & \dfrac{1}{c_0 \sin\beta_0 \sin\alpha_0^*} \end{bmatrix}. \tag{9-43}$$

In the absence of any microscopic nuclear displacement $\Delta \boldsymbol{x}^p$, the position vector of the nucleus Ip in the deformed crystal is given in terms of the same vector in the undeformed crystal by

$$\boldsymbol{r}_{Ip} = \boldsymbol{C}_r \boldsymbol{r}_{Ip}^c = \boldsymbol{C}_r \boldsymbol{r}_{Ip}^{c0} = \left(\boldsymbol{C}_r \boldsymbol{C}_{r0}^{-1}\right) \boldsymbol{r}_{Ip}^0. \tag{9-44}$$

The second equality in Eq. (9-44) follows from the relation $\boldsymbol{r}_{Ip}^c = \boldsymbol{r}_{Ip}^{c0}$ stating that relative coordinates are invariant under a purely macroscopic deformation for which $\Delta \boldsymbol{x}^p = \boldsymbol{0}$. The last equality is due to Eq. (A3-3). It follows from the definition that the (α, β) element of the matrix $\boldsymbol{C}_r \boldsymbol{C}_{r0}^{-1}$ is given by

$$(\boldsymbol{C}_r \boldsymbol{C}_{r0}^{-1})_{\alpha\beta} = \delta_{\alpha\beta} + \sigma_{\alpha\beta}, \tag{9-45}$$

where α and β represent any of X, Y and Z axes. Accordingly, by expanding each element of $\boldsymbol{C}_r \boldsymbol{C}_{r0}^{-1}$ as a power series of the elements of $\Delta \boldsymbol{a}'$ and taking the linear terms, the matrix elements of the transformation from $\Delta \boldsymbol{a}'$ to $\boldsymbol{\sigma}'$,

$$\boldsymbol{\sigma}' = \boldsymbol{A} \Delta \boldsymbol{a}' \tag{9-46}$$

are obtained. Since both $\boldsymbol{C}_r$ and $\boldsymbol{C}_{r0}^{-1}$ are upper triangular, the product $\boldsymbol{C}_r \boldsymbol{C}_{r0}^{-1}$ is also upper triangular. Its six elements are related to the six independent components $\sigma_{\alpha\beta}$ of the strain tensor in the form[4]

$$\left(\boldsymbol{C}_r \boldsymbol{C}_{r0}^{-1}\right)_{11} = \frac{a}{a_0} = 1 + \sigma_{XX} \tag{9-47a}$$

$$\left(\boldsymbol{C}_r \boldsymbol{C}_{r0}^{-1}\right)_{12} = -\frac{a\cos\gamma_0}{a_0 \sin\gamma_0} + \frac{b\cos\gamma}{b_0 \sin\gamma_0} = \sigma_{XY} \tag{9-47b}$$

$$\left(\boldsymbol{C}_r \boldsymbol{C}_{r0}^{-1}\right)_{13} = \frac{a\cos\beta_0^*}{a_0 \sin\gamma_0 \sin\beta_0^*} + \frac{b\cos\gamma\cos\alpha_0^*}{b_0 \sin\gamma_0 \sin\alpha_0^*} + \frac{c\cos\beta}{c_0 \sin\beta_0 \sin\alpha_0^*} = \sigma_{XZ} \tag{9-47c}$$

$$\left(\boldsymbol{C}_r\boldsymbol{C}_{r0}^{-1}\right)_{22} = \frac{b\sin\gamma}{b_0\sin\gamma_0} = 1+\sigma_{YY} \tag{9-47d}$$

$$\left(\boldsymbol{C}_r\boldsymbol{C}_{r0}^{-1}\right)_{23} = \frac{b\sin\gamma\cos\alpha_0^*}{b_0\sin\gamma_0\sin\alpha_0^*} - \frac{c\sin\beta\cos\alpha^*}{c_0\sin\beta_0\sin\alpha_0^*} = \sigma_{YZ} \tag{9-47e}$$

$$\left(\boldsymbol{C}_r\boldsymbol{C}_{r0}^{-1}\right)_{33} = \frac{c\sin\beta\sin\alpha^*}{c_0\sin\beta_0\sin\alpha_0^*} = 1+\sigma_{ZZ}. \tag{9-47f}$$

By differentiating each of Eqs. (9-47a–f) with respect to the six cell constants, we obtain the coefficients of the transformation from the components of the vector $\boldsymbol{\Delta a}'$ into the strain components up to the first order in the forms

$$\sigma_{XX} = \Delta a/a_0 \tag{9-48a}$$

$$\sigma_{YY} = \frac{\Delta b}{b_0} + \cot\gamma_0\,\Delta\gamma \tag{9-48b}$$

$$\sigma_{ZZ} = \frac{\Delta c}{c_0} - \frac{\cot\alpha_0^*\,\Delta\alpha}{\sin\gamma_0\,\sin\beta_0^*} - \frac{\cot\beta_0^*\,\Delta\beta}{\sin\gamma_0\,\sin\alpha_0^*} - \frac{\cot\alpha_0\,\cot\beta_0\,\Delta\gamma}{\sin\gamma_0} \tag{9-48c}$$

$$\begin{aligned}\sigma_{YZ} = {} & \frac{\cot\alpha_0^*\,\Delta b}{b_0} - \frac{\cot\alpha_0^*\,\Delta c}{c_0} - \frac{\Delta\alpha}{\sin\gamma_0\,\sin\beta_0^*} + \frac{\cot\gamma_0\,\Delta\beta}{\sin\alpha_0^*} \\ & + \frac{\cos\gamma_0\,\cot\alpha_0^* - \cot\beta_0^*}{\sin\gamma_0}\Delta\gamma\end{aligned} \tag{9-48d}$$

$$\begin{aligned}\sigma_{ZX} = {} & \frac{\cot\beta_0^*\,\Delta a}{a_0\,\sin\gamma_0} + \frac{\cot\gamma_0\,\cot\alpha_0^*\,\Delta b}{b_0} + \frac{\cot\beta_0\,\Delta\alpha}{c_0\,\sin\alpha_0^*} \\ & - \frac{\Delta\beta}{\sin\alpha_0^*} - \cot\alpha_0^*\,\Delta\gamma\end{aligned} \tag{9-48e}$$

$$\sigma_{XY} = -\frac{\cot\gamma_0\,\Delta a}{a_0} + \frac{\cot\gamma_0\,\Delta b}{b_0} - \Delta\gamma. \tag{9-48f}$$

The matrix $\boldsymbol{A}$ in Eq. (9-46) is formed by arranging the coefficients of Δa, Δb, Δc, $\Delta\alpha$, $\Delta\beta$ and $\Delta\gamma$ in each of Eqs. (9-48a–f) rowwise. The result can be written as $\boldsymbol{A} = \boldsymbol{A}_1\boldsymbol{A}_2$. The matrix $\boldsymbol{A}_1$ is a (6×6) asymmetric matrix in the form

$$\boldsymbol{A}_1 = \begin{bmatrix} 1 & 0 & 0 & 0 & 0 & 0 \\ 0 & 1 & 0 & 0 & 0 & c \\ 0 & 0 & 1 & -r & -p & -pr \\ 0 & r & -r & -1 & c & cr-p \\ p & cr & b/s & 0 & -1 & -r \\ -c & c & 0 & 0 & 0 & -1 \end{bmatrix}, \tag{9-49a}$$

where $c = \cot \gamma_0$, $r = \cot \alpha_0^*$, $p = \cot \beta_0^*/\sin \gamma_0$, $b = \cot \beta_0$ and $s = \sin \alpha_0^*$. The matrix $\boldsymbol{A}_2$ is a diagonal matrix, the inverse of which is given by

$$\boldsymbol{A}_2^{-1} = \mathrm{diag}[a_0 \quad b_0 \quad c_0 \quad \sin\gamma_0 \sin\beta_0^* \quad \sin\alpha_0^* \quad 1]. \tag{9-49b}$$

The vector of cell constant corrections is then calculated by

$$\Delta\boldsymbol{a}' = \boldsymbol{A}^{-1}\boldsymbol{\sigma}' = \boldsymbol{A}_2^{-1}\boldsymbol{A}_1^{-1}\boldsymbol{\sigma}'. \tag{9-50}$$

The inverse of the matrix $\boldsymbol{A}_1$ can be analytically expressed in the form

$$\boldsymbol{A}_1^{-1} = \sin^2\gamma_0 \begin{bmatrix} 1+c^2 & 0 & 0 & 0 & 0 & 0 \\ c^2 & 1 & 0 & 0 & 0 & c \\ 0 & 0 & 0 & 0 & 0 & 0 \\ pc & -pc & 0 & 0 & 0 & p \\ 0 & 0 & 0 & 0 & 0 & 0 \\ -c & c & 0 & 0 & 0 & -1 \end{bmatrix}$$

$$+\sin^2\beta_0 \begin{bmatrix} 0 & 0 & 0 & 0 & 0 & 0 \\ 0 & 0 & 0 & 0 & 0 & 0 \\ b^2 & a^2 & s^2 & -as & bs & -ab \\ bq & a(s-bp) & ws & (bp-s)s & qs & -aq \\ -b/s & arb & bs & -ab & -1 & r \\ 0 & 0 & 0 & 0 & 0 & 0 \end{bmatrix},$$

where

$$q = (pr - c)s = \cot\gamma_0^*/\sin\beta_0, \quad a = rs = \cos\alpha_0^*$$

and

$$w = bc - rs = \cos\alpha_0/(\sin\beta_0 \sin\gamma_0).$$

On correcting the macroscopic structure, the nuclear Cartesian coordinates should also be changed by such amounts as to relax the stress due to the changes in the cell constants. These changes are calculated from $\boldsymbol{Q}^p$ in Eq. (9-39) by the transformation

$$\Delta\boldsymbol{x}^p = \boldsymbol{L}^x\boldsymbol{Q}^p \tag{9-51}$$

Adding the elements of the left sides of Eqs. (9-50) and (9-51) to the cell constants and the nuclear Cartesian coordinates of the initial structure, respectively, we obtain the crystal structure in the next cycle of iteration. On repeating the iteration until all the linear terms in the expansion in Eq. (9-40) vanish practically, the crystal structure optimized with respect to all the degrees of freedom of the asymmetric unit is obtained.

The matrix of the coefficients of the quadratic terms in Eq. (9-40) for this structure, *i.e.*, the Hessian matrix with respect to the elements of $\boldsymbol{\sigma}$, divided by the volume of the unit cell V gives the elastic tensor $\boldsymbol{C}$ of the crystal in the absence of any thermal motion of the nuclei,[2)]

$$\boldsymbol{C} = \left(\boldsymbol{F}^{\sigma\sigma} - \tilde{\boldsymbol{F}}^{x\sigma}\boldsymbol{L}^x\boldsymbol{\Lambda}^{-1}\tilde{\boldsymbol{L}}^x\boldsymbol{F}^{x\sigma}\right)/V. \tag{9-52}$$

Table 9.1 Elastic constants and structure parameters of aspirin crystal.

Elastic constants	calculated	observed	Structure parameters	unit	calculated	observed
C_{11}	10.18	10.76	Lattice constants			
C_{22}	8.14	12.30	a	(nm)	1.1430	1.1434
C_{33}	9.24		b	(nm)	0.6591	0.6390
C_{44}	3.25	4.57	c	(nm)	1.1395	1.1169
C_{55}	2.13	2.72	β	(°)	95.68	97.70
C_{66}	3.86	3.77				
C_{12}	5.49	5.88	Volume	(nm^3)	0.8542	0.8087
C_{15}	−1.73	−0.03	Center of mass[a]			
C_{25}	0.19	0.64	X_g	(nm)	0.2538	0.2557
C_{46}	−0.22	−0.06	Y_g	(nm)	0.2623	0.2602
			Z_g	(nm)	0.0296	0.0223
Young's modulus[b]						
E_V	7.1	7.45	Eulerian angles[c]			
Deviations of atomic coordinates			ϕ	(°)	−44.89	−44.64
Average (nm)		0.0148	θ	(°)	57.37	54.47
Maximum (nm)		0.0220	χ	(°)	61.10	60.92

[a] Coordinates of the center of mass of molecule 1

[b] Ref 6.

[c] Eulerian angles of the inertial axes of molecule [1]. ϕ, θ and χ are defined according to E.B. Wilson, J.C. Decius and P. C. Cross, *Molecular Vibrations*, McGraw-Hill, New York (1955).

The elements of $\boldsymbol{C}$ are called the elastic constants. In Table 9.1, the elastic constants of aspirin calculated as described above[5] are compared with the experimental values obtained from the data for Brillouin scattering. According to York *et al.*, Young's modulus of aspirin crystal derived from the calculated elastic constants was shown to agree well with the experimental values obtained from the bending rod method.[6]

9.3 Crystal Vibrations

9.3.1 Normal Vibrations and Vibrational Spectra of Molecular Crystals

In the discussion in section 9.2, it has been assumed that the translational symmetry of the crystal is retained during the deformation. In other words, the microscopic distortion of the crystal has been assumed to be homogeneous. Now we consider the case of inhomogeneous microscopic distortion, that is, the case where the translational symmetry of the crystal is kept only at the equilibrium structure. Even in this case, one can factor the dynamical matrices by utilizing the translational symmetry at equilibrium, and can reduce the number of variables from infinity to a finite number corresponding to the degrees of freedom of nuclear motion in the asymmetric unit.[1] The principle of this factoring is similar to that used for separating the variables in the case of an isolated molecule having an n-fold axis of rotation with n greater than two.

The three vector products of the unit cell vectors divided by the unit cell volume are given in the form

$$\boldsymbol{a}^* = (\boldsymbol{b} \times \boldsymbol{c})/V, \quad \boldsymbol{b}^* = (\boldsymbol{c} \times \boldsymbol{a})/V, \quad \boldsymbol{c}^* = (\boldsymbol{a} \times \boldsymbol{b})/V. \tag{9-53}$$

The vectors $\boldsymbol{a}^*$, $\boldsymbol{b}^*$ and $\boldsymbol{c}^*$ are called the unit cell vectors in the reciprocal space or the reciprocal unit cell vectors briefly. The parallelepiped defined by $\boldsymbol{a}^*$, $\boldsymbol{b}^*$ and $\boldsymbol{c}^*$ is called a reciprocal unit cell. The components of the reciprocal unit cell vectors have the dimension of reciprocal length. From the property of the scalar triple product of three vectors, it is easily verified that

$$\begin{bmatrix} \tilde{\boldsymbol{a}}^* \\ \tilde{\boldsymbol{b}}^* \\ \tilde{\boldsymbol{c}}^* \end{bmatrix} \begin{bmatrix} \boldsymbol{a} & \boldsymbol{b} & \boldsymbol{c} \end{bmatrix} = \begin{bmatrix} \tilde{\boldsymbol{a}}^* \boldsymbol{a} & \tilde{\boldsymbol{a}}^* \boldsymbol{b} & \tilde{\boldsymbol{a}}^* \boldsymbol{c} \\ \tilde{\boldsymbol{b}}^* \boldsymbol{a} & \tilde{\boldsymbol{b}}^* \boldsymbol{b} & \tilde{\boldsymbol{b}}^* \boldsymbol{c} \\ \tilde{\boldsymbol{c}}^* \boldsymbol{a} & \tilde{\boldsymbol{c}}^* \boldsymbol{b} & \tilde{\boldsymbol{c}}^* \boldsymbol{c} \end{bmatrix} = \begin{bmatrix} 1 & 0 & 0 \\ 0 & 1 & 0 \\ 0 & 0 & 1 \end{bmatrix}. \tag{9-54}$$

Now let us take all the combinations of three integers K_a, K_b and K_c which satisfy the condition

$$0 \le K_a < N_a, \quad 0 \le K_b < N_b, \quad 0 \le K_c < N_c$$

and number them consecutively. The numbering is arbitrary except that the combination $K_a = K_b = K_c = 0$ corresponds to $K = 1$. For example, one may take

$$K = 1 + K_a + N_a K_b + N_a N_b K_c, \tag{9-55}$$

in the same way as the case of Eq. (9-5). For each combination of K_a, K_b and K_c, a vector $\boldsymbol{q}_K$ is introduced as a linear combination of the reciprocal unit cell vectors in the form

$$\boldsymbol{q}_K = \frac{K_a}{N_a}\boldsymbol{a}^* + \frac{K_b}{N_b}\boldsymbol{b}^* + \frac{K_c}{N_c}\boldsymbol{c}^*. \tag{9-56}$$

Note that $\boldsymbol{q}_1$ is a zero vector from the definition. If we place the starting points of all the other $\boldsymbol{q}_K$ vectors at the origin of the reciprocal space, called the Γ point, their tips will be homogeneously distributed within a reciprocal unit cell. The reciprocal unit cell is then divided into two parts of equal volume, *a* and *b*, so that two vectors specified by the integers K_a, K_b and K_c and by $N_a - K_a$, $N_b - K_b$ and $N_c - K_c$ terminate in different parts from each other. Such a choice is attained as $0 \le K_a \le N_a - 1$, $0 \le K_b \le N_b - 1$ and $0 \le K_c \le N_c/2$, or simply as $1 \le K \le N/2$ if the numbering of Eq. (9-55) is employed. In a general case, the choice is made straightforwardly by using a domain in the reciprocal space known as the first Brillouin zone, which is defined to be bounded by the bisector planes of the lines joining the Γ point with the origins of adjacent reciprocal unit cells. The boundaries of the Brillouin zones projected on the *ac* plane of an orthorhombic crystal and that of a monoclinic crystal are shown in Fig. 9.3. Note that the shape of the Brillouin zone becomes complicated if there are face angles other than $\pi/2$. If we allow for negative values of K_a, K_b and K_c, and take $\boldsymbol{q}_K$'s terminating within the first Brillouin zone, the parts *a* and *b* can be separated by a plane passing the Γ point. Because of the three-dimensional periodicity of the reciprocal space, the set of $\boldsymbol{q}_K$'s taken for the first Brillouin zone is equivalent to those for a reciprocal unit cell.

Taking the scalar product of the vectors $2\pi\tilde{\boldsymbol{q}}_K$ and $\boldsymbol{r}_l$ (Eq. (9-4)) and simplifying it with the help of Eq. (9-54), we obtain a dimensionless quantity θ_{Kl}, which may be regarded as an angle in radian in the form

$$2\pi\tilde{\boldsymbol{q}}_K \boldsymbol{r}_l = \theta_{Kl} = 2\pi\left(\frac{K_a l_a}{N_a} + \frac{K_b l_b}{N_b} + \frac{K_c l_c}{N_c}\right). \tag{9-57}$$

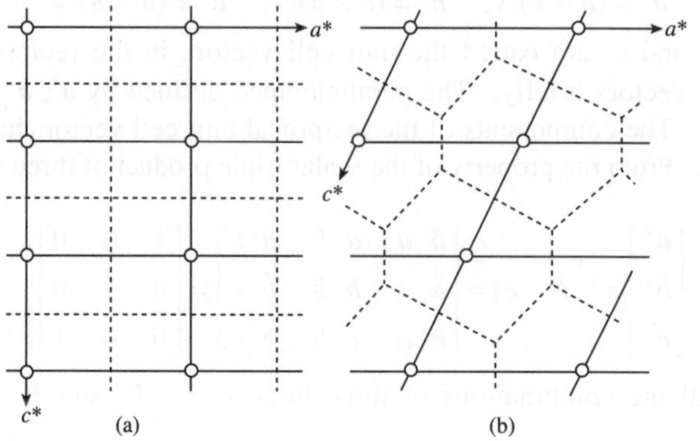

Fig. 9.3 Projections of Brillouin zone boundaries (dotted lines) onto the a^*c^* planes of orthorhombic (a) and monoclinic (b) lattices. Solid lines show reciprocal lattice planes.

A pair of symmetry Cartesian displacement coordinates is then defined by

$$\Delta\boldsymbol{x}(\boldsymbol{q}_K)_c = (2/N)^{1/2}\sum_{I=1}^{N}\Delta\boldsymbol{x}_I\cos\theta_{KI} \tag{9-58a}$$

$$\Delta\boldsymbol{x}(\boldsymbol{q}_K)_s = (2/N)^{1/2}\sum_{I=1}^{N}\Delta\boldsymbol{x}_I\sin\theta_{KI} \tag{9-58b}$$

for each of the vectors $\boldsymbol{q}_K$ which terminate in the part a, and it is assumed that any nuclear displacement in the crystal is given as a certain linear combination of these variables.[7] The explicit form of the linear combination will be given later. This assumption is equivalent to assuming that the equivalent parallelepiped blocks consisting of $N_a N_b N_c$ unit cells exhibit exactly the same nuclear motions as each other.[8] The set of symmetry Cartesian displacement coordinates defined by Eqs. (9-58a,b) has the same form as the internal symmetry coordinates introduced for isolated symmetric top molecules having an n-fold rotational axis, Eq. (4-107) in Chapter 4. The difference is that the periodicity introduced in a crystal by restricting the variables in the forms of Eqs. (9-58a,b) is three-dimensional and is a conventional assumption for deducing a tractable formulation, whereas a symmetric top molecule has a real one-dimensional periodicity. As in the case of an n-fold axis, the coordinates defined by Eq. (9-58b) vanish if the argument of the sine, $2\pi\tilde{\boldsymbol{q}}_K\boldsymbol{r}_I$ this time, is 0 or π.

In analogy with the case of an n-fold axis, the orthogonality of the transformation of Eqs. (9-58a,b) is proved as follows. Multiplying both sides of Eqs. (9-58a) and (9-58b) by $\cos\theta_{KJ}$ and $\sin\theta_{KJ}$, respectively, and summing over all $\boldsymbol{q}_K$'s, we have

$$\begin{aligned}
&\sum_{K\in a}\{\Delta\boldsymbol{x}(\boldsymbol{q}_K)_c\cos\theta_{KJ} + \Delta\boldsymbol{x}(\boldsymbol{q}_K)_s\sin\theta_{KJ}\} \\
&= (2/N)^{1/2}\sum_{I=1}^{N}\Delta\boldsymbol{x}_I\sum_{K\in a}(\cos\theta_{KI}\cos\theta_{KJ} + \sin\theta_{KI}\sin\theta_{KJ}) \\
&= (2/N)^{1/2}\sum_{I=1}^{N}\Delta\boldsymbol{x}_I\sum_{K\in a}\cos(\theta_{KI} - \theta_{KJ}).
\end{aligned} \tag{9-59}$$

By using Eq. (9-57) and the addition theorem, cos $(\theta_{KI} - \theta_{KJ})$ is expanded as

$$\begin{aligned}\cos(\theta_{KI} - \theta_{KJ}) &= \cos 2\pi \frac{K_a(I_a - J_a)}{N_a} \cos 2\pi \frac{K_b(I_b - J_b)}{N_b} \cos 2\pi \frac{K_c(I_c - J_c)}{N_c} \\ &- \sin 2\pi \frac{K_a(I_a - J_a)}{N_a} \sin 2\pi \frac{K_b(I_b - J_b)}{N_b} \cos 2\pi \frac{K_c(I_c - J_c)}{N_c} \\ &- \sin 2\pi \frac{K_a(I_a - J_a)}{N_a} \cos 2\pi \frac{K_b(I_b - J_b)}{N_b} \sin 2\pi \frac{K_c(I_c - J_c)}{N_c} \\ &- \cos 2\pi \frac{K_a(I_a - J_a)}{N_a} \sin 2\pi \frac{K_b(I_b - J_b)}{N_b} \sin 2\pi \frac{K_c(I_c - J_c)}{N_c}.\end{aligned}$$

On decomposing the sum of this expansion with respect to K into those with respect to K_a, K_b and K_c, the first term is factored as

$$\sum_{K_a=0}^{N_a-1} \cos 2\pi \frac{K_a(I_a - J_a)}{N_a} \sum_{K_b=0}^{N_b-1} \cos 2\pi \frac{K_b(I_b - J_b)}{N_b} \sum_{K_c=0}^{<N_b/2} \cos 2\pi \frac{K_c(I_c - J_c)}{N_c}.$$

The sums of the other three terms vanish because of the presence of factors involving sines. Hence we obtain

$$\sum_K{}' \cos(\theta_{KI} - \theta_{KJ}) = \delta_{IJ}\, N/2, \tag{9-60}$$

where the prime implies the sum over half the first Brillouin zone taken as part a above. Substituting Eq. (9-60) into Eq. (9-59), we have

$$\boldsymbol{\Delta x}_I = (2/N)^{1/2} \sum_{K=1}^{N/2} \{\boldsymbol{\Delta x}(\boldsymbol{q}_K)_c \cos\theta_{KI} + \boldsymbol{\Delta x}(\boldsymbol{q}_K)_s \sin\theta_{KI}\}. \tag{9-61}$$

Equation (9-61) gives the inverse transformation of Eqs. (9-58a,b), *i.e.*, the explicit form of the linear combinations of the symmetry Cartesian displacement coordinates representing the Cartesian displacements of nuclei. According to Eqs. (9-57) and (9-61), the vector $\boldsymbol{\Delta x}_I$ is left invariant on adding any integer multiples of N_a, N_b and N_c to I_a, I_b and I_c, respectively. It is thus understood that use of the symmetry Cartesian displacement coordinates in Eqs. (9-58a,b) brings about the dynamic periodicity for the crystal of infinite size.

The vector $\boldsymbol{q}_K$ is regarded to have the following geometrical meaning. In the basic parallelepiped containing N (= $N_aN_bN_c$) unit cells, the lattice vector (Eq. (9-4)) of the unit cell farthest from the crystal origin is given by

$$\boldsymbol{r}_N = (N_a - 1)\boldsymbol{a} + (N_b - 1)\boldsymbol{b} + (N_c - 1)\boldsymbol{c}.$$

In Fig. 9.4. let the points O and P be the vertices at both ends of a diagonal passing the body center of the basic parallelepiped. and the point O be the origin of the Cartesian coordinate system. Consider three vectors along the *a, b* and *c* axes defined by

$$\overrightarrow{\mathrm{OA}} \equiv \boldsymbol{r}_K^a = (N_a / K_a)\boldsymbol{a}, \quad \overrightarrow{\mathrm{OB}} \equiv \boldsymbol{r}_K^b = (N_b / K_b)\boldsymbol{b}, \quad \overrightarrow{\mathrm{OC}} \equiv \boldsymbol{r}_K^c = (N_c / K_c)\boldsymbol{c},$$

and a plane P determined by the triangle ABC. The wave vector $\boldsymbol{q}_K$ is perpendicular to two

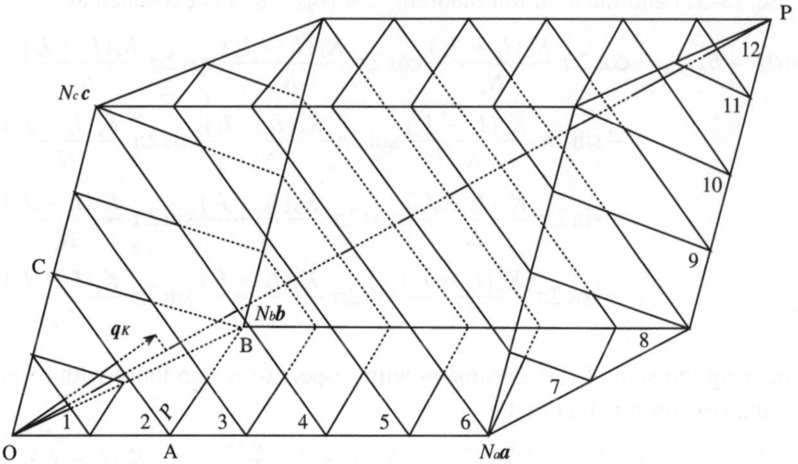

Fig. 9.4 Nodal planes for the wave vector $\boldsymbol{q}_K$ partitioning the basic parallelepiped crystal into $2(K_a + K_b + K_c)$ layers: $K_a = 3$, $K_b = 1$, $K_c = 2$.

vectors $\overrightarrow{\mathrm{AB}} = \boldsymbol{r}_K^b - \boldsymbol{r}_K^a$ and $\overrightarrow{\mathrm{AC}} = \boldsymbol{r}_K^c - \boldsymbol{r}_K^a$ both lying on the plane P, since we have

$$\tilde{\boldsymbol{q}}_K(\boldsymbol{r}_K^b - \boldsymbol{r}_K^a) = \tilde{\boldsymbol{b}}^*\boldsymbol{b} - \tilde{\boldsymbol{a}}^*\boldsymbol{a} = 0 \quad \text{and} \quad \tilde{\boldsymbol{q}}_K(\boldsymbol{r}_K^c - \boldsymbol{r}_K^a) = \tilde{\boldsymbol{c}}^*\boldsymbol{c} - \tilde{\boldsymbol{a}}^*\boldsymbol{a} = 0$$

from Eqs. (9-54) and (9-56). Furthermore, $\overrightarrow{\mathrm{AB}}$ and $\overrightarrow{\mathrm{AC}}$ are not parallel to each other. Hence $\boldsymbol{q}_K$ is perpendicular to P itself. The length of the normal from the origin to P is given as the projection of $\boldsymbol{r}_K{}^a$ on $\boldsymbol{q}_K$., *i.e.*,

$$\tilde{\boldsymbol{r}}_K^a\boldsymbol{q}_K / q_K = \tilde{\boldsymbol{a}}\boldsymbol{a}^* / q_K = 1/q_K, \tag{9-62a}$$

where $q_K = |\boldsymbol{q}_K|$. If the integers N_a, N_b and N_c are sufficiently large, the deviation of the origin of the unit cell N from point P is negligibly small compared to the size of the parallelepiped. The projection of the diagonal $\overline{\mathrm{OP}}$ on $\boldsymbol{q}_K$ is then approximately given by

$$\tilde{\boldsymbol{r}}_N\boldsymbol{q}_K / q_K = (K_a + K_b + K_c)/q_K \tag{9-62b}$$

In this case, one may choose the vector $\boldsymbol{r}_l$ in Eq. (9-4) in such a way that the integers l_a, l_b and l_c satisfy the condition ,

$$0 \le f \cong l_a/N_a \cong l_b/N_b \cong l_c/N_c < 1$$

for a given real number f such that $0 \le f < 1$. On the increase of f from 0 to 1, the origin of the unit cell specified by the vector $\boldsymbol{r}_l$ moves along the diagonal $\overline{\mathrm{OP}}$ from O to P. Calculating the scalar product $\theta_{Kl} = 2\pi\tilde{\boldsymbol{q}}_K\boldsymbol{r}_l$ from Eq. (9-57), and substituting it into Eq. (9-61) along with a special set of symmetry Cartesian displacement coordinates corresponding to Eq. (9-58a), *i.e.*,

$$\Delta\boldsymbol{x}(\boldsymbol{q}_K)_c = (N/2)^{1/2}\Delta\boldsymbol{x}, \quad \Delta\boldsymbol{x}(\boldsymbol{q}_{K'})_c = \boldsymbol{0}, \quad (K' \neq K), \quad \text{and} \quad \Delta\boldsymbol{x}(\boldsymbol{q}_K)_s = \boldsymbol{0},$$

we obtain

$$\Delta x_I = \Delta x \cos\theta_{KI} = \Delta x \cos\{2\pi f(K_a + K_b + K_c)\}. \tag{9-63}$$

According to Eq. (9-63), the microscopic distortion associated with the symmetry Cartesian displacement coordinate $\Delta x(q_K)_c$ is such that the nuclear displacements in the unit cells along the diagonal $\overline{\mathrm{OP}}$ change periodically from O to P, forming a series of $K_a + K_b + K_c$ loops of standing waves along the diagonal as the solution of Newton's equation of motion. The nuclear displacements in unit cells along the *a, b* and *c* axes form K_a, K_b and K_c loops of the same standing wave, respectively. Since $q_K \perp P$, the plane P is a nodal plane of this standing wave and the separation between the nodal planes, q_K^{-1}, is the wavelength of the standing wave. For this reason, the vector q_K is called the wave vector.

The potential function in terms of the symmetry Cartesian displacement coordinates has a great advantage in the separability of variables. On substituting Eq. (9-61) into Eq. (9-6), the linear part is written as

$$(2/N)^{1/2}\sum_I\left[\sum_K\{\Delta\tilde{x}(q_K)_c\cos\theta_{KI} + \Delta\tilde{x}(q_K)_s\sin\theta_{KI}\}\right]F_I^x$$

$$= (2/N)^{1/2}\sum_K\left\{\Delta\tilde{x}(q_K)_c\sum_I\cos\theta_{KI} + \Delta\tilde{x}(q_K)_s\sum_I\sin\theta_{KI}\right\}F_1^x$$

$$= (2/N)^{1/2}\left\{\sum_K\Delta\tilde{x}(q_K)_c\delta_{K1}N\right\}F_1^x = (2N)^{1/2}\Delta\tilde{x}(q_1)_cF_1^x. \tag{9-64a}$$

The quadratic part of Eq. (9-6) becomes

$$\frac{1}{2}\sum_I\sum_J\frac{2}{N}\sum_K\Big\{\Delta\tilde{x}(q_K)_cF_{IJ}^{xx}\sum_{K'}\Delta x(q_{K'})_c\cos\theta_{KI}\cos\theta_{K'J}$$
$$+\,\Delta\tilde{x}(q_K)_sF_{IJ}^{xx}\sum_{K'}\Delta x(q_{K'})_c\sin\theta_{KI}\cos\theta_{K'J}$$
$$+\,\Delta\tilde{x}(q_K)_cF_{IJ}^{xx}\sum_{K'}\Delta x(q_{K'})_s\cos\theta_{KI}\sin\theta_{K'J}$$
$$+\,\Delta\tilde{x}(q_K)_sF_{IJ}^{xx}\sum_{K'}\Delta x(q_{K'})_s\sin\theta_{KI}\sin\theta_{K'J}\Big\}$$
$$=\frac{1}{N}\sum_K\Big\{\Delta\tilde{x}(q_K)_c\sum_{K'}\Big(\sum_I\sum_JF_{IJ}^{xx}\cos\theta_{KI}\cos\theta_{K'J}\Big)\Delta x(q_{K'})_c$$
$$+\,\Delta\tilde{x}(q_K)_s\sum_{K'}\Big(\sum_I\sum_JF_{IJ}^{xx}\sin\theta_{KI}\cos\theta_{K'J}\Big)\Delta x(q_{K'})_c$$
$$+\,\Delta\tilde{x}(q_K)_c\sum_{K'}\Big(\sum_I\sum_JF_{IJ}^{xx}\cos\theta_{KI}\sin\theta_{K'J}\Big)\Delta x(q_{K'})_s$$
$$+\,\Delta\tilde{x}(q_K)_s\sum_{K'}\Big(\sum_I\sum_JF_{IJ}^{xx}\sin\theta_{KI}\sin\theta_{K'J}\Big)\Delta x(q_{K'})_s\Big\}. \tag{9-64b}$$

By referring to Eqs. (9-8) and (9-9) based on the translational symmetry of the crystal, the double sum in the first parentheses in the right side of Eq. (9-64b) is rewritten in the form

$$\begin{aligned}\sum_I\sum_J F^{xx}_{IJ}\cos\theta_{KI}\cos\theta_{K'J} &= \sum_L F^{xx}_{1L}\sum_I \cos\theta_{KI}\cos(\theta_{K'I}+\theta_{K'L})\\ &= \sum_L F^{xx}_{1L}\sum_I \cos\theta_{KI}\left(\cos\theta_{K'I}\cos\theta_{K'L}-\sin\theta_{K'I}\sin\theta_{K'L}\right)\\ &= \sum_L F^{xx}_{1L}\left(\cos\theta_{K'L}\sum_I\cos\theta_{KI}\cos\theta_{K'I}-\sin\theta_{K'L}\sum_I\cos\theta_{KI}\sin\theta_{K'I}\right)\\ &= (N/2)\sum_L F^{xx}_{1L}\cos\theta_{K'L}\,\delta_{KK'}.\end{aligned} \tag{9-65a}$$

The quantities in the other parentheses are similarly calculated as

$$\sum_I\sum_J F^{xx}_{IJ}\sin\theta_{KI}\cos\theta_{K'J} = (N/2)\sum_L F^{xx}_{1L}\sin\theta_{K'L}\,\delta_{KK'} \tag{9-65b}$$

$$\begin{aligned}\sum_I\sum_J F^{xx}_{IJ}\cos\theta_{KI}\sin\theta_{K'J} &= \sum_L F^{xx}_{1L}\sum_I\cos\theta_{KI}\left(\sin\theta_{K'I}\cos\theta_{K'L}-\cos\theta_{K'I}\sin\theta_{K'L}\right)\\ &= -(N/2)\sum_L F^{xx}_{1L}\sin\theta_{K'L}\,\delta_{KK'}\end{aligned} \tag{9-65c}$$

and

$$\sum_I\sum_J F^{xx}_{IJ}\sin\theta_{KI}\sin\theta_{K'J} = (N/2)\sum_L F^{xx}_{1L}\cos\theta_{K'L}\,\delta_{KK'}. \tag{9-65d}$$

Now we define a $6n$-dimensional vector of the symmetry Cartesian displacement coordinates in Eqs. (9-58a,b) for the case that $K \neq 1$ in the form

$$\Delta\boldsymbol{x}(\boldsymbol{q}_K) = \left[\Delta\tilde{\boldsymbol{x}}(\boldsymbol{q}_K)_c \quad \Delta\tilde{\boldsymbol{x}}(\boldsymbol{q}_K)_s\right]^{\mathrm{T}} \tag{9-66}$$

and abbreviate $\Delta\boldsymbol{x}(\boldsymbol{q}_1)_c$ as $\Delta\boldsymbol{x}(\boldsymbol{q}_1)$. Correspondingly, the potential energy matrix for the symmetry Cartesian displacement coordinates associated with $\boldsymbol{q}_K$ may simply be expressed, by combining Eqs. (9-65a–d) with Eq. (9-64b), as

$$\boldsymbol{F}^{xx}_K = \begin{bmatrix}\boldsymbol{F}^{xx}_{Kc} & -\boldsymbol{F}^{xx}_{Ks}\\ \boldsymbol{F}^{xx}_{Ks} & \boldsymbol{F}^{xx}_{Kc}\end{bmatrix} \equiv \begin{bmatrix}\sum_L \boldsymbol{F}^{xx}_{1L}\cos\theta_{KL} & -\sum_L \boldsymbol{F}^{xx}_{1L}\sin\theta_{KL}\\ \sum_L \boldsymbol{F}^{xx}_{1L}\sin\theta_{KL} & \sum_L \boldsymbol{F}^{xx}_{1L}\cos\theta_{KL}\end{bmatrix}. \tag{9-67}$$

In practical calculations, the sum with respect to L is taken over such unit cells surrounding the unit cell 1 that $|\boldsymbol{r}_L|$ is less than an appropriate limit, $r_{\max}$. By pre- and postmultiplying Eq. (9-67) by $\Delta\tilde{\boldsymbol{x}}(\boldsymbol{q}_K)$ and $\Delta\boldsymbol{x}(\boldsymbol{q}_K)$ in Eq. (9-66), respectively, and summing the result with respect to K over half the first Brillouin zone, the quadratic part of the potential energy is expressed in terms of the symmetry Cartesian displacements. By combining this result with Eq. (9-64a), the potential energy per unit cell is given in the form

$$V = \Delta\tilde{\boldsymbol{x}}(\boldsymbol{q}_1)\boldsymbol{F}^x_1 + \tfrac{1}{2}\sum_K \Delta\tilde{\boldsymbol{x}}(\boldsymbol{q}_K)\boldsymbol{F}^{xx}_K\Delta\boldsymbol{x}(\boldsymbol{q}_K). \tag{9-68}$$

In the first term in the right side of Eq. (9-68), no symmetry Cartesian displacement coordinates other than those with $K = 1$ appear. This result corresponds to the fact that only the totally symmetric coordinates appear in the linear terms of the potential function of an isolated molecule expanded at a non-equilibrium position.

If all the symmetry Cartesian displacement coordinates $\Delta\boldsymbol{x}(\boldsymbol{q}_K)$ except for those with $K = 1$ are set equal to zero, it follows from Eq. (9-61) that

$$\Delta\boldsymbol{x}_1 = \Delta\boldsymbol{x}_2 = \cdots = \Delta\boldsymbol{x}_N.$$

The Cartesian displacements of nuclei in this case keep the translational symmetry of the crystal. Thus the components of $\Delta\boldsymbol{x}(\boldsymbol{q}_1)$ are the variables representing the microscopic homogeneous deformation of the crystal. As mentioned in section 9.2, the search for equilibrium structures of molecular crystals is carried out by using these variables together with the macroscopic variables σ. The potential function expanded in powers of the symmetry Cartesian displacements by placing the nuclei at their equilibrium positions contains no linear terms, and Eq. (9-68) is reduced to the form

$$V = \tfrac{1}{2}\sum_K \Delta\tilde{\boldsymbol{x}}(\boldsymbol{q}_K)\boldsymbol{F}_K^{xx}\Delta\boldsymbol{x}(\boldsymbol{q}_K). \tag{9-69}$$

In contrast to the quadratic part of Eq. (9-6) which has a double sum over I and J, Eq. (9-69) has a sum over only a single index K. No terms containing the variable factors $\Delta\boldsymbol{x}(\boldsymbol{q}_K)$ and $\Delta\boldsymbol{x}(\boldsymbol{q}_{K'})$ $(K' \neq K)$ simultaneously appear in it. In this case, the Hessian matrix of the potential function with respect to the symmetry Cartesian displacements is written as a direct sum of diagonal blocks as shown in Fig. 9.5. The equations of motion can thus be set up for each wave vector $\boldsymbol{q}_K$ by using a finite number of variables, *i.e.*, the $3n$ or $6n$ components of $\Delta\boldsymbol{x}(\boldsymbol{q}_K)$, as given by

$$\boldsymbol{m}_K\,\Delta\ddot{\boldsymbol{x}}(\boldsymbol{q}_K) = -\boldsymbol{F}_K^{xx}\Delta\boldsymbol{x}(\boldsymbol{q}_K), \tag{9-70}$$

where $\boldsymbol{m}_K$ is a diagonal matrix with the dimensions $(3n \times 3n)$ for $K = 1$ and $(6n \times 6n)$ for $K > 1$ in the forms

$$\boldsymbol{m}_1 = \mathrm{diag}[m_1 \quad m_1 \quad m_1 \quad m_2 \quad m_2 \quad m_2 \quad \cdots \quad m_n \quad m_n \quad m_n], \tag{9-71a}$$

and

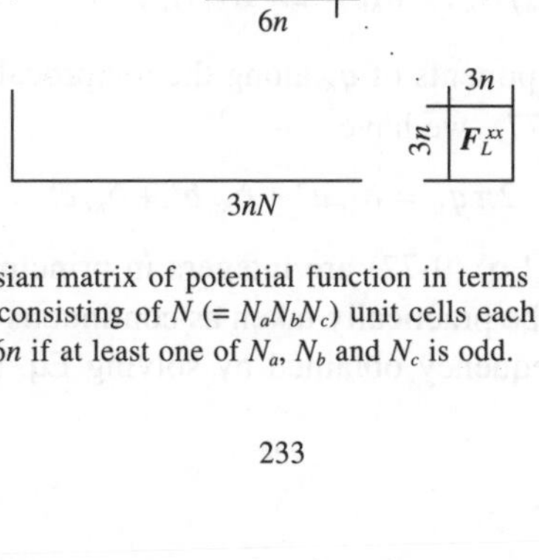

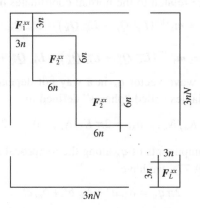

Fig. 9.5 Block diagonalized Hessian matrix of potential function in terms of symmetry Cartesian displacement coordinates of a crystal consisting of N (= $N_aN_bN_c$) unit cells each containing n nuclei. The dimension of the last block $\boldsymbol{F}_L^{xx}$ is $6n$ if at least one of N_a, N_b and N_c is odd.

$$\boldsymbol{m}_K = \mathrm{diag}[\boldsymbol{m}_1 \quad \boldsymbol{m}_1], \tag{9-71b}$$

respectively. In analogy with the case of isolated molecules, we introduce the mass-weighted symmetry Cartesian displacement coordinates in the form

$$\Delta\boldsymbol{x}_m(\boldsymbol{q}_K) = \begin{bmatrix} \Delta\boldsymbol{x}_m(\boldsymbol{q}_K)_c \\ \Delta\boldsymbol{x}_m(\boldsymbol{q}_K)_s \end{bmatrix} = \begin{bmatrix} \boldsymbol{m}_1^{1/2}\,\Delta\boldsymbol{x}(\boldsymbol{q}_K)_c \\ \boldsymbol{m}_1^{1/2}\,\Delta\boldsymbol{x}(\boldsymbol{q}_K)_s \end{bmatrix}, \tag{9-72}$$

and rewrite the potential energy in Eq. (9-69) as

$$V = \tfrac{1}{2}\sum_K \Delta\tilde{\boldsymbol{x}}_m(\boldsymbol{q}_K)\,\boldsymbol{F}_K^{mxx}\,\Delta\boldsymbol{x}_m(\boldsymbol{q}_K).$$

The Hessian matrix $\boldsymbol{F}_K^{mxx}$ is given by

$$\boldsymbol{F}_K^{mxx} = \boldsymbol{m}_K^{-1/2}\,\boldsymbol{F}_K^{xx}\,\boldsymbol{m}_K^{-1/2} = \begin{bmatrix} \boldsymbol{F}_{Kc}^{mxx} & -\boldsymbol{F}_{Ks}^{mxx} \\ \boldsymbol{F}_{Ks}^{mxx} & \boldsymbol{F}_{Kc}^{mxx} \end{bmatrix}, \tag{9-73}$$

where $\boldsymbol{F}_{Kc}^{mxx} = \boldsymbol{m}_1^{-1/2}\,\boldsymbol{F}_{Kc}^{xx}\,\boldsymbol{m}_1^{-1/2}$ and $\boldsymbol{F}_{Ks}^{mxx} = \boldsymbol{m}_1^{-1/2}\,\boldsymbol{F}_{Ks}^{xx}\,\boldsymbol{m}_1^{-1/2}$. [9)]
The mass-weighted Cartesian displacements can be transformed into the normal coordinates by an orthogonal transformation

$$\begin{bmatrix} \Delta\boldsymbol{x}_m(\boldsymbol{q}_K)_c \\ \Delta\boldsymbol{x}_m(\boldsymbol{q}_K)_s \end{bmatrix} = \begin{bmatrix} \boldsymbol{L}_{Kc}^{mx} & -\boldsymbol{L}_{Ks}^{mx} \\ \boldsymbol{L}_{Ks}^{mx} & \boldsymbol{L}_{Kc}^{mx} \end{bmatrix} \begin{bmatrix} \boldsymbol{Q}_K^a \\ \boldsymbol{Q}_K^b \end{bmatrix} = \boldsymbol{L}_K^{mx}\,\boldsymbol{Q}_K, \tag{9-74}$$

which is obtained by diagonalizing the real symmetric matrix $\boldsymbol{F}_K^{mxx}$ defined in Eq. (9-73) as

$$\begin{bmatrix} \tilde{\boldsymbol{L}}_{Kc}^{mx} & \tilde{\boldsymbol{L}}_{Ks}^{mx} \\ -\tilde{\boldsymbol{L}}_{Ks}^{mx} & \tilde{\boldsymbol{L}}_{Kc}^{mx} \end{bmatrix} \begin{bmatrix} \boldsymbol{F}_{Kc}^{mxx} & -\boldsymbol{F}_{Ks}^{mxx} \\ \boldsymbol{F}_{Ks}^{mxx} & \boldsymbol{F}_{Kc}^{mxx} \end{bmatrix} \begin{bmatrix} \boldsymbol{L}_{Kc}^{mx} & -\boldsymbol{L}_{Ks}^{mx} \\ \boldsymbol{L}_{Ks}^{mx} & \boldsymbol{L}_{Kc}^{mx} \end{bmatrix} = \begin{bmatrix} \boldsymbol{\Lambda}_K & \boldsymbol{0} \\ \boldsymbol{0} & \boldsymbol{\Lambda}_K \end{bmatrix}. \tag{9-75}$$

The appearance of doubly degenerate eigenvalues in Eq. (9-75) can be confirmed in the same way as the case of Eq. (4-131). For each wave vector $\boldsymbol{q}_K$, the symmetry Cartesian displacement coordinates are related to the normal coordinates by

$$\Delta\boldsymbol{x}(\boldsymbol{q}_K)_c = \boldsymbol{m}_1^{-1/2}(\boldsymbol{L}_{Kc}^{mx}\,\boldsymbol{Q}_K^a - \boldsymbol{L}_{Ks}^{mx}\,\boldsymbol{Q}_K^b) = \boldsymbol{L}_{Kc}^{x}\,\boldsymbol{Q}_K^a - \boldsymbol{L}_{Ks}^{x}\,\boldsymbol{Q}_K^b \tag{9-76a}$$

$$\Delta\boldsymbol{x}(\boldsymbol{q}_K)_s = \boldsymbol{m}_1^{-1/2}(\boldsymbol{L}_{Ks}^{mx}\,\boldsymbol{Q}_K^a + \boldsymbol{L}_{Kc}^{mx}\,\boldsymbol{Q}_K^b) = \boldsymbol{L}_{Ks}^{x}\,\boldsymbol{Q}_K^a + \boldsymbol{L}_{Kc}^{x}\,\boldsymbol{Q}_K^b. \tag{9-76b}$$

In order to specify the wave vector $\boldsymbol{q}_K$ in a way not depending on N_a, N_b and N_c, we introduce three angular variables called "phases" defined by

$$\delta_{Ka} = 2\pi\,K_a/N_a, \quad \delta_{Kb} = 2\pi\,K_b/N_b, \quad \delta_{Kc} = 2\pi\,K_c/N_c. \tag{9-77}$$

The phases represent the components of $\boldsymbol{q}_K$ along the reciprocal unit cell vectors, $\boldsymbol{a}^*$, $\boldsymbol{b}^*$ and $\boldsymbol{c}^*$. From Eqs. (9-56) and (9-77), we have

$$2\pi\,\boldsymbol{q}_K = \delta_{Ka}\,\boldsymbol{a}^* + \delta_{Kb}\,\boldsymbol{b}^* + \delta_{Kc}\,\boldsymbol{c}^*. \tag{9-78}$$

The indices K_a, K_b and K_c in Eq. (9-77) are integers in principle, but N_a, N_b and N_c are so large that δ_{Ka}, δ_{Kb} and δ_{Kc} can be practically taken as continuous variables between 0 and 2π. Accordingly, each normal frequency obtained by solving Eq. (9-70) can be regarded as a

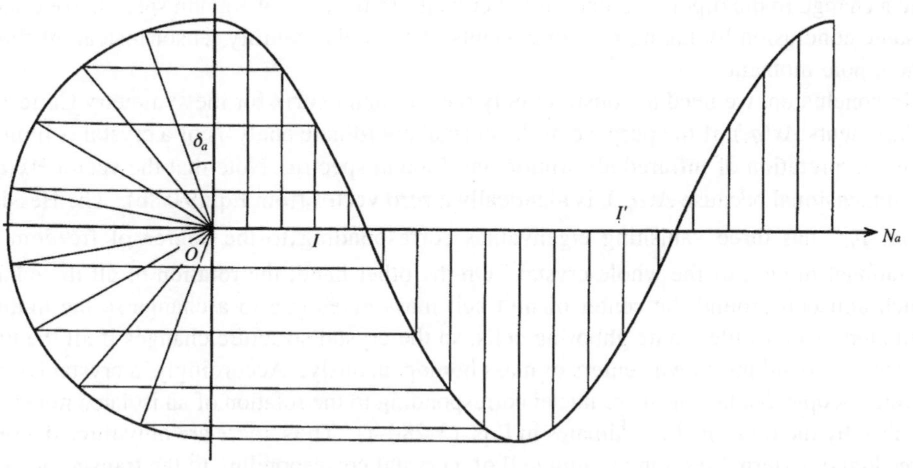

Fig. 9.6 Displacement amplitudes of equivalent nuclei in unit cells along the a-axis. (Phase difference between neighboring cells is δ_a.)

continuous function of the three components of the wave vector defined in the range $[0, 2\pi]$. The curve showing the dependence of a normal frequency on δ_{Ka}, δ_{Kb} or δ_{Kc} is called a dispersion curve. Similarly, the dependence of a normal frequency on two components is shown by a dispersion surface.

For a normal frequency of a crystal to be observable in infrared absorption and/or Raman spectra, a condition must be met by the wave vector of the normal frequency. From the definition of the symmetry Cartesian displacement coordinates in Eqs. (9-58a,b), the amplitudes of the translationally equivalent nuclei in successive unit cells along the a-axis are given by a cosine or sine function taken at every integer multiple of a constant phase difference $2\pi/N_a$ as shown in Fig. 9.6. If $K_a \neq 0$ and $K_b = K_c = 0$, an integer I' can be chosen to satisfy the condition

$$\boldsymbol{r}_{I'} = \boldsymbol{r}_I + (N_a/2\,K_a)\boldsymbol{a}$$

so that we have

$$2\pi\,\tilde{\boldsymbol{q}}_{Ka}\,\boldsymbol{r}_{I'} = \frac{2\pi\,K_a}{N_a}\tilde{\boldsymbol{a}}^*\left(\boldsymbol{r}_I + \frac{N_a}{2\,K_a}\boldsymbol{a}\right) = 2\pi\,\tilde{\boldsymbol{q}}_{Ka}\,\boldsymbol{r}_I + \pi\,. \tag{9-79}$$

According to Eq. (9-79), an increase in the symmetry Cartesian displacement coordinate $\Delta\boldsymbol{x}(\boldsymbol{q}_K)$ displaces all the nuclei in the unit cell I in the direction opposite to the displacements of the corresponding nuclei in another unit cell I'. If at least one of K_a, K_b and K_c is not zero, the basic parallelepiped can be partitioned into $2(K_a + K_b + K_c)$ layers of the same thichness $1/2\ q_K$ by a set of equidistant parallel planes including P in Fig. 9.4. The layers are then numbered in the ascending order from the origin along the diagonal $\overline{\mathrm{OP}}$, and unit cells I and I' are taken from odd- and even-numbered layers, respectively. In this case, the dipole moment changes arising from the nuclear displacements in the unit cells I and I' cancel each other out completely, giving rise to no changes in the dipole moment in the crystal as a whole. Hence it follows that all of K_a, K_b and K_c must be zero in order for an increase in $\Delta\boldsymbol{x}(\boldsymbol{q}_K)$ to cause a change in the dipole moment of the crystal. In the case of Raman spectra, we obtain the same conclusion by taking the components of the polarizability tensor instead of those of the dipole moment.

In conclusion, we need to construct only the Hessian matrix for the symmetry Cartesian displacements $\Delta \boldsymbol{x}(\boldsymbol{q}_1)$, if the purpose of the normal coordinate analysis of a crystal is limited to the interpretation of infrared absorption and Raman spectra. Note that the vector $\Delta \boldsymbol{x}(\boldsymbol{q}_1)$ is $3n$ dimensional because $\Delta \boldsymbol{x}(\boldsymbol{q}_1)_s$ is identically a zero vector from Eq. (9-58b). The Hessian matrix $\boldsymbol{F}_1^{xx}$ has three vanishing eigenvalues corresponding to the degree of freedom of translational motion of the whole crystal. On the other hand, the rotation of all the nuclei in each unit cell around the center of unit cell mass gives rise to a change in the mutual orientation of molecules in neighboring cells, so the crystal structure changes if all the unit cells rotate around their own centers of mass homogeneously. Accordingly, a crystal has no free microscopic displacements of nuclei corresponding to the rotation of an isolated molecule described by the rotational coordinates in Eqs. (3-28d–f). Thus, there are only three degrees of freedom of external motion per unit cell of a crystal corresponding to the translation, and the degrees of freedom of vibrations in the subspace spanned by the symmetry Cartesian displacement coordinates associated with the wave vector $\boldsymbol{q}_1 = \boldsymbol{0}$ are $3n - 3$. The dispersion curves for the three translational modes starting from zero at the Γ point are called the acoustic branches, while those of the remaining $3n - 3$ modes starting from non-zero frequencies are called the optical branches.

If a crystal can be subjected to any symmetry operations other than pure translations, the $3n - 3$ normal vibrations with $\boldsymbol{q}_1$ are classified into irreducible representations on the basis of their symmetry properties as in the case of isolated molecules.[10] In this case, the selection rule for infrared absorption and Raman spectra is deduced from the character table for the point group isomorphous to the factor group of the crystal. For a molecular crystal containing Z equivalent molecules per unit cell, each normal mode of the molecule in the isolated state corresponds to Z crystal modes. In the case of hexagonal, rhombohedral, cubic and some other crystal systems which have more-than-two-fold rotational and/or helical axes, some of these Z crystal modes are degenerate; hence the number of distinguishable normal frequencies derived from a normal mode of the isolated molecule is smaller than Z.

The intermolecular force arising from the interaction between nonbonded atoms and the hydrogen bond often gives rise to appreciable differences between the frequencies of the crystal modes arising from a single normal mode of the isolated molecule. Such a frequency difference is called a crystal field splitting, which often offers useful information for quantitative estimation of the interaction energy. There are two types of the crystal field splitting, *i.e.*, correlation field splitting and static field splitting. Correlation field splitting arises from the difference in the phase of vibration among equivalent molecules in a unit cell, while static field splitting is due to the lowering of the symmetry in the environment of the vibrating groups on crystallization. The crystal modes among which correlation field splitting is observed should belong to different irreducible representations of the factor group, while those showing static field splitting may belong to the same irreducible representation. Static field splitting is seen for degenerate modes of the isolated molecule and in principle, may occur even in the case where $Z = 1$. A degenerate pair of modes in the isolated state can belong in the crystalline state, where the degeneracy is removed, to either the same irreducible representation as each other or not. Large static field splitting may arise if one component of the initially degenerate pair couples with another mode which enters the same irreducible representation as that component on the lowering of symmetry. In the case where the symmetry of the isolated molecule is maintained in the crystal, correlation field splitting arises purely from the intermolecular interaction.

Since the interaction between adjacent molecules in a crystal is generally small, we rarely observe correlation field splittings exceeding the experimental uncertainty. If certain equivalent groups of equivalent molecules happen to form a close contact with each other, the nonbonded atom–atom interaction between the contacting groups would give rise to a large correlation field splitting from which reliable information on the interaction potential can be extracted.

The calculated and the experimental values of the lattice frequencies and the CO_2^- torsional frequencies of glycine and alanine crystals[11–14] are shown in Table 9.2. For flexible molecules having skeletal single bonds, the torsional modes have comparable frequencies with

Table 9.2 Lattice frequencies and CO_2 torsional frequencies of amino acid crystals (cm^{-1}).

α-glycine		L-alanine		DL-alanine	
observed	calculated	observed	calculated	observed	calculated
a_g		a		a_1	
194	202	190	219	180	205
178	184		185	145	170
155	164	137	145	115	136
	143	112	129	102	122
109	118	103	84		79
74	73	47	42		40
51	68		37		
a_u		b_1		b_1	
226	218	192	206		211
	185	157	176	150	164
	149	138	151	122	141
	122	112	119	117	107
	68	103	114	90	93
	43		86	72	77
b_g		b_2		a_2	
	210		224	180	204
178	166	144	162	160	171
163	167	112	143	145	148
109	136	86	104	115	111
84	114	73	80	98	105
74	101	48	74	90	84
51	75				70
b_u		b_3		b_2	
	225	190	224		214
175	185	137	169	150	171
	167	128	143	115	122
140	137	103	124	98	106
91	96	98	99	90	77
		47	79	72	74

Table 9.3 Alkyl deformation frequencies (cm^{-1}) of amino acid crystals.

molecule	vibrational form	symmetry	observed	calculated
α-glycine	CH_2 bending	a_g	1441	1435
		b_g	1457	1446
α-glycine-*C*-d_2	CD_2 bending	a_g	1039	1012
		b_g	1055	1019
L-alanine	CH_3 asymmetric deformation	a	1460	1475
		b_1	1460	1475
		b_2	1481	1489
		b_3	1480	1489
DL-alanine	CH_3 asymmetric deformation	a_1	1460	1480
		a_2	1461	1480
		b_1	1482	1502
		b_2	1482	1502
DL-alanine-α,β-d_4	CD_3 asymmetric deformation	a_1	1052	1048
		a_2	1053	1048
		b_1	1065	1067
		b_2	1065	1066

the lattice modes, giving rise to appreciable mixing between them. Each of these amino acid crystals contains four molecules per unit cell, so that the crystal vibrations with $\boldsymbol{q}_K = \mathbf{0}$ are classified into four irreducible representations according to the combination of the vibrational phases among the four molecules. The number of calculated frequencies per irreducible representation equals the degrees of freedom of the external motion and the CO_2^- torsion minus the number of translational motions of the whole crystal.

Although not all of these theoretically predicted frequencies have been identified in the observed spectra, the frequency ranges of the observed bands assignable to external and torsional modes correspond well with the calculated ranges. If the model function for the intermolecular potential energy is unreasonable, the calculated frequency ranges may deviate systematically from those observed, and the number of calculated frequencies in each irreducible representation may differ from that predicted theoretically.

The calculated deformation frequencies of the alkyl groups in glycine and alanine crystals are compared with the observed values in Table 9.3. For the CH_2 bending frequencies of the α-glycine crystal and the CH_3 asymmetric deformation frequencies of the L- and DL-alanine crystals, appreciable correlation field splittings due to the nonbonded atom–atom interaction between adjacent alkyl groups are observed in polarized Raman spectra. The calculation reproduced well the pattern and frequency difference in each of these splittings.

9.3.2 Anisotropic Temperature Factors in X-ray Diffraction by Crystals

In X-ray crystallographic measurements, the interference of X-rays scattered by electrons distributed around periodically located nuclei is observed. Hence the diffraction pattern of the scattered X-rays must contain useful information on the nuclear motion in the crystal force field.[15] Suppose that a parallelepiped block of a crystal considered in section 9. 2 is irradiated with the plane wave X-ray propagating along the unit vector $\boldsymbol{e}_0$. The X-ray scattered in the direction of the unit vector $\boldsymbol{e}_1$ is specified in terms of the scattering vector defined by

$$\boldsymbol{k} = (\boldsymbol{e}_1 - \boldsymbol{e}_0)/\lambda_0 . \qquad (9\text{-}80)$$

where λ_0 is the wavelength of the X-ray. The amplitude of X-ray scattered by an isolated atom p is measured in terms of the atomic structure factor, f_p, which is roughly proportional to the number of electrons carried by p, *i.e.*, the atomic number, since electrons are responsible for X-ray scattering. The values of f_p for most elements are given in the literature as a slowly varying function of $|\boldsymbol{k}|$[16]. The intensity of the scattered X-ray per unit cell is given in the form

$$I = N^{-1} I_e \left| \sum_{J=1}^{N} \sum_{p=1}^{n} f_p \exp\left(2\pi i \tilde{\boldsymbol{k}} \boldsymbol{r}_{Jp}\right) \right|^2 = N^{-1} I_e \sum_{J,p} \sum_{J',p'} f_p f_{p'}^{\ *} \exp\left\{2\pi i \tilde{\boldsymbol{k}} \left(\boldsymbol{r}_{Jp} - \boldsymbol{r}_{J'p'}\right)\right\}, \qquad (9\text{-}81)$$

where I_e is the scattering intensity of a single electron, Derivation of Eq. (9-81) is not given here, since it is mathematically analogous to the case of coherent neutron scattering. The latter is discussed in Appendix 7 together with the incoherent neutron scattering which has no counterpart in X-ray scattering. The position vector of the nucleus Jp, $\boldsymbol{r}_{Jp}$, can be decomposed into three parts as

$$\boldsymbol{r}_{Jp} = \boldsymbol{r}_J + \boldsymbol{r}_{1p}^0 + \Delta\boldsymbol{r}_{Jp}, \qquad (9\text{-}82)$$

where $\boldsymbol{r}_J$ is the position vector of the origin of the Jth cell, $\boldsymbol{r}_{1p}^0$ is the equilibrium position vector of the nucleus $1p$, and $\Delta\boldsymbol{r}_{Jp}$ is the displacement vector of the nucleus Jp. Then we can rewrite Eq. (9-81) as

$$I = N^{-1} I_e \sum_{J,J'} \exp\left\{2\pi i \tilde{\boldsymbol{k}} \left(\boldsymbol{r}_J - \boldsymbol{r}_{J'}\right)\right\} \sum_{p,p'} F_p F_{p'}^{\ *} \exp\left\{2\pi i \tilde{\boldsymbol{k}} \left(\Delta\boldsymbol{r}_{Jp} - \Delta\boldsymbol{r}_{J'p'}\right)\right\}, \qquad (9\text{-}83)$$

where

$$F_p = f_p \exp\left(2\pi i \tilde{\boldsymbol{k}} \boldsymbol{r}_{1p}^0\right) \qquad (9\text{-}84)$$

is the atomic structure factor of the pth nucleus in a unit cell. In order to compare the calculated intensities with the observed data, the nuclear displacements $\Delta\boldsymbol{r}_{Jp}$ in Eq. (9-83) must be averaged over the whole crystal and over the time required for the measurement. The averaged expression can be derived as follows. Let the displacement of the nucleus Jp be denoted in terms of the normal coordinates as

$$\Delta\boldsymbol{r}_{Jp} = \sum_{K} \left(\boldsymbol{L}_{Jp,K}^{Xa} \boldsymbol{Q}_K^a + \boldsymbol{L}_{Jp,K}^{Xb} \boldsymbol{Q}_K^b\right) = \sum_{K,j} \left(\boldsymbol{l}_{Jp,Kj}^{Xa} Q_{Kj}^a + \boldsymbol{l}_{Jp,Kj}^{Xb} Q_{Kj}^b\right), \qquad (9\text{-}85)$$

where $\boldsymbol{L}_{Jp,K}^{Xa}$ and $\boldsymbol{L}_{Jp,K}^{Xb}$ are the $(3 \times 3n)$ matrices in the form

$$\boldsymbol{L}_{Jp,K}^{Xu} = \left[\boldsymbol{l}_{Jp,K1}^{Xu} \quad \boldsymbol{l}_{Jp,K2}^{Xu} \quad \cdots \quad \boldsymbol{l}_{Jp,K3n}^{Xu}\right], \qquad (u = a \text{ or } b)$$

which relates the displacement $\Delta\boldsymbol{r}_{Jp}$ to the vectors $\boldsymbol{Q}_K{}^a$ and $\boldsymbol{Q}_K{}^b$, respectively. Explicit forms of the vectors $\boldsymbol{l}^{Xu}_{Jp,Kj}$ will be given later. The exponential factor in the sum with respect to p and p' in Eq. (9-83) is rewritten, with the help of Eq. (9-85), as

$$\exp\{2\pi i\tilde{\boldsymbol{k}}(\Delta\boldsymbol{r}_{Jp}-\Delta\boldsymbol{r}_{J'p'})\} = \exp\left[2\pi i\tilde{\boldsymbol{k}}\sum_{K,j}\left\{(\boldsymbol{l}^{Xa}_{Jp,Kj}-\boldsymbol{l}^{Xa}_{J'p',Kj})Q^a_{Kj}+(\boldsymbol{l}^{Xb}_{Jp,Kj}-\boldsymbol{l}^{Xb}_{J'p',Kj})Q^b_{Kj}\right\}\right]$$
$$= \prod_{K,j}\exp\{2\pi i\tilde{\boldsymbol{k}}(\boldsymbol{l}^{Xa}_{Jp,Kj}-\boldsymbol{l}^{Xa}_{J'p',Kj})Q^a_{Kj}\}\exp\{2\pi i\tilde{\boldsymbol{k}}(\boldsymbol{l}^{Xb}_{Jp,Kj}-\boldsymbol{l}^{Xb}_{J'p',Kj})Q^b_{Kj}\}. \tag{9-86}$$

Under the harmonic approximation, a normal coordinate Q_{Kj} with the normal frequency ω_{Kj} satisfies the relation[17,18)]

$$\langle\exp(2\pi i A Q_{Kj})\rangle = \exp(-2\pi^2 A^2\langle Q_{Kj}{}^2\rangle)$$
$$= \exp\left\{-\frac{hA^2}{4c\omega_{Kj}}\coth\left(\frac{hc\omega_{Kj}}{2kT}\right)\right\}, \tag{9-87}$$

where A is an arbitrary constant. The first equality in Eq. (9-87) is proved in Appendix 4, while the second equality follows from Eq. (5-122). Let the constant A in Eq. (9-87) be $A=\tilde{\boldsymbol{k}}(\boldsymbol{l}^{Xa}_{Jp,Kj}-\boldsymbol{l}^{Xa}_{J'p',Kj})$. It then follows that

$$A^2 = \left\{\tilde{\boldsymbol{k}}(\boldsymbol{l}^{Xa}_{Jp,Kj}-\boldsymbol{l}^{Xa}_{J'p',Kj})\right\}^2 = \tilde{\boldsymbol{k}}(\boldsymbol{l}^{Xa}_{Jp,Kj}-\boldsymbol{l}^{Xa}_{J'p',Kj})(\tilde{\boldsymbol{l}}^{Xa}_{Jp,Kj}-\tilde{\boldsymbol{l}}^{Xa}_{J'p',Kj})\boldsymbol{k}$$
$$= \tilde{\boldsymbol{k}}(\boldsymbol{l}^{Xa}_{Jp,Kj}\tilde{\boldsymbol{l}}^{Xa}_{Jp,Kj}-\boldsymbol{l}^{Xa}_{Jp,Kj}\tilde{\boldsymbol{l}}^{Xa}_{J'p',Kj}-\boldsymbol{l}^{Xa}_{J'p',Kj}\tilde{\boldsymbol{l}}^{Xa}_{Jp,Kj}+\boldsymbol{l}^{Xa}_{J'p',Kj}\tilde{\boldsymbol{l}}^{Xa}_{J'p',Kj})\boldsymbol{k} \tag{9-88}$$
$$= \tilde{\boldsymbol{k}}(\boldsymbol{l}^{Xa}_{Jp,Kj}\tilde{\boldsymbol{l}}^{Xa}_{Jp,Kj}-2\boldsymbol{l}^{Xa}_{Jp,Kj}\tilde{\boldsymbol{l}}^{Xa}_{J'p',Kj}+\boldsymbol{l}^{Xa}_{J'p',Kj}\tilde{\boldsymbol{l}}^{Xa}_{J'p',Kj})\boldsymbol{k},$$

since $\tilde{\boldsymbol{k}}\boldsymbol{l}^{Xa}_{Jp,Kj}\tilde{\boldsymbol{l}}^{Xa}_{J'p',Kj}\boldsymbol{k} = \tilde{\boldsymbol{k}}\boldsymbol{l}^{Xa}_{J'p',Kj}\tilde{\boldsymbol{l}}^{Xa}_{Jp,Kj}\boldsymbol{k}$.

By referring to Eqs. (9-87) and (9-88), the average of the product of exponential factors involving Q^a_{Kj} and Q^b_{Kj} in the third formula of Eq. (9-86) is written as

$$\langle\exp\{2\pi i\tilde{\boldsymbol{k}}(\boldsymbol{l}^{Xa}_{Jp,Kj}-\boldsymbol{l}^{Xa}_{J'p',Kj})Q^a_{Kj}\}\rangle\langle\exp\{2\pi i\tilde{\boldsymbol{k}}(\boldsymbol{l}^{Xb}_{Jp,Kj}-\boldsymbol{l}^{Xb}_{J'p',Kj})Q^b_{Kj}\}\rangle$$
$$= \exp\left\{-h\tilde{\boldsymbol{k}}\left(\boldsymbol{l}^{Xa}_{Jp,Kj}\tilde{\boldsymbol{l}}^{Xa}_{Jp,Kj}+\boldsymbol{l}^{Xb}_{Jp,Kj}\tilde{\boldsymbol{l}}^{Xb}_{Jp,Kj}+\boldsymbol{l}^{Xa}_{J'p',Kj}\tilde{\boldsymbol{l}}^{Xa}_{J'p',Kj}+\boldsymbol{l}^{Xb}_{J'p',Kj}\tilde{\boldsymbol{l}}^{Xb}_{J'p',Kj}\right.\right.$$
$$\left.\left.-2\boldsymbol{l}^{Xa}_{Jp,Kj}\tilde{\boldsymbol{l}}^{Xa}_{J'p',Kj}-2\boldsymbol{l}^{Xb}_{Jp,Kj}\tilde{\boldsymbol{l}}^{Xb}_{J'p',Kj}\right)\boldsymbol{k}\frac{\coth(hc\omega_{Kj}/2kT)}{4c\omega_{Kj}}\right\} \tag{9-89}$$

According to Eqs.(9-61) and (9-76a,b), the Cartesian displacements of nuclei in the Jth unit cell are given in terms of the normal coordinates as

$$\Delta\boldsymbol{x}_J = \left(\frac{2}{N}\right)^{1/2}\sum_{K=1}^{N/2}\{(\boldsymbol{L}^x_{Kc}\boldsymbol{Q}^a_K-\boldsymbol{L}^x_{Ks}\boldsymbol{Q}^b_K)\cos\theta_{KJ}+(\boldsymbol{L}^x_{Ks}\boldsymbol{Q}^a_K+\boldsymbol{L}^x_{kc}\boldsymbol{Q}^b_K)\sin\theta_{KJ}\}$$
$$= (2/N)^{1/2}\boldsymbol{m}_1{}^{-1/2}\sum_{K=1}^{N/2}\{(\boldsymbol{L}^{mx}_{Kc}\cos\theta_{KJ}+\boldsymbol{L}^{mx}_{Ks}\sin\theta_{KJ})\boldsymbol{Q}^a_K+(\boldsymbol{L}^{mx}_{Kc}\sin\theta_{KJ}-\boldsymbol{L}^{mx}_{Ks}\cos\theta_{KJ})\boldsymbol{Q}^b_K\}. \tag{9-90}$$

Let us partition the $(3n \times 3n)$ matrices $\boldsymbol{L}_{Kd}^{mx}$ ($d = c$ or s) in Eq. (9-90) into $n \times 3n$ blocks of (3×1) vectors, called atomic polarization vectors, in the form

$$\boldsymbol{L}_{Kd}^{mx} = \begin{bmatrix} (\boldsymbol{l}_{Kd}^{mx})_{11} & (\boldsymbol{l}_{Kd}^{mx})_{12} & \cdots & (\boldsymbol{l}_{Kd}^{mx})_{1,3n} \\ (\boldsymbol{l}_{Kd}^{mx})_{21} & (\boldsymbol{l}_{Kd}^{mx})_{22} & \cdots & (\boldsymbol{l}_{Kd}^{mx})_{2,3n} \\ \vdots & \vdots & \ddots & \vdots \\ (\boldsymbol{l}_{Kd}^{mx})_{n1} & (\boldsymbol{l}_{Kd}^{mx})_{n2} & \cdots & (\boldsymbol{l}_{Kd}^{mx})_{n,3n} \end{bmatrix}. \tag{9-91}$$

Then the vectors $\boldsymbol{l}_{Jp,Kj}^{Xa}$ and $\boldsymbol{l}_{Jp,Kj}^{Xb}$ introduced in Eq. (9-85) can be related to certain atomic polarization vectors in the pth row in the right side of Eq. (9-91) as

$$\boldsymbol{l}_{Jp,Kj}^{Xa} = (2/N m_p)^{1/2}\left\{(\boldsymbol{l}_{Kc}^{mx})_{pj}\cos\theta_{KJ} + (\boldsymbol{l}_{Ks}^{mx})_{pj}\sin\theta_{KJ}\right\} \tag{9-92a}$$

$$\boldsymbol{l}_{Jp,Kj}^{Xb} = (2/N m_p)^{1/2}\left\{(\boldsymbol{l}_{Kc}^{mx})_{pj}\sin\theta_{KJ} - (\boldsymbol{l}_{Ks}^{mx})_{pj}\cos\theta_{KJ}\right\}. \tag{9-92b}$$

By using Eqs. (9-92a,b), the sum of the first two terms in the parentheses between $\tilde{\boldsymbol{k}}$ and $\boldsymbol{k}$ in the right side of Eq. (9-89) is rewritten as

$$\begin{aligned} &\boldsymbol{l}_{Jp,Kj}^{Xa}\tilde{\boldsymbol{l}}_{Jp,Kj}^{Xa} + \boldsymbol{l}_{Jp,Kj}^{Xb}\tilde{\boldsymbol{l}}_{Jp,Kj}^{Xb} \\ &= \frac{2}{N m_p}\Big[\left\{(\boldsymbol{l}_{Kc}^{mx})_{pj}\cos\theta_{KJ} + (\boldsymbol{l}_{Ks}^{mx})_{pj}\sin\theta_{KJ}\right\}\left\{(\tilde{\boldsymbol{l}}_{Kc}^{mx})_{pj}\cos\theta_{KJ} + (\tilde{\boldsymbol{l}}_{Ks}^{mx})_{pj}\sin\theta_{KJ}\right\} \\ &\quad + \left\{(\boldsymbol{l}_{Kc}^{mx})_{pj}\sin\theta_{KJ} - (\boldsymbol{l}_{Ks}^{mx})_{pj}\cos\theta_{KJ}\right\}\left\{(\tilde{\boldsymbol{l}}_{Kc}^{mx})_{pj}\sin\theta_{KJ} - (\tilde{\boldsymbol{l}}_{Ks}^{mx})_{pj}\cos\theta_{KJ}\right\}\Big] \\ &= \frac{2}{N m_p}\left\{(\boldsymbol{l}_{Kc}^{mx})_{pj}(\tilde{\boldsymbol{l}}_{Kc}^{mx})_{pj} + (\boldsymbol{l}_{Ks}^{mx})_{pj}(\tilde{\boldsymbol{l}}_{Ks}^{mx})_{pj}\right\}. \end{aligned} \tag{9-93}$$

Note that the index J disappears from the final expression of Eq. (9-93), whereby a programable formula for the subsequent calculation is obtained. Similarly, the sum of the third and fourth terms in the same parentheses in Eq. (9-89) is calculated by replacing p in Eq. (9-93) with p'. The sum of the fifth and sixth terms is calculated as

$$\begin{aligned} &-2\left(\boldsymbol{l}_{Jp,Kj}^{Xa}\tilde{\boldsymbol{l}}_{J'p',Kj}^{Xa} + \boldsymbol{l}_{Jp,Kj}^{Xb}\tilde{\boldsymbol{l}}_{J'p',Kj}^{Xb}\right) \\ &= \frac{-4}{N(m_p m_{p'})^{1/2}}\Big[\left\{(\boldsymbol{l}_{Kc}^{mx})_{pj}(\tilde{\boldsymbol{l}}_{Kc}^{mx})_{p'j} + (\boldsymbol{l}_{Ks}^{mx})_{pj}(\tilde{\boldsymbol{l}}_{Ks}^{mx})_{p'j}\right\}\cos(\theta_{KJ} - \theta_{KJ'}) \\ &\quad - \left\{(\boldsymbol{l}_{Kc}^{mx})_{pj}(\tilde{\boldsymbol{l}}_{Ks}^{mx})_{p'j} - (\boldsymbol{l}_{Ks}^{mx})_{pj}(\tilde{\boldsymbol{l}}_{Kc}^{mx})_{p'j}\right\}\sin(\theta_{KJ} - \theta_{KJ'})\Big]. \end{aligned} \tag{9-94}$$

The second term in brackets in Eq. (9-94) vanishes on taking the sum with respect to p and p' because the exchange of these indices always produces a pair of canceling terms. The first term including the cosine factor is related to a (3×3) subblock $\boldsymbol{\Sigma}_{pp'}^{x}(T)$ of the mean square amplitude matrix per unit cell, $\boldsymbol{\Sigma}^{x}(T)$, which describes the correlation between the motions of nuclei in a unit cell just as in the case of isolated molecules in Eq. (5-123). Each subblock of $\boldsymbol{\Sigma}^x(T)$ may be expressed as the sum of the contributions from the normal modes belonging to individual wave vectors,

$$\boldsymbol{\Sigma}^x_{pp'}(T) = \sum_{K=1}^{N} \boldsymbol{\Sigma}^x_{pp'}(T)_K \ .$$

Since the normal coordinates Q^a_{Kj} and Q^b_{Kj} are equivalent to each other with respect to mean square amplitudes, we may write, in analogy with Eq. (5-122) for an isolated molecule, the squared average in the form

$$\left\langle Q^{a\ 2}_{Kj} \right\rangle^T = \left\langle Q^{b\ 2}_{Kj} \right\rangle^T = \Sigma^Q_{Kj,Kj,}(T).$$

By referring to Eqs. (5-122) and (9-94), the contribution from the wave vector $\boldsymbol{q}_K$ to $\boldsymbol{\Sigma}^x_{pp'}(T)$ is given by

$$\begin{aligned}\boldsymbol{\Sigma}^x_{pp'}(T)_K &= \left\langle \Delta \boldsymbol{r}_{Jp}\, \Delta \tilde{\boldsymbol{r}}_{Jp'} \right\rangle^T_K = \sum_j \left\{ \boldsymbol{l}^{Xa}_{Jp,Kj} \left\langle Q^{a\ 2}_{Kj} \right\rangle^T \tilde{\boldsymbol{l}}^{Xa}_{Jp',Kj} + \boldsymbol{l}^{Xb}_{Jp,Kj} \left\langle Q^{b\ 2}_{Kj} \right\rangle^T \tilde{\boldsymbol{l}}^{Xb}_{Jp',Kj} \right\} \\ &= \sum_j \left(\boldsymbol{l}^{Xa}_{Jp,Kj}\, \tilde{\boldsymbol{l}}^{Xa}_{Jp',Kj} + \boldsymbol{l}^{Xb}_{Jp,Kj}\, \tilde{\boldsymbol{l}}^{Xb}_{Jp',Kj} \right) \Sigma^Q_{Kj,Kj}(T) \qquad (9\text{-}95) \\ &= h \sum_j \boldsymbol{D}^{Kj}_{pp'} \coth\left(hc\omega_{Kj}/2kT\right) \Big/ \left\{ 4\pi^2 N \left(m_p m_{p'}\right)^{1/2} c\omega_{Kj} \right\},\end{aligned}$$

where $\boldsymbol{D}^{Kj}_{pp'}$ is a (3×3) matrix defined by

$$\boldsymbol{D}^{Kj}_{pp'} = \left(\boldsymbol{l}^{mx}_{Kc}\right)_{pj} \left(\tilde{\boldsymbol{l}}^{mx}_{Kc}\right)_{p'j} + \left(\boldsymbol{l}^{mx}_{Ks}\right)_{pj} \left(\tilde{\boldsymbol{l}}^{mx}_{Ks}\right)_{p'j} \ . \qquad (9\text{-}96)$$

Let us introduce the dimensionless scalar quantities

$$W_p = 2\pi^2 \tilde{\boldsymbol{k}} \left\{ \sum_K \boldsymbol{\Sigma}^x_{pp}(T)_K \right\} \boldsymbol{k} \qquad (9\text{-}97a)$$

and

$$U^{JJ'}_{pp'} = 4\pi^2 \tilde{\boldsymbol{k}} \left\{ \sum_K \boldsymbol{\Sigma}^x_{pp'}(T)_K \cos(\theta_{KJ} - \theta_{KJ'}) \right\} \boldsymbol{k}. \qquad (9\text{-}97b)$$

The parameter W_p is called the Debye–Waller factor.[19,20] The sum with respect to p and p' in Eq. (9-83) is now thermally averaged, by using Eqs. (9-86), (9-89), (9-93~95) and (9-97a,b), and by expanding the exponential factor including $U^{JJ'}_{pp'}$. The result is given in the form

$$\begin{aligned}&\sum_{p,p'} F_p F_{p'}^* \exp(-2W_p) \exp(-2W_{p'}) \exp(U^{JJ'}_{pp'}) \\ &= \sum_{p,p'} F_p F_{p'}^* \exp(-2W_p) \exp(-2W_{p'}) \left\{ 1 + \sum_{n=1}^{\infty} (U^{JJ'}_{pp'})^n / n! \right\}.\end{aligned} \qquad (9\text{-}98)$$

Furthermore, by introducing the temperature-dependent structure factor,

$$F^W_p = F_p \exp(-2W_p), \qquad (9\text{-}99)$$

the total intensity is given by

$$I = I_0 + I_1 + I_2 + \cdots \tag{9-100a}$$

where

$$I_0 = N^{-1} I_e \sum_{J,J'} \exp\{2\pi i \tilde{\boldsymbol{k}}(\boldsymbol{r}_J - \boldsymbol{r}_{J'})\} \sum_p F_p^W \sum_{p'} F_{p'}^{W*} \tag{9-100b}$$

and

$$I_n = N^{-1} I_e \sum_{J,J'} \exp\{2\pi i \tilde{\boldsymbol{k}}(\boldsymbol{r}_J - \boldsymbol{r}_{J'})\} \sum_{p,p'} F_p^W F_{p'}^{W*} (U_{pp'}^{JJ'})^n / n!. \tag{9-100c}$$

In Eq. (9-100a), the scattering due to the term I_0 is called the Laue–Bragg scattering while that due to I_n is called the temperature diffuse scattering of the nth order. For I_0 in Eq. (9-100b), the indices J and J' do not appear in the second and third sums, so that the first sum can be calculated separately as

$$\sum_{J,J'} \exp\{2\pi i \tilde{\boldsymbol{k}}(\boldsymbol{r}_J - \boldsymbol{r}_{J'})\} = \frac{\sin^2(\pi k_a N_a)}{\sin^2(\pi k_a)} \frac{\sin^2(\pi k_b N_b)}{\sin^2(\pi k_b)} \frac{\sin^2(\pi k_c N_c)}{\sin^2(\pi k_c)}. \tag{9-101}$$

Derivation of Eq. (9-101), which is known as the Laue function, is given in Appendix 5. The Laue function divided by NV^{-1} gives unity when integrated over a reciprocal unit cell, but diminishes to zero at the limit $N \to \infty$ unless $\boldsymbol{k}$ is very near one of the reciprocal lattice vectors defined by

$$\boldsymbol{h}_{hkl} \equiv h\boldsymbol{a}^* + k\boldsymbol{b}^* + l\boldsymbol{c}^*, \tag{9-102a}$$

where h, k and l are the Miller indices. According to Eqs. (9-54) and (A3-2), we can rewrite Eq. (9-102a) as

$$\boldsymbol{h}_{hkl} = [\boldsymbol{a}^* \quad \boldsymbol{b}^* \quad \boldsymbol{c}^*]\boldsymbol{k}_c = \tilde{\boldsymbol{C}}_{r0}{}^{-1}\boldsymbol{k}_c, \tag{9-102b}$$

where $\boldsymbol{k}_c$ is the vector of Miller indices defined by

$$\boldsymbol{k}_c = [h \quad k \quad l]^{\mathrm{T}}. \tag{9-103}$$

From Eq. (9-101), the Laue function can be regarded as a periodic δ-function in the three-dimensional reciprocal space; see Appendix 5. Because of this property, the Laue–Bragg scattering gives rise to distinct spots of diffraction at the reciprocal lattice points where it holds that

$$\boldsymbol{k} - \boldsymbol{h}_{hkl} = \boldsymbol{0}. \tag{9-104}$$

The Debye–Waller factor defined by Eq. (9-97a) is now expressed, with the help of Eqs. (9-104) and (9-102b), in terms of the Miller indices in the form

$$\begin{aligned} W_p &= 2\pi^2 \tilde{\boldsymbol{h}} \left\{ \sum_K \boldsymbol{\Sigma}_{pp}^x(T)_K \right\} \boldsymbol{h} = \tilde{\boldsymbol{k}}_c \boldsymbol{B}_p \boldsymbol{k}_c \\ &= B_{11}h^2 + B_{22}k^2 + B_{33}l^2 + 2B_{12}hk + 2B_{13}hl + 2B_{23}kl. \end{aligned} \tag{9-105}$$

The coefficients B_{ij} are the anisotropic temperature factors, which constitute a part of parameter set for fitting the experimental data in X-ray crystallographic analyses. By using the well-known relation between the hyperbolic and the exponential functions (Eq. (5-114)), the anisotropic temperature factors in Eq. (9-105) are given as the components of the tensor in the form

$$\boldsymbol{B}_p = \frac{h}{N m_p} \boldsymbol{C}_{r0}^{-1} \left(\sum_{K,j} \frac{\boldsymbol{D}_{pp}^{Kj}}{2 c \omega_{Kj}} \frac{1 + x_{Kj}}{1 - x_{Kj}} \right) \tilde{\boldsymbol{C}}_{r0}^{-1}, \tag{9-106}$$

where

$$x_{Kj} = \exp(-h c \omega_{Kj} / k T). \tag{9-107}$$

In actual calculation, the sum with respect to K in Eq. (9-106) is taken over a set of those wave vectors $\boldsymbol{q}_K$ which point from the origin toward the homogeneously distributed lattice points in a half volume of the first Brillouin zone specified as part a in section 9.3.1. In this process, a difficulty arises from the fact that certain terms with vanishing denominators inevitably enter the sum. For small ω_{Kj}, the quantity x_{Kj} in Eq. (9-107) is expanded as

$$x_{Kj} = 1 - \frac{h c \omega_{Kj}}{k T} + \frac{1}{2} \left(\frac{h c \omega_{Kj}}{k T} \right)^2 - \frac{1}{3!} \left(\frac{h c \omega_{Kj}}{k T} \right)^3 + \cdots \tag{9-108}$$

so that a term with small ω_{Kj} takes the divergent form

$$\frac{\boldsymbol{D}_{pp}^{Kj}}{2 c \omega_{Kj}} \frac{1 + x_{Kj}}{1 - x_{Kj}} = \frac{k T \boldsymbol{D}_{pp}^{Kj}}{h c^2 \omega_{Kj}^2} \left\{ 1 + \frac{1}{12} \left(\frac{h c \omega_{Kj}}{k T} \right)^2 + \cdots \right\}. \tag{9-109}$$

This difficulty can be avoided by replacing the divergent part of the sum in Eq. (9-106) with a definite integral of a function of the frequency from zero to a certain frequency ω_D.[21] The integrand is the product of the right side of Eq. (9-109), in which ω_{Kj} is replaced by a continuous variable ω, and a distribution function $g(\omega)$ playing the role of a weight. Here we must note that the elements of $\boldsymbol{D}_{pp}^{Kj}$ are related to ω through the dependence on K. The functional forms of $g(\omega)$ and $\boldsymbol{D}_{pp}^{Kj}$ are determined as follows.

As proved later, the frequency of an acoustic mode is proportional to the phase of the wave vector defined by Eq. (9-77) if $|\boldsymbol{q}_K| \ll 1$,[21] as given by

$$\omega_{Ki} = u_{ai}\,\delta_{Ka} + u_{bi}\,\delta_{Kb} + u_{ci}\,\delta_{Kc}. \tag{9-110}$$

This approximation is well known as the Debye approximation. Representing the phase space in a polar coordinate system, we have

$$\delta_{Ka} = \delta_K \sin\theta_K \cos\phi_K, \quad \delta_{Kb} = \delta_K \sin\theta_K \sin\phi_K \quad \text{and} \quad \delta_{Kc} = \delta_K \cos\theta_K.$$

The length of the vector $[\delta_{Ka} \quad \delta_{Kb} \quad \delta_{Kc}]^{\mathrm{T}}$ is given by

$$\delta_K = \delta_{Ka} \sin\theta_K \cos\phi_K + \delta_{Kb} \sin\theta_K \sin\phi_K + \delta_{Kc} \cos\theta_K. \tag{9-111}$$

By using δ_K, Eq. (9-110) is rewritten as

$$\omega_{Ki} = C(\theta_K, \phi_K)\,\delta_K. \tag{9-112}$$

The coefficients in Eqs. (9-110) and (9-112) are related to each other by

$$C(\theta_K, \phi_K) = u_{ai} / \sin\theta_K \cos\phi_K = u_{bi} / \sin\theta_K \sin\phi_K = u_{ci} / \cos\theta_K. \tag{9-113}$$

Now we assume that the Debye approximation is valid for acoustic modes with frequencies below ω_D. The phase corresponding to the frequency ω is represented by a point in the phase space located on a closed surface depicted by the tip of a radial vector whose length depends on the direction as given by

$$\delta_D = \omega_D / C(\theta, \phi). \tag{9-114}$$

The volume of the domain surrounded by such an iso-frequency surface is calculated by integrating the volume element $\delta^2 \sin\theta \, d\theta \, d\phi \, d\delta$ of the phase space and making use of Eq. (9-114).[7] The result is written in the form

$$\begin{aligned} V_D &= \int_0^{2\pi}\int_0^{\pi}\int_0^{\delta_D} \delta^2 \sin\theta \, d\theta \, d\phi \, d\delta = \int_0^{2\pi}\int_0^{\pi} (\delta_D{}^3/3) \sin\theta \, d\theta \, d\phi \\ &= (\omega_D{}^3/3) \int_0^{2\pi}\int_0^{\pi} C(\theta,\phi)^{-3} \sin\theta \, d\theta \, d\phi. \end{aligned} \tag{9-115}$$

The double integral in the last formula of Eq. (9-115) is evaluated numerically by replacing the integral with a sum in the form

$$S \equiv \int_0^{2\pi}\int_0^{\pi} \frac{\sin\theta}{C(\theta,\phi)^3} d\theta \, d\phi = \frac{\pi^2}{nm} \sum_{i=1}^{n} \sum_{j=1}^{m} \sin\frac{\pi i}{n} \Big/ C\left(\frac{\pi i}{n}, \frac{2\pi j}{m}\right)^3. \tag{9-116}$$

Since there are $3N$ acoustic modes homogeneously distributed in the whole volume $(2\pi)^3$ of the phase space, the number of acoustic modes with the frequencies between ω and $\omega + d\omega$ is given by

$$g(\omega) d\omega = \{3N/(2\pi)^3\} (\partial V_D / \partial \omega) d\omega = \{3N/(2\pi)^3\} S \omega^2 d\omega.$$

The distribution function is thus given by

$$g(\omega) = \{3N/(2\pi)^3\} S \omega^2 = a \omega^2. \tag{9-117}$$

If the wave vector $\boldsymbol{q}_K$ is taken along the $\boldsymbol{a}^*$-axis, it follows from Eqs. (9-4), (9-57) and (9-77) that the angle $\theta_{KI} = 2\pi \tilde{\boldsymbol{q}}_K \boldsymbol{r}_I$ can always be represented as $\theta_{KI} = l_a \delta_{ka}$. In the vicinity of the origin of the first Brillouin zone, the dynamical matrix for such $\boldsymbol{q}_K$ can be expanded in powers of δ_{Ka} as

$$\boldsymbol{F}_K^{mxx} = \begin{bmatrix} \boldsymbol{F}_{1c}^{mxx} + \sum_{n=1}^{\infty} \boldsymbol{F}'_{ca,2n} \delta_{Ka}{}^{2n} & -\sum_{n=1}^{\infty} \boldsymbol{F}'_{sa,2n-1} \delta_{Ka}{}^{2n-1} \\ \sum_{n=1}^{\infty} \boldsymbol{F}'_{sa,2n-1} \delta_{Ka}{}^{2n-1} & \boldsymbol{F}_{1c}^{mxx} + \sum_{n=1}^{\infty} \boldsymbol{F}'_{ca,2n} \delta_{Ka}{}^{2n} \end{bmatrix}. \tag{9-118}$$

By referring to Eqs. (9-67) and (9-73), and remembering the relations

$$\sin x = x - \tfrac{1}{3!} x^3 + \tfrac{1}{5!} x^5 - \cdots \quad \text{and} \quad \cos x = 1 - \tfrac{1}{2} x^2 + \tfrac{1}{4!} x^4 - \cdots.$$

the matrices of the coefficients in the perturbing terms in the diagonal and the off-diagonal subblocks in the right side of Eq. (9-118) are given by

$$\boldsymbol{F}'_{ca,2n} = (-1)^n \sum_{L} \{L_a{}^{2n}/(2n)!\} \boldsymbol{m}_1{}^{-1/2} \boldsymbol{F}_{1L}^{xx} \boldsymbol{m}_1{}^{-1/2}$$

and

$$\boldsymbol{F}'_{sa,2n-1} = (-1)^{n-1} \sum_{L} \{ L_a{}^{2n-1}/(2n-1)! \} \boldsymbol{m}_1{}^{-1/2} \boldsymbol{F}_{1L}^{xx} \boldsymbol{m}_1{}^{-1/2},$$

respectively, where L_a is determined for a given L according to the rule of counting the unit cells as illustrated by I_a and I in Eq. (9-5). The dynamical matrix in the form of Eq. (9-118) is diagonalized, within the framework of the first order perturbation theory, by taking the terms with the smallest powers in δ_{Ka} in each subblock. The result is written in the form

$$\tilde{\boldsymbol{L}}_K^{mx} \begin{bmatrix} \boldsymbol{F}_{1c}^{mxx} + \boldsymbol{F}'_{ca,2}\delta_{Ka}{}^2 & -\boldsymbol{F}'_{sa,1}\delta_{Ka} \\ \boldsymbol{F}'_{sa,1}\delta_{Ka} & \boldsymbol{F}_{1c}^{mxx} + \boldsymbol{F}'_{ca,2}\delta_{Ka}{}^2 \end{bmatrix} \boldsymbol{L}_K^{mx} = \begin{bmatrix} \boldsymbol{\Lambda}_1 + \boldsymbol{\Lambda}'_a\delta_{Ka}{}^2 & \boldsymbol{0} \\ \boldsymbol{0} & \boldsymbol{\Lambda}_1 + \boldsymbol{\Lambda}'_a\delta_{Ka}{}^2 \end{bmatrix}, \quad (9\text{-}119)$$

where the matrix of the eigenvectors has the form

$$\boldsymbol{L}_K^{mx} = \begin{bmatrix} \boldsymbol{L}_{1c}^{mx} & \boldsymbol{0} \\ \boldsymbol{0} & \boldsymbol{L}_{1c}^{mx} \end{bmatrix} \begin{bmatrix} \boldsymbol{E} + \boldsymbol{P}'_{ca}\delta_{Ka}{}^2 & -\boldsymbol{P}'_{sa}\delta_{Ka} \\ \boldsymbol{P}'_{sa}\delta_{Ka} & \boldsymbol{E} + \boldsymbol{P}'_{ca}\delta_{Ka}{}^2 \end{bmatrix}. \quad (9\text{-}120)$$

Equation (9-119) corresponds to Eq. (9-75). Note that the perturbing terms in the diagonal and the off-diagonal subblocks in the right sides of Eqs. (9-119) and (9-120) are quadratic and linear in δ_{Ka}, respectively. This result is confirmed as follows. By writing $\boldsymbol{L}_K^{mx}$ in terms of a small matrix $\boldsymbol{P}_K$ in the form $\boldsymbol{L}_K^{mx} = \boldsymbol{L}_1^{mx}(\boldsymbol{E} + \boldsymbol{P}_K)$,
we have

$$\tilde{\boldsymbol{L}}_K^{mx}\boldsymbol{L}_K^{mx} = (\boldsymbol{E} + \tilde{\boldsymbol{P}}_K)\tilde{\boldsymbol{L}}_1^{mx}\boldsymbol{L}_1^{mx}(\boldsymbol{E} + \boldsymbol{P}_K) = \boldsymbol{E} + \tilde{\boldsymbol{P}}_K + \boldsymbol{P}_K + \tilde{\boldsymbol{P}}_K\boldsymbol{P}_K. \quad (9\text{-}121)$$

In order for $\boldsymbol{L}_K^{mx}$ to be orthogonal to the first order of approximation, the terms $\tilde{\boldsymbol{P}}_K + \boldsymbol{P}_K$ in the last formula of Eq. (9-121) must vanish, *i.e.*, $\boldsymbol{P}_K$ must be antisymmetric. On partitioning $\boldsymbol{P}_K$ into four blocks as

$$\boldsymbol{P}_K = \begin{bmatrix} \boldsymbol{P}_{Kc} & -\boldsymbol{P}_{Ks} \\ \boldsymbol{P}_{Ks} & \boldsymbol{P}_{Kc} \end{bmatrix}, \quad (9\text{-}122a)$$

the transpose of $\boldsymbol{P}_K$ is given by

$$\tilde{\boldsymbol{P}}_K = \begin{bmatrix} \tilde{\boldsymbol{P}}_{Kc} & \tilde{\boldsymbol{P}}_{Ks} \\ -\tilde{\boldsymbol{P}}_{Ks} & \tilde{\boldsymbol{P}}_{Kc} \end{bmatrix} = \begin{bmatrix} -\boldsymbol{P}_{Kc} & \boldsymbol{P}_{Ks} \\ -\boldsymbol{P}_{Ks} & -\boldsymbol{P}_{Kc} \end{bmatrix}. \quad (9\text{-}122b)$$

It then follows that $\boldsymbol{P}_{Kc}$ is antisymmetric and $\boldsymbol{P}_{Ks}$ is symmetric. By expressing the dynamical matrix $\boldsymbol{F}_K^{mxx}$ as $\boldsymbol{F}_1^{mxx} + \delta\boldsymbol{F}_K^{mxx}$, the congruence transformation in Eq. (9-119) is rewritten in the form

$$\begin{aligned} &(\boldsymbol{E} - \boldsymbol{P}_K)\tilde{\boldsymbol{L}}_1^{mx}(\boldsymbol{F}_1^{mxx} + \delta\boldsymbol{F}_K^{mxx})\boldsymbol{L}_1^{mx}(\boldsymbol{E} + \boldsymbol{P}_K) \\ &= \boldsymbol{\Lambda}_1 - \boldsymbol{P}_K\boldsymbol{\Lambda}_1 + \boldsymbol{\Lambda}_1\boldsymbol{P}_K + \tilde{\boldsymbol{L}}_1^{mx}(\delta\boldsymbol{F}_K^{mxx})\boldsymbol{L}_1^{mx} + \text{higher order terms.} \end{aligned} \quad (9\text{-}123)$$

The fourth term in the right side of Eq. (9-123) is rewritten, by introducing the abbreviations, $\delta\boldsymbol{\Lambda}_{Kc} = \tilde{\boldsymbol{L}}_{1c}^{mx}\boldsymbol{F}'_{ca,2}\boldsymbol{L}_{1c}^{mx}\delta_{Ka}{}^2$ and $\delta\boldsymbol{\Lambda}_{Ks} = \tilde{\boldsymbol{L}}_{1c}^{mx}\boldsymbol{F}'_{sa,1}\boldsymbol{L}_{1c}^{mx}\delta_{Ka}$, in the form

$$\tilde{L}_1^{mx}(\delta F_K^{mxx})L_1^{mx} = \delta \Lambda_K = \begin{bmatrix} \delta \Lambda_{Kc} & -\delta \Lambda_{Ks} \\ \delta \Lambda_{Ks} & \delta \Lambda_{Kc} \end{bmatrix}. \tag{9-124}$$

By substituting Eqs. (9-122a) and (9-124) into Eq. (9-123), the first order terms in the right side can be rewritten in the partitioned form as

$$\begin{bmatrix} -P_{Kc}\Lambda_1 & P_{Ks}\Lambda_1 \\ -P_{Ks}\Lambda_1 & -P_{Kc}\Lambda_1 \end{bmatrix} + \begin{bmatrix} \Lambda_1 P_{Kc} & -\Lambda_1 P_{Ks} \\ \Lambda_1 P_{Ks} & \Lambda_1 P_{Kc} \end{bmatrix} + \begin{bmatrix} \delta \Lambda_{Kc} & -\delta \Lambda_{Ks} \\ \delta \Lambda_{Ks} & \delta \Lambda_{Kc} \end{bmatrix}. \tag{9-125}$$

From the requirement that the off-diagonal terms in the (1,1) submatrix of Eq. (9-125) must vanish, it follows that

$$-(P_{Kc})_{ij}(\Lambda_1)_{jj} + (\Lambda_1)_{ii}(P_{Kc})_{ij} + (\delta\Lambda_{Kc})_{ij} = 0, ,$$

or

$$(P_{Kc})_{ij} = (\delta\Lambda_{Kc})_{ij} \Big/ \left\{(\Lambda_1)_{ii} - (\Lambda_1)_{jj}\right\}.$$

Since the elements of $\delta\Lambda_{Kc}$ are quadratic in δ_{Ka} and those of Λ_1 do not involve δ_{ka}, the elements of P_{Kc} must be quadratic in δ_{Ka}. From a similar argument on the (2,1) submatrix of Eq. (9-125), it is concluded that the elements of P_{Ks} must be linear in δ_{Ka} just as those of $\delta\Lambda_{Ks}$. Obviously, the diagonal elements of Eq. (9-125) depends quadratically on δ_{Ka} in the same way as that of $\delta\Lambda_{Kc}$.

Since eigenvalues of acoustic modes are zero for $K = 1$, the δ_{ka}-dependence of an acoustic eigenvalue is written as $(\Lambda_K)_{ii} = \varepsilon_i' \, \delta_{Ka}{}^2$. As given in Eq. (4-14), a normal frequency is proportional to the square root of the corresponding eigenvalue. A similar argument holds for the change in q_K along the b^*- and c^*-axes. Hence the frequency of an acoustic mode changes in proportion to δ_{Ka}, δ_{Ka} and δ_{Ka} on the increase of $|q_K|$ along a^*, b^* and c^*, respectively, from 0 to a certain amount below which terms of order higher than quadratic can be neglected. Since the effects of small and continuous changes must be additive in general, the behavior of an acoustic branch near the origin of the first Brillouin zone is described by Eq. (9-110) for a wave vector q_K as stated in the Debye approximation.

The phase dependence of the eigenvalue of an optical mode is given by

$$(\Lambda_K)_{jj} = (\Lambda_1)_{jj} + \varepsilon_j' \delta_{Ka}{}^2,$$

which involves the non-vanishing zeroth-order term in contrast to the case of acoustic modes. Then an optical frequency depends on the phase quadratically as given by

$$(\omega_K)_{jj} = (\omega_1)_{jj} + (2\pi c)^{-1}(\varepsilon_j'/2)\delta_{Ka}{}^2,$$

and the dispersion curve of an optical mode intersects the ordinate at the right angle.

The dependence of the matrix D_{pp}^{Kj} on δ_{Ka} is obtained as follows. From Eq. (9-119), the transformation matrix L_k^{mx} can be written as

$$L_K^{mx} = \begin{bmatrix} L_{1c}^{mx} + L'_{ca}\delta_{Ka}{}^2 & -L'_{sa}\delta_{Ka} \\ L'_{sa}\delta_{Ka} & L_{1c}^{mx} + L'_{ca}\delta_{Ka}{}^2 \end{bmatrix}.$$

From the definition in Eq. (9-96), we have

$$\boldsymbol{D}_{pp}^{Kj} = \left\{ (\boldsymbol{l}_{1c}^{mx})_{pj} + (\boldsymbol{l}'_{ca})_{pj} \delta_{Ka}{}^2 \right\} \left\{ \left(\tilde{\boldsymbol{l}}_{1c}^{mx}\right)_{pj} + \left(\tilde{\boldsymbol{l}}'_{ca}\right)_{pj} \delta_{Ka}{}^2 \right\} + (\boldsymbol{l}'_{sa}\delta_{Ka})\left(\tilde{\boldsymbol{l}}'_{sa}\delta_{Ka}\right)$$
$$= \boldsymbol{D}_{pp}^{1j} + \boldsymbol{C}_{pj}^{a} \delta_{Ka}{}^2 + (\boldsymbol{l}'_{ca})_{pj} \left(\tilde{\boldsymbol{l}}'_{ca}\right)_{pj} \delta_{Ka}{}^4, \qquad (9\text{-}126)$$

where

$$\boldsymbol{C}_{pj}^{a} = (\boldsymbol{l}_{1c}^{mx})_{pj}\left(\tilde{\boldsymbol{l}}'_{ca}\right)_{pj} + (\boldsymbol{l}'_{ca})_{pj}\left(\tilde{\boldsymbol{l}}_{1c}^{mx}\right)_{pj} + (\boldsymbol{l}'_{sa})_{pj}\left(\tilde{\boldsymbol{l}}'_{sa}\right)_{pj}.$$

The terms up to quadratic in Eq. (9-126) and similar quadratic terms with respect to δ_{Kb} and δ_{Kc} are then collected to give

$$\boldsymbol{D}_{pp}^{Kj} = \boldsymbol{D}_{pp}^{1j} + \boldsymbol{C}_{pj}^{a} \delta_{Ka}{}^2 + \boldsymbol{C}_{pj}^{b} \delta_{Kb}{}^2 + \boldsymbol{C}_{pj}^{c} \delta_{Kc}{}^2 = \boldsymbol{D}_{pp}^{1j} + \boldsymbol{C}_{pj}\omega_{Kj}{}^2, \qquad (9\text{-}127)$$

where

$$\boldsymbol{C}_{pj} = \frac{1}{3}\left(\frac{\boldsymbol{C}_{pj}^{a}}{u_{aj}{}^2} + \frac{\boldsymbol{C}_{pj}^{b}}{u_{bj}{}^2} + \frac{\boldsymbol{C}_{pj}^{c}}{u_{cj}{}^2} \right).$$

Substitute Eq. (9-127) into Eq. (9-109) and omit higher-than-second order terms in ω. Multiplying the result by $g(\omega)$ in Eq. (9-117) and integrating, we have

$$\frac{3NSkT}{(2\pi)^3 hc^2} \int_0^{\omega_D} \left(\boldsymbol{D}_{pp}^{1j} + \boldsymbol{C}_{pj}\,\omega^2\right)\left(1 + \frac{h^2c^2\omega^2}{12k^2T^2}\right) d\omega$$
$$= \frac{3NSkT\omega_D}{(2\pi)^3 hc^2}\left\{ \boldsymbol{D}_{pp}^{1j} + \frac{\omega_D{}^2}{3}\left(\boldsymbol{C}_{pj} + \frac{\boldsymbol{D}_{pp}^{1j} h^2c^2}{12k^2T^2} \right) \right\}. \qquad (9\text{-}128)$$

For the modes with frequency smaller than ω_D in each of the three acoustic branches, the contributions to the elements of $\boldsymbol{B}_p$ are calculated by using Eq. (9-128) in place of the sum with respect to K in the right side of Eq. (9-106). The result is given by

$$\boldsymbol{B}_p = \frac{3SkT\omega_D}{8\pi\, m_p c^2} \boldsymbol{C}_{r0}{}^{-1} \sum_j \left\{ \boldsymbol{D}_{pp}^{1j} + \frac{\omega_D{}^2}{3}\left(\boldsymbol{C}_{pj} + \frac{\boldsymbol{D}_{pp}^{1j} h^2c^2}{12k^2T^2} \right) \right\} \tilde{\boldsymbol{C}}_{r0}{}^{-1}. \qquad (9\text{-}129)$$

The coefficients u_{aj}, u_{bj} and u_{cj} necessary for this calculation are obtained from the slope of the corresponding dispersion curves along δ_a, δ_b and δ_c, respectively. The contribution of the other modes is summed up by using Eq. (9-106), in which the factor $2/N$ is replaced by inverse the number of modes taken in the sum.

In order to calculate the summation in Eq. (9-129) accurately, normal coordinate analyses must be carried out for a fairly large number of wave vectors in the first Brillouin zone. Kitagawa and Miyazawa calculated the anisotropic temperature factors for the carbon atom of orthorhombic polyethylene crystal at 100 K and 298 K.[22] This crystal contains two repeating units (four methylene groups) in a unit cell, and the polymer chain is parallel with the c-axis, so that there are four lattice modes per repeating unit consisting of a rotatory mode around the c-axis and three translatory modes, and each wave vector is associated with eight lattice modes. Based on the symmetry of the reciprocal lattice, one-eighth of the first Brillouin zone was scanned by limiting the phases δ_a, δ_b and δ_c within the range $[0, \pi]$. By setting δa, δb = 10° + 20n° (n = 0–8), and δ_c = 0°, 5°, 10°, 20°, 40°, 60° + 30n° (n = 0–4), 810 grid points were taken in the phase space, and the dynamical matrix at each point was

Table 9.4 Temperature factor of carbon atom in orthorhombic crystal of polyethylene (Å^2)[a].

Temp.	B_{aa}	B_{ab}	B_{bb}	B_{cc}
100 K	1.23	− 0.20	1.14	0.37
298 K	3.31	− 0.54	3.00	0.84
$\Delta B/\Delta T$ (Å^{2}K^{-1})[b]	0.0094	− 0.0017	0.0094	0.0024

[a] Å = 10^{-10} m

[b] The temperature gradients of the mean square displacements of the carbon atom agree well with the experimental data by Y. Aoki, A. Chiba and M. Kaneko, *J. Phys. Soc. Jpn.*, **27**, 1579 (1969).

diagonalized. The frequencies and the elements of the eigenvectors at four successive points along the δ_c axis were fitted to a polynomial, an even-power quartic function near the Γ point and a cubic function in the other region, and used for generating the corresponding data at 20′ intervals of δ_c by interpolation. In this way, 540 grid points were taken along the δ_c axis, and *ca.* 350,000 (≈ 9 × 9 × 540 × 8) freqeuncies were generated. These data were then put into Eqs. (9-106) and (9-129), and yielded the results shown in Table 9.4.

Since the expansion of $\exp(hc\omega_{Kj}/kT)$ in Eq. (9-108) is not valid unless $hc\omega_{Kj} \ll kT$, Eq. (9-129) cannot be used when the temperature is nearly 0 K. This problem is solved by using the Debye function defined in the form

$$D_n = (n/u^n)\int_0^u y^n/(e^y - 1)\,dy, \tag{9-130}$$

and replacing the sum with respect to K in Eq. (9-106) with an integral in the form

$$\begin{aligned}\sum_K \frac{D_{pp}^{Kj}}{2c\omega_{Kj}}\frac{1+x_{Kj}}{1-x_{Kj}} &\rightarrow \int_0^{\omega_D} g(\omega)\frac{\left(D_{pp}^{1j} + C_{pj}\omega^2\right)}{2c\omega}\frac{1+e^{-hc\omega/kT}}{1-e^{-hc\omega/kT}}d\omega \\ &= \frac{3NS}{8\pi^3 c}\int_0^{\omega_D}\left(D_{pp}^{1j}\,\omega + C_{pj}\,\omega^3\right)\left(\frac{1}{e^{hc\omega/kT}-1}+\frac{1}{2}\right)d\omega \\ &= \frac{3NS}{8\pi^3 c}\left[\left(\frac{kT}{hc}\right)^2 D_{pp}^{1j}\left\{D_1(u_D)+\frac{\omega_D{}^2}{4}\right\}+\left(\frac{kT}{hc}\right)^4 C_{pj}\left\{D_3(u_D)+\frac{\omega_D{}^4}{8}\right\}\right],\end{aligned} \tag{9-131}$$

where $g(\omega)$ is taken from Eq. (9-117), and

$$u_D = hc\omega_D/kT. \tag{9-132}$$

The Debye function $D_3(u_D)$ has been introduced first in relation to calculation of thermodynamic quantities of crystals, and has proved to reproduce experimental data quite accurately.[21] Remarks on numerical calculation of the Debye function are given in Appendix 6.

9.3.3 Rigid Molecule Approximation

Calculations involved in lattice dynamics of molecular crystals can be simplified if the molecule can be regarded approximately as "rigid". Here, a rigid molecule implies a molecule in which any nuclear displacements may be taken as negligible. Aromatic hydrocarbons are typical examples of rigid molecules. The procedure of calculation in rigid molecule approximation will be explained below according to the formulation by Shimanouchi and Harada.[23)]

Two types of Cartesian coordinate systems are used for describing the nuclear displacements in the lth unit cell. One is the crystal coordinate system in which the Cartesian displacements are given as a vector $\boldsymbol{\Delta x}_l$ defined in Eq. (9-3), while the other is the coordinate system in terms of the principal axes of inertia of individual molecules. On assuming that each unit cell contains Z molecules, the vector of the nuclear displacements in the latter system, $\boldsymbol{\Delta x}_l^P$, is given by

$$\boldsymbol{\Delta x}_l^P = [\boldsymbol{\Delta \tilde{x}}_{l1}^P \quad \boldsymbol{\Delta \tilde{x}}_{l2}^P \quad \cdots \quad \boldsymbol{\Delta \tilde{x}}_{lZ}^P]^{\mathrm{T}}, \tag{9-133}$$

where $\boldsymbol{\Delta x}_{lk}^P$ is the subvector consisting of the nuclear displacements of the kth molecule. Analogously, the vector $\boldsymbol{\Delta x}_l$ can be partitioned into the subvectors, $\boldsymbol{\Delta x}_{l1}$, $\boldsymbol{\Delta x}_{l2}$, $\cdots$, $\boldsymbol{\Delta x}_{lZ}$. Let $(\boldsymbol{L}^P_{\mathrm{mol}})_k$ be the matrix of transformation from $\boldsymbol{\Delta x}_k$ to $\boldsymbol{\Delta x}_{lk}^P$., *i.e.*,

$$\boldsymbol{\Delta x}_{lk}^P = (\boldsymbol{L}^P_{\mathrm{mol}})_k\, \boldsymbol{\Delta x}_{lk}. \tag{9-134}$$

If the kth molecule contains n_k nuclei, the matrix $(\boldsymbol{L}^P_{\mathrm{mol}})_k$ is expressible as the direct sum of n_k identical (3×3) matrices obtained by diagonalizing the tensor of inertia of the kth molecule in the crystal system as shown in Eq. (3-22). Thus, replacing $\boldsymbol{L}^P$ in Eq. (3-25) with $(\boldsymbol{L}^P_{\mathrm{mol}})_k$ and collecting over the nuclei in the kth molecule, we obtain Eq. (9-134). The zeroth order normal coordinates of the kth molecule in the isolated state, $\boldsymbol{Q}_{\mathrm{in},lk}$, is obtained by diagonalizing the intramolecular dynamical matrix in the principal system as

$$\begin{aligned}\boldsymbol{\Delta \tilde{x}}_{lk}^P\, \boldsymbol{F}_{\mathrm{in},k}^{Pxx}\, \boldsymbol{\Delta x}_{lk}^P &= \left(\boldsymbol{\tilde{Q}}_{\mathrm{in},lk}\, \boldsymbol{\tilde{L}}_{\mathrm{in},k}^{Px}\right) \boldsymbol{F}_{\mathrm{in},k}^{Pxx} \left(\boldsymbol{L}_{\mathrm{in},k}^{Px}\, \boldsymbol{Q}_{\mathrm{in},lk}\right)\\ &= \boldsymbol{\tilde{Q}}_{\mathrm{in},lk}\, \boldsymbol{\Lambda}_{\mathrm{in},k}\, \boldsymbol{Q}_{\mathrm{in},lk}.\end{aligned} \tag{9-135}$$

It follows from Eq. (9-135) that

$$\boldsymbol{Q}_{\mathrm{in},lk} = \left(\boldsymbol{L}_{\mathrm{in},k}^{Px}\right)^{-1} \boldsymbol{\Delta x}_{lk}^P. \tag{9-136a}$$

The external coordinates of the kth molecule is defined in the principal system as described in Eq. (3-28a–f), and are given in a vector notation analogous to Eq. (3-32), *i.e.*,

$$\boldsymbol{\Delta R}_{\mathrm{ex},lk} = \boldsymbol{B}_{\mathrm{ex}}\, \boldsymbol{\Delta x}_{lk}^P. \tag{9-136b}$$

Combining $\boldsymbol{Q}_{\mathrm{in},lk}$ and $\boldsymbol{\Delta R}_{\mathrm{ex},lk}$ into a single vector, $\boldsymbol{Q}_{lk}$, we have

$$\boldsymbol{Q}_{lk} = \begin{bmatrix}\boldsymbol{Q}_{\mathrm{in},lk}\\ \boldsymbol{\Delta R}_{\mathrm{ex},lk}\end{bmatrix} = \begin{bmatrix}\left(\boldsymbol{L}_{\mathrm{in},k}^{Px}\right)^{-1}\\ \boldsymbol{B}_{\mathrm{ex}}\end{bmatrix} \boldsymbol{\Delta x}_{lk}^P = \begin{bmatrix}\left(\boldsymbol{L}_{\mathrm{in},k}^{Px}\right)^{-1}\\ \boldsymbol{B}_{\mathrm{ex}}\end{bmatrix} (\boldsymbol{L}^P_{\mathrm{mol}})_k\, \boldsymbol{\Delta x}_{lk}. \tag{9-137}$$

Introducing the matrices $\boldsymbol{L}_{\mathrm{in},k}^x = \left(\boldsymbol{\tilde{L}}^P_{\mathrm{mol}}\right)_k \boldsymbol{L}_{\mathrm{in},k}^{Px}$ and $\boldsymbol{D}_{\mathrm{ex},k}^x = \left(\boldsymbol{\tilde{L}}^P_{\mathrm{mol}}\right)_k \boldsymbol{D}_{\mathrm{ex},k}^{Px}$, where $\boldsymbol{D}_{\mathrm{ex},k}^{Px}$ is constructed for the principal system according to Eq. (3-38), and noting that $(\boldsymbol{L}^P_{\mathrm{mol}})_k$ is orthogonal, we can invert Eq. (9-137) as

$$\boldsymbol{\Delta x}_{lk} = \boldsymbol{L}_{\mathrm{in},k}^x\, \boldsymbol{Q}_{\mathrm{in},lk} + \boldsymbol{D}_{\mathrm{ex},k}^x\, \boldsymbol{\Delta R}_{\mathrm{ex},lk} = \boldsymbol{L}_k^x\, \boldsymbol{Q}_{lk}. \tag{9-138}$$

Let $\boldsymbol{F}_{lk,Jl}^{xx}$ be the subblock of the Hessian matrix with respect to the Cartesian displacements representing the interaction between the kth molecule in the Ith cell and the lth molecule in the Jth cell. The total potential function of the crystal is then given by

$$V = \tfrac{1}{2}\sum_{I,k}\sum_{J,l} \Delta\tilde{\boldsymbol{x}}_{Ik}\, \boldsymbol{F}^{xx}_{Ik,Jl}\, \Delta\boldsymbol{x}_{Jl}. \tag{9-139}$$

The term in Eq. (9-139) with $Jl = Ik$ represents the potential function of the molecule Ik buried in a hypothetical crystal in which the nuclei in any molecule other than Ik are fixed at their equilibrium positions. This potential is rewritten in terms of the zeroth-order normal coordinates as

$$\begin{aligned} V_{Ik} &= \tfrac{1}{2}\Delta\tilde{\boldsymbol{x}}_{Ik}\, \boldsymbol{F}^{xx}_{Ik,Ik}\, \Delta\boldsymbol{x}_{Ik} \\ &= \tfrac{1}{2}\begin{bmatrix}\tilde{\boldsymbol{Q}}_{\text{in},Ik} & \tilde{\boldsymbol{R}}_{\text{ex},Ik}\end{bmatrix}\begin{bmatrix}\tilde{\boldsymbol{L}}^{x}_{\text{in},k} \\ \tilde{\boldsymbol{D}}^{x}_{\text{ex},k}\end{bmatrix}\boldsymbol{F}^{xx}_{Ik,Ik}\begin{bmatrix}\boldsymbol{L}^{x}_{\text{in},k} & \boldsymbol{D}^{x}_{\text{ex},k}\end{bmatrix}\begin{bmatrix}\boldsymbol{Q}_{\text{in},Ik} \\ \boldsymbol{R}_{\text{ex},Ik}\end{bmatrix} \\ &= \tfrac{1}{2}\tilde{\boldsymbol{Q}}_{Ik}\, \boldsymbol{F}^{QQ}_{Ik,Ik}\, \boldsymbol{Q}_{Ik}. \end{aligned} \tag{9-140}$$

By partitioning $\boldsymbol{F}^{xx}_{Ik,Ik}$ into the contributions from V_{in} [Eq. (9-35)] and V_{ex} [Eq. (9-32)] as $\boldsymbol{F}^{xx}_{Ik,Ik} = (\boldsymbol{F}^{xx}_{\text{in}})_{Ik,Ik} + (\boldsymbol{F}^{xx}_{\text{ex}})_{Ik,Ik}$, the matrix $\boldsymbol{F}^{QQ}_{Ik,Ik}$ can be rewritten in the form

$$\begin{aligned} \boldsymbol{F}^{QQ}_{Ik,Ik} &= \begin{bmatrix}\tilde{\boldsymbol{L}}^{x}_{\text{in},k} \\ \tilde{\boldsymbol{D}}^{x}_{\text{ex},k}\end{bmatrix}\left\{(\boldsymbol{F}^{xx}_{\text{in}})_{Ik,Ik} + (\boldsymbol{F}^{xx}_{\text{ex}})_{Ik,Ik}\right\}\begin{bmatrix}\boldsymbol{L}^{x}_{\text{in},k} & \boldsymbol{D}^{x}_{\text{ex},k}\end{bmatrix} \\ &= \begin{bmatrix}\boldsymbol{\Lambda}_{\text{in},Ik} + \tilde{\boldsymbol{L}}^{x}_{\text{in},k}(\boldsymbol{F}^{xx}_{\text{ex}})_{Ik,Ik}\boldsymbol{L}^{x}_{\text{in},k} & \tilde{\boldsymbol{L}}^{x}_{\text{in},k}(\boldsymbol{F}^{xx}_{\text{ex}})_{Ik,Ik}\boldsymbol{D}^{x}_{\text{ex},k} \\ \tilde{\boldsymbol{D}}^{x}_{\text{ex},k}(\boldsymbol{F}^{xx}_{\text{ex}})_{Ik,Ik}\boldsymbol{L}^{x}_{\text{in},k} & \tilde{\boldsymbol{D}}^{x}_{\text{ex},k}(\boldsymbol{F}^{xx}_{\text{ex}})_{Ik,Ik}\boldsymbol{D}^{x}_{\text{ex},k}\end{bmatrix}. \end{aligned} \tag{9-141}$$

Usually, the intramolecular forces exerted in rigid molecules are far stronger than intermolecular forces of any types. Hence the diagonal elements of the (1,1) subblock in the right side of Eq. (9-141) are much larger than any other elements in the whole matrix. On diagonalizing such a matrix $\boldsymbol{F}^{QQ}_{Ik,Ik}$, we obtain two clearly distinguishable types of eigenvectors, one contributed mainly by $\boldsymbol{Q}_{\text{in},Ik}$ (internal modes) and the other by $\boldsymbol{R}_{\text{ex},Ik}$ (external modes). Correspondingly, the six lowest eigenvalues of $\boldsymbol{F}^{QQ}_{Ik,Ik}$ must be well approximated by the eigenvalues of the (6×6) matrix

$$(\boldsymbol{F}^{QQ}_{\text{ex}})_{Ik,Ik} = \tilde{\boldsymbol{D}}^{x}_{\text{ex},k}(\boldsymbol{F}^{xx}_{\text{ex}})_{Ik,Ik}\boldsymbol{D}^{x}_{\text{ex},k}.$$

Let the vectors of external coordinates of all the molecules in the Ith unit cell be merged into a single vector in the form

$$\Delta\boldsymbol{R}_{\text{ex},I} = \begin{bmatrix}\Delta\tilde{\boldsymbol{R}}_{\text{ex},I1} & \Delta\tilde{\boldsymbol{R}}_{\text{ex},I2} & \cdots & \Delta\tilde{\boldsymbol{R}}_{\text{ex},IZ}\end{bmatrix}^{\text{T}}. \tag{9-142}$$

In the rigid molecule approximation, we define the external symmetry coordinates associated with the wave vector $\boldsymbol{q}_K$ by using Eqs. (9-58a,b) in which $\Delta\boldsymbol{x}_I$ is replaced by $\Delta\boldsymbol{R}_{\text{ex},I}$, *i.e.*,

$$\Delta\boldsymbol{R}_{\text{ex}}(\boldsymbol{q}_K)_c = (2/N)^{1/2}\sum_{I=1}^{N}\Delta\boldsymbol{R}_{\text{ex},I}\cos\theta_{KI} \tag{9-143a}$$

$$\Delta \boldsymbol{R}_{\text{ex}}(\boldsymbol{q}_K)_s = (2/N)^{1/2} \sum_{l=1}^{N} \Delta \boldsymbol{R}_{\text{ex},l} \sin \theta_{Kl}. \tag{9-143b}$$

The subsequent procedure of the rigid molecule approximation in lattice dynamics is the same as that employed in dealing with all the degrees of freedom of nuclear motion. If the unit cell has any symmetry elements, the dynamical matrix for external symmetry coordinates can be factored, irrespective of whether Z molecules are all equivalent or not, into appropriate subblocks according to the symmetry of the wave vector. The rigid molecule approximation was proved to be useful for interpreting vibrational spectra due to lattice vibrations of uracil,[23] benzene and naphthalene[24]. The dispersion curves of naphthalene were calculated for both the rigid molecule approximation and the overall normal coordinate analysis of intra- and intermolecular vibrations. The maximum deviation in lattice frequencies between the two calculations was within 10 cm^{-1}.[25] The rigid molecule approximation is particularly advantageous in calculations of anisotropic temperature factors and thermodynamic functions, since these quantities are neither affected sensitively by small frequency differences nor receive much contribution from high-frequency intramolecular modes. When the intramolecular distortion is absent, the first term in the second formula of Eq. (9-138), $\boldsymbol{L}^x_{\text{in},lk} \boldsymbol{Q}_{\text{in},lk}$, is dropped, and the Cartesian displacements and external coordinates are related to each other by $\Delta \boldsymbol{x}_{lk} = \boldsymbol{D}^x_{\text{ex},k} \Delta \boldsymbol{R}_{\text{ex},lk}$. The anisotropic temperature factors under the rigid molecule approximation are then derived by transforming the mean square amplitude matrix for the external coordinates to that of nuclear displacements by

$$\begin{aligned} \boldsymbol{\Sigma}^x_{kk}(T) = \langle \Delta \boldsymbol{x}_k \Delta \tilde{\boldsymbol{x}}_k \rangle^T &= \boldsymbol{D}^x_{\text{ex},k} \langle \Delta \boldsymbol{R}_{\text{ex},k} \Delta \tilde{\boldsymbol{R}}_{\text{ex},k} \rangle^T \tilde{\boldsymbol{D}}^x_{\text{ex},k} \\ &= \boldsymbol{D}^x_{\text{ex},k} \boldsymbol{\Sigma}^{\text{ex}}_{kk}(T) \tilde{\boldsymbol{D}}^x_{\text{ex},k}. \end{aligned} \tag{9-144}$$

Since the external coordinates consist of three translational and three rotational coordinates, the (6×6) mean square amplitude matrix for the external coordinates of the kth molecule can be partitioned into four (3×3) matrices as

$$\boldsymbol{\Sigma}^{\text{ex}}_{kk}(T) = \begin{bmatrix} \boldsymbol{\Sigma}^{TT}_{kk}(T) & \boldsymbol{\Sigma}^{TR}_{kk}(T) \\ \boldsymbol{\Sigma}^{RT}_{kk}(T) & \boldsymbol{\Sigma}^{RR}_{kk}(T) \end{bmatrix}. \tag{9-145}$$

The matrices $\boldsymbol{\Sigma}^{TT}_{kk}(T)$, $\boldsymbol{\Sigma}^{RR}_{kk}(T)$ and $\boldsymbol{\Sigma}^{RT}_{kk}(T)$ correspond to the $\boldsymbol{T}$, $\boldsymbol{L}$ and $\boldsymbol{S}$ tensors, respectively, which have been widely used in crystallography for estimating the anisotropic temperature factor under the rigid molecule approximation. The tensors $\boldsymbol{T}$ and $\boldsymbol{L}$ (with the original name $\boldsymbol{\omega}$) were introduced first by Cruickshank for the purpose of estimating the effect of thermal motions of nuclei on the internuclear distances.[26] Later, Trueblood and Schomaker pointed out the necessity of $\boldsymbol{S}$ for correctly describing the nuclear motion of rigid molecules in crystals.[27]

9.3.4 Temperature Diffuse Scattering of X-ray by Crystals

The intensity of the inelastic scattering of X-ray by a crystal is given by Eq. (9-100c), in which the cell indices J and J' appear in the exponential factor involving the lattice vectors as well as in $U_{pp'}^{JJ'}$. With the help of Euler's formula for cosine, Eq. (9-97b) is rewritten as

$$\begin{aligned} U_{pp'}^{JJ'} &= 2\pi^2 \tilde{k} \sum_{K=1}^{N} \boldsymbol{\Sigma}_{pp'}^{x}(T)_K \{ e^{2\pi i \tilde{q}_K (r_J - r_{J'})} + e^{-2\pi i \tilde{q}_K (r_J - r_{J'})} \} k \\ &= \sum_K H_{K,pp'}\, e^{2\pi i \tilde{q}_K (r_J - r_{J'})} \end{aligned} \tag{9-146}$$

where

$$H_{K,pp'} = 4\pi^2 \,\tilde{k}\, \boldsymbol{\Sigma}_{pp'}^{x}(T)_K \,\boldsymbol{k}. \tag{9-147}$$

The second equality in Eq. (9-146) is justified since there must be such a pair of indices K and K' in the first Brillouin zone that the relations $\boldsymbol{q}_{K'} = -\boldsymbol{q}_K$ and $\boldsymbol{\Sigma}_{pp'}^{x}(T)_K = \boldsymbol{\Sigma}_{pp'}^{x}(T)_{K'}$ hold simultaneously. Let us remember that the polynomial theorem is expressed in the form

$$\left(\sum_{i=1}^{l} a_i \right)^n = n! \sum_{n_1=0}^{n} \frac{a_1^{\,n_1}}{n_1!} \sum_{n_2=0}^{n-n_1} \frac{a_2^{\,n_2}}{n_2!} \cdots \sum_{n_{l-1}=0}^{n'_{l-1}} \frac{a_{l-1}^{\,n_{l-1}}}{n_{l-1}!} \frac{a_l^{\,n'_l}}{n'_l!}, \tag{9-148}$$

where

$$n'_i = n - \sum_{k=1}^{i-1} n_k \quad (2 \le i \le l). \tag{9-149}$$

Then the nth power of Eq. (9-146) is given by

$$\left(U_{pp'}^{JJ'} \right)^n = n! \sum_{\sum n_K = n} \left\{ \prod_K \frac{(H_{K,pp'})^{n_K}}{n_K!} \right\} e^{2\pi i \left(\sum_K n_K \tilde{q}_K \right)(r_J - r_{J'})}, \tag{9-150}$$

where the first sum in the right side is taken over those combinations of non-negative integers n_1 through n_K the sum of which is n. Substituting Eq. (9-150) into Eq. (9-100c) and using Eq. (A5-13), we may collect the exponential factor involving J and J', and replace the result with the δ-function in the form

$$\sum_{J,J'} \exp\left\{ 2\pi i \left(\tilde{k} + \sum_K n_K \tilde{q}_K \right)(r_J - r_{J'}) \right\} = NV^{-1} \delta\left(\boldsymbol{k} + \sum_K n_K \boldsymbol{q}_K - \boldsymbol{h}_\lambda \right), \tag{9-151}$$

where V is the unit cell volume, and $\boldsymbol{h}_\lambda$ is reciprocal lattice vector with suffix λ representing a set of the Miller indices, h, k and l. If n is 1 in Eq. (9-150), the sums with respect to i in both sides of Eq. (9-151) are replaced by $\boldsymbol{q}_K$. This case represents a first order temperature diffuse scattering, the intensity of which is given by

$$I_1 = I_e V^{-1} \sum_{p,p'} F_p^W F_{p'}^{W*} \sum_K H_{K,pp'}\, \delta(\boldsymbol{k} + \boldsymbol{q}_K - \boldsymbol{h}_\lambda). \tag{9-152}$$

The condition for a non-vanishing first order temperature diffuse scattering is

$$\boldsymbol{k} + \boldsymbol{q}_K - \boldsymbol{h}_\lambda = \boldsymbol{0}, \tag{9-153}$$

which differs from the condition for the Laue–Bragg scattering in Eq. (9-104) by $\boldsymbol{q}_K$. Accordingly, only the modes with such $\boldsymbol{q}_K$ as to satisfy Eq. (9-153) contribute to the sum with respect to K in Eq. (9-152).[28] Equation (9-153) assures the conservation of momentum during the process of scattering. In order for the energy to be conserved, the frequency of the scattered X-ray must be smaller than the frequency of the incident X-ray by the amount ω_{Kj}. However, since the frequencies of X-rays are about 10^5 times larger than the frequencies of nuclear vibrations, the frequency changes associated with the inelastic scattering of X-rays are mostly negligible. Since the wave vector $\boldsymbol{q}_K$ is usually small and continuously distributed in the first Brillouin zone, the terms I_1, I_2, $\cdots$ give diffuse backgrounds around the spots due to the Laue–Bragg scattering.[29] The intensities of higher-than-first order temperature diffuse scatterings are given by[18,30]

$$I_n = I_e V^{-1} \sum_{p,p'} F_p^W F_{p'}^{W*} \sum_{\sum n_K = n} \left\{ \prod_K \frac{\left(H_{K,pp'}\right)^{n_K}}{n_K!} \right\} \delta\left(\boldsymbol{k} + \sum_K n_K \boldsymbol{q}_K - \boldsymbol{h}_\lambda \right) \tag{9-154}$$

with the condition for the scattering direction

$$\boldsymbol{k} + \sum_K n_K \boldsymbol{q}_K - \boldsymbol{h}_\lambda = \boldsymbol{0}. \tag{9-155}$$

The scattering directions satisfying the condition for non-vanishing I_1 and I_2 are illustrated in Fig. 9.7, where wave vectors are taken in a two-dimensional reciprocal space. Equation (9-154) shows that the intensities of temperature diffuse scattering can be calculated by using the same quantities as those used in the calculation of the anisotropic temperature factors. In practical calculation, an asymmetric unit of the first Brillouin zone is partitioned into an appropriate number of grid compartments according to the experimental resolution, and each compartment is given an address in the array for intensity storage. The calculated intensities, *i.e.*, the coefficients of δ-functions in Eqs. (9-152) and (9-154), are accumulated in the array according to the address specification using the wave vector or the sum of wave vectors satisfying Eq. (9-155).

Experimental and theoretical data of temperature diffuse scattering were compared with each other for the crystal of hexamethylenetetramine (hexamine, $C_6H_{12}N_4$).[31,32] The calculated

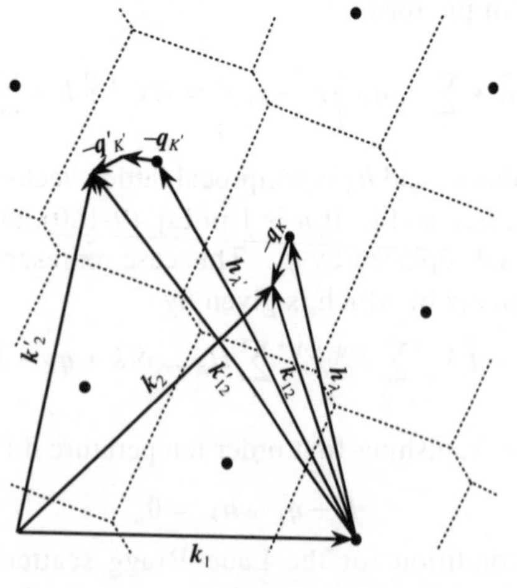

Fig. 9.7 Relations among the scattering vectors ($\boldsymbol{k}_{12}$, $\boldsymbol{k}'_{12}$), reciprocal lattice vectors ($\boldsymbol{h}_\lambda$, $\boldsymbol{h}'_\lambda$) and wave vectors ($\boldsymbol{q}_K$, $\boldsymbol{q}'_K$, $\boldsymbol{q}'_{K'}$) in one-phonon and two-phonon temperature diffuse scatterings.

intensities were obtained by using the rigid molecular model with isotropic intermolecular pairwise forces between the nearest and the second nearest neighbors. Despite the simplicity of the models, the calculation was successful in reproducing the positions of the maxima and minima and the relative magnitudes of the scattering intensities.

As seen from Eqs. (9-104) and (9-153), the temperature diffuse scattering for $\boldsymbol{q}_K = \mathbf{0}$ overlaps with the Laue–Bragg scattering. The calculation of temperature diffuse scattering intensities based on a reasonable potential model is thus important for evaluating the correction terms in the precise measurement of intensities of the Laue-Bragg scattering.

9.3.5 Thermodynamic Functions

The thermodynamic functions of molecular crystals can be calculated from the partition function just as in the case of isolated molecules described in Chapter 6. The difference is that the translational and rotational motions of each molecule are replaced by vibrational motions in crystals because of the large restoring force against deviation of the position and orientation of the molecules. Thus we need to consider only the vibrational degrees of freedom for the nuclear motion in a molecular crystal, unless certain disordered phases involving rotating molecules must be dealt with. Among the thermodynamic functions of a molecular crystal, those having the dimension of an energy are divided into static and dynamic parts, of which only the latter is taken up in this subsection. The calculation of the static energy is discussed in section 9. 4.

The formulae for the thermodynamic functions used for molecular crystals have the same form as those for isolated molecules.[15] For molecular crystals, it is convenient to evaluate a thermodynamic function per unit cell first, then divide it by the number of molecules in a unit cell to obtain the value per molecule. By substituting Eq. (6-32) for crystal vibrations into Eq. (6-20), the dynamic part of the Helmholtz free energy of a crystal per unit cell is derived as

$$\begin{aligned} A^C &= -\frac{kT}{N}\ln\left(\prod_{K,j}\frac{e^{-hc\omega_{Kj}/2kT}}{1-e^{-hc\omega_{Kj}/kT}}\right) = -\frac{kT}{N}\sum_{K,j}\ln\left\{2\sinh\left(\frac{hc\omega_{Kj}}{2kT}\right)\right\} \\ &= \frac{hc}{2N}\sum_{K,j}\omega_{Kj} + \frac{kT}{N}\sum_{K,j}\ln(1-e^{-hc\omega_{Kj}/kT}), \end{aligned} \tag{9-156a}$$

where the summation index i runs over all the normal modes asociated with a given wave vector $\boldsymbol{q}_K$ and K runs over all the wave vectors in the first Brillouin zone. Here the dynamic part of free energy implies the free energy due to the thermal motion of nuclei, *i.e.*, the total free energy minus the static part, which is the equilibrium value of the potential function of the crystal. In this definition, the total free energy corresponds to the atomic heat of formation of the crystal. Subtracting the sum of the dynamic and static free energies of the constituent molecules of the unit cell from this total free energy gives the lattice free energy of the crystal. Similarly, the dynamic part of internal energy per unit cell of a crystal is given by

$$U^C = hc\sum_{K,i}\omega_{Ki}\coth(hc\omega_{Ki}/2kT)\Big/2N. \tag{9-156b}$$

The static part of internal energy is the same as that of free energy.

The entropy and the heat capacity at constant volume per unit cell are

$$S^C = \frac{k}{N}\sum_{K,i}\left[\frac{hc\omega_{Ki}}{2kT}\coth\left(\frac{hc\omega_{Ki}}{2kT}\right) - \ln\left\{2\sinh\left(\frac{hc\omega_{Ki}}{2kT}\right)\right\}\right] \tag{9-156c}$$

and

$$C_v^C = (k/N)\sum_{K,i}(hc\omega_{Ki}/2kT)^2/\sinh^2(hc\omega_{Ki}/2kT), \tag{9-156d}$$

respectively. Equations (9-156b–d) are the equivalent forms of Eqs. (6-34c), (6-35d) and (6-38c), respectively, rewritten by using the well-known relations between exponential and hyperbolic functions. Since the normal frequencies belonging to the three acoustic branches approach zero when $\boldsymbol{q}_K$ approaches $\mathbf{0}$, certain terms in Eqs. (9-156a–d) diverge on the limit that $\omega_{ki} \to 0$, and cause serious problems in practical calculations. The definite integrals of thermodynamic quantities weighted by a distribution function from zero to an upper bound frequency is useful for avoiding this difficulty. By using the same distribution function and the range of integration as used for calculating the Debye–Waller factor, the diverging part of the second sum in the last formula of Eq. (9-156a) is replaced by a definite integral which can be evaluated by integrating by parts in the form

$$\begin{aligned} 3a\int_0^{\omega_D}\omega^2\ln(1-e^{-hc\omega/kT})\,d\omega &= 3a(kT/hc)^3\int_0^{u_D}u^2\ln(1-e^{-u})\,du \\ &= a\frac{\omega_D{}^3}{u_D{}^3}\left\{u_D{}^3\ln(1-e^{-u_D}) - \int_0^{u_D}\frac{u^3\,e^{-u}}{1-e^{-u}}du\right\}, \end{aligned} \tag{9-157}$$

where $u = hc\omega/kT$ and $u_D = hc\omega_D/kT$. Numerical calculation of Eq. (9-157) can be carried out efficiently by introducing the Debye function[21)] defined by Eq. (9-131) and the Debye temperature,

$$\Theta = hc\omega_D/k = u_D T. \tag{9-158}$$

The contribution of the acoustic modes with the frequencies below ω_D to the Helmholtz free energy is written as

$$A_I^C = \frac{akT\omega_D{}^3}{9N}\left\{\frac{9\Theta}{8T} + 3\ln(1-e^{-\Theta/T}) - D_3\left(\frac{\Theta}{T}\right)\right\}. \tag{9-159a}$$

Contributions from the same low-frequency acoustic modes to the thermodynamic functions U^C, S^C and C_v^C are calculated by substituting Eq. (9-159a) into Eqs. (6-11), (6-9) and (6-36), respectively. The results are

$$U_I^C = (akT\omega_D{}^3/N)\{(1/8) + D_3(\Theta/T)\} \tag{9-159b}$$

$$S_I^C = (ak\omega_D{}^3/9N)\{4D_3(\Theta/T) - 3\ln(1-e^{-\Theta/T})\} \tag{9-159c}$$

and

$$C_{v,\mathrm{I}}^{C} = \left(ak\omega_{\mathrm{D}}^{3}/3N\right)\left\{4D_3(\Theta/T) - 3(\Theta/T)\left(e^{\Theta/T} - 1\right)^{-1}\right\}. \tag{9-159d}$$

In deriving C_v^C from Eq. (6-36), use has been made of the relation

$$\frac{d}{dT}D_3(\Theta/T) = \frac{\Theta}{T^2}\left\{\frac{3T}{\Theta}D_3(\Theta/T) - \frac{3}{e^{\Theta/T} - 1}\right\}. \tag{9-160}$$

The heat capacity $C_{v,\mathrm{I}}^{C}$ is often given in the form

$$C_{v,\mathrm{I}}^{C} = \left(ak\omega_{\mathrm{D}}^{3}/N\right)(T/\Theta)^{3}\int_{0}^{\Theta/T} x^{4}e^{x}\left(e^{x} - 1\right)^{-2}dx, \tag{9-161}$$

which can be shown to be equivalent to Eq. (9-159d) by integrating by parts. It is noted from Eqs. (159a–d) that various important thermodynamic functions can be evaluated once the Debye function is calculated. The contributions to F^C, U^C, S^C and C_v^C from normal modes with frequencies higher than ω_{D} are caculated by Eqs. (9-156a–d) and added to those contributions from Eqs. (159a–d) to give the total thermodynamic functions. In earlier works, the characteristic frequency ω_{D} was usually taken to be high enough to leave no lattice frequencies outside the range of the integral, thus enabling estimation of thermodynamic functions in the absence of efficient computing facilities. Recent progress in computer technology is making it easier to carry out the calculation over a wide region of the dispersion surfaces so that the first Brillouin zone can be reasonably partitioned into the integration and the summation regions from the standpoints of efficiency and accuracy.

The heat capacity of orthorhombic polyethylene crystal at constant volume was calculated in the temperature range 0 – 150 K by Kitagawa and Miyazawa by using the same set of lattice frequencies as used for the calculation of anisotropic temperature factors.[33] The discrepancy between the calculated heat capacity and experimental data is almost negligible below 100 K, but the former starts to deviate slightly downwards from the latter on increase of temperature above 100 K. Kobayashi and Tadokoro calculated the thermodynamic functions of triclinic and orthorhombic polyethylene crystals, and discussed the stability of these polymorphic forms.[34] By referring to the experimetal data of Chang[35] and the data of the heat of fusion of polyethylene measured by Atkinson and Richardson,[36] the enthalpy and the Gibbs free energy of liquid polyethylene were successfully estimated.

9.3.6 Cross Section of Inelastic Scattering of Neutrons by Crystals

The distribution of normal frequencies of the lattice modes in an asymmetric unit of the first Brillouin zone must be known to calculate the anisotropic temperature factors and thermodynamic quantities of a crystal. On the other hand, little information on the detailed structure of dispersion curves of the lattice modes is available from the observed data of these quantities, since each datum involves contributions from many modes having different frequencies. Alternatively, detailed information as to the frequency resolution of lattice modes is available from the cross section of inelastic scattering of slow neutrons by crystals. Neutron scattering by atomic nuclei in a crystal is classified into elastic and inelastic scatterings according to whether an energy transfer between the crystal and neutrons takes place or not. The situation is analogous to the case of X-ray scattering except that there are four types of neutron scatterings distinguished by two criteria, energy transfer and coherency, *i.e.*, coherent elastic, coherent inelastic, incoherent elastic and incoherent inelastic scatterings. It is shown in Appendix 7 that the difference in coherency arises from stochastic distribution of isotopes and nuclear spin states among translationally equivalent sites in a crystal. Coherent elastic and coherent inelastic neutron scatterings correspond to the Laue–Bragg scattering and the temperature diffuse scattering of X-ray, respectively. In contrast to the case of X-ray, the energy of incident neutrons is comparable to phonon energies, so that the energy difference between the incident and the scattered neutrons after inelastic collision with nuclei can be easily detected.

Quantum mechanically, a crystal exposed to a stationary beam of neutrons can be regarded to be equivalent to a crystal interacting with a neutron, if the wavefunction of the neutron is represented by a superposition of incident and scattered waves. The interaction between the neutron and a nucleus Jp in the crystal is described by the Fermi pseudopotential in terms of the δ-function in the form

$$V_{Jp} = (2\pi\hbar^2/m)b_{Jp}\delta(\boldsymbol{r}_\mathrm{n} - \boldsymbol{r}_{Jp}), \tag{9-162a}$$

where m and $\boldsymbol{r}_\mathrm{n}$ are the mass and the position vector of a neutron, respectively, and b_{Jp} is a constant representing the scattering power of the nucleus at the site Jp. The constant b_{Jp} has the dimension of length, and is called the scattering length of nucleus Jp. The potential in Eq. (9-162a) states that the energy of interaction remains zero as long as the neutron and the nucleus occupy different positions, but increases abruptly to infinity if they collide with each other. Summing up Eq. (9-162a) over all nuclei in the crystal, we obtain the total potential of interaction between a neutron and the crystal in the form

$$V_\mathrm{int} = \sum_{J,p} V_{Jp} = (2\pi\hbar^2/m)\sum_{J,p} b_{Jp}\delta(\boldsymbol{r}_\mathrm{n} - \boldsymbol{r}_{Jp}). \tag{9-162b}$$

The intensity of a scattered neutron beam is measured in terms of the differential cross section, $(d\sigma/d\Omega)$, where $d\sigma$ is the number of neutrons scattered within the solid angle $d\Omega$ per unit time divided by the unit flux of incident neutrons; a flux means the number of particles passing a unit area per unit time. As shown in Appendix 7, a differential cross section of neutron scattering consists of the contributions from coherent and incoherent scatterings in the form

$$\frac{d\sigma}{d\Omega} = \frac{d\sigma_\mathrm{coh}}{d\Omega} + \frac{d\sigma_\mathrm{incoh}}{d\Omega}. \tag{9-163}$$

Let $\boldsymbol{k}_1$ and $\boldsymbol{k}_2$ be the momentum vectors of incident and scattered neutrons. Then the coherent cross section is given by

$$(d\sigma_\mathrm{coh}/d\Omega) = (|\boldsymbol{k}_2|/|\boldsymbol{k}_1|)\sum_{J,p}\sum_{J',p'}\langle b_p\rangle\langle b_{p'}\rangle M^*_{Jp} M_{J'p'}, \tag{9-164a}$$

where $\langle b_p\rangle$ is the isotopic average of b_{Jp} throughout N unit cells in the crystal, while the incoherent cross section is given by

$$(d\sigma_\mathrm{incoh}/d\Omega) = (|\boldsymbol{k}_2|/|\boldsymbol{k}_1|)\sum_{J,p}\sigma_p^\mathrm{incoh}|M_{Jp}|^2. \tag{9-164b}$$

The coefficient σ_p^incoh depends on averages of b_{Jp} and b_{Jp}^2 and spin of nucleus p as given by

Eq. (A7-39b). In Eqs. (9-164a,b), M_{Jp} is a definite integral over the configuration space of the crystal in the form

$$M_{Jp} = \int_{c} \psi_2^* \exp\left(2\pi i\tilde{\boldsymbol{k}}_{12}\boldsymbol{r}_{Jp}\right)\psi_1 \, d\boldsymbol{r}_c, \tag{9-165}$$

where $\boldsymbol{k}_{12} = \boldsymbol{k}_1 - \boldsymbol{k}_2$ is called a scattering vector, and ψ_1 and ψ_2 are the wave functions of initial and final states of the crystal (see Appendix 7). The coherent cross section in Eq. (9-164a) takes formally the same form as Eq. (9-81) for X-ray scattering intensities, except that a factor involving absolute values of the initial and final momenta is required for neutrons. Hence the formulae for coherent elastic and coherent inelastic scattering cross sections of neutrons are derived in a way analogous to the cases of Laue–Bragg scattering and temperature diffuse scattering intensities, respectively, of X-rays. The formulae for incoherent cross sections are derived as follows.

Within the framework of the harmonic approximation, the total wave function of the nuclear motion of a crystal is a product of harmonic oscillator wave functions of individual normal coordinates in the forms

$$\psi_1 = \prod_{K,j} \left| v_{Kaj} \right\rangle \left| v_{Kbj} \right\rangle \quad \text{and} \quad \psi_2 = \prod_{K,j} \left| v'_{Kaj} \right\rangle \left| v'_{Kbj} \right\rangle,$$

and nuclear displacements can be expressed as linear combinations of the normal coordinates.[37] Thus, substituting Eqs. (9-82) and (9-85) into Eq. (9-165) gives

$$M_{Jp} = \exp\left\{2\pi i\tilde{\boldsymbol{k}}_{12}\left(\boldsymbol{r}_J + \boldsymbol{r}_{1p}^0\right)\right\} \prod_{K,j} \left\{ \left\langle v'_{Kaj} \right| \exp\left(iF_{Jp,Kj}^a Q_{Kj}^a\right) \left| v_{Kaj} \right\rangle \left\langle v'_{Kbj} \right| \exp\left(iF_{Jp,Kj}^b Q_{Kj}^b\right) \left| v_{Kbj} \right\rangle \right\}, \tag{9-166}$$

where

$$F_{Jp,Kj}^u = 2\pi \tilde{\boldsymbol{k}}_{12} \boldsymbol{l}_{Jp,Kj}^{Xu} \qquad (u = a \text{ or } b). \tag{9-167}$$

In Eq. (9-166), $\left| v_{Kaj} \right\rangle$ and $\left| v'_{Kaj} \right\rangle$ are single variable wave functions of the normal coordinate Q_{Kj}^a, while $\left| v_{Kbj} \right\rangle$ and $\left| v'_{Kbj} \right\rangle$ are those of Q_{Kj}^b. Let us take the case in which the state $\left| v'_{Kaj} \right\rangle$ is s quanta above or below the state $\left| v_{Kaj} \right\rangle$, and let the matrix element of $\exp\left(iF_{Jp,Kj}^a Q_{Kj}^a\right)$ in that case be expressed by

$$\left(I_{Jp,Kj}^a\right)_{n,n+s} = \left\langle n + s \right| \exp\left(iF_{Jp,Kj}^a Q_{Kj}^a\right) \left| n \right\rangle, \tag{9-168}$$

where $s \geq -n$. Since the population of the initial state $|n\rangle$ depends on the temperature, the thermal average of the product of Eq. (9-168) and its complex conjugate is necessary for calculating $\left|M_{Jp}\right|^2$. The weight for the initial state of the lattice mode K_j, is given by the distribution function for vibrational levels of a harmonic oscillator in the form

$$\left(w_{Kj}\right)_n = \left(1 - x_{Kj}\right) x_{Kj}{}^n, \tag{9-169}$$

where x_{Kj} is defined in Eq. (9-107).

The thermal average of the probabilities for transitions of the vibrational levels associated with the normal coordinate Q_{Kj}^a by s quanta is given by

$$\left(J^a_{Jp,Kj}\right)_s = \sum_{n=0}^{\infty}\left(w_{Kj}\right)_n \left(I^a_{Jp,Kj}\right)^*_{n,n+s}\left(I^a_{Jp,Kj}\right)_{n,n+s}. \tag{9-170}$$

In Appendix 8, the averaged probability is calculated to be

$$\left(J^a_{Jp,Kj}\right)_s = \exp\left\{-\frac{{t^a_{Jp,Kj}}^2}{2}\frac{1+x_{Kj}}{1-x_{Kj}}\right\}\frac{I_s\left\{{t^a_{Jp,Kj}}^2/2\sinh\left(hc\omega_{Kj}/2kT\right)\right\}}{\exp\left(-shc\omega_{Kj}/2kT\right)}, \tag{9-171}$$

where $I_s(x)$ is the modified Bessel function of the sth order,[38] and $t^a_{Jp,Kj}$ is a dimensionless quantity defined by

$$t^a_{Jp,Kj} = \left(h/4\pi^2 c\omega_{Kj}\right)^{1/2} F^a_{Jp,Kj}. \tag{9-172}$$

On scattering for which the transition probability is given by Eq. (9-171), the energy transferred from the neutron to the crystal is $s\omega_{Kj}$. Since the same argument holds for the normal coordinate Q^b_{Kj}, the crystal gains the energy $s\omega_{Kj}$ whenever the net increase in the vibrational quanta related to Q^a_{Kj} and Q^b_{Kj} is s. Hence the combined contribution from Q^a_{Kj} and Q^b_{Kj} to the probability of such an energy transfer is given by

$$\begin{aligned}\left(J_{p,Kj}\right)_s &= \sum_{r=-\infty}^{\infty}\left(J^a_{Jp,Kj}\right)_{s-r}\left(J^b_{Jp,Kj}\right)_r \\ &= \exp\left\{-\frac{{t^a_{Jp,Kj}}^2+{t^b_{Jp,Kj}}^2}{2}\frac{1+x_{Kj}}{1-x_{Kj}}\right\} x_{Kj}^{-s/2}\, I_s\left(\frac{{t^a_{Jp,Kj}}^2+{t^b_{Jp,Kj}}^2}{x_{Kj}^{-1/2}-x_{Kj}^{1/2}}\right),\end{aligned} \tag{9-173}$$

where the second equality is proved in Appendix 8. Note that the suffix J is dropped from the joint probability $(J_{p,Kj})_s$ in Eq. (9-173), since it follows from Eqs. (9-172), (9-167), (9-94) and (9-96) that

$$\begin{aligned}{t^a_{Jp,Kj}}^2+{t^b_{Jp,Kj}}^2 &= \frac{h\left({F^a_{Jp,Kj}}^2+{F^b_{Jp,Kj}}^2\right)}{4\pi^2 c\omega_{Kj}} = \frac{h\tilde{\boldsymbol{k}}_{12}}{c\omega_{Kj}}\left(\boldsymbol{l}^{Xa}_{Jp,Kj}\tilde{\boldsymbol{l}}^{Xa}_{Jp,Kj}+\boldsymbol{l}^{Xb}_{Jp,Kj}\tilde{\boldsymbol{l}}^{Xb}_{Jp,Kj}\right)\boldsymbol{k}_{12} \\ &= \frac{2h\tilde{\boldsymbol{k}}_{12}\boldsymbol{D}^{Kj}_{pp}\boldsymbol{k}_{12}}{Nm_p c\omega_{Kj}}.\end{aligned} \tag{9-174}$$

In Eq. (9-173), the argument of the modified Bessel function is so small for a usual condition that the approximation

$$I_s(x) \approx (x/2)^{|s|}/(|s|!) \tag{9-175}$$

can be adopted. When s is zero, no exchange of energy occurs between the neutron and the crystal. The scattering of neutrons in such a case is called elastic scattering, the probability of which is given by

$$\left(J_{p,Kj}\right)_0 = \exp\left\{-\frac{{t^a_{Jp,Kj}}^2+{t^b_{Jp,Kj}}^2}{2}\frac{1+x_{Kj}}{1-x_{Kj}}\right\}. \tag{9-176a}$$

Substituting Eqs. (9-175) and (9-176a) into Eq. (9-173) and noting that $I_0(x) \approx 1$ for $x \ll 1$, we have

$$\left(J_{p,Kj}\right)_s = \left(J_{p,Kj}\right)_0 \left\{\left(t_{Jp,Kj}^{a}{}^2 + t_{Jp,Kj}^{b}{}^2\right)/2\left(1-x_{Kj}\right)\right\}^s / s!. \tag{9-176b}$$

The total transition probability for an elastic scattering is obtained, by comparing Eq. (9-174) with Eqs. (9-95) and (9-97a), in terms of diagonal subblocks of the mean square amplitude matrix in the form

$$\prod_{K,j}\left(J_{p,Kj}\right)_0 = \exp\left\{-4\pi^2\,\tilde{\boldsymbol{k}}_{12}\sum_K \boldsymbol{\Sigma}_{pp}^{x}(T)_K\,\boldsymbol{k}_{12}\right\} = \exp(-2W_p). \tag{9-177}$$

The scattering of neutrons accompanied by an energy transfer between colliding particles is called inelastic scattering. In the case when the energy is transferred from the nucleus to the neutron, the scattering is called up-scattering, while the case of transfer from the neutron to the nucleus is called down-scattering. Both the up- and down-scatterings are classified into single-, two-, three-phonon processes and so on, according to how many vibrational quanta are participating in the energy transfer. The numerical calculation is facilitated by introducing a dimensionless scalar quantity

$$H_{Kj,p}^{+} \equiv \left(J_{p,Kj}\right)_1 = \frac{h\,\tilde{\boldsymbol{k}}_{12}\boldsymbol{D}_{pp}^{Kj}\boldsymbol{k}_{12}}{m_p\,Nc\omega_{Kj}}\,\frac{1}{1-x_{Kj}} \tag{9-178a}$$

and

$$H_{Kj,p}^{-} = x_{Kj}\,H_{Kj,p}^{+} = \frac{h\,\tilde{\boldsymbol{k}}_{12}\boldsymbol{D}_{pp}^{Kj}\boldsymbol{k}_{12}}{m_p\,Nc\omega_{Kj}}\,\frac{x_{Kj}}{1-x_{Kj}} \tag{9-178b}$$

corresponding to gain and loss, respectively, of one quantum of the normal mode *Kj* by the crystal.[7] The coefficient x_{Kj} in Eq. (9-178b) enters as the Boltzmann factor between upper and lower levels. Then the differential cross section of single-phonon incoherent inelastic neutron scattering is calculated by

$$\frac{d\sigma_{\text{incoh}}}{d\Omega}\left(\pm\omega_{Kj}\right) = \frac{|\boldsymbol{k}_1|}{|\boldsymbol{k}_2|}\,\sigma_p^{\text{incoh}}\,H_{Kj,p}^{\pm}\exp(-2W_p), \tag{9-179a}$$

where + and − represents down- and up-scattering cross sections, respectively. The differential cross section for the multiphonon scattering involving *r* quanta of mode *Kj*, *s* quanta of mode *K'j'*, ... is given by

$$\frac{d\sigma_{\text{incoh}}}{d\Omega}\left(\pm r\omega_{Kj} \pm s\omega_{K'j'} \pm\cdots\right) = \frac{|\boldsymbol{k}_1|}{|\boldsymbol{k}_2|}\sum_p \sigma_p^{\text{incoh}}\,\frac{\left(H_{Kj,p}^{\pm}\right)^r}{r!}\,\frac{\left(H_{K'j',p}^{\pm}\right)^s}{s!}\cdots\exp\left(-2W_p\right). \tag{9-179b}$$

To simulate the scattering intensity spectrum, the whole spectral range is partitioned into small segments corresponding to the experimental resolution, and each segment is attributed to an address in the intensity storage array. Both sides of Eq. (9-179a) or (9-179b) are calculated and the summands in the right side are accumulated in the address specified by the freqeuncy or frequency sum in the left side. Thus incoherent inelastic scattering of neutrons offers detailed information on the energy distribution of lattice vibrations, but cannot tell about spatial distribution of wave vectors because of its incoherency. In this sense, incoherent inelastic neutron scattering is complementary to temperature diffuse scattering of X-ray as a tool in lattice dynamics. There is a close analogy in intensity formulae between the two methods. Compare Eq. (9-179) with Eq. (9-154). The constants $H^{\pm}_{Kj,p}$ introduced here correspond to the constant $H_{K,pp'}$ introduced for temperature diffuse scattering of X-ray. If the scattering vectors are formally taken to be common, it follows from Eqs. (9-147) and (9-178a,b) that

$$H_{K,pp'} = \sum_j \left(H^{+}_{Kj,p} + H^{-}_{Kj,p}\right).$$

Distinction between up- and down-scatterings is necessary for incoherent inelastic neutron scattering in order to fully utilize its advantage in frequency resolution. Theoretically, coherent inelastic scattering is advantageous in both the energetic and spatial resolutions. It is practically difficult, however, to apply this method to light hydrogen nuclei for which $\sigma^{\text{incoh}} \gg \sigma^{\text{coh}}$. The reverse is true for many other nuclear species including D, C, O, N, *etc.*, so that the progress in coherent inelastic scattering of neutrons is expected to serve greatly to lattice dynamics of organic molecular crystals in the future.

Since the scattering intensities in Eqs. (9-179a,b) involve scalar products of scattering vector and the atomic polarization vectors, certain information as to the anisotropy in the frequency distribution may be available from appropriate experiments on oriented samples. In stretched polymer samples, the chain axes are uniaxially oriented along the stretching direction. By taking this direction as the *Z*-axis, and setting the scattering vector perpendicular to it, only the *X*- and *Y*-components of the polarization vectors contribute to the scattering intensity. A polymer sample uniaxially oriented along the *Z*-axis has no regularity as to the orientation in the *XY* plane. In this case, the mean value of the orientation dependent factor in the right side of Eq. (9-174) becomes

$$\left\langle \tilde{\boldsymbol{k}}_{12}\left\{\left(\boldsymbol{l}^{mx}_{Kc}\right)_{pj}\left(\tilde{\boldsymbol{l}}^{mx}_{Kc}\right)_{pj} + \left(\boldsymbol{l}^{mx}_{Ks}\right)_{pj}\left(\tilde{\boldsymbol{l}}^{mx}_{Ks}\right)_{pj}\right\}\boldsymbol{k}_{12}\right\rangle = \tfrac{1}{2}|\boldsymbol{k}_{12}|^2\left\{\left(D^{Kj}_{pp}\right)_{XX} + \left(D^{Kj}_{pp}\right)_{YY}\right\}$$
$$= \tfrac{1}{2}|\boldsymbol{k}_{12}|^2\left\{\left(l^{mx}_{Kc,X}\right)_{pj}^{\,2} + \left(l^{mx}_{Kc,Y}\right)_{pj}^{\,2} + \left(l^{mx}_{Ks,X}\right)_{pj}^{\,2} + \left(l^{mx}_{Ks,Y}\right)_{pj}^{\,2}\right\} \qquad (9\text{-}180)$$

Similarly, for the scattering vector along the stretching direction of the sample, we have the orientation dependent factor along the *Z*-axis in the form

$$\tfrac{1}{2}\left(D^{Kj}_{pp}\right)_{ZZ} = \tfrac{1}{2}\left\{\left(l^{mx}_{Kc,Z}\right)_{pj}^{\,2} + \left(l^{mx}_{Ks,Z}\right)_{pj}^{\,2}\right\}. \qquad (9\text{-}181)$$

Accordingly, by defining the quantities

$$H^{+}_{Kj,p}(\perp) = \frac{h|\boldsymbol{k}_{12}|^2\left\{\left(D^{Kj}_{pp}\right)_{XX} + \left(D^{Kj}_{pp}\right)_{YY}\right\}}{2m_p\, Nc\omega_{Kj}\left(1 - x_{Kj}\right)} \qquad (9\text{-}182a)$$

and

$$H^{+}_{Kj,p}(//) = \frac{h|\boldsymbol{k}_{12}|^2\left(D^{Kj}_{pp}\right)_{ZZ}}{2m_p\, Nc\omega_{Kj}\left(1 - x_{Kj}\right)}, \qquad (9\text{-}182b)$$

and introducing the corresponding symbols for the up-scattering,

$$H^{-}_{Kj,p}(\perp) = x_{Kj}\, H^{+}_{Kj,p}(\perp) \qquad \text{and} \qquad H^{-}_{Kj,p}(//) = x_{Kj}\, H^{+}_{Kj,p}(//),$$

the anisotropic scattering cross sections for uniaxially oriented polymer samples can be calculated. The theoretical cross sections of polyethylene calculated by Kitagawa and Miyazawa successfully reproduced the peak positions and relative intensities in the experimental anisotropic cross sections reported by Myers, Summerfield and King.[39)]

With the recent progress in inelastic neutron scattering spectroscopy, high quality spectra of molecular crystals of relatively complex organic compounds have been accumulating. Heyward *et al.* measured the inelastic neutron scattering spectrum of acetanilide polycrystals and simulated the spectra with molecular mechanics and molecular dynamics calculations involving both intra- and intermolecular vibrations.[40)] The space group is *Pbca*, $Z = 8$ and each molecule consists of 19 atoms. The CHARMm potential function including the Urey–Bradley type 1–3 interaction was used.[41)] In the molecular mechanics calculation, the normal coordinate analyses were made at 125 points in one-eighth of the first Brillouin zone, generating $3 \times 19 \times 8 \times 125 \times 8 = 456{,}000$ frequencies. The polycrystal spectrum was obtained as the average of three spectra calculated for the scattering vectors along the three unit cell axes. The single, two- and three-phonon processes were taken into account. In both the observed and simulated spectra, the strongest peak due to the methyl torsional mode appears at 145 cm^{-1}. Many other prominent peaks in the observed spectra were well reproduced by the simulation. As seen from this result, inelastic neutron scattering is a very powerful tool for estimating potential functions of molecular crystals, since the spectra can be simulated without introducing any adjustable parameters which control the intensities.

The advantages and shortcomings of molecular mechanics and of molecular dynamics in simulating inelastic neutron scattering spectra are summarized as follows. At present, molecular mechanics formulation is hardly extended beyond the harmonic approximation, but the method can be applied to infinitetly large system in principle. In molecular dynamics calculation, on the other hand, the sample size is severely restricted by the computer efficiency but anharmonic effects can be easily incorporated without much change in the program. The roles of molecular mechanics and molecular dynamics are thus complementary, so that the results obtained by the two methods must be carefully compared with each other in order that the reliability of the whole calculation is ascertained.

9.4 Lattice Energies of Static Crystals

The lattice energy of a molecular crystal is defined as the energy required for separating all the molecules contained in a specified amount of the crystal into the isolated state. If all the nuclei in both isolated and crystallized molecules are assumed to be at rest, the lattice energy is given by the difference between the potential minimum of the isolated molecule and that of the crystal per molecule. Accordingly, if V_{in} in Eq. (9-35) is taken to be the potential function of the isolated molecule expanded at the equilibrium, the constant term V_{in}^0 is zero so that minus the constant term V^0 in Eq. (9-36) divided by the number of molecules per unit cell represents the static lattice energy per molecule. The lattice energy of a real crystal estimable from the observed heat of sublimation consists of the static and dynamic lattice energies. The latter, which has been treated in subsection 9.3.4 is the difference in the energy of thermal motion of nuclei per molecule between the crystal and the isolated state. This section deals with the calculation of static lattice energy.

If the molecular structure in the isolated state is not much different from that in the crystal, the lattice energy can be regarded to arise mainly from the nonbonded atom–atom interaction between neighboring molecules. If the crystal involves intermolecular hydrogen bonds or metal coordination bonds, the potential function to describe the dependence of these bond energies on the related local structures should be added. The sum of these energies of intermolecular interaction minus the increment of the intramolecular energy due to the molecular distortion in the crystal gives the lattice energy. The distortion of a molecule on bringing it from the isolated state into the crystal is denoted in terms of the normal coordinates by $\boldsymbol{Q}^{\rho}$ in Eq. (9-39), and leads to the increment of the intramolecular energy given by

$$V_{\text{in}} = \tfrac{1}{2}\tilde{\boldsymbol{Q}}^{\rho}\,\boldsymbol{\Lambda}\,\boldsymbol{Q}^{\rho}. \tag{9-183}$$

If the electronic structure of a molecule is not much affected by the states of aggregation, the difference in V_{in}^{0} between the crystal and the gas phase may be small, and the static lattice energy arises from the energy of intermolecular nonbonded interaction, V_{ex}^{0}. How far the summation limit should be taken in calculating V_{ex}^{0} is an important problem in applying molecular mechanics to crystals. In a space where N_d particles per unit volume are homogeneously distributed, the number of particles contained in a shell between two spheres of radii r and $r + \mathrm{d}r$ is $4\pi r^2 N_d\,\mathrm{d}r$. All crystals are electrically neutral in a normal state, so that a unit cell carries a vanishing net charge as a whole, but the dipole moment of a unit cell does not necessarily vanish.

As shown in Eq. (7-18d), the interaction energy between two dipoles separated by the distance r is proportional to r^{-3}. On multiplying this energy by the number of particles and summing over the spherical shells of homogeneous thickness, the total energy of dipole–dipole interaction is estimated to be roughly proportional to

$$1 + 2^{-1} + 3^{-1} + \cdots + n^{-1} + \cdots.$$

It is well known that this series diverges. If the charge distribution in a unit cell is described by any moments whose orders are higher than the first, the sum of the electrostatic interaction converges in principle, but the convergence is very slow when only a simple sum is computed. Various techniques to accelerate the convergence have been developed. A rigorous method of calculating the sum of pairwise interaction energies inversely proportional to the distance by utilizing the periodicity of the crystal was proposed first by Ewald,[42] and extended to cases including the multipoles by Nijboer and DeWette.[43] Williams applied this method to general inverse power potentials of the type $q_s q_t / r_{st}^{\,n}$.[44]

Derivation of the acceleration formulae starts with dividing the lattice sum of the form $\frac{1}{2}\sum_{s,t} q_s q_t\, r_{st}^{-n}$ per unit cell into two parts as

$$S_n = \tfrac{1}{2}\sum_{s=1}^{n'N}\sum_{t=1}^{n'} q_s\, q_t\, r_{st}^{-n}\varphi(r_{st}) + \tfrac{1}{2}\sum_{s=1}^{n'N}\sum_{t=1}^{n'} q_s\, q_t\, r_{st}^{-n}\{1-\varphi(r_{st})\}, \tag{9-184}$$

where $\varphi(r)$ is called the convergence function, and transforming the second sum into a sum over the lattice points in the reciprocal space. In Eq. (9-184), the index t runs over a unit cell involving n' nuclei, and s over the whole crystal involving N unit cells. By attributing an adequate value to a parameter in $\varphi(r)$, the convergence in each sum in Eq. (9-184) can be made quick enough to be used in practical calculation. Nijboer and DeWette[43] introduced an efficient convergence function defined in terms of an incomplete gamma function of the second kind $\Gamma(a, b)$ and a gamma function $\Gamma(x)$ as

$$\varphi(r) = \Gamma(n/2, K^2\pi r^2)/\Gamma(n/2). \tag{9-185}$$

Following Bertaut,[45] let us introduce the Patterson function expressed in terms of the net atomic charge q_s in place of the atomic structure factor f_s used in X-ray crystallography,[46] *i.e.*,

$$\boldsymbol{P}(\boldsymbol{r}) = \sum_{s=1}^{n'N}\sum_{t=1}^{n'} q_s\, q_t\, \delta(\boldsymbol{r} - \boldsymbol{r}_{st}). \tag{9-186}$$

Derivation of Eq. (9-186) from the definition of $P(\boldsymbol{r})$ as the self convolution of electron density is given in Appendix 9 (Eq. (A9-6)). By using Eq. (9-186) and the property of the δ-function, the lattice sum in Eq. (9-184) can be rewritten as

$$S_n = \tfrac{1}{2} \int_{\text{crys}} \{P(\boldsymbol{r}) - P(\boldsymbol{r})\delta(\boldsymbol{r})\}\varphi(r) r^{-n} d\boldsymbol{r} + \tfrac{1}{2} \int_{\text{crys}} \{P(\boldsymbol{r}) - P(\boldsymbol{r})\delta(\boldsymbol{r})\}\{1 - \varphi(r)\} r^{-n}\, d\boldsymbol{r}, \tag{9-187}$$

where the integrals must be taken over whole crystal in order that long range interaction is taken into account. Note that the definition of δ-function gives

$$\int \delta(\boldsymbol{r} - \boldsymbol{r}_{st}) f(r) d\boldsymbol{r} = f(r_{st}) \tag{9-188a}$$

and

$$\int \delta(\boldsymbol{r} - \boldsymbol{r}_{st})\delta(\boldsymbol{r}) f(r) d\boldsymbol{r} = f(r_{ss}) = f(0), \tag{9-188b}$$

whence the operator $P(\boldsymbol{r})\delta(\boldsymbol{r})$ in Eq. (9-187) has the role of removing the case $s = t$ from the sums in Eq. (9-184). Let the first and second terms in the right side of Eq. (9-187) be called I_1 and I_2, respectively, and let I_2 be divided into two parts I_{2a} and I_{2b} in the forms

$$I_{2a} = \tfrac{1}{2} \int_{\text{crys}} P(\boldsymbol{r})\{1 - \varphi(r)\} r^{-n}\, d\boldsymbol{r} \tag{9-189a}$$

and

$$I_{2b} = -\tfrac{1}{2} \int_{\text{crys}} P(\boldsymbol{r})\delta(\boldsymbol{r})\{1 - \varphi(r)\} r^{-n}\, d\boldsymbol{r}. \tag{9-189b}$$

The integral I_1 is rewritten, by replacing $P(\boldsymbol{r})$ with the last formula of Eq. (9-186) and using Eqs. (9-185) and (9-188a,b), in the form

$$\begin{aligned} I_1 &= \tfrac{1}{2} \sum_{s=1}^{n'N}\sum_{t=1}^{n'} \int_{\text{crys}} q_s q_t \{\delta(\boldsymbol{r} - \boldsymbol{r}_{st}) - \delta(\boldsymbol{r} - \boldsymbol{r}_{st})\delta(\boldsymbol{r})\} \frac{\Gamma(n/2, K^2\pi r^2)}{\Gamma(n/2)} r^{-n}\, d\boldsymbol{r} \\ &= \frac{1}{2\,\Gamma(n/2)} \sum_{s=1}^{n'N}\sum_{t=1}^{n'} (1 - \delta_{st}) q_s q_t\, \Gamma(n/2, K^2\pi r_{st}^{\,2}) r_{st}^{\,-n}. \end{aligned} \tag{9-190}$$

In calculating the integral I_{2a} for a vibrating crystal, we cannot assume the translational symmetry among equivalent unit cells. In this case, individual motions of all nuclei in the crystal may be easily taken into account by introducing the total structure factor given by

$$F_C(\boldsymbol{h}) = \sum_{t=1}^{n'N} q_t \exp\left(2\pi i \tilde{\boldsymbol{h}} \boldsymbol{r}_t\right)$$
$$= \sum_{t=1}^{n'} q_t \sum_{L=1}^{N} \exp\left\{2\pi i \tilde{\boldsymbol{h}} (\boldsymbol{r}_L + \boldsymbol{r}_t^0)\right\} \exp\left(2\pi i \tilde{\boldsymbol{h}}\, \Delta \boldsymbol{r}_{Lt}\right). \tag{9-191}$$

The expression of the Patterson function in Eq. (9-186) is now modified so as to involve all nuclei in the crystal, and subjected to the three-dimensional Fourier transform defined in Appendix 10. The result is given in terms of $F_C(\boldsymbol{h})$ as

$$\mathrm{FT}_3\{P(\boldsymbol{r})\} = N^{-1} \sum_{s=1}^{n'N} \sum_{t=1}^{n'N} q_s q_t \int_{\text{crys}} \delta(\boldsymbol{r} - \boldsymbol{r}_{st}) \exp\left(2\pi i \tilde{\boldsymbol{h}} \boldsymbol{r}\right) d\boldsymbol{r}$$
$$= N^{-1} \sum_{s=1}^{n'N} \sum_{t=1}^{n'N} q_s q_t \exp\left(2\pi i \tilde{\boldsymbol{h}} \boldsymbol{r}_{st}\right) = N^{-1} \left|F_C(\boldsymbol{h})\right|^2. \tag{9-192}$$

where V is the unit cell volume. By substituting Eq. (9-192) into the second Parseval theorem for real functions[47] given in the form

$$\int f(\boldsymbol{r}) g(\boldsymbol{r}) d\boldsymbol{r} = \int \mathrm{FT}_3\{f(\boldsymbol{r})\} \mathrm{FT}_3\{g(-\boldsymbol{r})\} d\boldsymbol{h},$$

where $f(\boldsymbol{r})$ and $g(\boldsymbol{r})$ stand for $\{1 - P(\boldsymbol{r})\} r^{-n}$ and $P(\boldsymbol{r})$, respectively, and noting that $P(\boldsymbol{r}) = P(-\boldsymbol{r})$ as shown in Eq. (A9-12), the integral I_{2a} in Eq. (9-189a) is written in the form

$$I_{2a} = \frac{1}{2N} \sum_{\lambda} \int_{\text{rec}} \left|F_C(\boldsymbol{h})\right|^2 \mathrm{FT}_3\left\{\frac{1 - \varphi(\boldsymbol{r})}{r^n}\right\} d\boldsymbol{h}. \tag{9-193a}$$

According to Eq. (A10-4) in Appendix 10, Eq. (9-193a) is further rewritten as

$$I_{2a} = \frac{\pi^{n-\frac{3}{2}} \int_{\text{rec}} \left|F_C(\boldsymbol{h})\right|^2 h^{n-3} \Gamma\{(-n+3)/2, \pi h^2/K^2\} d\boldsymbol{h}}{2N\,\Gamma(n/2)}. \tag{9-193b}$$

The quantity $\left|F_C(\boldsymbol{h})\right|^2$ is expanded in powers of nuclear displacements in Appendix 11. Substituting the constant term in the expansion into Eq. (9-193b) and using the property of δ-function, we have

$$I_{2a} = \frac{\pi^{n-\frac{3}{2}}}{2V\,\Gamma(n/2)} \sum_{\lambda} \left|F(\boldsymbol{h}_\lambda)_0\right|^2 h_\lambda^{\,n-3} \Gamma\left\{\frac{-n+3}{2}, \frac{\pi h_\lambda^{\,2}}{K^2}\right\}, \tag{9-194}$$

where $\boldsymbol{h}_\lambda$ is the reciprocal lattice vector pointing to the λth lattice point in the reciprocal lattice. Here the sum with respect to λ may be divided into two parts corresponding to $\boldsymbol{h}_\lambda = \boldsymbol{0}$ and $\boldsymbol{h}_\lambda \neq \boldsymbol{0}$. Since $F(\boldsymbol{0})_0 = \sum_{\text{cell}} q_t$, it follows from Eq. (A11-17) in Appendix 11 that the contribution from the case where $\boldsymbol{h}_\lambda = \boldsymbol{0}$ to I_{2a} is given by

$$I_{2a}(\boldsymbol{h}_\lambda = \boldsymbol{0}) = \frac{\pi^{n/2}}{2V\,\Gamma(n/2)} \left(\sum_{t=1}^{n'} q_t\right)^2 K^{n-3} \frac{2}{n-3}. \tag{9-195}$$

Finally, according to Eq. (A11-23), the integral I_{2b} in Eq. (9-189b) is rewritten as

$$I_{2b} = -\tfrac{1}{2}\left\{\Gamma\left(\frac{n}{2}\right)\right\}^{-1} \frac{2K^n \pi^{n/2}}{n} \sum_{t=1}^{n'} (q_t^2). \tag{9-196}$$

By summing up the contributions from Eq. (9-194) with $\boldsymbol{h}_\lambda \neq \boldsymbol{0}$ and Eqs. (9-195) and (9-196), the reciprocal sum is written in the form

$$I_2 = \frac{\pi^{n/2}}{2\Gamma(n/2)}\left\{\frac{\pi^{(n-3)/2}}{V}\sum_{\boldsymbol{h}_\lambda \neq \boldsymbol{0}} |F(\boldsymbol{h}_\lambda)|^2 h_\lambda^{\,n-3}\,\Gamma\left(\frac{3-n}{2}, \frac{\pi h_\lambda^{\,2}}{K^2}\right)\right.$$
$$\left. + \frac{K^{n-3}}{V}\frac{2}{n-3}\left(\sum_{t=1}^{n'} q_t\right)^2 - \frac{2K^n}{n}\left(\sum_{t=1}^{n'} q_t^{\,2}\right)\right\}, \tag{9-197}$$

The total lattice sum S_n is the sum of Eqs. (9-190) and (9-197). How to partition S_n into the direct and the reciprocal sums depends on the choice of the convergence parameter K. Note that the structure factor involved in Eq. (9-197) is obtained by a single sum over nuclei per unit cell while the pairwise interaction in Eq. (9-190) must be summed up by a double loop over nuclei. From this reason, Eq. (9-197) is much more efficient than Eq. (9-190). The advantage of reciprocal sum in saving computing time is reduced in the calculation of dynamical matrix since we need to calculate $n(n+1)/2$ matrix elements separately in both direct and reciprocal sums. The intermolecular interaction energy due to the nth inverse power potential is obtained by subtracting from $I_1 + I_2$ the intramolecular part of S_n,

$$I_{\text{in}} = \tfrac{1}{2}\sum_{t}\sum_{s'} q_{s'} q_t r_{s't}^{\;-n},$$

where the index s' runs over the nuclei in the molecule containing the nucleus t. In the case of Coulomb energy of an electrically neutral crystal, the sum of q_t over a unit cell vanishes and the second term in the braces in the right side of Eq. (9-197) disappears.

In searching for equilibrium crystal structures by the Newton–Raphson method, we need to calculate the first and second derivatives of the lattice sum S_n. The second derivatives of S_n must also be calcualted in the normal coordinate analysis of lattice vibrations. In this case, partitioning of S_n into direct and reciprocal sums is necessary not only for assuring the convergent lattice sum but for interpreting vibrational spectra of ionic crystals or crystals containing polar molecules.[1)] As shown in Appendix 11, certain terms in the dynamical matrix elements of such a crystal consist of projections of nuclear displacement vector on the wave vector $\boldsymbol{q}_K$. These projections change with the direction of $\boldsymbol{q}_K$ but are not affected by $|\boldsymbol{q}_K|$. Hence the differences between optical normal frequencies ω_{Kj} and $\omega_{K'j}$ due to any difference in directions between $\boldsymbol{q}_K$ and $\boldsymbol{q}_{K'}$ such that $|\boldsymbol{q}_K| = |\boldsymbol{q}_{K'}|$ remain finite at the limit $|\boldsymbol{q}_K| \to 0$, *i.e.*, at the Γ point. For a given normal mode, the largest projection appears when the nuclear displacements are parallel with $\boldsymbol{q}_K$ while the projection disappears when the displacements are perpendicular to $\boldsymbol{q}_K$. The normal modes with parallel and perpendicular displacements give rise to longitudinal and transverse waves, respectively, of nuclear motions in the crystal. The splitting of this type is thus called TO (transverse-optical)-LO (longitudinal-optical) splitting. Experimental data of TO-LO splittings are available from infrared reflection spectra(e.g., $KNiF_3$,[48)] $RbMnF_3$,[49)] $CsMgCl_3$[50)] and HCOONa[51)]) or polarized Raman spectra(e.g., ZnO[52)]) of single crystals. The magnitudes of TO-LO splittings for crystals with rigid nuclear charges are calculated from the corresponding difference in eigenvalues of dynamical matrices with and without the first term in the right side of Eq. (A11-14). Analytical expressions for the first and second derivatives of the direct sum in S_n with respect to the Cartesian displacements of nuclei are given by Williams.[44] Pietila and Rasmussen examined the effect of convergence accelleration on the inverse power potentials of the types r^{-1}, r^{-6}, r^{-9} and r^{-12}, by using the molecular mechanics program CFF coded according to Williams' formula.[53] In the case of argon and ethane crystals, the recommended value of the convergence constant K for r^{-6} is 2.0 nm^{-1} for the summation limit 0.6 nm with an estimated truncation error less than 0.2%.

The structure parameters at the equilibrium and the lattice energies of a few simple amino acids were calculated from the same empirical potential function as used for interpreting vibrational spectra by the normal coordinate analyses. The summation limit for the Coulomb potential was taken to be 4 nm with the convergence constant K = 0.5 nm^{-1}. The large summation limit is necessary for attaining sufficient convergence for L-alanine crystal in which each unit cell has a residual dipole moment. Zero point energies were roughly estimated from the normal frequencies ω_i with the wave vector **0** calculated as described in section 9.3.1 in the form

$$V_0 = \frac{1}{2}\sum_i \omega_i / Z. \qquad (9\text{-}198)$$

The results are shown in Table 9.5. Gaffney *et al.* calculated the energy difference between the non-dissociated and the zwitter-ionic structures of glycine by the *ab initio* MO method.[54] These authors estimated the heat of sublimation of α-glycine crystal to zwitter-ionic molecules

Table 9.5 Structures and lattice energies of amino acid crystals.

	α-Glycine		L-Alanine		DL-Alanine	
	observed[a]	calculated	observed[b]	calculated	observed[c]	calculated
Lattice constants						
a (nm)	0.5105	0.5120	0.6032	0.6368	1.2060	1.2133
b (nm)	1.1969	1.2036	1.2343	1.2127	0.605	0.6204
c (nm)	0.5465	0.5569	0.5784	0.5816	0.582	0.5829
Center of mass						
X_g (nm)	0.0679	0.0769	0.3184	0.3331	0.1621	0.1636
Y_g (nm)	0.1411	0.1357	0.1641	0.1642	0.1662	0.1619
Z_g (nm)	−0.0090	−0.0056	0.2702	0.2575	0.1700	0.1703
Eulerian angles[d]						
ϕ (°)	8.64	3.16	31.05	36.33	115.42	124.77
θ (°)	48.39	46.40	10.95	9.91	11.96	11.67
χ (°)	77.64	73.56	52.57	44.37	120.08	134.17
$\lvert \Delta r \rvert$ (nm)[e]						
Average		0.013		0.022		0.013
Maximum		0.022		0.029		0.021
Lattice energies (kJ·mol^{-1})		−216.0		−194.4		−192.1

[a] P.G. Jensson and K. Kvick, *Acta Crystallogr.*, **B28**, 1827 (1972).
[b] H.J. Simpson and R.E. Marsh, *Acta Crystallogr.*, **20**, 550 91966).
[c] J. Donohue, *J. Am. Chem. Soc.*, **72**, 949 (1950).
[d] Defined in the same way as in footnote [a] of Table 9.1.
[e] Δr is the distance between the observed and calculated nuclear positions.

in the isolated state to be 218 kJ mol^{-1}, by summing the *ab initio* energy difference and the experimental heat of sublimation from the mass spectrum. This value agrees well with the lattice energy calculated by molecular mechanics in Table 9.5.

The calculation of the lattice energy by molecular mechanics may also be used in discussing the stability of the crystalline solid solutions. The two-component system of benzoic acid (BA) and *p*-fluorobenzoic acid (PFB) is crystallized as a solid solution containing 30–86% PFB from an aqueous solution of the constant total concentration 0.029 mol·dm^{-3} (PFB 20–70%). Yamamoto *et al.* found the crystal structure of this solid solution to be different from that of either BA or PFB, and examined the dependence of the lattice energy of the three types of crystal structures on the composition by using an empirical intermolecular potential function.[55)]

The molecular model in a mixed crystal was constructed by placing the fluorine and the hydrogen atoms at the points 1.364 Å and 1.0 Å apart, respectively, from the benzene ring along the radial direction. These molecular models were placed in the unit cell according to the crystal structures of BA, PFB and the mixed crystals determined by X-ray analysis, and the energy due to the intermolecular nonbonded atom–atom interaction was calculated. An exp–6 type nonbonded atom–atom potential function was used for describing the energy due to the dispersion and the exchange repulsion interactions. The same parameters for C, H, O and N as adopted in the calculation of amino acid crystals were used, while those for the fluorine atom were determined so as to reproduce the crystal structure of PFB. The total

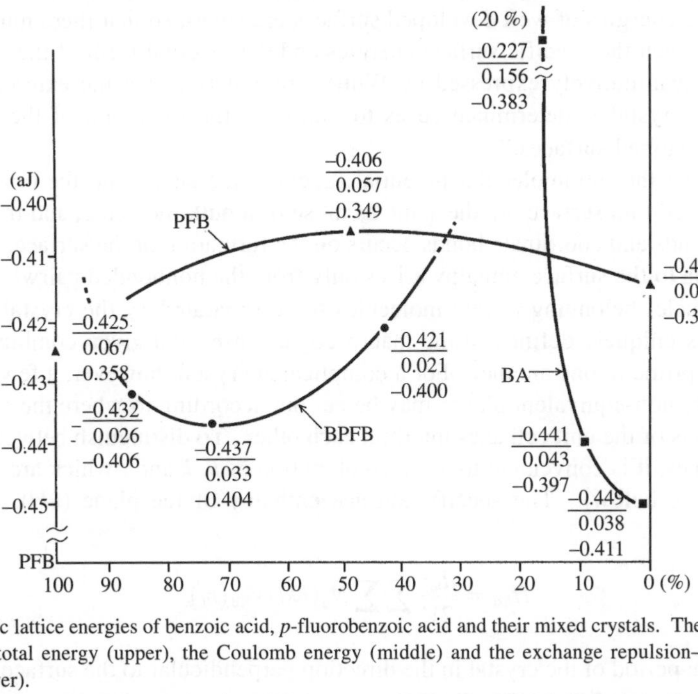

Fig. 9.8 Static lattice energies of benzoic acid, *p*-fluorobenzoic acid and their mixed crystals. The numerals indicate the total energy (upper), the Coulomb energy (middle) and the exchange repulsion–dispersion energy (lower).
(Reproduced with permission from N. Yamamoto, T. Taga and K. Machida, *Acta Crystallogr.*, **B45**, 162 (1989), p.166 (Fig. 5))

energy was calculated by summing up all the pairwise interaction energies each multiplied by the occupation ratios of the sites of interacting nuclei, whereby the contribution from any type of interacting atom pairs randomly distributing in the mixed crystal was taken into account. The dependence of the lattice enthalpy on the composition is shown in Fig. 9.8. The most stable structure predicted by the calculation is the mixed type in the range of composition between 30 and 90 %, and the structures of pure crystals of BA and PFB near 0 and 100 % of PFB, respectively, in good agreement with the experiment.

9.5 Surface Energies

Generally, a molecule situated near the surface of any molecular crystal has more or less excess energy compared to molecules situated in the bulk of the crystal. The reason for such a difference is that a bulk molecule is stabilized by the interaction with neighboring molecules in all directions around it while a molecule near the surface lacks partners for interaction in the direction of nearby surfaces. Accordingly, any creation of a new surface by cutting a molecular crystal requires some energy corresponding to this excess energy of surface molecules. This energy is called the surface energy, and the surface energy per unit area of a given crystal surface is called the specific surface energy. For an anisotropic crystal, crystal surfaces with different combinations of the Miller indices have different specific surface energies in general. If the crystal has a special symmetry, exchange of the Miller indices in limited ways may leave the specific surface energy invariant. As naturally expected, the specific surface energies of well developed surfaces are small, so that there must be a close correlation between the specific surface energies and the external form of the crystal. This correlation is quantitatively expressed by Wulff's rule stating that the external form of a slowly formed crystal is determined so as to minimize the total sum of the surface free energies of developed surfaces.[56]

Suppose that the intramolecular potential function, the shape and the orientation of a molecule exposed to a surface are the same as those of a bulk molecule, and that no fission of hydrogen bonds and coordinate bonds occurs on the formation of the surface. In this case, the contribution to the surface enthalpy arises only from the nonbonded pairwise interaction between the nuclei belonging to two molecules to be separated by the crystal plane.[57] A crystal plane is uniquely defined in the macroscopic sense by a given combination of the Miller indices prime to one another. For a complicated crystal, however, a few parallel but microscopically non-equivalent planes may be cut out according to where the surface plane and the segments of the unit cell axes intersect each other. To distinguish between such non-equivalent planes, it is convenient to use a combination of h, k and l which are not prime to one another (see below). The specific surface enthalpy of the plane (hkl), H_{hkl}, is then calculated by

$$H_{hkl} = \frac{d_{hkl}}{2V} \sum_i \sum_j N_{ij}(hkl)\, V_{\mathrm{NB}}(r_{ij}), \tag{9-199}$$

where d_{hkl} is the period of the crystal in the direction perpendicular to the surface in question, and V is the volume of the unit cell. The sum with respect to i is taken over the nuclei belonging to the molecules in the basic unit cell, and the sum with respect to j over all the nuclei belonging to any molecule which is separated from the molecule containing i by an intermolecular distance less than the summation limit r_{max}. To define the intermolecular distance, the representative nucleus of each molecule in the asymmetric unit is chosen desirably from the central part of the molecule. The intermolecular distance is then taken to be the internuclear distance between the representative nuclei of the two molecules. The factor $N_{ij}(hkl)$ in Eq. (9-199) is the number of equivalent nuclear pairs i–j separated by the surface plane, and is determined by using the following equations.

$$N_{ij}(hkl) = N_j(hkl) - N_i(hkl) \qquad (\text{if } N_j(hkl) \geq N_i(hkl)) \tag{9-200a}$$

$$N_{ij}(hkl) = 0 \qquad (\text{if } N_j(hkl) < N_i(hkl)) \tag{9-200b}$$

where

$$N_i(hkl) = \text{INT}\{F_{hkl}(\boldsymbol{r}_i^c)\} - \delta\{F_{hkl}(\boldsymbol{r}_i^c)\} \tag{9-201}$$

$$F_{hkl}(\boldsymbol{r}_i^c) = h'X_i^c + k'Y_i^c + l'Z_i^c - 1/P \tag{9-202}$$

and

$$h' = h/P, \qquad k' = k/P, \qquad l' = l/P \tag{9-203}$$

In Eq. (9-201), INT(F) is the integer part of the real number F, $\delta(F)$ is a step function which is 0 if F is positive, and 1 if F is negative. In Eq. (9-203), P is the greatest common divisor of h, k and l, and h', k' and l' are the Miller indices in the macroscopic sense. Now we introduce a notation

$$\xi(X) = PX + 1 \tag{9-204}$$

and write

$$a_i = \xi\{F_{hkl}(\boldsymbol{r}_i^c)\}/h \tag{9-205a}$$

$$b_i = \xi\{F_{hkl}(\boldsymbol{r}_i^c)\}/k \tag{9-205b}$$

$$c_i = \xi\{F_{hkl}(\boldsymbol{r}_i^c)\}/l \tag{9-205c}$$

Then the equation of the plane which has intercepts a_i, b_i and c_i on the a-, b- and c-axes, respectively, in the crystal coordinate system is given by

$$hX^c + kY^c + lZ^c = \xi\{F_{hkl}(\boldsymbol{r}_i^c)\}. \tag{9-206}$$

On substituting X_i^c, Y_i^c and Z_i^c into Eq. (9-206) and using Eqs. (9-204) and (9-205a–c), we obtain the same expression as Eq. (9-203), whence the nucleus i lies on the plane given by Eq. (9-206).

On the other hand, let N be an arbitrary integer, and the plane with the intercepts on the a-, b- and c-axes given by

$$\left.\begin{aligned} a(N) &= (N + 1/P)/h' \\ b(N) &= (N + 1/P)/k' \\ c(N) &= (N + 1/P)/l' \end{aligned}\right\} \tag{9-207}$$

be called P_N. Then the equation of the plane P_N is given by

$$hX^c + kY^c + lZ^c = \xi(N). \tag{9-208}$$

From Eq. (9-207), $1/P$ is the relative coordinate of the intersection between the plane P_0 and the axis for which the Miller index is 1 in the macroscopic sense. If $P = 1$, the plane P_{-1} passes through the origin. Changing N in Eq. (9-208) from $-\infty$ to $+\infty$ yields a set of equidistant parallel planes, whereby the space is divided into infinite layers of an equal thickness, as shown in Fig. 9.9.

On denoting the layer between the planes P_N and P_{N+1} by N, it follows from Eqs. (9-201)–(9-203) that the layer contaning the nucleus i is given by N_i. If $\delta\{F_{hkl}(\boldsymbol{r}_i^c)\}$ in Eq. (9-201) is omitted, the layer containing i turns out to be $N_i - 1$ for a negative $F_{hkl}(\boldsymbol{r}_i^c)$ on taking the integer part of the latter, giving two layers named N_0 sparated by the boundary plane on which the origin lies. From the above consideration, it is concluded that $N_{ij}(hkl)$ in Eq. (9-200a) is the number of boundary planes given by Eq. (9-208). Obviously, $N_{ij}(hkl)$ is the same as the number of nuclear pairs equivalent to i–j separated by one of the set of planes P_N, for example P_0 (see Fig. 9.9).

The double sum in Eq. (9-199) gives the total interaction energy between the nuclei in a unit cell exposed to the surface and in all the unit cells constituting a semifinite pillar behind the exposed unit cell and those in the hemispace at the opposite side of the boundary plane. Since this sum multiplied by the number of exposed unit cells in one side of the boundary plane, N_t, gives the total surface energy, E_t, the original double sum which is E_t/N_t must give the surface energy per exposed unit cell. The exposed area of the surface unit cell S is the volume divided by the thickness which is the period along the line perpendicular to the surface plane, V/d_{hkl}. Since the surface plane is formed at both sides of the boundary, the double sum divided by $2S$ gives the specific surface energy as shown in Eq. (9-199).[57]

Often the crystal habit of a given compound depends considerably on the kind of solvent from which the material is crystallized. The crystalline forms of aspirin obtained from a few kinds of solvents by Watanabe *et al.* are shown in Fig. 9.10.[58] These authors have confirmed

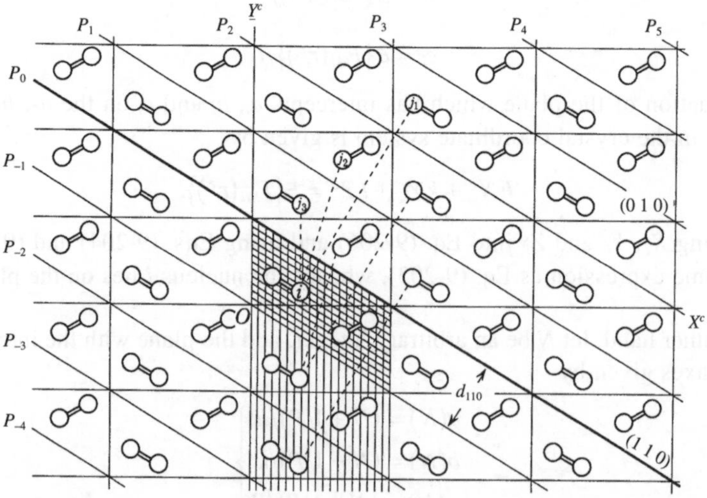

Fig. 9.9 Illustrative nonbonded nuclear pairs taken into account in the calculation of surface energies. There are three i–j_1 pairs, two i–j_2 pairs and an i–j_3 pair separated by the (1 1 0) plane P_0.

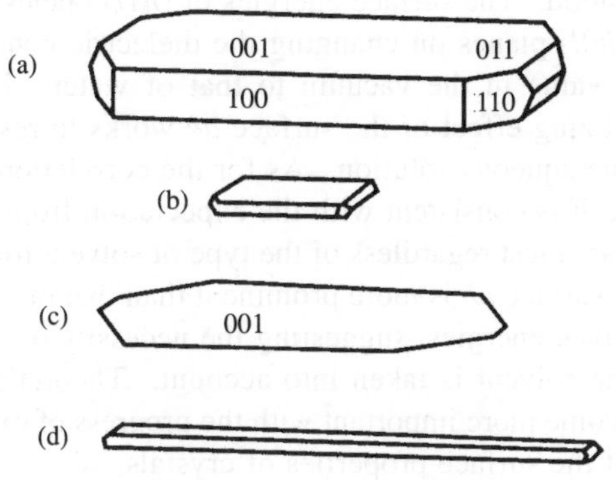

Fig. 9.10 Habits of aspirin crystals crystallized from various solvents.
(a) indices of developed crystal planes, (b) crystallized from ethanol, (c) from water, (d) from dioxane. (Reproduced with permission from A. Watanabe, Y. Yamaoka and K. Takada, *Chem. Pharm. Bull.* (Tokyo), **30**, 2958 (1982), p. 2958 (Fig. 1)).

that the rates of dissolution of these crystals are affected by the crystal habit. The dependence of the rate of regression of the crystal surface on the Miller indices on dissolving a large single crystal has been confirmed by direct measurements with the micrometer.[57] The observed rates of regression of various surfaces are shown together with the calculated values of specific surface energies in Table 9.6.

Kim *et al.* calculated the Coulomb energy terms in Eq. (9-199) for the dielectric constants of the vacuum and of water with the purpose of estimating the difference in specific surface energy between the vacuum and the aqueous solution. Since a surface with a large rate of regression tends to diminish the area of adjacent surfaces quickly, the order of the rate of regression and that of the diminishing rate of the area are expected to be inverse to each other. It is energetically preferred that surfaces with larger surface energies disappear more quickly. Hence it is reasonable that the rates of regression of surfaces *ab* and *bc* are in the order inverse to the order of specific surface energies, but the large surface energies of (*h*1*l*) planes constituting the most quickly regressing (in other works, the most slowly diminishing)

Table 9.6 Rate of regression on dissolution and specific surface energies of aspirin crystal.

Plane	Rate of regression ($nm \cdot s^{-1}$)[a)]				Surface energy (mJm^{-2})[b)]	
	26 °C	36 °C	46 °C	(*hkl*)	$D = 1.0$	$D = 78.56$
				(110)	237.8	208.0
ac[c)]	254	465	1001	(010)	394.9	296.3
				(011)	236.9	182.5
bc	187	334	736	(100)	131.5	127.2
ab	121	210	348	(001)	163.6	157.5

[a)] Apparatus 2 in USP XX was used.
[b)] D : Dielectric constant
[c)] The apparent *ac* plane consists of the (1 1 0) and the (0 1 1) planes, and the true (0 1 0) plane does not develop.

surface *ac* cannot be understood. The surface energies of (*h*1*l*) planes decrease much more appreciably than those of (*h*0*l*) planes on changing the dielectric constant in the Coulomb interaction terms from the value in the vacuum to that of water. From this result it is suggested that a large stabilizing effect of the surface *ac* works to resist the diminution of the area of this surface in the aqueous solution. As for the correlation between the surface energy and the crystal habit, it is consistent with the expectation from Wulff's rule that the area of the surface *ac* is the smallest regardless of the type of solvent for crystallization. The fact that the development of surface *ab* is more prominent than that of surface *bc* contradicts, however, the calculated surface energies, suggesting the necessity of a theory in which the effect of interaction with the solvent is taken into account. Theoretical role of molecular mechanics is expected to become more important with the progress of experimental technique in quantitative evaluation of the surface properties of crystals.

References

1) M. Born and K. Huang, *Dynamical Theory of Crystal Lattices*, Oxford University Press, London (1954).
2) Y. Shiro and T. Miyazawa, *Bull. Chem. Soc. Jpn.*, **44**, 2371 (1971).
3) K. Machida and Y. Kuroda, *Bull. Chem. Soc. Jpn.*, **54**, 1343 (1981).
4) J.L. Schlenker, G.V. Gibbs and M.B. Boisen, Jr., *Acta Crystallogr.*, **A34**, 52 (1978).
5) Y. Kim, K. Machida, T. Taga and K. Osaki, *Chem. Pharm. Bull.* (*Tokyo*), **33**, 2641 (1985).
6) R. J. Roberts, R. C. Rowe and P. York, *Powder Technol.*, **65**, 139 (1991).
7) T. Kitagawa and T. Miyazawa, *Adv. Polymer Sci.*, **9**, 336 (1972).
8) M. Born and T. von Karman, *Physik. Z.*, **13**, 297 (1912).
9) T. Miyazawa, Y. Ideguchi and K. Fukushima, *J. Chem. Phys.*, **38**, 2709 (1963).
10) P.M.A. Sherwood, *Vibrational Spectroscopy of Solids*, Cambridge University Press, London (1972).
11) K. Machida, A. Kagayama and Y. Kuoroda, *Bull. Chem. Soc. Jpn.*, **54**, 1348 (1981).
12) K. Machida, A. Kagayama, Y. Saito and T. Uno, *Spectrochim. Acta*, **34**, 909 (1978).
13) K. Machida, A. Kagayama and Y. Saito, *J. Raman Spectrosc.*, **7**, 188 (1978).
14) K. Machida, A. Kagayama and Y. Saito, *J. Raman Spectrosc.*, **8**,133 (1979).
15) A.I. Kitaigorodsky, *Molecular Crystals and Molecules*, Academic Press, New York (1973).
16) A. J. C. Wilson, editor, *International Tables for X-Ray Crystallography, Vol. C, Mathematical, Physical and Chemical Tables*, Kluwer Academic Publishers, Dordrecht, The Netherlands (1992).
17) V.H. Ott, *Ann. Physik*, **23**, 169 (1935).
18) M. Born and K. Sarginson, *Proc. Roy. Soc.* (London), **A179**, 69 (1941).
19) P.P. Debye, *Ann. Physik.*, **43**, 49 (1914).
20) I. Waller, *Z. Physik.*, **17**, 398 (1923); **51**, 213 (1928).
21) P.P. Debye, *Ann. Physik*, **39**, 789 (1912).
22) T. Kitagawa and T. Miyazawa, *Rept. Progr. Polymer Phys. Jpn.*, **11**, 219 (1968).
23) T. Shimanouchi and I. Harada, *J. Chem. Phys.*, **41**, 2651 (1964).
24) I. Harada and T. Shimanouchi, *J. Chem. Phys.*, **44**, 2016 (1966).
25) G.S. Pawley and S.J. Cyvin, *J. Chem. Phys.*, **52**, 4073 (1970).
26) D.W.J. Cruickshank, *Acta Crystallogr.*, **9**, 754, 757 (1956).
27) V. Schomaker and K.N. Trueblood, *Acta Crystallogr.*, **B24**, 63 (1968).
28) M. Born, *Proc. Roy. Soc.* (London), **A180**, 397 (1942).
29) J.L. Amoros and M. Amoros, *Molecular Crystals: Their Transforms and Diffuse Scattering*, John Wiley & Sons, New York (1968).
30) Y. Kashiwase and J. Harada, *J. Phys. Soc. Jpn.*, **35**, 1711 (1973).
31) G.S. Pawley, *Acta Crystallogr.*, **A25**, 702 (1969).
32) W.G. Ferrier, J.T. McMullan and D.C. Sutherland, *J. Phys.* C, *Solid State Phys.*, **6**, 1489 (1973).
33) T. Kitagawa and T. Miyazawa, *Bull. Chem. Soc. Jpn.*, **43**, 372 (1970).
34) M. Kobayashi and H. Tadokoro, *Macromolecules*, **8**, 897 (1975).
35) S.S. Chang, *J. Res. Nat'l. Bur. Stand. Sect. A*, **78**, 387 (1974).
36) C.M.L. Atkinson and M.J. Richardson, *Trans. Faraday Soc.*, **65**, 1749,1764 (1969).
37) R. Weinstock, *Phys. Rev.*, **65**, 1 (1944).
38) A.C. Zemach and R.J. Glauber, *Phys. Rev.*, **101**, 118 (1956).
39) W. Myers, G.C. Summerfield and J.S. King, *J. Chem. Phys.*, **44**, 184 (1966).

40) R.L. Heyward, H.D. Middendorf, U. Wanderlingh and J.C. Smith, *J. Chem. Phys.*, **102**, 5525 (1995).
41) C.L. Brooks III, M. Karplus and B.M. Pettitt, *Advances in Chemical Physics, Vol. 61*, John Wiley, New York (1988).
42) P.P. Ewald, *Ann. Phys.*, **54**, 519, 557 (1917).
43) B.R.A. Nijboer and F.W. DeWette, *Physica*, **23**, 309 (1957).
44) D.E. Williams, *Acta Crystallogr.*, **A27**, 452 (1971), **A28**, 629 (1972).
45) F. Bertaut, *J. Phys. Radium*, **13**, 499 (1952).
46) A.L. Patterson, *Phys. Rev.*, **46**, 372 (1934); *Z. Krist.*, **90**, 517, 543 (1935).
47) E. Butkov, *Mathematical Physics*, Addison-Wesley Publishing Co., Reading (1968), Chapter 7.
48) M. Balkanski, P. Moch and M.K. Teng, *J. Chem. Phys.*, **46**, 1621 (1967).
49) I. Nakagawa, *Spectrochim. Acta*, **29A**, 1451 (1973).
50) Y. Morioka and I. Nakagawa, *Bull. Chem. Soc. Jpn.*, **51**, 2467 (1978).
51) I. Tajima, H. Takahashi and K. Machida, *Spectrochim. Acta*, **37A**, 905 (1981).
52) T.C. Damen, S.P.S. Porto and B. Tell, *Phys. Rev.*, **142**, 570 (1966).
53) L. -O. Pietilä and Kj. Rasmussen, *J. Comput. Chem.*, **5**, 252 (1984).
54) J.S. Gaffney, R.C. Pierce and L. Friedman, *J. Am. Chem. Soc.*, **99**, 4293 (1977).
55) N. Yamamoto, T. Taga and K. Machida, *Acta Crystallogr.*, **B45**, 162 (1989).
56) G. Wulff, *Z. Krist.*, **34**, 449 (1901).
57) Y. Kim, M. Matsumoto and K. Machida, *Chem. Pharm. Bull.* (*Tokyo*), **33**, 4125 (1985).
58) A. Watanabe, Y. Yamaoka and K. Takada, *Chem. Pharm. Bull.* (*Tokyo*), **30**, 2958 (1982).

Appendix

Appendix 1 The Cosine Theorem in Spherical Trigonometry

Suppose that three points A, B, and C are taken on a sphere centered at a point O in such a way that

$$\angle \mathrm{AOB} = \theta_1, \quad \angle \mathrm{AOC} = \theta_2, \quad \angle \mathrm{BOC} = \theta_3.$$

Let D and E be the points of intersection between the tangential plane of the sphere at A and the straight lines OB and OC, respectively (see Fig. A1.1). From the nature of the tangential plane, angles ∠OAD and ∠OAE are both right angles, whence the dihedral angle τ between planes OAD and OAE is equal to angle ∠DAE. Applying the cosine theorem of plane trigonometry to triangles DEO and DEA, we have

$$\overline{\mathrm{DE}}^2 = \overline{\mathrm{OD}}^2 + \overline{\mathrm{OE}}^2 - 2\,\overline{\mathrm{OD}} \cdot \overline{\mathrm{OE}} \cos\theta_3 \tag{A1-1}$$

and

$$\overline{\mathrm{DE}}^2 = \overline{\mathrm{AD}}^2 + \overline{\mathrm{AE}}^2 - 2\,\overline{\mathrm{AD}} \cdot \overline{\mathrm{AE}} \cos\tau, \tag{A1-2}$$

respectively. Subtracting Eq. (A1-1) from Eq. (A1-2) gives

$$0 = \left(\overline{\mathrm{OD}}^2 - \overline{\mathrm{AD}}^2\right) + \left(\overline{\mathrm{OE}}^2 - \overline{\mathrm{AE}}^2\right) - 2\,\overline{\mathrm{OD}} \cdot \overline{\mathrm{OE}} \cos\theta_3 + 2\,\overline{\mathrm{AD}} \cdot \overline{\mathrm{AE}} \cos\tau. \tag{A1-3}$$

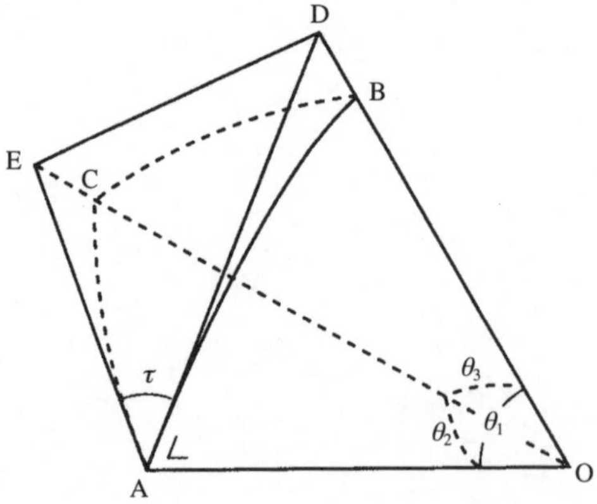

Fig. A1.1 Spherical triangle ABC and plane triangle ADE contacting at a common apex A, ∠ BAC = ∠ DAE.

From Pythagoras' theorem applied to right triangles ODE and OEA, it follows that

$$\overline{\mathrm{OD}}^2 - \overline{\mathrm{AD}}^2 = \overline{\mathrm{OE}}^2 - \overline{\mathrm{AE}}^2 = \overline{\mathrm{OA}}^2. \tag{A1-4}$$

Substituting Eq. (A1-4) into Eq. (A1-3), and dividing the result by $2\overline{\mathrm{OD}}\cdot\overline{\mathrm{OE}}$, we have

$$\frac{\overline{\mathrm{OA}}}{\overline{\mathrm{OD}}}\cdot\frac{\overline{\mathrm{OA}}}{\overline{\mathrm{OE}}} - \cos\theta_3 + \frac{\overline{\mathrm{AD}}}{\overline{\mathrm{OD}}}\cdot\frac{\overline{\mathrm{AE}}}{\overline{\mathrm{OE}}}\cos\tau = 0. \tag{A1-5}$$

Since it holds from the definitions that

$$\overline{\mathrm{OA}}/\overline{\mathrm{OD}} = \cos\theta_1, \quad \overline{\mathrm{OA}}/\overline{\mathrm{OE}} = \cos\theta_2,$$

$$\overline{\mathrm{AD}}/\overline{\mathrm{OD}} = \sin\theta_1, \quad \overline{\mathrm{AE}}/\overline{\mathrm{OE}} = \sin\theta_2,$$

Eq. (A1-5) is rewritten as

$$\cos\theta_3 - \cos\theta_1\cos\theta_2 = \sin\theta_1\sin\theta_2\cos\tau. \tag{A1-6}$$

By appropriate substitutions, Eqs. (2-17), (3-132) and (9-11) are derived from Eq. (A1-6).

Appendix 2 Matrix Elements of the Coordinates and Momenta with the Wave Functions for Harmonic Oscillators

The matrix elements of the coordinate x are obtained as follows. A recursion formula for the normalization constant of the wave function of harmonic oscillators is derived from the definition in the form

$$N_n = \left\{\frac{1}{\sqrt{\pi}}\frac{1}{2^n n!}\right\}^{1/2} = \left(\frac{1}{2n}\right)^{1/2} N_{n-1} = \{2(n+1)\}^{1/2} N_{n+1}. \tag{A2-1}$$

The recursion formula for the Hermite polynomials in Eq. (4-61) is written as

$$x\,H_n(x) = \tfrac{1}{2}H_{n+1}(x) + n\,H_{n-1}(x). \tag{A2-2}$$

By multiplying Eq. (A2-2) by N_n $\exp(-x^2/2)$ and substituting Eq. (A2-1), the recursion formula for the wave function is obtained in the form

$$x|n\rangle = \left(\frac{n+1}{2}\right)^{1/2}|n+1\rangle + \left(\frac{n}{2}\right)^{1/2}|n-1\rangle. \tag{A2-3}$$

Premultiplying Eq. (A2-3) by $\langle n-1|$ and $\langle n+1|$, integrating and using the orthonormality of the wave functions, $\langle n \mid n \rangle = \delta_{mn}$ we obtain

$$\langle n-1|x|n\rangle = (n/2)^{1/2} \tag{A2-4a}$$

and

$$\langle n+1|x|n\rangle = \{(n+1)/2\}^{1/2} \tag{A2-4b}$$

respectively. The form of the right side of Eq. (A2-3) shows that the matrix elements other than Eqs. (A2-4a,b) are all vanishing. The matrix elements of x^2 are calculated by using the fact that the matrix elements of the product AB are the elements of the product of matrices of A and B, that is,

$$\begin{aligned}\langle n|x^2|n\rangle &= \langle n|x|n-1\rangle\langle n-1|x|n\rangle + \langle n|x|n+1\rangle\langle n+1|x|n\rangle \\ &= \frac{n}{2} + \frac{n+1}{2} = n + \frac{1}{2}.\end{aligned}$$

The matrix elements of the momentum conjugate to x are calculated as follows. The wave function expressed as a product of a Gaussian distribution function and a Hermite polynomial

$$\psi(x;n) = N_n \exp(-x^2/2) H_n(x) \tag{A2-5}$$

is differentiated to give

$$\psi'(x;n) = N_n \exp(-x^2/2)\{H_n'(x) - x\,H_n(x)\}. \tag{A2-6}$$

Substituting Eqs. (4-63) and (4-61) into the first and the second term in the braces in the right side of Eq. (A2-6), and using Eq. (A2-1), we obtain

$$\begin{aligned}\psi'(x;n) &= N_n \exp(-x^2/2)\left[2n\,H_{n-1}(x) - \{\tfrac{1}{2} H_{n+1}(x) + n\,H_{n-1}(x)\}\right] \\ &= (n/2)^{1/2}\psi(x;n-1) - \{(n+1)/2\}^{1/2}\psi(x;n+1).\end{aligned} \tag{A2-7}$$

By using the operator p_x, Eq. (A2-7) is written also as

$$\frac{i\,p_x}{\hbar}|n\rangle = \left(\frac{n}{2}\right)^{1/2}|n-1\rangle - \left(\frac{n+1}{2}\right)^{1/2}|n+1\rangle\,. \tag{A2-8}$$

Manipulation of Eq. (A2-8) in the same way as Eq. (A2-3) leads to

$$\langle n-1|\frac{i\,p_x}{\hbar}|n\rangle = \left(\frac{n}{2}\right)^{1/2}$$

and

$$\langle n+1|\frac{i\,p_x}{\hbar}|n\rangle = -\left(\frac{n+1}{2}\right)^{1/2}.$$

The matrix elements of $p_x^2/\hbar^2$ are calculated, by using the matrix elements of the product as in the case of x^2, to be

$$\begin{aligned}\langle n|\frac{p_x^{\,2}}{\hbar^2}|n\rangle &= -\langle n|\frac{i\,p_x}{\hbar}|n-1\rangle\langle n-1|\frac{i\,p_x}{\hbar}|n\rangle - \langle n|\frac{i\,p_x}{\hbar}|n+1\rangle\langle n+1|\frac{i\,p_x}{\hbar}|n\rangle \\ &= \frac{n}{2} + \frac{n+1}{2} = n + \frac{1}{2}.\end{aligned} \tag{A2-9}$$

The non-vanishing off-diagonal elements of the matrix of x^2 are

$$\begin{aligned}\langle n|x^2|n-2\rangle &= \langle n|x|n-1\rangle\langle n-1|x|n-2\rangle \\ &= \{n(n-1)\}^{1/2}/2\end{aligned} \tag{A2-10a}$$

and

$$\langle n|x^2|n+2\rangle = \langle n|x|n+1\rangle\langle n+1|x|n+2\rangle$$
$$= \{(n+1)(n+2)\}^{1/2}/2. \qquad \text{(A2-10b)}$$

By using Eqs. (A2-10a,b), the diagonal element of the matrix of x^4 is calculated to be

$$\langle n|x^4|n\rangle = \{n(n-1) + (2n+1)^2 + (n+1)(n+2)\}/4$$
$$= \frac{6}{4}\left(n^2 + n + \frac{1}{2}\right) = \frac{6}{4}\left(n + \frac{1}{2}\right)^2 + \frac{3}{8}. \qquad \text{(A2-11)}$$

Equation (A2-11) appears in the perturbation theory of anharmonic molecular vibrations for calculating the matrix elements of the second-order Hamiltonians $H'_{2;3,3}$ and $H'_{2;4,4}$. In this case, the constant term 3/8 in Eq. (A2-11) is customarily omitted since it does not affect the transition energy to be observed in the experiment.

Appendix 3 The Matrix of Inverse Transformation of $\boldsymbol{r} = \boldsymbol{C}\boldsymbol{r}_c$

In the Cartesian coordinate system, the unit cell vectors $\boldsymbol{a}$, $\boldsymbol{b}$ and $\boldsymbol{c}$ are given by

$$\boldsymbol{a} = [a \quad 0 \quad 0]^{\mathrm{T}}, \qquad \text{(A3-1a)}$$

$$\boldsymbol{b} = [b\cos\gamma \quad b\sin\gamma \quad 0]^{\mathrm{T}} \qquad \text{(A3-1b)}$$

and

$$\boldsymbol{c} = [c\cos\beta \quad -c\sin\beta\cos\alpha^* \quad c\sin\beta\cos\alpha^*]^{\mathrm{T}}. \qquad \text{(A3-1c)}$$

Arranging the (3 × 1) column vectors $\boldsymbol{a}$, $\boldsymbol{b}$ and $\boldsymbol{c}$ in a row, we obtain a (3 × 3) matrix

$$\boldsymbol{C}_r = [\boldsymbol{a} \quad \boldsymbol{b} \quad \boldsymbol{c}] = \begin{bmatrix} a_X & b_X & c_X \\ a_Y & b_Y & c_Y \\ a_Z & b_Z & c_Z \end{bmatrix}, \qquad \text{(A3-2)}$$

and Eq. (9-10) is rewritten as

$$\boldsymbol{r}_{Ip} = X^c_{Ip}\boldsymbol{a} + Y^c_{Ip}\boldsymbol{b} + Z^c_{Ip}\boldsymbol{c} = [\boldsymbol{a} \quad \boldsymbol{b} \quad \boldsymbol{c}]\boldsymbol{r}^c_{Ip}. \qquad \text{(A3-3)}$$

The determinant of the triangular matrix is the product of the diagonal elements given by

$$|\boldsymbol{C}_r| = abc\sin\gamma\sin\beta\sin\alpha^*. \qquad \text{(A3-4)}$$

For the (3 × 3) matrix in Eq. (A3-2), Cramer's rule can be written as

$$\boldsymbol{C}_r^{-1} = \begin{bmatrix} b_Y c_Z - b_Z c_Y & b_Z c_X - b_X c_Z & b_X c_Y - b_Y c_X \\ c_Y a_Z - c_Z a_Y & c_Z a_X - c_X a_Z & c_X a_Y - c_Y a_X \\ a_Y b_Z - a_Z b_Y & a_Z b_X - a_X b_Z & a_X b_Y - a_Y b_X \end{bmatrix} |\boldsymbol{C}_r|^{-1}$$

$$= \begin{bmatrix} (\boldsymbol{b}\times\boldsymbol{c})^{\mathrm{T}} \\ (\boldsymbol{c}\times\boldsymbol{a})^{\mathrm{T}} \\ (\boldsymbol{a}\times\boldsymbol{b})^{\mathrm{T}} \end{bmatrix} |\boldsymbol{C}_r|^{-1}. \tag{A3-5}$$

From Eq. (9-10), the non-zero elements of $\boldsymbol{C}_r^{-1}$ are calculated to be

$$\left(C_r^{-1}\right)_{11} = (\boldsymbol{b}\times\boldsymbol{c})_X/|\boldsymbol{C}_r| = bc\sin\gamma\sin\beta\sin\alpha^*/|\boldsymbol{C}_r| = a^{-1} \tag{A3-6a}$$

$$\left(C_r^{-1}\right)_{12} = (\boldsymbol{b}\times\boldsymbol{c})_Y/|\boldsymbol{C}_r| = -bc\cos\gamma\sin\beta\sin\alpha^*/|\boldsymbol{C}_r| = -\cos\gamma/(a\sin\gamma) \tag{A3-6b}$$

$$\left(C_r^{-1}\right)_{13} = (\boldsymbol{b}\times\boldsymbol{c})_Z/|\boldsymbol{C}_r| = bc(-\cos\gamma\sin\beta\cos\alpha^* - \cos\beta\sin\gamma)/|\boldsymbol{C}_r|$$

$$= \frac{-\cos\gamma\sin\beta(\cos\beta\cos\gamma - \cos\alpha) - \cos\beta\sin^2\gamma\sin\beta}{a\sin^2\gamma\sin^2\beta\sin\alpha^*}$$

$$= \frac{-\cos\beta + \cos\gamma\cos\alpha}{a\sin^2\gamma\sin\beta\sin\alpha^*} \tag{A3-6c}$$

$$\left(C_r^{-1}\right)_{22} = (\boldsymbol{c}\times\boldsymbol{a})_Y/|\boldsymbol{C}_r| = ac\sin\beta\sin\alpha^*/|\boldsymbol{C}_r| = 1/b\sin\gamma \tag{A3-6d}$$

$$\left(C_r^{-1}\right)_{23} = (\boldsymbol{c}\times\boldsymbol{a})_Z/|\boldsymbol{C}_r| = -(-ac\sin\beta\cos\alpha^*)/|\boldsymbol{C}_r| = \cos\alpha^*/(b\sin\gamma\sin\alpha^*) \tag{A3-6e}$$

$$\left(C_r^{-1}\right)_{33} = (\boldsymbol{a}\times\boldsymbol{b})_Z/|\boldsymbol{C}_r| = ab\sin\gamma/|\boldsymbol{C}_r| = 1/(c\sin\beta\sin\alpha^*). \tag{A3-6f}$$

By analogy with the case of α^*, the dihedral angle between the ab and the bc planes, β^*, and that between the ac and the bc planes, γ^*, satisfy the relations,

$$\cos\beta^* = (\cos\gamma\cos\alpha - \cos\beta)/(\sin\gamma\sin\alpha) \tag{A3-7a}$$

and

$$\cos\gamma^* = (\cos\alpha\cos\beta - \cos\gamma)/(\sin\alpha\sin\beta), \tag{A3-7b}$$

respectively. Since the choice of the three axes is arbitrary for the triclinic system, the cell volume must be written in analogous ways to Eq. (A3-4) in the forms

$$|\boldsymbol{C}_r| = abc\,\sin\alpha\,\sin\beta^*\,\sin\gamma \tag{A3-8}$$

and

$$|\boldsymbol{C}_r| = abc\,\sin\alpha\,\sin\beta\,\sin\gamma^*. \tag{A3-9}$$

Combining Eqs. (A3-4, 8 and 9), we have

$$\frac{\sin\alpha}{\sin\alpha^*} = \frac{\sin\beta}{\sin\beta^*} = \frac{\sin\gamma}{\sin\gamma^*}. \tag{A3-10}$$

By using Eqs. (A3-7a,b) and (A3-10), Eq. (A3-6c) is rewritten as

$$\left(C_r^{-1}\right)_{13} = \cos\beta^*/(a\sin\gamma\sin\beta^*).$$

Now we define the quantities a^*, b^* and c^* in terms of the unit cell vectors $\boldsymbol{a}$, $\boldsymbol{b}$ and $\boldsymbol{c}$ in the forms

$$a^* = |\boldsymbol{a}^*| = |\boldsymbol{b}\times\boldsymbol{c}|/|\boldsymbol{C}_r| = \sin\alpha/(a\sin\beta\sin\gamma\sin\alpha^*) \tag{A3-11a}$$

$$b^* = |\boldsymbol{b}^*| = |\boldsymbol{c}\times\boldsymbol{a}|/|\boldsymbol{C}_r| = \sin\beta/(b\sin\gamma\sin\alpha\sin\beta^*) \tag{A3-11b}$$

$$c^* = |\boldsymbol{c}^*| = |\boldsymbol{a}\times\boldsymbol{b}|/|\boldsymbol{C}_r| = \sin\gamma/(c\sin\alpha\sin\beta\sin\gamma^*). \tag{A3-11c}$$

Then the quantities a^*, b^*, c^*, α^*, β^* and γ^* are called the cell constants of the reciprocal unit cell. In analogy with the relation

$$\boldsymbol{b}\cdot\boldsymbol{c} = bc\cos\alpha,$$

the cell constants of the reciprocal unit cell satisfy the relation

$$\boldsymbol{b}^*\cdot\boldsymbol{c}^* = b^*c^*\cos\alpha^*. \tag{A3-12}$$

Substituting Eqs. (A3-11a,b) into Eq. (A3-12), and rearranging the result with reference to the formula of the vector triple product, we have

$$\begin{aligned}\cos\alpha^* &= (\boldsymbol{c}\times\boldsymbol{a})\cdot(\boldsymbol{a}\times\boldsymbol{b})/|\boldsymbol{C}_r|^2 = \boldsymbol{a}\cdot\{\boldsymbol{b}\times(\boldsymbol{c}\times\boldsymbol{a})\}/|\boldsymbol{C}_r|^2\\ &= \boldsymbol{a}\cdot\{\boldsymbol{c}(\boldsymbol{b}\cdot\boldsymbol{a})-\boldsymbol{a}(\boldsymbol{b}\cdot\boldsymbol{c})\}/|\boldsymbol{C}_r|^2 = \{(\boldsymbol{a}\cdot\boldsymbol{c})(\boldsymbol{b}\cdot\boldsymbol{a})-(\boldsymbol{a}\cdot\boldsymbol{a})(\boldsymbol{b}\cdot\boldsymbol{c})\}/|\boldsymbol{C}_r|^2\\ &= a^2bc(\cos\beta\cos\gamma-\cos\alpha)/|\boldsymbol{C}_r|^2.\end{aligned} \tag{A3-13}$$

The second equalities in Eqs. (A3-11b,c) lead to the relation

$$|\boldsymbol{C}_r| = |\boldsymbol{c}\times\boldsymbol{a}|/b^* = ac\sin\beta/b^* = |\boldsymbol{a}\times\boldsymbol{b}|/c^* = ab\sin\gamma/c^*. \tag{A3-14}$$

Substituting Eq. (A3-14) into Eq. (A3-13), we obtain Eq. (9-11),

$$\cos\alpha^* = (\cos\beta\cos\gamma-\cos\alpha)/(\sin\beta\sin\gamma).$$

By comparing Eq. (9-11) with Eq. (A1-5), it is seen that the dihedral angle formed by the *ab* and *ac* planes containing the positive directions of the *b*- and *c*-axes, respectively, is the supplementary angle of α^*.

Appendix 4 Average of the Exponential Function of a Random Variable

For a set of random variables x distributed around the origin according to the normal distribution, it can be shown that

$$\langle\exp(2\pi i x)\rangle = \exp(-2\pi^2\langle x^2\rangle). \tag{A4-1}$$

Using the expansion

$$e^z = 1 + z + \frac{1}{2}z^2 + \cdots + \frac{1}{n!}z^n + \cdots$$

and noting that $\langle x^n \rangle = 0$ for an odd n if x is distributed in the positive and the negative sides of the origin homogeneously, we have

$$\langle e^{2\pi ix} \rangle = 1 - 2\pi^2 \langle x^2 \rangle + \frac{2}{3}\pi^4 \langle x^4 \rangle - \cdots + (-1)^n \frac{(2\pi)^{2n}}{(2n)!} \langle x^{2n} \rangle + \cdots. \quad \text{(A4-2a)}$$

Note that $\langle x^{2n} \rangle$ appears in the $(n + 1)$th term of the expansion in Eq. (A4-2a). On the other hand, the right side of Eq. (A4-1) is expanded as

$$e^{-2\pi^2 \langle x^2 \rangle} = 1 - 2\pi^2 \langle x^2 \rangle + 2\pi^4 \langle x^2 \rangle^2 - \cdots + (-1)^n \frac{(2\pi^2)^n}{n!} \langle x^2 \rangle^n + \cdots. \quad \text{(A4-2b)}$$

The Gaussian distribution function with the dispersion σ is written in the form

$$g(x) = (1/\sigma\sqrt{2\pi})\exp(-x^2/2\sigma^2)$$

so that the average of x^{2n} is given by

$$\begin{aligned} \langle x^{2n} \rangle &= \frac{1}{\sigma\sqrt{2\pi}} \int_{-\infty}^{\infty} x^{2n} \exp\left(-\frac{x^2}{2\sigma^2}\right) dx = \frac{2}{\sigma\sqrt{2\pi}} \frac{(2n-1)!!}{2^{n+1}} \sqrt{(2\sigma^2)^{2n+1}\pi} \\ &= \frac{(2n-1)!!}{2^n} 2^n \sigma^{2n}. \end{aligned} \quad \text{(A4-3)}$$

Substituting Eq. (A4-3) into the right side of Eq. (A4-2a) and using the relation

$$\frac{(2n-1)!!}{(2n)!} = \frac{(2n-1)(2n-3)\cdots 3\cdot 1}{(2n)(2n-1)\cdots 2} = \frac{1}{2n(2n-2)\cdots 2} = \frac{1}{2^n n!},$$

we have

$$\begin{aligned} (-1)^n \frac{(2\pi)^{2n}}{(2n)!} \langle x^{2n} \rangle &= (-1)^n \frac{(2\pi)^{2n}}{(2n)!} (2n-1)!!\sigma^{2n} \\ &= (-1)^n \frac{(2\pi)^{2n}}{2^n n!} \sigma^{2n} = (-1)^n \frac{(2\pi^2)^n}{n!} \sigma^{2n}. \end{aligned} \quad \text{(A4-4)}$$

As confirmed by putting $n = 1$ in Eq. (A4-3), it holds that $\langle x^2 \rangle = \sigma^2$ for the normal distribution. Accordingly, the $(n + 1)$th term of the right side of Eq. (A4-2b) is given by

$$(-1)^n \frac{(2\pi^2)^n}{n!} \langle x^2 \rangle^n = (-1)^n \frac{(2\pi^2)^n}{n!} \sigma^{2n}. \quad \text{(A4-5)}$$

The coincidence between Eqs. (A4-4) and (A4-5) proves that the left sides of Eqs. (A4-2a) and (A4-2b) are equal to each other.

Next it shall be proved that the thermal average of the squared wave functions of a harmonic oscillator is a Gaussian distribution function with the mean squared amplitude as the dispersion. Taking the sum of squares of the wave functions in Eq. (4-65) weighted by the Boltzmann factors, we obtain the thermally averaged distribution in the form

$$g(q) = \sum_{n=0}^{\infty} \frac{e^{-q^2}}{2^n n! \sqrt{\pi}} \{H_n(q)\}^2 e^{-hc(n+1/2)\omega/kT}. \tag{A4-6}$$

Let us define the function

$$S(q) = \sum_{n=0}^{\infty} \frac{1}{2^n n!} \{H_n(q)\}^2 e^{-hcn\omega/kT}. \tag{A4-7}$$

Differentiating Eq. (A4-7) with respect to q and using Eq. (4-63), we have

$$S'(q) = \sum_{n=1}^{\infty} \frac{2H_n(q)}{2^n n!} H_n'(q) e^{-hcn\omega/kT} = \sum_{n=1}^{\infty} \frac{2n\, H_n(q)}{2^{n-1} n!} H_{n-1}(q) e^{-hcn\omega/kT}. \tag{A4-8}$$

Multiply Eq. (A4-7) by $2q$, and substitute Eqs. (4-61) and (A4-8) successively to obtain

$$\begin{aligned} 2q\, S(q) &= 2q + \sum_{n=1}^{\infty} \frac{1}{2^n n!} \{H_{n+1}(q) + 2n\, H_{n-1}(q)\} H_n(q) e^{-hcn\omega/kT} \\ &= \tfrac{1}{2}(e^{hc\omega/kT} + 1) S'(q) \end{aligned} \tag{A4-9}$$

or

$$\frac{S'(q)}{S(q)} = \frac{d}{dq} \ln S(q) = \frac{4q}{e^{hc\omega/kT} + 1}. \tag{A4-10}$$

Integrating Eq. (A4-10) leads to

$$\ln S(q) = 2q^2/(e^{hc\omega/kT} + 1) + C, \tag{A4-11}$$

where C is the constant of integration. Solving Eq. (A4-11) for $S(q)$ and substituting the result back into Eq. (A4-6), we obtain

$$\begin{aligned} g(q) &= \pi^{-1/2} e^{-hc\omega/2kT} \exp\left(\frac{2q^2}{e^{hc\omega/kT} + 1} - q^2 + C \right) \\ &= \pi^{-1/2} e^{C - hc\omega/2kT} \exp\left(-\frac{1 - e^{-hc\omega/kT}}{1 + e^{-hc\omega/kT}} q^2 \right) \\ &= \pi^{-1/2} e^{C - hc\omega/2kT} \exp\left(-\frac{q^2}{\coth(hc\omega/2kT)} \right). \end{aligned} \tag{A4-12}$$

From Eq. (A4-12) it follows that a normal coordinate under the harmonic potential satisfies Eq. (A4-1). The constant of integration determined so as to normalize $g(q)$ is given in the form

$$C = \frac{hc\omega}{2kT} - \frac{1}{2} \ln\left\{ \coth\left(\frac{hc\omega}{2kT} \right) \right\}.$$

Appendix 5 Laue Function and Periodic Delta Function

The sum over J and J' in Eq. (9-101) is rewritten as

$$\sum_{J,J'} \exp\{2\pi i \tilde{\boldsymbol{k}}(\boldsymbol{r}_J - \boldsymbol{r}_{J'})\} = \sum_J \exp(2\pi i \tilde{\boldsymbol{k}} \boldsymbol{r}_J) \sum_{J'} \exp(-2\pi i \tilde{\boldsymbol{k}} \boldsymbol{r}_{J'}). \quad \text{(A5-1)}$$

The scattering vector is expressed in terms of the components along the reciprocal unit cell vectors as

$$\boldsymbol{k} = k_a \boldsymbol{a}^* + k_b \boldsymbol{b}^* + k_c \boldsymbol{c}^*. \quad \text{(A5-2)}$$

Taking the scalar product of $\boldsymbol{k}$ and the lattice vector $\boldsymbol{r}_J$ given by replacing I in Eq. (9-4) with J, and making use of Eq. (9-54), we have

$$\tilde{\boldsymbol{k}} \boldsymbol{r}_J = k_a J_a + k_b J_b + k_c J_c. \quad \text{(A5-3)}$$

The sum over J in Eq. (A5-1) is factored into three factors in the form

$$\sum_{J=1}^{N} \exp(2\pi i \tilde{\boldsymbol{k}} \boldsymbol{r}_J) = \sum_{J_a=0}^{N_a-1} \exp(2\pi i k_a J_a) \sum_{J_b=0}^{N_b-1} \exp(2\pi i k_b J_b) \sum_{J_c=0}^{N_c-1} \exp(2\pi i k_c J_c). \quad \text{(A5-4)}$$

The first factor in the right side of Eq. (A5-4) becomes

$$\begin{aligned} \sum_{J_a=0}^{N_a-1} \exp(2\pi i k_a J_a) &= \frac{1 - e^{2\pi i k_a N_a}}{1 - e^{2\pi i k_a}} = \frac{e^{2\pi i k_a N_a/2}}{e^{2\pi i k_a/2}} \frac{e^{-2\pi i k_a N_a/2} - e^{2\pi i k_a N_a/2}}{e^{-2\pi i k_a/2} - e^{2\pi i k_a/2}} \\ &= e^{2\pi i k_a (N_a - 1)/2} \frac{\sin(2\pi k_a N_a/2)}{\sin(2\pi k_a/2)}. \end{aligned} \quad \text{(A5-5)}$$

By taking the product of Eq. (A5-5) and its complex conjugate, Eq. (A5-1) is rewritten as

$$\begin{aligned} \sum_{J=1}^{N} \sum_{J'=1}^{N} \exp\{2\pi i \tilde{\boldsymbol{k}}(\boldsymbol{r}_J - \boldsymbol{r}_{J'})\} &= L(\boldsymbol{k}, N) \\ &= L_1(k_a, N_a) L_1(k_b, N_b) L_1(k_c, N_c), \end{aligned} \quad \text{(A5-6)}$$

where $L_1(k_a, N_a)$ is a periodic function of k_a in the form

$$L_1(k_a, N_a) = \sin^2(\pi k_a N_a)/\sin^2(\pi k_a). \quad \text{(A5-7)}$$

It follows from Eq. (A5-7) that

$$\lim_{\mathrm{Mod}(k_a,1)\to 0} L_1(k_a, N_a) = N_a^{\,2}. \quad \text{(A5-8)}$$

The function $L_1(k_a, N_a)$ oscillates N_a times within its period, 1. If N_a is a large number, the submaxima near the origin are approximately given by

$$L_1(m/2N_a, N_a) \approx 4(N_a/m\pi)^2, \quad \text{(A5-9)}$$

where m is an odd integer such that $m << N_a$. The ratio of the peak height of the first submaximum ($m = 1$) to the main peak is $4/\pi^2 \approx 0.4$, but the peak heights of the successive maxima quickly decrease with the increase of m. In order to integrate the function $L_1(k_a,N_a)$ over a period, we rewrite the numerator according to

$$\sin^2(\pi N_a k_a) = \tfrac{1}{2}\{1 - \cos(2\pi N_a k_a)\} \tag{A5-10}$$

and introduce the abbreviations

$$\cos(\pi k_a M) = C_M \quad \text{and} \quad \sin(\pi k_a M) = S_M .$$

By using the addition theorem for cosines and sines successively, we then have the relation

$$\begin{aligned} C_{2N} &= C_{2N-1} C_1 - S_{2N-1} S_1 = C_{2N-2} C_1^2 - S_{2N-2} S_1 C_1 - S_{2N-1} S_1 \\ &= C_{2N-2}(C_1^2 + S_1^2) - 2S_{2N-1} S_1 = C_{2N-2} - 2S_{2N-1} S_1 \\ &= C_{2N-4} - 4S_{2N-3} S_1 - 4C_{2N-2} S_1^2 \\ &= C_{2N-6} - 6S_{2N-5} S_1 - 8C_{2N-4} S_1^2 - 4C_{2N-2} S_1^2 = \cdots \\ &= C_{2N-2i} - 2i\, S_{2N-2i+1} S_1 - 4\sum_{k=1}^{i-1} k\, C_{2N-2k} S_1^2 = \cdots \\ &= C_2 - 2(N-1) S_2 S_1 - 4\sum_{k=1}^{N-2} k\, C_{2N-2k} S_1^2 \\ &= C_0 - 2N S_1^2 - 4\sum_{k=1}^{N-1} k\, C_{2N-2k} S_1^2 . \end{aligned} \tag{A5-11}$$

By using Eqs. (A5-10) and (A5-11) and noting that $C_0 = 1$, the integral is obtained as

$$\int_0^1 \frac{\sin^2(\pi N_a k_a)}{\sin^2(\pi k_a)} dk_a = N_a \int_0^1 dk_a + 2\sum_{k=1}^{N_a-1} k \int_0^1 \cos\{2\pi (N_a - k)k_a\} dk_a = N_a . \tag{A5-12}$$

This result means that the function $L_1(k_a,N_a)/N_a$ behaves like a periodic δ-function satisfying

$$\int_x^{x+1} \delta(k_a)\, dk_a = 1,$$

where x is any real number, on the limit that $N_a \to \infty$. The same results are obtained with respect to k_b and k_c. Furthermore, the integral of $dk_a dk_b dk_c$ over a period of each variable in an oblique coordinate system corresponds to the three-dimensional integral of a volume element in the reciprocal space over a reciprocal unit cell, *i.e.*,

$$\int_0^1\int_0^1\int_0^1 dk_a\, dk_b\, dk_c = \int_{\text{rec}} d\boldsymbol{k} = V^* .$$

where V^* is volume of the reciprocal unit cell. Accordingly, for sufficiently large N, we may write

$$\sum_{J=1}^{N}\sum_{J'=1}^{N} \exp\{2\pi i \tilde{\boldsymbol{k}}(\boldsymbol{r}_J - \boldsymbol{r}_{J'})\} = L(\boldsymbol{k}, N) = N V^* \delta(\boldsymbol{k} - \boldsymbol{h}_{hkl}), \tag{A5-13}$$

where $\boldsymbol{h}_{hkl}$ is a reciprocal lattice vector defined by Eq. (9-102a).

Appendix 6 Evaluation of the Debye Function

For a given positive argument u, the Debye function can be numerically calculated by using either the expansion

$$\frac{n}{u^n}\int_0^u \frac{y^n}{e^y-1}dy = 1-\frac{nu}{2(n+1)}+n\sum_{k=1}^{\infty}\frac{B_{2k}\,u^{2k}}{(2k+n)(2k)!} \qquad \text{(A6-1a)}^{1)}$$

or the double expansion

$$\frac{n}{u^n}\int_u^\infty \frac{y^n}{e^y-1}dy = n\sum_{k=1}^{\infty}\left\{\frac{e^{-ku}}{k}\sum_{l=0}^{n}\frac{n!}{(n-l)!(ku)^l}\right\}. \qquad \text{(A6-1b)}^{1)}$$

In Eq. (A6-1a), B_{2k} is the Bernoulli number defined as the coefficients of the expansion

$$y/(e^y-1) = \sum_{k'=0}^{\infty} B_{k'}\,y^{k'}/k'!. \qquad \text{(A6-2)}$$

It follows from Eq. (A6-2) that

$$y = (e^y-1)\sum_{k'=0}^{\infty} B_{k'}\frac{y^{k'}}{k'!} = \left(y+\frac{y^2}{2}+\frac{y^3}{3!}+\cdots\right)\left(B_0+B_1\,y+\frac{B_2\,y^2}{2}+\cdots\right)$$

$$= B_0\,y+\left(\frac{B_0}{2}+B_1\right)y^2+\left(\frac{B_0}{3!}+\frac{B_1}{2}+\frac{B_2}{2}\right)y^3+\cdots,$$

whence we have $B_0 = 1$ and $(B_0/2) + B_1 = 0$, so that $B_1 = -1/2$ and

$$B_n = -n!\sum_{l=1}^{n} B_{n-l}/\{(l+1)!(n-l)!\}. \qquad \text{(A6-3)}$$

To prove Eq. (A6-1a), let us transfer the second term in the right side of Eq. (A6-2), $B_1 y$, to the left side, and write the result in terms of the hyperbolic cotangent in the form

$$\frac{y}{e^y-1}+\frac{y}{2} = \frac{y}{2}\frac{e^y+1}{e^y-1} \equiv \frac{y}{2}\coth\frac{y}{2} = 1+\sum_{n=2}^{\infty}\frac{B_n\,y^n}{n!}. \qquad \text{(A6-4)}$$

Since the function defined by the third formula in Eq. (A6-4) is a product of two odd functions, *i.e.*, an even function, the terms with odd powers of y in the series in the right side must vanish. Multiply both sides of Eq. (A6-2) by y^{n-1} and integrate term by term to give

$$\int_0^u \frac{y^n}{e^y-1}dy = \frac{u^n}{n}-\frac{u^{n+1}}{2(n+1)}+\sum_{k'=2}^{\infty}\frac{B_{k'}\,u^{n+k'}}{(k'+n)k'!}. \qquad \text{(A6-5)}$$

Multiplying Eq. (A6-5) by n/u^n and rewriting the index k' as $2k$, we obtain Eq. (A6-1a).

We can prove Eq. (A6-1b) by substituting the expansion

$$(1-x)^{-1} = 1+x+x^2+x^3+\cdots$$

with $x = e^{-y}$ into the integrand of the Debye function, obtaining

$$y^n(e^y-1)^{-1} = y^n\,e^{-y}(1-e^{-y})^{-1} = y^n\sum_{k=1}^{\infty}e^{-ky}. \qquad \text{(A6-6)}$$

Each term in the right side of Eq. (A6-6) is integrated by parts successively to give

$$\begin{aligned}
\int_u^\infty y^n e^{-ky}\,dy &= \left|\frac{y^n e^{-ky}}{-k}\right|_u^\infty + \frac{n}{k}\int_u^\infty y^{n-1} e^{-ky}\,dy \\
&= \frac{u^n e^{-ku}}{k} + \frac{n}{k}\left(\frac{u^{n-1} e^{-ku}}{k} + \frac{n-1}{k}\int_u^\infty y^{n-2} e^{-ky}\,dy\right) \\
&= \frac{u^n e^{-ku}}{k} + \frac{n u^{n-1} e^{-ku}}{k^2} + \frac{n(n-1)u^{n-2} e^{-ku}}{k^3} + \cdots + \frac{n!}{k^n}\int_u^\infty e^{-ky}\,dy \\
&= n!\,e^{-ku}\sum_{l=0}^{n}\frac{u^{n-l}}{(n-l)!\,k^{l+1}}
\end{aligned} \tag{A6-7}$$

Integrating Eq. (A6-6) term by term with the help of Eq. (A6-7) and multiplying the result by n/u^n, we obtain Eq. (A6-1b). Subtracting Eq. (A6-1b) from $D_n(\infty)$ gives $D_n(u)$. The former is related to the gamma function through a definite integral in the form

$$\int_0^\infty x^{n-1} e^{-kx}\,dx = k^{-n}\int_0^\infty u^{n-1} e^{-u}\,du = k^{-n}\,\Gamma(n). \tag{A6-8}$$

Substitute $k = 1, 2, \cdots, \infty$ into Eq. (A6-8) and sum up to obtain

$$\int_0^\infty x^{n-1}\sum_{k=1}^{\infty} e^{-kx}\,dx = \int_0^\infty x^{n-1}\left(e^x - 1\right)^{-1} dx = \left(\sum_{k=1}^{\infty} k^{-n}\right)\Gamma(n). \tag{A6-9}$$

The series in the right side of Eq. (A6-9) is the sum of inverse powers of positive integers, and is known to converge for $n > 1$. This sum can be extended to the case where n is a continuous variable, and in this case becomes a continuous function of n called Riemann's zeta function, $\zeta(n)$. For an even integer $2l$, it holds that

$$\zeta(2l) = (-1)^{l+1} 2^{2l-1} \pi^{2l} B_{2l}/(2l)! \tag{A6-10}$$

To prove Eq. (A6-10), we apply first the residue theorem to the complex function cot z. According to the theory of function of complex variables, cot z has simple poles at $z = 0, \pm\pi, \pm 2\pi, \cdots$. Here a simple pole of a function $f(z)$ means a singularity a such that $(z - a)f(z)$ is analytic near and at a. Noting that sin $k\pi$ is zero for an integer k, we rewrite cot z as

$$\cot z = \frac{\cos z}{\sin(z - k\pi + k\pi)} = \frac{\cos z}{\sin(z - k\pi)\cos k\pi}.$$

The residue at the pole $z = k\pi$ is given by

$$\lim_{z\to k\pi}(z - k\pi)\cot z = \lim_{z\to k\pi}\frac{\cos z}{\cos k\pi}\,\frac{z - k\pi}{\sin(z - k\pi)} = 1.$$

Let us apply the residue theorem to the complex integral along a circle centered at the origin. If the radius R of the circle is such that $n\pi < R < (n + 1)n$, we have

$$\frac{1}{2\pi i}\int_C \frac{\cot z}{z-x}dz = \cot x + \sum_{k=-n}^{n} \frac{1}{k\pi - x}, \tag{A6-11}$$

where $x \neq k\pi$ and $|x| < R$. The integral in the left side of Eq. (A6-11) is rewritten as

$$\int_C \frac{\cot z}{z-x}dz = \int_C \frac{\cot z}{z}dz + x\int_C \frac{\cot z}{z(z-x)}dz. \tag{A6-12}$$

The integrand of the first integral in the right side of Eq. (A6-12) is an even function of z, so that its integral along a path centrosymmetric with respect to the origin vanishes. The second integral becomes zero as $R \to \infty$, since $|\cot z|$ approaches 1 as $|z| \to \infty$, implying that $|\cot z| < 2$ for sufficiently large R, and

$$|z(z-x)| = |z||z-x| \geq |z|(|z| - |x|) = R(R - |x|),$$

so that

$$\lim_{R\to\infty}\left|\int_C \frac{\cot z}{z(z-x)}dz\right| \leq \lim_{R\to\infty} \frac{2}{R(R-|x|)}\int_C ds = \lim_{R\to\infty}\frac{4\pi}{R-|x|} = 0.$$

On the other hand, the summation limit n in the right side of Eq. (A6-11) becomes ∞ as $R \to \infty$, whence we have

$$\cot x = -\sum_{k=-n}^{n}\frac{1}{k\pi - x} = \frac{1}{x} + \sum_{k=1}^{\infty}\left(\frac{1}{x-k\pi} + \frac{1}{x+k\pi}\right)$$

or

$$\begin{aligned} x\cot x &= 1 + x\sum_{k=1}^{\infty}\left(\frac{2x}{x^2 - k^2\pi^2}\right) = 1 - 2\sum_{k=1}^{\infty}\frac{(x/k\pi)^2}{1-(x/k\pi)^2} \\ &= 1 - 2\sum_{k=1}^{\infty}\left(\frac{x}{k\pi}\right)^2\left\{1 + \left(\frac{x}{k\pi}\right)^2 + \left(\frac{x}{k\pi}\right)^4 + \cdots\right\} \\ &= 1 - 2\sum_{k=1}^{\infty}\sum_{l=1}^{\infty}\left(\frac{x}{k\pi}\right)^{2l} = 1 - 2\sum_{l=1}^{\infty}\left(\sum_{k=1}^{\infty}\frac{1}{k^{2l}}\right)\frac{x^{2l}}{\pi^{2l}}. \end{aligned} \tag{A6-13}$$

On the other hand, substituting $y = 2ix$ and $n = 2l$ into Eq. (A6-4) gives

$$ix\coth ix = 1 + \sum_{l=1}^{\infty}\frac{B_{2l}(2ix)^{2l}}{(2l)!} = 1 - \sum_{l=1}^{\infty}\frac{(-1)^{l+1}2^{2l}B_{2l}}{(2l)!}x^{2l}. \tag{A6-14}$$

Finally, noting that

$$x\cot x = ix\frac{e^{ix/2} + e^{-ix/2}}{e^{ix/2} - e^{-ix/2}} = ix\coth ix,$$

and equating the coefficients of x^{2l} in Eqs. (A6-13) and (A6-14), we obtain Eq. (A6-10).

Equations (A6-1a) and (A6-1b) show good performance for small and large values of u, respectively. For the range $0 \leq u < 1.0$, Eq. (A6-1a) with six to eight terms gives eight significant digits, while Eq. (A6-1b) with k up to 15 gives similar accuracy for u greater than 0.9. A five significant digit table of $D_3(u)$ was pulished by Beattie for u from 0.0 to 24.0

with the increment of 0.01.[2)]

References

1) I.A. Stegun, in *Handbook of Mathematical Functions*, ed. M. Abramowitz and I.A. Stegun, p. 997, Dover Publications, New York (1965).
2) J.A. Beattie, *J. Math. Phys.*, **6**, 1 (1926).

Appendix 7 Coherent and Incoherent Neutron Scattering by Crystals

The Schrödinger equation for a joint system of a neutron and a molecular crystal is written as

$$\left(-\frac{\hbar^2}{2m}\nabla^2 + H_c + V_{int}\right)\Psi = E\Psi, \tag{A7-1}$$

where m is the neutron mass, H_c is the crystal Hamiltonian and V_{int} is the interaction potential between the neutron and crystal. The scalar operator $\nabla^2 \equiv \tilde{\nabla}\nabla$ is operated only on the neutron coordinates. The crystal is taken as an assembly of coupled harmonic oscillators, for which the Hamiltonian has been solved to give eigenfunctions ψ_j satisfying

$$H_c\,\psi_j(\boldsymbol{r}_c) = E_j\,\psi_j(\boldsymbol{r}_c). \tag{A7-2}$$

Then any solution of Eq. (A7-1) is expressed as a linear combination of eigenfunctions of the crystal Hamiltonian given in the form

$$\Psi(\boldsymbol{r}_n, \boldsymbol{r}_c) = \sum_j \phi_j(\boldsymbol{r}_n)\psi_j(\boldsymbol{r}_c), \tag{A7-3}$$

where $\boldsymbol{r}_n$ is the position vector of the neutron. Substituting Eqs. (A7-2) and (A7-3) into Eq. (A7-1) gives

$$-\frac{\hbar^2}{2m}\sum_j\left(\nabla^2 + 4\pi^2 k_j^2\right)\phi_j\,\psi_j + V_{int}\Psi = 0, \tag{A7-4}$$

where

$$4\pi^2 k_j^2 = 2m\left(E - E_j\right)/\hbar^2. \tag{A7-5}$$

Premultiply Eq. (A7-4) by ψ_l^* and integrate over the crystal coordinates. Using the orthonormality of ψ_l, we have

$$\frac{\hbar^2}{2m}\left(\nabla^2 + 4\pi^2 k_l^2\right)\phi_l(\boldsymbol{r}_n) = \int_c \psi_l^*(\boldsymbol{r}_c)V_{int}(\boldsymbol{r}_n, \boldsymbol{r}_c)\Psi(\boldsymbol{r}_n, \boldsymbol{r}_c)\,d\boldsymbol{r}_c. \tag{A7-6a}$$

If a neutron with the wave function $\phi_1(\boldsymbol{r}_n)$does not interact with the crystal (*i.e.*, $V_{int} = 0$), Eq. (A7-6a) becomes the homogeneous equation

$$\frac{\hbar^2}{2m}\left(\nabla^2 + 4\pi^2 k_1^2\right)\phi_1(\boldsymbol{r}_n) = 0, \tag{A7-6b}$$

which has a solution in the form

$$\phi_1(\boldsymbol{r}_n) = \exp\left(-2\pi i\,\tilde{\boldsymbol{k}}_1\,\boldsymbol{r}_n\right), \tag{A7-7}$$

where $|\boldsymbol{k}_1| = k_1$, so that $\tilde{\boldsymbol{\nabla}}\boldsymbol{\nabla}\ \phi_1 = 4\pi^2 i^2\, \tilde{\boldsymbol{k}}_1 \boldsymbol{k}_1\, \phi_1 = -4\pi^2 k_1^2\, \phi_1$.

Next let us introduce a function of $\boldsymbol{r}_{\rm n}$ defined by

$$\chi_l(\boldsymbol{r}_{\rm n}) = (2m/\hbar^2) \int_{\rm n} G_l(\boldsymbol{r}_{\rm n} - \boldsymbol{r}') \left\{ \int_{\rm c} \psi_l^*(\boldsymbol{r}_{\rm c})\, V_{\rm int}(\boldsymbol{r}', \boldsymbol{r}_{\rm c})\, \Psi(\boldsymbol{r}', \boldsymbol{r}_{\rm c})\, d\boldsymbol{r}_{\rm c} \right\} d\boldsymbol{r}'. \qquad \text{(A7-8)}$$

Function $G_l(\boldsymbol{r})$ is a solution of the differential equation involving a three-dimensional δ-function in the form

$$\left(\nabla^2 + 4\pi^2 k_l^2\right) G_l(\boldsymbol{r}) = \delta(\boldsymbol{r}), \qquad \text{(A7-9)}$$

and is known as the Green's function. Operating $\hbar^2\left(\nabla^2 + k_l^2\right)/2m$ on both sides of Eq.(A7-8) and using the property of δ-function, we have

$$\begin{aligned} \frac{\hbar^2}{2m}\left(\nabla^2 + 4\pi^2 k_l^2\right)\chi(\boldsymbol{r}_{\rm n}) &= \int_{\rm n} \delta(\boldsymbol{r}_{\rm n} - \boldsymbol{r}') \int_{\rm c} \psi_l^*(\boldsymbol{r}_{\rm c}) V_{\rm int}(\boldsymbol{r}', \boldsymbol{r}_{\rm c}) \Psi(\boldsymbol{r}', \boldsymbol{r}_{\rm c})\, d\boldsymbol{r}_{\rm c}\, d\boldsymbol{r}' \\ &= \int_{\rm c} \psi_l^*(\boldsymbol{r}_{\rm c}) V_{\rm int}(\boldsymbol{r}_{\rm n}, \boldsymbol{r}_{\rm c}) \Psi(\boldsymbol{r}_{\rm n}, \boldsymbol{r}_{\rm c})\, d\boldsymbol{r}_{\rm c}, \end{aligned} \qquad \text{(A7-10)}$$

which indicates that $\chi_l(\boldsymbol{r}_{\rm n})$ is a particular solution of Eq. (A7-6). The form of function $G_l(\boldsymbol{r})$ is determined as follows.

The Fourier transform of a three-dimensional Gaussian distribution function

$$f(\boldsymbol{h}) = e^{-\pi a^2 \left(h_x^2 + h_y^2 + h_z^2\right)} \equiv e^{-\pi a^2 \tilde{\boldsymbol{h}} \boldsymbol{h}} \qquad \text{(A7-11)}$$

is given by

$$F(\boldsymbol{r}_{\rm n}) = \int_{-\infty}^{\infty}\int_{-\infty}^{\infty}\int_{-\infty}^{\infty} e^{-\pi a^2 \left(h_x^2 + h_y^2 + h_z^2\right)} e^{-2\pi i \left(x_{\rm n} h_x + y_{\rm n} h_y + z_{\rm n} h_z\right)} dh_x\, dh_y\, dh_z, \qquad \text{(A7-12)}$$

where $x_{\rm n}$, $y_{\rm n}$ and $z_{\rm n}$ are the components of vector $\boldsymbol{r}_{\rm n}$. The integral is factored into three one-dimensional integrals with respect to h_x, h_y and h_z, of which the integral with respect to h_x is given by

$$F_x(x_{\rm n}) = \int_{-\infty}^{\infty} e^{-\pi a^2 h_x^2 - 2\pi i x_{\rm n} h_x}\, dh_x = \int_{-\infty}^{\infty} e^{-\pi\left(a h_x + i\frac{x_{\rm n}}{a}\right)^2} e^{-\pi\left(\frac{x_{\rm n}}{a}\right)^2} dh_x. \qquad \text{(A7-13a)}$$

Introducing a new variable t defined by

$$t = a h_x + i\frac{x_{\rm n}}{a},$$

so that $dt = a dh_x$, we can rewrite Eq. (A7-13a) as

$$F_x(x_{\rm n}) = \frac{1}{a} e^{-\pi\left(\frac{x_{\rm n}}{a}\right)^2} \int_{-\infty}^{\infty} e^{-\pi t^2} dt = \frac{1}{a} e^{-\pi\left(\frac{x_{\rm n}}{a}\right)^2}, \qquad \text{(A7-13b)}$$

since $\int_{-\infty}^{\infty} e^{-\pi t^2} dt = 1$. Similar arguments hold for the integrals with respect to h_y and h_z. Hence we have

$$F(\boldsymbol{r}_n) = \frac{1}{a} e^{-\frac{\pi}{a^2}(x_n{}^2 + y_n{}^2 + z_n{}^2)} = \frac{1}{a} e^{-\frac{\pi}{a^2}\tilde{\boldsymbol{r}}_n \boldsymbol{r}_n}. \tag{A7-14}$$

Since the limits of Eqs. (A7-12) and (A7-14) at $a \to 0$ are equivalent to each other, we obtain, by noting that

$$\lim_{a\to 0} e^{-\pi a^2(h_x{}^2 + h_y{}^2 + h_z{}^2)} = e^0 = 1,$$

two different definitions of the δ-function in the forms

$$\delta(\boldsymbol{r}_n) = \lim_{a\to 0} \frac{1}{a} e^{-\pi \tilde{\boldsymbol{r}}_n \boldsymbol{r}_n / a^2} = \int_{\boldsymbol{h}} e^{-2\pi i \boldsymbol{r}_n \boldsymbol{h}}\, d\boldsymbol{h}. \tag{A7-15}$$

On the other hand, the Green's function $G_l(\boldsymbol{r})$ may be regarded as the Fourier transform of a function in the form

$$G_l(\boldsymbol{r}) = \int_{\boldsymbol{h}} g_l(\boldsymbol{h}) e^{-2\pi i \boldsymbol{r}_n \boldsymbol{h}}\, d\boldsymbol{h}. \tag{A7-16}$$

Substituting Eqs. (A7-15) and (A7-16) into Eq. (A7-9) and noting that

$$\nabla^2 g_l(\boldsymbol{h}) e^{-2\pi i \boldsymbol{r}_n \boldsymbol{h}} = g_l(\boldsymbol{h}) \nabla^2 e^{-2\pi i \boldsymbol{r}_n \boldsymbol{h}} = -4\pi^2 h^2 g_l(\boldsymbol{h}) e^{-2\pi i \boldsymbol{r}_n \boldsymbol{h}},$$

we have

$$4\pi^2 \int_{\boldsymbol{h}} (k_l^2 - h^2) g_l(\boldsymbol{h}) e^{-2\pi i \boldsymbol{r}_n \boldsymbol{h}}\, d\boldsymbol{h} = \delta(\boldsymbol{r}) \equiv \int_{\boldsymbol{h}} e^{-2\pi i \boldsymbol{r}_n \boldsymbol{h}}\, d\boldsymbol{h},$$

from which it follows that

$$g_l(\boldsymbol{h}) = \left\{4\pi^2 (k_l^2 - h^2)\right\}^{-1}. \tag{A7-17}$$

Substituting Eq. (A7-17) into Eq. (A7-16) and performing the integration in the polar coordinate system, we have

$$\begin{aligned}
G_l(\boldsymbol{r}_n) &= (4\pi^2)^{-1} \int_{\boldsymbol{h}} (k_l^2 - h^2)^{-1} e^{-2\pi i \tilde{\boldsymbol{r}}_n \boldsymbol{h}}\, d\boldsymbol{h} \\
&= (4\pi^2)^{-1} \int_0^{2\pi}\!\int_0^{\pi} \left\{ \int_0^{\infty} (k_l^2 - h^2)^{-1} e^{-2\pi i r_n h \cos\theta}\, h^2 dh \right\} \sin\theta\, d\theta\, d\phi \\
&= \frac{1}{2\pi} \int_{-1}^{1}\!\int_0^{\infty} \frac{e^{2\pi i r_n h t}}{k_l^2 - h^2} h^2 dh\, dt = \frac{1}{4\pi^2} \int_0^{\infty} \frac{h\left(e^{2\pi i r_n h} - e^{-2\pi i r_n h}\right)}{i r_n (k_l^2 - h^2)} dh \qquad \text{(A7-18)} \\
&= \frac{1}{4\pi^2} \left\{ \int_0^{\infty} \frac{h e^{2\pi i r_n h}}{i r_n (k_l^2 - h^2)} dh - \int_0^{-\infty} \frac{h e^{2\pi i r_n h}}{i r_n (k_l^2 - h^2)} dh \right\} \\
&= \frac{1}{4\pi^2} \int_{-\infty}^{\infty} \frac{h e^{2\pi i r_n h}}{i r_n (k_l^2 - h^2)} dh.
\end{aligned}$$

Note that the variable θ in the third formula in Eq. (A7-18) has been replaced by $t = -\cos\theta$ in the fourth formula. The integral in the last formula in Eq. (A7-18) can be evaluated as a

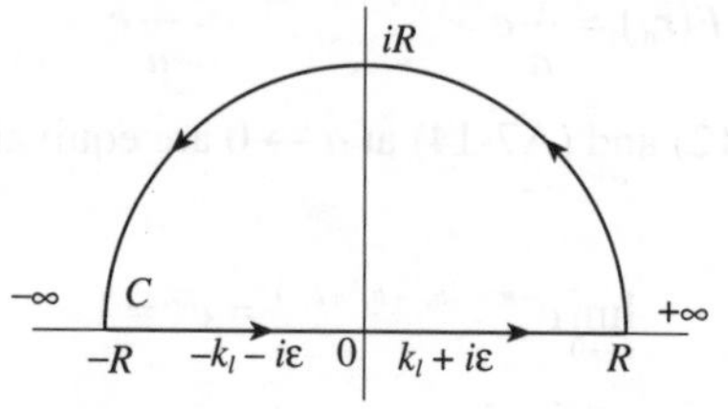

Fig. A7.1 Path of integration in the complex plane for the Green's function in Eq. (A7-18).

portion on the real axis of a complex integral along the closed path C taken as shown in Fig. A7.1. To avoid ambiguity in choosing the detour around the singularities $h = \pm k_l$, the constant k_l in the integrand is replaced by $k_l + i\varepsilon$, and the limit as $\varepsilon \to 0$ is taken after the integration. The real factor $e^{-2\pi r_n \varepsilon}$ appearing in the numerator represents the effect of certain dissipation of the scattering energy due to the crystal imperfection. The parameter ε must be positive in order that the residue $e^{2\pi i r_n k_l} e^{-2\pi r_n \varepsilon}$ does not diverge even if $r_n \to 0$; hence the singularity $h = k_l + i\varepsilon$ is in the interior of C, while $h = -k_l - i\varepsilon$ is not. Decomposing the last formula in Eq. (A7-18) into partial fractions, replacing k_l with $k_l + i\varepsilon$ and noting that the integral along the semicircle vanishes as $|h| \to \infty$, we have

$$\begin{aligned} G_l(\boldsymbol{r}_\text{n}) &= \lim_{\varepsilon\to 0} \frac{i}{8\pi^2 r_\text{n}} \int_C \left(\frac{e^{2\pi i r_\text{n} h}}{h + k_l + i\varepsilon} + \frac{e^{2\pi i r_\text{n} h}}{h - k_l - i\varepsilon} \right) dh \\ &= -\frac{2\pi}{8\pi^2 r_\text{n}} \lim_{\varepsilon\to 0} e^{2\pi i r_\text{n}(k_l + i\varepsilon)} = -\frac{1}{4\pi} \frac{e^{2\pi i |\boldsymbol{r}_\text{n}| k_l}}{|\boldsymbol{r}_\text{n}|}. \end{aligned} \tag{A7-19}$$

Substituting Eq. (A7-19) into Eq. (A7-8) gives a particular solution of Eq. (A7-6a) in the form

$$\chi_l(\boldsymbol{r}_\text{n}) = -\frac{2m}{4\pi\hbar^2} \int_\text{n} \frac{e^{2\pi i k_l |\boldsymbol{r}_\text{n} - \boldsymbol{r}'|}}{|\boldsymbol{r}_\text{n} - \boldsymbol{r}'|} \left\{ \int_\text{c} \psi_l^*(\boldsymbol{r}_\text{c}) V_\text{int}(\boldsymbol{r}', \boldsymbol{r}_\text{c}) \Psi(\boldsymbol{r}', \boldsymbol{r}_\text{c}) d\boldsymbol{r}_\text{c} \right\} d\boldsymbol{r}'. \tag{A7-20}$$

The sum of Eq. (A7-7) and Eq. (A7-20) is a general solution of Eq. (A7-6a). In this case, Eq. (A7-7) represents the incident neutron incoming as a plane wave, while Eq. (A7-20) represents the scattered neutron outgoing asymptotically as a spherical wave. Since the distance $|\boldsymbol{r}_\text{n}|$ at which we observe the scattering intensity is much larger than the distance $|\boldsymbol{r}'|$ at which the neutron–crystal interaction is effective, the portion involving $|\boldsymbol{r}_\text{n} - \boldsymbol{r}'|$ in Eq. (A7-20) may be approximated as

$$\begin{aligned} \frac{\exp(2\pi i k_l |\boldsymbol{r}_\text{n} - \boldsymbol{r}'|)}{|\boldsymbol{r}_\text{n} - \boldsymbol{r}'|} &= \frac{\exp\left[2\pi i k_l \{(\tilde{\boldsymbol{r}}_\text{n} - \tilde{\boldsymbol{r}}')(\boldsymbol{r}_\text{n} - \boldsymbol{r}')\}^{1/2}\right]}{|\boldsymbol{r}_\text{n} - \boldsymbol{r}'|} \\ &\approx \frac{1}{r_\text{n}} \exp\left\{ 2\pi i k_l \left(r_\text{n} - \frac{\tilde{\boldsymbol{r}}_\text{n} \boldsymbol{r}'}{r_\text{n}} \right) \right\} = \frac{e^{2\pi i k_l r_\text{n}}}{r_\text{n}} e^{-2\pi i \tilde{\boldsymbol{k}}_2 \boldsymbol{r}'}, \end{aligned} \tag{A7-21}$$

where $\boldsymbol{k}_2 = k_l(\boldsymbol{r}_\text{n}/r_\text{n})$ is the momentum vector of the scattered neutron directed from the scattering point to the detector. Now let $\psi_2(\boldsymbol{r}_\text{c})$ be the wave function of the crystal after scattering of a neutron, and let Eq. (A7-21) with $l = 1$ be substituted into Eq. (A7-20).

According to the first approximation of Born, which is valid for a system involving weakly interacting particles, we can replace the wavefunction $\Psi(\boldsymbol{r}', \boldsymbol{r}_c)$ in Eq. (A7-20) with that of the initial state, $\phi_1(\boldsymbol{r}_n)\psi_1(\boldsymbol{r}_c)$. obtaining

$$\begin{aligned}\chi_2(\boldsymbol{r}_n) &= -\frac{m e^{2\pi i k_2 r_n}}{2\pi\hbar^2 r_n}\int_n e^{2\pi i \boldsymbol{k}_2 \boldsymbol{r}'}\int_c \psi_2^*(\boldsymbol{r}_c)V_{int}(\boldsymbol{r}',\boldsymbol{r}_c)\Psi(\boldsymbol{r}',\boldsymbol{r}_c)\,d\boldsymbol{r}_c d\boldsymbol{r}' \\ &= -\frac{m e^{2\pi i k_2 r_n}}{2\pi\hbar^2 r_n}\int_n e^{2\pi i(\boldsymbol{k}_2-\boldsymbol{k}_1)\boldsymbol{r}'}\int_c \psi_2^*(\boldsymbol{r}_c)V_{int}(\boldsymbol{r}',\boldsymbol{r}_c)\psi_1(\boldsymbol{r}_c)\,d\boldsymbol{r}_c d\boldsymbol{r}'.\end{aligned} \tag{A7-22}$$

Employing the Fermi pseudopotential in Eq. (9-162b) as $V_{int}(\boldsymbol{r}', \boldsymbol{r}_c)$, we can rewrite Eq. (A7-22) as

$$\chi_2(\boldsymbol{r}_n) = -\frac{e^{2\pi i k_2 r_n}}{r_n}\int_c \psi_2^*(\boldsymbol{r}_c)\sum_{J=1}^{N}\sum_{p=1}^{n} b_{Jp}e^{2\pi i \boldsymbol{k}_{12}\boldsymbol{r}_{Jp}}\psi_1(\boldsymbol{r}_c)\,d\boldsymbol{r}_c, \tag{A7-23}$$

where $\boldsymbol{k}_{12} = \boldsymbol{k}_2 - \boldsymbol{k}_1$ is the scattering vector representing the momentum change of a neutron by the scattering. The integrand in Eq. (A7-23) involves the neutron coordinates only in the form $\boldsymbol{r}_n/r_n$ in the definition of $\boldsymbol{k}_2$. Hence the neutron coordinates appearing in the integral in the polar coordinate system are only θ and ϕ, and the integral can be expressed as

$$f(\theta,\phi) = \sum_{J=1}^{N}\sum_{p=1}^{n} b_{Jp}\int_c \psi_2^*(\boldsymbol{r}_c)e^{2\pi i \boldsymbol{k}_{12}\boldsymbol{r}_{Jp}}\psi_1(\boldsymbol{r}_c)\,d\boldsymbol{r}_c\,. \tag{A7-24}$$

The function $f(\theta, \phi)$ is called the scattering amplitude. Consider a small area dS on the surface of a sphere of radius R centered at the scattering point. The number of scattered neutrons passing through dS in time Δt is

$$|\chi_2(\boldsymbol{R})|^2 v_2\,\Delta t\,dS = |f(\theta,\phi)|^2\, v_2\,\Delta t\,dS/R^2, \tag{A7-25a}$$

where $\boldsymbol{R}$ is the vector defined by R, θ and ϕ in the polar coordinate system, and v_2 is the speed of scattered neutrons. Note that neutrons in the volume $v_2\Delta t dS$ pass the surface in time Δt. On the other hand, the number of incident neutrons passing unit area with the speed v_1 in time Δt is

$$|\phi_1|^2 v_1\,\Delta t = v_1\,\Delta t. \tag{A7-25b}$$

When $\Delta t = 1$, Eq. (A7-25b) gives the incident flux. The number of neutrons scattered in unit time in the direction within a solid angle $d\Omega$ per unit flux is called the scattering cross section, and is denoted by $d\sigma$. Introducing the solid angle spanned by dS, $d\Omega = dS/R^2$, into Eq. (A7-25a) and dividing the result by Eq. (A7-25b), we obtain

$$d\sigma = |f(\theta,\phi)|^2(v_2/v_1)d\Omega = |f(\theta,\phi)|^2(k_2/k_1)d\Omega, \tag{A7-26a}$$

since the ratio in speeds is equal to the ratio in momenta if the mass does not change. The cross section per unit solid angle in the direction of θ and ϕ,

$$\frac{d\sigma}{d\Omega} = |f(\theta,\phi)|^2\,\frac{k_2}{k_1} \tag{A7-26b}$$

is called the differential cross section, which is partitioned into coherent and incoherent parts as given by Eqs. (9-164a,b). The theoretical basis for this partitioning is given as follows according to the treatment by Izyumov and Chernoplekov.[1)]

Combining Eqs. (A7-24) and (9-165), we can rewrite the scattering amplitudes in the form

$$f(\theta,\phi) = \sum_{J,p} b_{Jp} M_{Jp}. \tag{A7-27}$$

The square of the absolute value of Eq. (A7-27) gives the probability of neutron scattering associated with the transition $\psi_1 \rightarrow \psi_2$ in the form

$$|f(\theta,\phi)|^2 = \sum_{J,p} \sum_{J',p'} b_{Jp} b^*_{J'p'} M_{Jp} M^*_{J'p'}. \tag{A7-28}$$

To take into account the effect of stochastic distribution of b_{Jp} over J, let us define the average of the product $b_{Jp} b^*_{J'p'}$ such that the relation

$$\sum_{J,p} \sum_{J',p'} b_{Jp} b^*_{J'p'} M_{Jp} M^*_{J'p'} = \sum_{p,p'} \langle b_p b^*_{p'} \rangle \sum_{J,J'} M_{Jp} M_{J'p'} \tag{A7-29}$$

holds. We shall take up first only the isotopic averages, leaving the spin averages to later consideration. To rewrite the average in a more explicit form, we introduce an index called the occupation number, P^s_{Jp}, which is defined as $P^s_{Jp} = 1$ if the site Jp is occupied by the isotope s, and otherwise $P^s_{Jp} = 0$. Let b^s_p be the scattering length of isotope s of the element p. The scattering length of the site Jp can be written as $b_{Jp} = b^s_p P^s_{Jp}$, so that Eq. (A7-23) becomes

$$\sum_{J,p} \sum_{J',p'} b_{Jp} b^*_{J'p'} M_{Jp} M^*_{J'p'} = \sum_{p,p'} \sum_{s,s'} b^s_p b^{s'}_{p'} \sum_{J,J'} P^s_{Jp} P^{s'}_{J'p'} M_{Jp} M^*_{J'p'}. \tag{A7-30}$$

The averaged product of occupation numbers, $\langle P^s_p P^{s'}_{p'} \rangle$, is then defined so as to satisfy the relation

$$\sum_{J,p,J',p'} P^s_{Jp} P^{s'}_{J'p'} M_{Jp} M^*_{J'p'} = \sum_{p,p'} \langle P^s_p P^{s'}_{p'} \rangle \sum_{J,J'} M_{Jp} M^*_{J'p'}. \tag{A7-31}$$

Combining Eqs. (A7-29), (A7-30) and (A7-31), we have

$$\langle b_p b^*_{p'} \rangle = \sum_{s,s'} b^s_p b^{s'}_{p'} \langle P^s_p P^{s'}_{p'} \rangle. \tag{A7-32}$$

If isotopes are stochastically distributed in the crystal, the average $\langle P^s_p P^{s'}_{p'} \rangle$ can be calculated from the definition of P^s_{Jp} in the form

$$\langle P^s_p P^{s'}_{p'} \rangle = \frac{1}{N} \sum_{J} P^s_{Jp} \left\{ \frac{1}{N} \sum_{J'} P^{s'}_{J'p'} (1 - \delta_{JJ'} \delta_{pp'}) + \delta_{ss'} \delta_{JJ'} \delta_{pp'} \right\}, \tag{A7-33}$$

where N is the total number of sites in the crystal. The factor $(1 - \delta_{JJ'}\delta_{pp'})$ in Eq. (A7-33) implies that different isotpes s and s' cannot occupy the same site Jp at once. The second term in braces in Eq. (A7-33) states that $P^{s}_{J'p'}$ becomes unity if $P^s_{Jp} = 1$ and s', J' and p' are the same as s, J and p, respectively. For large N, it holds that

$$\sum_{J,p} P_{Jp}^s \Big/ N = c_s,$$

where c_s is the natural abundance of the isotope s, whence Eq. (A7-33) is rewritten as

$$\begin{aligned}\langle P_p^s P_{p'}^{s'} \rangle &= c_s\, c_{s'} (1 - \delta_{JJ'} \delta_{pp'}) + c_s\, \delta_{ss'} \delta_{JJ'} \delta_{pp'} \\ &= c_s\, c_{s'} + c_s (\delta_{ss'} - c_{s'}) \delta_{JJ'} \delta_{pp'}.\end{aligned} \tag{A7-34}$$

Substitute Eq. (A7-34) into Eq. (A7-32), rearrange the right side and use the fact that the averages of scattering lengths and squared scattering lengths in a random assembly of isotopes are given by $\langle b_p \rangle = \sum_s b_p^s c_s$ and $\langle {b_p}^2 \rangle = \sum_s b_p^{s2} c_s$, respectively. The result is

$$\begin{aligned}\langle b_p\, b_{p'}^* \rangle &= \sum_{s,s'} b_p^s b_{p'}^{s'} \{ c_s\, c_{s'} + c_s (\delta_{ss'} - c_{s'}) \delta_{JJ'} \delta_{pp'} \} \\ &= \sum_s b_p^s c_s \sum_{s'} b_{p'}^{s'} c^{s'} + \left(\sum_s b_p^{s2} c_s - \sum_s b_p^s c_s \sum_{s'} b_{p'}^{s'} c^{s'} \right) \delta_{JJ'} \delta_{pp'} \\ &= \langle b_p \rangle \langle b_{p'} \rangle + \left(\langle {b_p}^2 \rangle - \langle b_p \rangle \langle b_{p'} \rangle \right) \delta_{JJ'} \delta_{pp'}.\end{aligned} \tag{A7-35}$$

The above discussion concerns only the averaging of scattering lengths of isotopes. The effect of stochastic distribution of spin states of neutrons and nuclei may be roughly taken into account in the following way. If the nucleus p has a spin $\boldsymbol{I}_p$, its spin state is distributed among $2I_p + 1$ levels, and nuclei in different spin levels interact with neutrons in different ways. In addition, a neutron may have a spin of 1/2 or –1/2. In this case, the scattering length b_{Jp} in Eq. (9-162b) is replaced as

$$b_{Jp} \to b_{Jp} + B_p\, \tilde{\mathbf{s}}_{\mathrm{n}} \boldsymbol{I}_p, \tag{A7-36}$$

where B_p is a constant specific to the element at the site p but is independent of its spin state. In accordance with Eq. (A7-36), the average of products $\langle b_p b_{p'}^* \rangle$ in Eq. (A7-35) must also be changed. Noting that terms linear in $\langle \tilde{\mathbf{s}}_{\mathrm{n}} \boldsymbol{I}_p \rangle$ vanish since positive and negative spins are equally distributed for a given spin operator and that $|\mathbf{s}_{\mathrm{n}}|^2 = 1/4$ for neutrons and $\langle {I_p}^2 \rangle = \langle I_p (I_p + 1) \rangle$ for any angular momentum operator, we have

$$\begin{aligned}\langle b_p b_{p'}^* \rangle &\to \langle b_p b_{p'}^* \rangle + 2\, B_p \langle b_p \tilde{\mathbf{s}}_{\mathrm{n}} \boldsymbol{I}_p \rangle + {B_p}^2 \langle \tilde{\mathbf{s}}_{\mathrm{n}} \boldsymbol{I}_p\, \tilde{\mathbf{s}}_{\mathrm{n}} \boldsymbol{I}_{p'} \rangle \delta_{JJ'} \\ &= \langle b_p b_{p'}^* \rangle + \tfrac{1}{4} {B_p}^2 \langle I_p (I_p + 1) \rangle \delta_{JJ'} \delta_{pp'}.\end{aligned} \tag{A7-37}$$

Modifying Eq. (A7-35) according to Eq. (A7-37), substituting the result into Eq. (A7-29) and then into Eq. (A7-28) gives

$$|f(\theta,\phi)|^2 = [\langle b_p \rangle \langle b_{p'} \rangle + \{ \langle {b_p}^2 \rangle - \langle b_p \rangle \langle b_{p'} \rangle + \tfrac{1}{4} {B_p}^2 \langle I_p (I_p + 1) \rangle \} \delta_{JJ'} \delta_{pp'}] \sum_{J,p} \sum_{J',p'} M_{Jp} M_{J'p'}^*. \tag{A7-38}$$

Now we define the bound atom coherent and incoherent cross sections of the element p in the forms

$$\sigma_p^{\mathrm{coh}} = \langle b_p \rangle^2 \tag{A7-39a}$$

and

$$\sigma_p^{\mathrm{incoh}} = \left(\langle b_p{}^2 \rangle - \langle b_p \rangle \langle b_{p'} \rangle\right) + \tfrac{1}{4}\langle B^2 I(I+1) \rangle. \tag{A7-39b}$$

Substituting Eq. (A7-38) into Eq. (A7-26b) and comparing the result with Eq. (9-163) gives Eqs. (9-164a,b).

Reference

1) Yu. A. Izyumov and N. A. Chernoplekov, *Neutron Spectroscopy*, translated by J. Buchner, Consultants Bureau, A division of Plenum Publishing Corp., New York (1994).

Appendix 8 Evaluation of the Integral $(I^a_{Jp,Kj})_{n,n+s}$ and the Thermal Average of the Square of Its Absolute Value $(J^a_{Jp,Kj})_s$

The discussion in this appendix is limited to a single normal mode, so that the quantities $(I^a_{Jp,Kj})_{n,n+s}$ and $(J^a_{Jp,Kj})_s$ in the text will be denoted by simplified notations $I_{n,n+s}$ and J_s, respectively. The Hermite polynomials are given as the coefficients in the expansion of the generating function in the form

$$e^{-z^2+2zx} = \sum_{n=0}^{\infty} H_n(x) z^n / n!. \tag{A8-1}$$

According to the theory of integrals of functions of complex variables,[1] a function $f(x)$ differentiable at every point within a closed curve can be expanded as a Taylor series in the form

$$f(x) = \sum_{n=0}^{\infty} \frac{f^{(n)}(z_0)}{n!}(x - z_0)^n, \tag{A8-2a}$$

where

$$f^{(n)}(z_0) \equiv \frac{d^n f}{dx^n}(z_0) = \frac{n!}{2\pi i}\int_C \frac{f(z)}{(z - z_0)^{n+1}}dz. \tag{A8-2b}$$

Comparing Eq. (A8-1) with Eq. (A8-2), we have

$$\frac{H_n(x)}{n!} = \frac{1}{2\pi i}\int_C \frac{e^{-z^2+2zx}}{z^{n+1}}dz. \tag{A8-3}$$

By substituting the wave functions of the harmonic oscillator, Eq. (4-65), into Eq. (9-168), the integral $I_{n,n+s}$ is written as

$$\begin{aligned} I_{n,n+s} &= \int_{-\infty}^{\infty} \psi^*(q; n+s)\exp(2itq)\psi(q;n)dq \\ &= (2^{2n+s}\pi(n+s)!n!)^{-1/2}\int_{-\infty}^{\infty} \exp(-q^2 + 2itq) H^*_{n+s}(q) H_n(q) dq, \end{aligned} \tag{A8-4}$$

where $t = t^{a}_{Jp,Kj}/2$.[2)] Substituting Eq. (A8-3) into Eq. (A8-4) gives

$$I_{n,n+s} = \frac{\{(n+s)!n!\}^{1/2}}{(2^{2n+s}\pi)^{1/2}(2\pi i)^2} \int_{-\infty}^{\infty}\int_C\int_C e^{-q^2+2itq}\frac{e^{-u^2+2uq}}{u^{n+1}}\frac{e^{-v^2+2vq}}{v^{n+s+1}}dudvdq$$
$$= \frac{\{(n+s)!n!\}^{1/2}e^{-t^2}}{(2^{2n+s}\pi)^{1/2}(2\pi i)^2} \int_{-\infty}^{\infty}\int_C\int_C e^{-\{q-(it+u+v)\}^2}\frac{e^{2u(it+v)}}{u^{n+1}}\frac{e^{2itv}}{v^{n+s+1}}dudvdq, \tag{A8-5}$$

where use has been made of the relation

$$\{q-(it+u+v)\}^2 = q^2 - t^2 + u^2 + v^2 - 2q(it+u+v) + 2u(it+v) + 2itv.$$

By changing the variable as $q-(it+u+v) = q'$, and remembering that the integral of $\exp(-x^2)$ from $-\infty$ to ∞ is $\sqrt{\pi}$, the integral factor in Eq. (A8-5) is rewritten as

$$\int_{-\infty}^{\infty}\int_C\int_C e^{-q'^2}\frac{e^{2u(it+v)}}{u^{n+1}}\frac{e^{2itv}}{v^{n+s+1}}dudvdq' = \sqrt{\pi}\int_C\int_C \frac{e^{2u(it+v)}}{u^{n+1}}\frac{e^{2itv}}{v^{n+s+1}}dudv. \tag{A8-6}$$

Integrating Eq. (A8-6) with respect to u, substituting the result into Eq. (A8-5) and introducing a new variable z by $v = itz$, we have

$$I_{n,n+s} = \left\{\frac{(n+s)!}{n!}\right\}^{1/2}\frac{e^{-t^2}}{(2^{2n+s})^{1/2}(2\pi i)}\int_C \frac{\{2(it+v)\}^n e^{-2itv}}{v^{n+s+1}}dv$$
$$= \left\{\frac{(n+s)!}{n!}\right\}^{1/2}\frac{2^{-s/2}e^{-t^2}(it)^{-s}}{(2\pi i)}\int_C \frac{(1+z)^n e^{-2t^2 z}}{z^{n+s+1}}dz. \tag{A8-7}$$

Exchanging the suffices $n+s$ and n in Eq. (A8-7) with each other and taking the complex conjugate, we have

$$I^*_{n+s,n} = \left\{\frac{n!}{(n+s)!}\right\}^{1/2}\frac{2^{s/2}e^{-t^2}(-it)^s}{(2\pi i)}\int_C \frac{(1+y)^{n+s}e^{-2t^2 y}}{y^{n+1}}dy. \tag{A8-8}$$

The thermal average of the transition probability is the weighted sum of the product of Eqs. (A8-7) and (A8-8):

$$J_s = \sum_{n=0}^{\infty} w^n I_{n,n+s}I^*_{n,n+s} = (1-x)\sum_{n=0}^{\infty} x^n I_{n,n+s}I^*_{n,n+s}, \tag{A8-9}$$

where $x = e^{-hc\omega/kT}$. Substituting Eqs. (A8-7) and (A8-8) into Eq. (A8-9), and introducing new variables ζ and η defined by $z = \zeta/(1-\zeta)$ and $y = \eta/(1-\eta)$, we have

$$\frac{J_s}{1-x} = \frac{e^{-2t^2}(-1)^s}{(2\pi i)^2}\sum_{n=0}^{\infty}x^n\int_C\int_C \frac{(1+z)^n(1+y)^{n+s}e^{-2t^2(z+y)}}{z^{n+s+1}y^{n+1}}dzdy$$
$$= \frac{e^{-2t^2}(-1)^s}{(2\pi i)^2}\int_C\int_C\sum_{n=0}^{\infty}x^n\frac{(1-\zeta)^{s-1}e^{-2t^2\left(\frac{\zeta}{1-\zeta}+\frac{\eta}{1-\eta}\right)}}{\zeta^{n+s+1}\eta^{n+1}(1-\eta)^{s+1}}d\zeta d\eta. \tag{A8-10}$$

Since $|x| \le 1$, it holds always that $|x/\zeta\eta| \le 1$, whence we have

$$\sum_{n=0}^{\infty}(x/\zeta\eta)^n = \{1-(x/\zeta\eta)\}^{-1}. \tag{A8-11}$$

Substituting Eq. (A8-11) into Eq. (A8-10) leads to

$$\frac{J_s}{1-x} = \frac{(-1)^s}{(2\pi i)^2}\int_C\int_C \frac{(1-\zeta)^{s-1} e^{-t^2\left(2+\frac{2\zeta}{1-\zeta}+\frac{2\eta}{1-\eta}\right)}}{\{\eta-(x/\zeta)\}\zeta^{s+1}(1-\eta)^{s+1}}\,d\zeta\, d\eta. \tag{A8-12}$$

The integral with respect to η can be taken to encircle no singular points other than $\eta = x/\zeta$, whence it follows from Cauchy's theorem[1] that

$$\frac{J_s}{1-x} = \frac{(-1)^s}{2\pi i}\int_C \frac{(1-\zeta)^{s-1} e^{-t^2\left(\frac{1+\zeta}{1-\zeta}+\frac{\zeta+x}{\zeta-x}\right)}}{\{1-(x/\zeta)\}^{s+1}\zeta^{s+1}}\,d\zeta. \tag{A8-13}$$

Let us rewrite Eq. (A8-13) using new variables ξ and a defined by

$$\xi = 2t^2(x-\zeta)/\{(1-x)(1-\zeta)\}. \tag{A8-14a}$$

and

$$a = 2t^2/(1-x). \tag{A8-14b}$$

Solving Eq. (A8-14a) for ζ, and using Eq. (A814b), we have

$$\zeta = (2t^2x - \xi + x\xi)/(2t^2 - \xi + x\xi) = (ax-\xi)/(a-\xi). \tag{A8-15}$$

Substituting Eqs. (A8-14b) and (A8-15) into relevant expressions in Eq. (A8-13) gives

$$t^2\left(\frac{1+\zeta}{1-\zeta}+\frac{\zeta+x}{\zeta-x}\right) = a(1+x) - \xi - a^2\frac{x}{\xi}, \quad d\zeta = -\frac{a(1-x)}{(a-\xi)^2}d\xi$$

and

$$\frac{(1-\zeta)^{s-1}}{(\zeta-x)^{s+1}} = \left(\frac{a}{-\xi}\right)^{s-1}\left\{\frac{a-\xi}{\xi(1-x)}\right\}^2,$$

whence it follows from Eq. (A8-13) that

$$J_s = \frac{(-i)^s e^{-a(1+x)}}{(2\pi i)x^{s/2}}\int_C \frac{e^{\xi-(ai\sqrt{x})^2/\xi}}{\xi^{s+1}}(ai\sqrt{x})^s\, d\xi. \tag{A8-16}$$

On introducing a new variable $z = 2a\sqrt{x}$, Eq. (A8-16) is rewritten as

$$J_s = \frac{(-i)^s e^{-a(1+x)}}{(2\pi i)x^{s/2}}\left(\frac{iz}{2}\right)^s\int_C \frac{e^{\xi-(iz/2)^2/\xi}}{\xi^{s+1}}\,d\xi. \tag{A8-17}$$

The right side of Eq. (A8-17) can be compactly expressed in terms of the Bessel function,[2] which is defined in terms of the generating function in the form

$$\exp\left\{\frac{z}{2}\left(x-\frac{1}{x}\right)\right\} = \sum_{n=-\infty}^{\infty} J_n(z)x^n. \qquad \text{(A8-18)}$$

The explicit form of $J_n(z)$ is obtained by expanding the left side of Eq. (A8-18) as

$$\begin{aligned}\exp\left\{\frac{z}{2}\left(x-\frac{1}{x}\right)\right\} &= \sum_{l=0}^{\infty}\frac{1}{l!}\left\{\frac{z}{2}\left(x-\frac{1}{x}\right)\right\}^l \\ &= \sum_{l=0}^{\infty}\frac{1}{l!}\left(\frac{z}{2}\right)^l \sum_{j=0}^{l}\frac{(-1)^j l!}{(l-j)!\,j!}x^{l-j}\left(\frac{1}{x}\right)^j = \sum_{l=0}^{\infty}\left(\frac{z}{2}\right)^l \sum_{j=0}^{l}\frac{(-1)^j x^{l-2j}}{(l-j)!\,j!}.\end{aligned} \qquad \text{(A8-19)}$$

By changing the index l to $s = l - 2j$, Eq. (A8-19) is rewritten as

$$\exp\left\{\frac{z}{2}\left(x-\frac{1}{x}\right)\right\} = \sum_{s=-\infty}^{\infty}\left\{\sum_{j=0}^{\infty}\frac{(-1)^j}{(s+j)!\,j!}\left(\frac{z}{2}\right)^{s+2j}\right\}x^s. \qquad \text{(A8-20)}$$

From Eqs. (A8-18) and (A8-20), it follows that

$$J_s(z) = (z/2)^s\sum_{j=0}^{\infty}(-1)^j (z/2)^{2j}/\{(s+j)!\,j!\}. \qquad \text{(A8-21)}$$

Applying Cauchy's integral formula to Eq. (A8-18), and introducing a new variable $\xi = zx/2$, we have

$$J_s(z) = \frac{1}{2\pi i}\int_C \frac{e^{(z/2)(x-1/x)}}{x^{s+1}}dx = \frac{1}{2\pi i}\left(\frac{z}{2}\right)^s\int_C \frac{e^{\xi-(z^2/4\xi)}}{\xi^{s+1}}d\xi. \qquad \text{(A8-22)}$$

The modified Bessel function of the sth order is defined by

$$I_s(z) = (z/2)^s\sum_{j=0}^{\infty}(z/2)^{2j}/\{(s+j)!\,j!\}. \qquad \text{(A8-23)}$$

It follows from Eqs. (A8-21) and (A8-23) that

$$(-i)^s J_s(iz) = I_s(z). \qquad \text{(A8-24)}$$

By combining Eqs. (A8-17), (A8-22) and (A8-24), we obtain

$$J_s = \frac{(-i)^s e^{-a(1+x)}}{x^{s/2}}J_s(iz) = \frac{e^{-2t^2(1+x)/(1-x)}}{x^{s/2}}I_s(z). \qquad \text{(A8-25)}$$

Equation (A8-25) is equivalent to Eq. (9-171) in the text, since

$$z = \frac{4t^2}{1-x}x^{1/2} = \frac{4t^2}{e^{hc\omega/2kT} - e^{-hc\omega/2kT}} = \frac{\left(t^a_{Jp.Kj}\right)^2}{2\sinh(hc\omega/2kT)}$$

from the definition of the variables z and t.[2)]

Equation (9-173) in the text is based on the addition theorem for Bessel functions stating that

$$J_s(x+y) = \sum_{n=-\infty}^{\infty} J_{s-n}(x)\, J_n(y). \tag{A8-26}$$

Equation (A8-26) is proved as follows. From the property of the exponential function, we have

$$\exp\left\{\frac{x+y}{2}\left(u-\frac{1}{u}\right)\right\} = \exp\left\{\frac{x}{2}\left(u-\frac{1}{u}\right)\right\}\exp\left\{\frac{y}{2}\left(u-\frac{1}{u}\right)\right\}. \tag{A8-27}$$

Substituting Eq. (A8-18) into both the sides of Eq. (A8-27) gives

$$\sum_{n=-\infty}^{\infty} J_n(x+y)u^n = \sum_{r=-\infty}^{\infty} J_r(x)u^r \sum_{s=-\infty}^{\infty} J_s(x)u^s\,. \tag{A8-28}$$

The right side of Eq. (A8-28) is rewritten as

$$\sum_{r=-\infty}^{\infty}\sum_{s=-\infty}^{\infty} J_r(x)\, J_s(x) u^{r+s} = \sum_{n=-\infty}^{\infty}\left\{\sum_{s=-\infty}^{\infty} J_{n-s}(x)\, J_s(x)\right\}u^n. \tag{A8-29}$$

Equating the coefficients of u^n in the left side of Eq. (A8-28) and the right side of Eq. (A8-29), we obtain Eq. (A8-26).

References

1) E Butkov, *Mathematical Physics*, Chap. 2, Addison-Wesley Publishing Co., Reading, Massachusetts (1973).
2) V.H. Ott, *Ann. Phys.*, **23**, 169 (1935).

Appendix 9 Structure Factor and Patterson Function for Coulomb Potential

In X-ray crystallography, the Patterson function is defined as

$$P(\boldsymbol{u}) = \int_{\text{cell}} \rho(\boldsymbol{r})\, \rho(\boldsymbol{r}+\boldsymbol{u})\, d\boldsymbol{r}, \tag{A9-1}$$

where $\rho(\boldsymbol{r})$ is the electron density.[1] The integration covers the space of a unit cell. Because of the periodicity of crystals, the electron density is expressed as a Fourier series in which each term is specified by a reciprocal lattice vector $\boldsymbol{h}_\lambda$ as given by

$$\rho(\boldsymbol{r}) = V^{-1}\sum_{\lambda} F(\boldsymbol{h}_\lambda)\exp\left(-2\pi i\, \tilde{\boldsymbol{h}}_\lambda \boldsymbol{r}\right), \tag{A9-2}$$

where the coefficients $F(\boldsymbol{h}_\lambda)$ are called structure factors. To determine each structure factor, we multiply both sides of Eq. (A9-2) by $\exp\left(2\pi i\, \tilde{\boldsymbol{h}}_{\lambda'} \boldsymbol{r}\right)$, integrate over a unit cell and use the relation

$$\int_{\text{cell}} \exp\left\{2\pi i\left(\tilde{\boldsymbol{h}}_{\lambda'} - \tilde{\boldsymbol{h}}_\lambda\right)\boldsymbol{r}\right\} d\boldsymbol{r} = \delta_{\lambda\lambda'}\int_{\text{cell}} d\boldsymbol{r} = \delta_{\lambda\lambda'}\, V, \tag{A9-3}$$

where V is the unit cell volume, obtaining

$$\int_{\text{cell}} \rho(\boldsymbol{r}) \exp\left(2\pi i \tilde{\boldsymbol{h}}_{\lambda'} \boldsymbol{r}\right) d\boldsymbol{r} = F(\boldsymbol{h}_{\lambda'}). \tag{A9-4}$$

Substituting Eq. (A9-2) into Eq. (A9-1) gives

$$P(\boldsymbol{u}) = V^{-2} \sum_{\lambda} \sum_{\lambda'} F(\boldsymbol{h}_{\lambda}) F(\boldsymbol{h}_{\lambda'}) \exp\left(-2\pi i \tilde{\boldsymbol{h}}_{\lambda'} \boldsymbol{u}\right) \int_{\text{cell}} \exp\left\{-2\pi i \left(\tilde{\boldsymbol{h}}_{\lambda} + \tilde{\boldsymbol{h}}_{\lambda'}\right) \boldsymbol{r}\right\} d\boldsymbol{r}. \tag{A9-5}$$

Comparing the integral in the right side of Eq. (A9-5) with Eq. (A9-3) and noting that $F(\boldsymbol{h}_{\lambda'})$ is the complex conjugate of $F(\boldsymbol{h}_{\lambda})$ if $\boldsymbol{h}_{\lambda'} = -\boldsymbol{h}_{\lambda}$, we have

$$P(\boldsymbol{u}) = V^{-1} \sum_{\lambda} \left|F(\boldsymbol{h}_{\lambda})\right|^2 \exp\left(2\pi i \tilde{\boldsymbol{h}}_{\lambda} \boldsymbol{u}\right). \tag{A9-6}$$

In dealing with the Coulomb potential in molecular mechanics, we consider the charge density instead of electron density. In addition, our model employs point charges instead of continuous charges distributed in space. In this case, it is convenient to represent the charge density in terms of a δ-function, *i.e.*,

$$\rho(\boldsymbol{r}) = \sum_{s=1}^{nN} q_s \, \delta(\boldsymbol{r} - \boldsymbol{r}_s), \tag{A9-7}$$

where q_s and $\boldsymbol{r}_s$ are the effective charge and the position vector of nucleus s, respectively, and the sum is taken over the whole crystal. The Coulomb potential for continuously distributed charges is given by

$$V_{\mathrm{C}} = \tfrac{1}{2} \int_{\boldsymbol{u}}^{\text{crys}} \int_{\boldsymbol{r}}^{\text{cell}} \{\rho(\boldsymbol{r})\, \rho(\boldsymbol{r} + \boldsymbol{u}) / |\boldsymbol{u}|\}\, d\boldsymbol{r}\, d\boldsymbol{u}. \tag{A9-8}$$

Substituting Eq. (A9-7) into Eq. (A9-8), we have

$$\begin{aligned} V_{\mathrm{C}} &= \tfrac{1}{2} \sum_{s=1}^{nN} \sum_{t=1}^{n} q_s q_t \int_{\boldsymbol{u}}^{\text{crys}} |\boldsymbol{u}|^{-1} \int_{\boldsymbol{r}}^{\text{cell}} \delta(\boldsymbol{r} - \boldsymbol{r}_t)\, \delta(\boldsymbol{r} + \boldsymbol{u} - \boldsymbol{r}_s)\, d\boldsymbol{r}\, d\boldsymbol{u} \\ &= \tfrac{1}{2} \sum_{s=1}^{nN} \sum_{t=1}^{n} q_s q_t \int_{\boldsymbol{u}}^{\text{crys}} |\boldsymbol{u}|^{-1} \delta(\boldsymbol{r}_t + \boldsymbol{u} - \boldsymbol{r}_s)\, d\boldsymbol{u} = \tfrac{1}{2} \sum_{s=1}^{nN} \sum_{t=1}^{n} q_s q_t / r_{st}, \end{aligned} \tag{A9-9}$$

where $r_{st} = |\boldsymbol{r}_t - \boldsymbol{r}_s|$. Thus the charge density in the form of Eq. (A9-7) gives the correct form of the Coulomb potential for point charges. Substituting Eq. (A9-7) into Eq. (A9-1) gives

$$\begin{aligned} P(\boldsymbol{u}) &= \sum_{s=1}^{nN} \sum_{t=1}^{n} q_s q_t \int_{\text{cell}} \delta(\boldsymbol{r} - \boldsymbol{r}_t)\, \delta(\boldsymbol{r} + \boldsymbol{u} - \boldsymbol{r}_s)\, d\boldsymbol{r} \\ &= \sum_{s=1}^{nN} \sum_{t=1}^{n} q_s q_t\, \delta(\boldsymbol{r}_s + \boldsymbol{u} - \boldsymbol{r}_t) = \sum_{s=1}^{nN} \sum_{t=1}^{n} q_s q_t\, \delta(\boldsymbol{u} - \boldsymbol{r}_{st}). \end{aligned} \tag{A9-10}$$

Similarly, the structure factor for point charges is obtained from Eqs. (A9-4) and (A9-7) as

$$F(\boldsymbol{h}_\lambda) = \int_{\text{cell}} \sum_{s=1}^{nN} q_s\, \delta(\boldsymbol{r} - \boldsymbol{r}_s) \exp\left(2\pi i \tilde{\boldsymbol{h}}_\lambda \boldsymbol{r}\right) d\boldsymbol{r}$$

$$= \sum_{s=1}^{n} q_s \exp\left(-2\pi i \tilde{\boldsymbol{h}}_\lambda \boldsymbol{r}_s\right). \tag{A9-11}$$

Changing variables in Eq. (A9-1) as $\boldsymbol{r} + \boldsymbol{u} = \boldsymbol{r}'$ and $\boldsymbol{r} = \boldsymbol{r}' - \boldsymbol{u}$, and noting that the integrals with respect to $\boldsymbol{r}$ and $\boldsymbol{r}'$ are equivalent to each other because of the periodicity of the crystal, we have

$$P(\boldsymbol{u}) = \int_{\text{cell}} \rho(\boldsymbol{r}')\, \rho(\boldsymbol{r}' - \boldsymbol{u})\, d\boldsymbol{r}' = P(-\boldsymbol{u}). \tag{A9-12}$$

Reference

1) F. Bertaut, *J. Phys. Radium*, **13**, 499 (1952).

Appendix 10 Three-dimensional Fourier Transform of the Function $\gamma(n, \pi r^2)/r^{2n}$

In this Appendix 10 as well as in the text, we follow the definition and the notation of the three dimensional Fourier transform of a function $f(\boldsymbol{r})$ and the inverse Fourier transform of the transformed function $F(\boldsymbol{h})$ adopted by Nijboer and De Wette[1] in the forms

$$\mathrm{FT}_3\{f(\boldsymbol{r})\} = \iiint f(\boldsymbol{r}) \exp\left(2\pi i \tilde{\boldsymbol{h}} \boldsymbol{r}\right) d\boldsymbol{r} = F(\boldsymbol{h}) \tag{A10-1a}$$

and

$$\mathrm{FT}_3\{F(\boldsymbol{h})\} = \iiint F(\boldsymbol{h}) \exp\left(-2\pi i \tilde{\boldsymbol{h}} \boldsymbol{r}\right) d\boldsymbol{h} = f(\boldsymbol{r}). \tag{A10-1b}$$

From the definition of the gamma function

$$\Gamma(z) = \int_0^\infty e^{-t} t^{z-1} dt \tag{A10-2a}$$

and of the incomplete gamma functions of the first kind,

$$\gamma(z, p) = \int_0^p e^{-t} t^{z-1} dt \tag{A10-2b}$$

and of the second kind

$$\Gamma(z, p) = \int_p^\infty e^{-t} t^{z-1} dt, \tag{A10-2c}$$

the convergence function for the integrals I_{2a} and I_{2b} in the text is written as

$$\{1 - \varphi(\boldsymbol{r})\} r^{-n} = \gamma\left(n/2, \pi K^2 r^2\right) / r^n \Gamma(n/2). \tag{A10-3}$$

The relation to be used for deriving Eq. (9-192b) from Eq. (9-192a) in the text is written as

$$\mathrm{FT}_3\left\{\frac{\gamma(n/2,\pi K^2 r^2)}{r^n}\right\} = \pi^{n-3/2} h^{n-3} \Gamma\left(\frac{-n+3}{2}, \frac{\pi h^2}{K^2}\right). \tag{A10-4}$$

Instead of proving Eq. (A10-4) directly, we shall take up the inverse transform of Eq. (A10-4) in the form

$$\mathrm{FT}_3\left\{\pi^{n-3/2} h^{n-3} \Gamma\left(\frac{-n+3}{2}, \frac{\pi h^2}{K^2}\right)\right\} = \frac{\gamma(n/2,\pi K^2 r^2)}{r^n}, \tag{A10-5}$$

and prove that

$$d(r^n F_1)/dr = d(r^n F_2)/dr, \tag{A10-6}$$

where F_1 and F_2 are the left and right sides of Eq. (A10-5) divided by $2\sqrt{\pi}$, respectively.

Let us first take up the quantity F_1 and rewrite it in a simpler form. Since the transformed function involves only a single variable h, it is advantageous to use the polar coordinate system. The vector $\boldsymbol{r}$ can always be taken along the polar without loss of generality. In this case the scalar product $\tilde{\boldsymbol{h}}\boldsymbol{r}$ is expressed as $hr\cos\theta$, whence we have

$$\begin{aligned}\mathrm{FT}_3\{F(h)\} &= \int_0^\infty\left[\int_0^\pi\left\{\int_0^{2\pi} F(h)\exp\left(-2\pi i\tilde{\boldsymbol{h}}\boldsymbol{r}\right)d\phi\right\}\sin\theta d\theta\right]h^2 dh \\ &= 2\pi\int_0^\infty F(h)\left[\int_0^\pi \exp(-2\pi i\,hr\cos\theta)\sin\theta d\theta\right]h^2 dh.\end{aligned} \tag{A10-7}$$

By changing the variable as $\zeta = -\cos\theta$, we have $d\zeta = \sin\theta\, d\theta$, so that Eq. (A10-7) is rewritten as

$$\begin{aligned}\mathrm{FT}_3\{F(h)\} &= 2\pi\int_0^\infty F(h)\left[\int_{-1}^{+1}\exp(2\pi i\,hr\zeta)d\zeta\right]h^2 dh \\ &= 2\pi\int_0^\infty F(h)\left(\frac{e^{2\pi i hr\zeta} - e^{-2\pi i hr\zeta}}{2\pi i\,hr}\right)h^2 dh \\ &= 2\pi\int_0^\infty F(h)\left(\frac{\sin 2\pi\,hr}{r}\right)h\,dh.\end{aligned} \tag{A10-8}$$

Taking the right side of Eq. (A10-4) as $F(h)$, using the definition of $\Gamma(z,p)$ in Eq. (A10-2c), and replacing the dummy variable from h to $t = 2\pi rh$, we have

$$\begin{aligned}F_1 &= \frac{2\pi^{n-2}}{2r}\int_0^\infty\left(h^{n-3}\sin 2\pi\,hr\int_{\pi h^2/K^2}^\infty e^{-x}x^{-(n-1)/2}dx\right)h\,dh \\ &= \frac{2\pi^{n-2}}{2r(2\pi r)^{n-1}}\int_0^\infty t^{n-2}\sin t\left(\int_{t^2/4\pi K^2 r^2}^\infty e^{-x}x^{-(n-1)/2}\,dx\right)dt.\end{aligned} \tag{A10-9}$$

Multiplying both sides of Eq. (A10-9) by r^n and exchanging the order of integration, we have

$$r^n F_1 = 2^{1-n}\pi^{-1}\int_0^\infty e^{-x}x^{-(n-1)/2}\left(\int_0^{2Kr\sqrt{\pi x}} t^{n-2}\sin t\,dt\right)dx. \qquad \text{(A10-10)}$$

Replacing t in Eq. (A10-10) with $y = t/2K\sqrt{\pi x}$, whereby we have $t = 2Ky\sqrt{\pi x}$ and $dt = 2K\sqrt{\pi x}\,dy$, leads to

$$\begin{aligned} r^n F_1 &= 2^{1-n}\pi^{-1}\int_0^\infty e^{-x}x^{-(n-1)/2}\left\{\int_0^r y^{n-2}\left(2K\sqrt{\pi x}\right)^{n-1}\sin\left(2Ky\sqrt{\pi x}\right)dy\right\}dx \\ &= \pi^{-1}\int_0^\infty e^{-x}x^{-(n-1)/2}\left\{\int_0^r K^{n-1}y^{n-2}(\pi x)^{(n-1)/2}\sin\left(2Ky\sqrt{\pi x}\right)dy\right\}dx. \end{aligned} \qquad \text{(A10-11)}$$

Recalling that an indefinite integral is related to its derivative by

$$d\left\{\int_0^r f(y)dy\right\}\Big/dr = f(r), \qquad \text{(A10-12)}$$

we differentiate Eq. (A10-11) with respect to r, and change the variable of the remaining integration from x to $u = x^{1/2}$, obtaining

$$\begin{aligned} \frac{d}{dr}(r^n F_1) &= \pi^{(n-3)/2}\int_0^\infty e^{-x}K^{n-1}r^{n-2}\sin\left(2Kr\sqrt{\pi x}\right)dx \\ &= 2\pi^{(n-3)/2}K^{n-1}r^{n-2}\int_0^\infty e^{-u^2}u\sin(2\pi^{1/2}Kru)du. \end{aligned} \qquad \text{(A10-13)}$$

By using the formula for the definite integral

$$\int_0^\infty e^{-x^2}x\sin bx\,dx = (\sqrt{\pi}/4)b\,e^{-b^2/4},$$

Eq. (A10-13) is finally led to

$$\frac{d(r^n F_1)}{dr} = 2\pi^{(n-3)/2}r^{n-2}\left(\frac{\sqrt{\pi}}{4}2\sqrt{\pi}\,r\,e^{-\pi K^2 r^2}\right) = \left(\pi K^2 r^2\right)^{(n-1)/2}e^{-\pi K^2 r^2}. \qquad \text{(A10-14)}$$

Next let us take up the quantity F_2. From Eq. (A10-5) and the definition of the function $\gamma(z, p)$ in Eq. (A10-2b), we have

$$r^n F_2 = \left(2\sqrt{\pi}\right)^{-1}\int_0^{\pi K^2 r^2} e^{-x}x^{(n-2)/2}\,dx. \qquad \text{(A10-15)}$$

By introducing a variable y such that $x = \pi K^2 y^2$, we have $dx = 2\pi Ky\,dy$ and $y = r$ when $x = \pi K^2 r^2$. Then Eq. (A10-15) is rewritten as

$$r^n F_2 = (2\sqrt{\pi})^{-1} \int_0^r e^{-\pi K^2 y^2} (\pi K^2 y^2)^{(n-2)/2} 2\pi K y \, dy$$

$$= \pi^{(n-1)/2} \int_0^r e^{-\pi K^2 y^2} (Ky)^{n-1} \, dy. \qquad \text{(A10-16)}$$

By referring to Eq. (A10-12), the derivative of Eq. (A10-16) with respect to r is obtained as

$$d(r^n F_2)/dr = \pi^{(n-1)/2} e^{-\pi K^2 r^2} (Kr)^{n-1},$$

which is equal to Eq. (A10-14).

The relation to be used in deriving Eq. (9-195) in the text,

$$\lim_{h \to 0} h^{n-3} \Gamma\left(\frac{3-n}{2}, \frac{\pi h^2}{K^2}\right) = \pi^{(3-n)/2} K^{n-3} \frac{2}{n-3}, \qquad \text{(A10-17)}$$

is proved as follows. From the definition of an incomplete gamma function of the second kind, Eq. (A10-2c), we have

$$\Gamma\left(\frac{3-n}{2}, \frac{\pi h^2}{K^2}\right) = \int_{\pi h^2/K^2}^{\infty} e^{-t} t^{-(1-n)/2} \, dt. \qquad \text{(A10-18)}$$

By introducing a new variable u such that $t = \pi u^2/K^2$, we have $dt = (2\pi u/K^2)du$. The lower limit of the integral in Eq. (A10-18) $t = \pi h^2/K^2$ corresponds to $u = h$. Then the integral in Eq.(A10-18) is rewritten as

$$\int_h^{\infty} e^{-\pi u^2/K^2} (\pi u^2/K^2)^{-(n-1)/2} (2\pi u/K^2) du = \left(\frac{\sqrt{\pi}}{K}\right)^{-(n-1)} \frac{2\pi}{K^2} \int_h^{\infty} e^{-\pi u^2/K^2} u^{2-n} \, du. \qquad \text{(A10-19)}$$

The integral in the right side of Eq. (A10-19) can be carried out by the integration by parts

$$\int_h^{\infty} f(u) g'(u) du = |f(u) g(u)|_h^{\infty} - \int_h^{\infty} f'(u) g(u) du,$$

where $f(u) = e^{-bu^2}$ ($b = \pi/K^2$), $g'(u) = u^{2-n}$, and $g(u) = u^{3-n}/(3-n)$. The result is

$$\int_h^{\infty} e^{-bu^2} u^{2-n} du = -e^{-bh^2} \frac{h^{3-n}}{3-n} + \frac{2b}{3-n} \int_h^{\infty} e^{-bu^2} u^{4-n} du. \qquad \text{(A10-20)}$$

Since the second term in the right side of Eq. (A10-20) yields only the terms of higher order than $3-n$ in h on further integration by parts, it must vanish when multiplied by h^{n-3} and the limit on $h \to 0$ is taken. Substituting the first term of Eq. (A10-20) into Eq. (A10-19) and then into Eq. (A10-18), and taking the limit of the product with h^{n-3} on $h \to 0$, we obtain Eq. (A10-17).

Finally, let us calculate the integral I_{2b} in Eq. (9-189b) in the text. By differentiating both sides of Eq. (A10-2b) with respect to p, and expanding the exponential factor in the right side, we have

$$d\gamma(z,p)/dp = e^{-p} p^{z-1} = p^{z-1} \sum_{n=0}^{\infty} (-p)^n / n! = \sum_{n=0}^{\infty} (-1)^n p^{n+z-1} / n!.$$

Integrating the last expression term by term leads to the power series expansion of the function $\gamma(z, p)$ in the form

$$\gamma(z,p) = \sum_{n=0}^{\infty}(-1)^n p^{z+n}/\{n!(z+n)\}. \tag{A10-21}$$

It follows that

$$\gamma(n/2, K^2\pi r^2) = \sum_{j=0}^{\infty}(-1)^j (K^2\pi r^2)^{(n/2)+j}/[j!\{(n/2)+j\}]. \tag{A10-22}$$

Substituting Eq. (A10-22) into Eq. (A10-3) and taking the limit at $r \to 0$, we have

$$\lim_{r\to 0}\{1-\varphi(r)\}r^{-n} = K^n \pi^{n/2}(2/n)/\Gamma(n/2). \tag{A10-23}$$

Using the property of δ-function together with Eq. (A10-23), and noting that

$$P(\mathbf{0}) = N\sum_{\text{cell}} q_t^{\,2},$$

we have Eq. (9-196) in the text.[2)]

References

1) B.R.A. Nijboer and F.W. De Wette, *Physica*, **23**, 309 (1957).
2) D.E. Williams, *Acta Crystallogr.*, **A27**, 452 (1971).

Appendix 11 Differentiation of the Coulomb Potential in Crystals and TO-LO Splitting

The distance-dependent part of the direct sum in Eq. (9-190) is written as

$$f(r) = \Gamma(n/2, K^2\pi r^2)r^{-n}. \tag{A11-1}$$

Substituting the arguments in parentheses in the right side of Eq. (A11-1) into the definition of incomplete gamma function of the second kind and changing the variable of integration from t to $s = t^{1/2}$, whereby $dt = 2s\,ds$, we have

$$\Gamma\left(\frac{n}{2}, K^2\pi r^2\right) = \int_{K^2\pi r^2}^{\infty} e^{-t}\, t^{(n/2)-1}\, dt = 2\int_{\sqrt{\pi}Kr}^{\infty} e^{-s^2} s^{n-1}\, ds \tag{A11-2}$$

Differentiating Eq. (A11-2) with respect to r gives

$$\frac{d}{dr}\Gamma\left(\frac{n}{2}, K^2\pi r^2\right) = -2\,e^{-\pi K^2 r^2}\left(\sqrt{\pi}K\right)^n r^{n-1}. \tag{A11-3}$$

Successive use of Eqs. (A11-1) and (A11-3) gives

$$f'(r) \equiv \frac{d}{dr}f(r) = -\frac{1}{r}\left\{n\,f(r) + 2\,e^{-\pi K^2 r^2}\left(\sqrt{\pi}K\right)^n\right\} \tag{A11-4a}$$

and

$$f''(r) \equiv \frac{d^2}{dr^2} f(r) = \frac{n+1}{r} f'(r) + 4e^{-\pi K^2 r^2}\left(\sqrt{\pi} K\right)^{n+2}. \tag{A11-4b}$$

The contribution from the direct sum to the elements of dynamical matrix is calculated by differentiating Eq. (A11-1) successively. The results are given by

$$\boldsymbol{\nabla} f(r) = (\boldsymbol{r}/r) f'(r) \tag{A11-5a}$$

and

$$\boldsymbol{\nabla} \tilde{\boldsymbol{\nabla}} f(r) = \boldsymbol{\nabla} (\tilde{\boldsymbol{r}}/r) f'(r) = \frac{\boldsymbol{r}\tilde{\boldsymbol{r}}}{r^2} f''(r) + \left(\boldsymbol{E} - \frac{\boldsymbol{r}\tilde{\boldsymbol{r}}}{r^2}\right)\frac{f'(r)}{r}. \tag{A11-5b}$$

Combining Eqs. (A11-4a,b) and (A11-5a,b) gives

$$\begin{aligned}\boldsymbol{\nabla} \tilde{\boldsymbol{\nabla}} f(r) = &\left\{\frac{(n+2)\boldsymbol{r}\tilde{\boldsymbol{r}}}{r^2} - \boldsymbol{E}\right\} \frac{n}{r^{n+2}} \Gamma\left(\frac{n}{2}, \pi K^2 r^2\right) \\ &+ 2(\sqrt{\pi} K)^n \frac{e^{-\pi K^2 r^2}}{r^2}\left\{\left(\frac{n+2}{r^2} - 2\pi K^2\right)\boldsymbol{r}\tilde{\boldsymbol{r}} - \boldsymbol{E}\right\}.\end{aligned} \tag{A11-6}$$

The contribution from the reciprocal sum to dynamical matrix elements is calculated as follows. The last exponential factor of the total structure factor in Eq. (9-191) may be expanded in powers of a small quantity $\Delta \boldsymbol{r}_{Lt}$ in the form

$$\begin{aligned}F_{\mathrm{C}}(\boldsymbol{h}) &= \sum_{L=1}^{N} \sum_{t=1}^{n} q_t \exp\{2\pi i \tilde{\boldsymbol{h}}(\boldsymbol{r}_L + \boldsymbol{r}_t^0)\}(1 + 2\pi i \tilde{\boldsymbol{h}} \Delta \boldsymbol{r}_{Lt} + \cdots) \\ &= F(\boldsymbol{h})_0 \sum_{L=1}^{N} \exp(2\pi i \tilde{\boldsymbol{h}} \boldsymbol{r}_L) + 2\pi i \sum_{L=1}^{N} \exp(2\pi i \tilde{\boldsymbol{h}} \boldsymbol{r}_L) \sum_{t=1}^{n} q_t \exp(2\pi i \tilde{\boldsymbol{h}} \boldsymbol{r}_t^0) \tilde{\boldsymbol{h}} \Delta \boldsymbol{r}_{Lt} + \cdots,\end{aligned} \tag{A11-7}$$

where $F(\boldsymbol{h})_0$ is the unit cell structure factor at the equilibrium defined by

$$F(\boldsymbol{h})_0 = \sum_{t=1}^{n} q_t \exp(2\pi i \tilde{\boldsymbol{h}} \boldsymbol{r}_t^0).$$

Expanding the product of Eq. (A11-7) and its complex conjugate up to terms quadratic in Cartesian displacements and using Eq. (A4-13), we have

$$|F_{\mathrm{C}}(\boldsymbol{h})|^2 = (N/V)\sum_{\lambda} |F(\boldsymbol{h})_0|^2 \delta(\boldsymbol{h} - \boldsymbol{h}_\lambda) + 4\pi^2 \sum_{s,t=1}^{n} q_s q_t \, e^{2\pi i \tilde{\boldsymbol{h}}(\boldsymbol{r}_s^0 - \boldsymbol{r}_t^0)} \sum_{L,M=1}^{N} e^{2\pi i \tilde{\boldsymbol{h}}(\boldsymbol{r}_L - \boldsymbol{r}_M)} \Delta \tilde{\boldsymbol{r}}_{Mt} \boldsymbol{h} \tilde{\boldsymbol{h}} \Delta \boldsymbol{r}_{Ls}. \tag{A11-8}$$

Let us introduce here complex symmetry Cartesian displacement coordinates in the form

$$\Delta \boldsymbol{r}_s(\boldsymbol{q}_K) = (2/N)^{1/2} \sum_{J=1}^{N} \Delta \boldsymbol{r}_{Js} \exp(2\pi i \tilde{\boldsymbol{q}}_K \boldsymbol{r}_J), \tag{A11-9a}$$

by subtracting Eq. (9-58b) times $i = \sqrt{-1}$ from Eq. (9-58a). Multiplying Eq. (A11-9a) by $\exp(-2\pi i \tilde{\boldsymbol{q}}_K \boldsymbol{r}_J)$, summing both sides over wave vectors in half the first Brillouin zone and noting that

$$(2/N)\sum_{K=1}^{N/2}\exp\{2\pi i\tilde{\boldsymbol{q}}_K(\boldsymbol{r}_L-\boldsymbol{r}_J)\}=\delta_{JL},$$

we obtain

$$\Delta\boldsymbol{r}_{Ls}=(2/N)^{1/2}\sum_K\exp(-2\pi i\boldsymbol{q}_K\boldsymbol{r}_L)\Delta\boldsymbol{r}_s(\boldsymbol{q}_K). \tag{A11-9b}$$

Substituting Eq. (A11-9b) into Eq. (A11-8), the sum with respect to L and M can be rewritten as

$$\sum_{L,M=1}^{N}e^{2\pi i\tilde{\boldsymbol{h}}(\boldsymbol{r}_L-\boldsymbol{r}_M)}\Delta\tilde{\boldsymbol{r}}_{Mt}\boldsymbol{h}\tilde{\boldsymbol{h}}\Delta\boldsymbol{r}_{Ls}=\frac{2}{N}\sum_K\left\{\sum_{L,M}e^{-2\pi i(\tilde{\boldsymbol{h}}-\tilde{\boldsymbol{q}}_K)(\boldsymbol{r}_L-\boldsymbol{r}_M)}\right\}\Delta\tilde{\boldsymbol{r}}_t(\boldsymbol{q}_K)\boldsymbol{h}\tilde{\boldsymbol{h}}\Delta\boldsymbol{r}_s(\boldsymbol{q}_K). \tag{A11-10}$$

Remembering Eq. (A4-13), we obtain

$$\begin{aligned}|F_C(\boldsymbol{h})|^2=(N/V)\Big\{&\sum_\lambda|F(\boldsymbol{h})_0|^2\delta(\boldsymbol{h}-\boldsymbol{h}_\lambda)+4\pi^2\sum_K\sum_{s,t}q_sq_t\,e^{2\pi i\tilde{\boldsymbol{h}}(\boldsymbol{r}_s^0-\boldsymbol{r}_t^0)}\\&\times\delta(\boldsymbol{h}-\boldsymbol{q}_K-\boldsymbol{h}_\lambda)\Delta\tilde{\boldsymbol{r}}_s(\boldsymbol{q}_K)\boldsymbol{h}\tilde{\boldsymbol{h}}\,\Delta\boldsymbol{r}_t(\boldsymbol{q}_K)\Big\}.\end{aligned} \tag{A11-11}$$

The constant factors in I_{2a} for $n=1$ are given by

$$\pi^{1-(3/2)}/\{2V\,\Gamma(1/2)\}=1/2\pi V \tag{A11-12a}$$

and

$$\Gamma(1,\pi h^2/K^2)=\int_{\pi h^2/K^2}^{\infty}e^{-t}\,dt=\exp(-\pi h^2/K^2). \tag{A11-12b}$$

The contribution from I_{2a} to the elements of dynamical matrix for $\boldsymbol{q}_K$ is given by

$$\begin{aligned}\boldsymbol{\nabla}_{Ks}\tilde{\boldsymbol{\nabla}}_{Kt}\,I_{2a}&=\frac{2\pi}{V}q_sq_t\int_h e^{-\pi(h/K)^2}h^{-2}\,e^{-2\pi i\tilde{\boldsymbol{h}}(\boldsymbol{r}_s^0-\boldsymbol{r}_t^0)}\,\boldsymbol{h}\tilde{\boldsymbol{h}}\,\delta(\boldsymbol{h}-\boldsymbol{q}_K-\boldsymbol{h}_\lambda)\,d\boldsymbol{h}\\&=\frac{2\pi}{V}q_sq_t\sum_\lambda\frac{e^{-\pi|\boldsymbol{h}_\lambda+\boldsymbol{q}_K|^2/K^2}}{|\boldsymbol{h}_\lambda+\boldsymbol{q}_K|^2}e^{-2\pi i(\tilde{\boldsymbol{h}}_\lambda+\tilde{\boldsymbol{q}}_K)(\boldsymbol{r}_s^0-\boldsymbol{r}_t^0)}(\boldsymbol{h}_\lambda+\boldsymbol{q}_K)(\tilde{\boldsymbol{h}}_\lambda+\tilde{\boldsymbol{q}}_K).\end{aligned} \tag{A11-13}$$

The sum with respect to λ in Eq. (A11-13) is taken over all the reciporocal lattice points. In this sum, the term with $\boldsymbol{h}_\lambda=\boldsymbol{0}$ becomes indefinite at the limit that $\boldsymbol{q}_K\to\boldsymbol{0}$. To avoid this difficulty, the term in question is written separately in the form

$$\begin{aligned}\frac{e^{-\pi|\boldsymbol{q}_K|^2/K^2}}{|\boldsymbol{q}_K|^2}\boldsymbol{q}_K\tilde{\boldsymbol{q}}_Ke^{-2\pi i\tilde{\boldsymbol{q}}_K(\boldsymbol{r}_s^0-\boldsymbol{r}_t^0)}&=\frac{\boldsymbol{q}_K\tilde{\boldsymbol{q}}_K}{|\boldsymbol{q}_K|^2}e^{-2\pi i\tilde{\boldsymbol{q}}_K(\boldsymbol{r}_s^0-\boldsymbol{r}_t^0)}\\&\quad-\frac{\boldsymbol{q}_K\tilde{\boldsymbol{q}}_K}{|\boldsymbol{q}_K|^2}\left(1-e^{-\pi|\boldsymbol{q}_K|^2/K^2}\right)e^{-2\pi i\tilde{\boldsymbol{q}}_K(\boldsymbol{r}_s^0-\boldsymbol{r}_t^0)}.\end{aligned} \tag{A11-14}$$

The values of dyadic elements in the first term in the right side of Eq. (A11-14) at the Γ point depend on the direction from which $\boldsymbol{q}_K$ approaches a zero vector. The portion of potential energy arising from this term is transformed from the Cartesian displacement space to the normal coordinate space through Eq. (9-85) as given by

$$\begin{aligned} V^a_{\mathrm{rec}} &= \tfrac{1}{2}(2\pi/V)\sum_{s,t} \Delta\tilde{\boldsymbol{r}}_{Is}\, \boldsymbol{e}_K\, q_s q_t\, e^{-2\pi i \tilde{\boldsymbol{q}}_K(\boldsymbol{r}^0_{Is}-\boldsymbol{r}^0_{Jt})}\, \tilde{\boldsymbol{e}}_K\, \Delta\boldsymbol{r}_{Jt} \\ &= \tfrac{1}{2}(2\pi/V)\tilde{\boldsymbol{Q}}^a_K \left\{ \sum_{s,t} \tilde{\boldsymbol{L}}^{Xa}_{Is,K}\, q_s q_t\, e^{-2\pi i \tilde{\boldsymbol{q}}_K(\boldsymbol{r}^0_{Is}-\boldsymbol{r}^0_{Jt})}\, \boldsymbol{e}_K \tilde{\boldsymbol{e}}_K\, \boldsymbol{L}^{Xa}_{Jt,K} \right\} \boldsymbol{Q}^a_K, \end{aligned} \tag{A11-15}$$

where $\boldsymbol{e}_K = \boldsymbol{q}_K/|\boldsymbol{q}_K|$ is the unit vector along the wave vector $\boldsymbol{q}_K$. The exponential factor in Eq. (A11-15) becomes 1 when $\boldsymbol{q}_K = \boldsymbol{0}$, so that the quantity in braces at the Γ point is given by

$$\sum_{s,t} \tilde{\boldsymbol{L}}^{Xa}_{Is,K}\, q_s q_t\, \boldsymbol{e}_K \tilde{\boldsymbol{e}}_K\, \boldsymbol{L}^{Xa}_{Jt,K} = \left(\sum_s q_s \tilde{\boldsymbol{L}}^{Xa}_{Is,K}\right) \boldsymbol{e}_K \tilde{\boldsymbol{e}}_K \left(\sum_t q_t \boldsymbol{L}^{Xa}_{Jt,K}\right). \tag{A11-16}$$

The vectors in parentheses in Eq. (A11-16) represent the changes in dipole moment on the nuclear displacements along the normal coordinates specified by $\boldsymbol{Q}^a_K$. Thus, the potential term V^a_{rec} in Eq. (A11-15) vanishes if the normal coordinate vector $\boldsymbol{Q}^a_K$ involves no infrared active modes, and TO-LO splittings are found only for infrared active modes. The merit of separating the term with $\boldsymbol{h}_\lambda = \boldsymbol{0}$ in Eq. (A11-13) from the others is that the second term in Eq. (A11-14) has the definite limit 0 as $\boldsymbol{q}_K$ approaches $\boldsymbol{0}$ from any direction, irrespective of whether the modes involved are infrared active or not. The indeterminacy of the first term in the right side of Eq. (A11-14) causing a TO-LO splitting arises from the factor $h^{-2}\boldsymbol{h}\tilde{\boldsymbol{h}}$ in the integrand in Eq. (A11-13). Since h^{-2} in Eq. (A11-13) is the special case of h_λ^{n-3} in Eq. (9-194) with $n = 1$, no TO-LO splitting occurs for any pairwise interaction with $n > 1$.

It is worth noting that definition of symmetry Cartesian displacement coordinates is not unique. Some authors defined the complex symmetry Cartesian displacements in the form[1)]

$$\Delta\boldsymbol{r}_t^{\,p}(\boldsymbol{q}_K)_\pm = (2/N)^{1/2} \sum_{I=1}^{N/2} \Delta\boldsymbol{r}_{It} \exp\{\pm 2\pi i \tilde{\boldsymbol{q}}_K(\boldsymbol{r}_I + \boldsymbol{r}_t^0)\}. \tag{A11-17}$$

The coefficients of the displacement of nucleus It in Eq. (A11-17) is different from the corresponding coefficients in the definition of complex symmetry Cartesian displacements derived from the real symmetry coordinates in Eqs. (9-58a,b), *i.e.*,

$$\begin{aligned} \Delta\boldsymbol{r}_t(\boldsymbol{q}_K)_\pm &= 2^{-1/2}\{\Delta\boldsymbol{r}_t(\boldsymbol{q}_K)_c \pm i\Delta\boldsymbol{r}_t(\boldsymbol{q}_K)_s\} \\ &= (2/N)^{1/2} \sum_{I=1}^{N/2} \Delta\boldsymbol{r}_{It} \exp(\pm 2\pi i \tilde{\boldsymbol{q}}_K \boldsymbol{r}_I) \end{aligned} \tag{A11-18}$$

by a constant factor $\exp(2\pi i \tilde{\boldsymbol{q}}_K \boldsymbol{r}_t^0)$. Thus, the dynamical matrix elements associated with the coordinate pair $\Delta\boldsymbol{r}_s(\boldsymbol{q}_K)_\pm$ and $\Delta\boldsymbol{r}_t(\boldsymbol{q}_K)_\pm$ correspond to Eq. (A11-13) divided by $e^{2\pi i \tilde{\boldsymbol{q}}_K(\boldsymbol{r}_t^0-\boldsymbol{r}_s^0)}$, *i.e.*,

$$\frac{2\pi}{V} q_s q_t \sum_\lambda \frac{e^{-\pi|\boldsymbol{h}_\lambda+\boldsymbol{q}_K|^2/K^2}}{|\boldsymbol{h}_\lambda+\boldsymbol{q}_K|^2} e^{-2\pi i \tilde{\boldsymbol{h}}_\lambda(\boldsymbol{r}_s^0-\boldsymbol{r}_t^0)} (\boldsymbol{h}_\lambda + \boldsymbol{q}_K)(\tilde{\boldsymbol{h}}_\lambda + \tilde{\boldsymbol{q}}_K). \tag{A11-19}$$

Obviously, the use of Eqs. (A11-17) and (A11-19) gives the same result as the use of Eqs. (A11-18) and (A11-13).

Reference

1) M. Born and K. Huang, *Dynamical Theory of Crystal Lattices*, p. 223, Eq. (24.5), Oxford University Press, London (1954).

Addendum

The following is a list of selected references to recent development of the theory and application of molecular mechanics, especially on the topics not fully discussed in the text. Each reference except for the books is followed by the parentheses in which the subjects, the methods employed and/or the materials studied are given.

Chapter 1

[Books and Reviews]

A 1 D. C. Rapaport, *The Art of Molecular Dynamics Simulations*, Cambridge University Press, Cambridge (1995).

A 2 G. D. Billing and K. V. Mikkelson, *Introduction to Molecular Dynamics and Chemical Kinetics*, John Wiley, New York (1996).

A 3 A. Gavezzotti (Editor), *Theoretical Aspects and Computer Modeling of the Molecular Solid State*, Wiley, Chichester (1997).

A 4 I. Pattersson and T. Liljefors, *Rev. Comput. Chem.*, **9**, 167 (1996). (Molecular mechanics; organic molecules).

A 5 M. E. Karpens and C. L. Brooks, III, *Modelling Protein Conformation by Molecular Mechanics and Dynamics*, in *Protein Structure Prediction*, ed. by M. J. E. Sternberg, pp. 229-261, IRL Press, Oxford (1996).

A 6 R. H. Stote, A. Dejaegere and M. Karplus, *Understanding Chem. React.*, **19**, 153 (1997). (Molecular mechanics and dynamics,; Enzymes)

Chapter 2

[New Force Fields]

A 7 N. Gresh, *J. Phys. Chem.*, **A101**, 8680 (1997). (SIBFA, Sum of interactions between fragments ab initio computed; water oligomers)

A 8 A. M. Schneider and P. Behrens, *Chem. Mater.* **10**, 679 (1998). (ESFF, Extensible systematic force field)

A 9 A. Hocquet and M. Langgard, *J. Mol. Model.*, **4**, 94 (1998).(MM+, a variation of MM2)

[Comparison of Force Fields]

A10 J. L. Assensio, M. Martin-Pastor and J. Jimenez-Barbero, *J. Mol. Struct.*, **395/396**, 245 (1997). (MM3, ESFF; Carbohydrates)

A11 P. Hobza, M. Kabelac, J. Sponer, P. Mejzlik and J. Vondrasek, *J. Comp. Chem.*, **18**, 1136 (1997). (AMBER, CFF95, CVFF, CHARMm, OPLS, POLTEV and quantum mechanical methods)

A12 N. Arona and B. Jayaram, *J. Comp. Chem.*, **18**, 1245 (1997). (AMBER, CHARMm, ECEPP, GROMOS, OPLS; H-bonded α-helices).

A13 J. P. Bowen and G. Liang, *Practical Application of Computer-Aided Drug Design*, ed. by P. S. Charifson, pp. 495-538, Dekker, New York 1997. (MM4, AMBER, CFF93, CHARMm, MMFF94)

A14 B. Chen, M. G. Martin and J. I. Siepmann, *J. Phys. Chem.* **B102**, 2578(1998). (OPLS, MMFF94; Thermodynamic properties).

A15 J. Gruza, J. Koca, S. Perez and A. Imberty, *J. Mol. Struct.*, **424**, 269 (1998). (AMBER, MM3)

A16 S. N. Rao, *Biophys. J.*, **74**, 3131 (1998); *Nucleosides Nucleotides*, **17**, 791 (1998). (AMBER, MM2)

A17 A. M. Rodriguez, H. A. Baldoni, F. Suvire, R. N. Vazquez, G. Zamarbide, R. D. Enriz, O. Farkas, A. Perczel, M. A. McAllister, L. L. Torday, J. G. Papp and I. G. Csizmadia, *J. Mol. Struct.*, **455**, 278 (1998). (MM+, AMBER, CHARMm, OPLS and quantum mechanical methods; Ramachandran map)

Chapter 3

[Internal coordinates]

A18 A. Blondel and M. Karplus, *J. Comp. Chem.*, **17**, 1132 (1996). (Derivatives of torsion and improper torsion angles, removal of singularity).

[Search for minimum-energy structure]

A19 M. Kerer and S. I. Stupp, *Comput. Chem.*, **22**, 345 (1998) (Generic Algorithm)

A20 D. Overlin, Jr., and H. A. Scheraga, *J. Comput. Chem.*, **19**, 71 (1998). (3-dimensional spline method).

A21 C. T. Klein, B. Mayer, G. Kohler and P. Wolschann, *J. Comput. Chem.*, **19**, 1470 (1998). (systematic stepwise variation method)

A22 C. Baysal,H. Meirovitch and I. M. Navan, *J. Comput. Chem.*, **20**, 354 (1999). (Truncated Newton method)

Chapter 4

[Quantum mechanically estimated force fields]

A23 Y. Karzazi, G. Vergoten and G. Sarpateanu, *Electron. J. Theor. Chem.*, **2**, 85 (1997). (Transferability)

A24 J. R. Maple, M. -J. Hwang, K. J. Jalkanen, T. P. Stockfisch and A. T. Hagler, *J. Comp. Chem.*, **19**, 430(1998). (Amides, peptides and related compounds).

A25 J. Baker and P. Pulay, *J. Comput. Chem.*, **19**, 1187 (1998). (Scaling; fluorocarbons).

A26 S. Samdal, *J. Mol. Struct.*, **440**, 165 (1998). (Scaling; Acetamide)

Chapter 5

[Anharmonic Force Field]

A27 C. Puzzarini, R. Tarroni and P. Palmieri, *Spectrochim. Acta, Part A*, **53A**, 1079 (1997). (Chlorofluoromethane).

A28 S. Dressler and W. Thiel, *Chem. Phys. Lett.*, **273**, 71 (1997). (Evaluation by density functional theory).

[Dynamic Structures of Molecules]

A29 G. M. Kuramshina, V. P. Spiridonov, A. G. Yagola and T. G. Strand, *J. Mol. Struct.*, **445**, 243 (1998). (Higher order approximation for $\langle \Delta r^n \rangle$)

Chapter 6

[Partition Function]

A30 A. Nino and C. Munos-Caro, *Comput. Chem.*, **21**, 143 (1997). (non-rigid molecules)

A31 A. D. Isaakson, *J. Chem. Phys.*, **108**, 9978(1998). (Removal of resonance effect from anharmonic contribution)

Chapter 7

[Quantum mechanical estimation of electric properties of molecules]

A32 E. Sigfridsson and U. Ryde, *J. Comput. Chem.*, **19**, 377 (1998). (atomic charges)

A33 A. Hinchliffe, J. J. Perez and H. J. S. Humberto, *Electron. J. Theor. Chem.*, **2**, 325 (1997). (dipole moments, polarizabilities by density functional theory)

Chapter 8

[Infrared Absorption Intensity]

A34 K. Palmo and S. Krimm, *J. Comput. Chem.*, **19**, 754(1998). (Introduction of internal coordinate dipole moments).

Chapter 9

[Elastic Constants]

A35 M. R. Sorino, J. A. Bruno and A. Batana, *Comput. Chem.*, **20**, 485 (1996). (Pressure dependence; ionic crystals).

[Crystal Structure]

A36 T. Yui, K. Miyawaki, M. Yada and K. Ogawa, Int. *J. Biol. Macromol.*, **21**, 243 (1997). (Combined use of X-ray Powder diffraction data and MM studies(MM3); Mannans).

[Temperature Diffuse Scattering]

A37 M. E. Wall, S. E. Ealick and S. M. Gruner, *Proc. Nat. Acad. Sci. USA*, **94**, 6180 (1997). (Crystal disorder; proteins).

[Inelastic Neutron Scattering]

A38 S. F. Parker, D. A. Braden, J. Tomkenson and B. S. Hudson, *J. Phys. Chem.*, **B102**, 5955(1998)(Octadecane)

A39 M. Barthes, H. N. Bordallo, J. Eckert, O. Maurus G. de Nunzio and J. Leon, *J. Phys. Chem.*, **B102**, 6177 (1998)(*N*-methylacetamide).

A40 R. Caciuffo, A. D. Esposti, M. S. Deleuze, D. A. Leigh A. Murphy, B. Paci, S. F. Parker and F. Zerbetto, *J. Chem. Phys.*, **109**, 11094 (1998). (Benzylic amide [2] catenane)

[Lattice Sum]

A41 H. Dufner, S. M. Kast, J. Brickmann and M. Schlenkrich, *J. Comput. Chem.*, **18**, 660 (1997). (Comparison between Ewald's method and the direct summation of shifted force potentials)

A42 Z. M. Chen, T. Cagin and W. A. Goddard III, *J. Comput. Chem.* **18**, 1365 (1997). (Ewald methods for van der Waals potential)

[Surface Energy]

A43 R. K. Marwala and B. S. Shah, *Indian J. Phys. A*, **71A**, 53 (1997). (Relation to Vicker's hardness; anthracene, phenanthrene, etc.)

Subject Index

(References to Chapters or Sections are given in boldface figures.)